Moduli Spaces of Riemann Surfaces

IAS/PARK CITY
MATHEMATICS SERIES

Volume 20

Moduli Spaces of Riemann Surfaces

Benson Farb
Richard Hain
Eduard Looijenga
Editors

American Mathematical Society
Institute for Advanced Study

John C. Polking, Series Editor
Benson Farb, Volume Editor
Richard Hain, Volume Editor
Eduard Looijenga, Volume Editor

IAS/Park City Mathematics Institute runs mathematics education programs that bring together high school mathematics teachers, researchers in mathematics and mathematics education, undergraduate mathematics faculty, graduate students, and undergraduates to participate in distinct but overlapping programs of research and education. This volume contains the lecture notes from the Graduate Summer School program held in July 2011.

2000 *Mathematics Subject Classification.* Primary 14-06, 14H10, 32G15, 55R40, 57S05.

Library of Congress Cataloging-in-Publication Data

Moduli spaces of Riemann surfaces / Benson Farb, Richard Hain, Eduard Looijenga, editors.
pages cm – (IAS/Park City mathematics series ; volume 20)
Includes bibliographical references.
ISBN 978-0-8218-9887-1 (alk. paper)
1. Moduli theory. 2. Riemann surfaces. I. Farb, Benson, II. Hain, Richard M. (Richard Martin), 1953– III. Looijenga, Eduard.

QA564.M644 2013
515′.93–dc23 2013007216

∞ The paper used in this book is acid-free and falls within the guidelines
established to ensure permanence and durability.
Visit the AMS home page at `http://www.ams.org/`

10 9 8 7 6 5 4 3 2 1   18 17 16 15 14 13

Contents

Preface

The IAS/Park City Mathematics Institute (PCMI) was founded in 1991 as part of the "Regional Geometry Institute" initiative of the National Science Foundation. In mid 1993 the program found an institutional home at the Institute for Advanced Study (IAS) in Princeton, New Jersey.

The IAS/Park City Mathematics Institute encourages both research and education in mathematics and fosters interaction between the two. The three-week summer institute offers programs for researchers and postdoctoral scholars, graduate students, undergraduate students, high school teachers, undergraduate faculty, and researchers in mathematics education. One of PCMI's main goals is to make all of the participants aware of the total spectrum of activities that occur in mathematics education and research. We wish to involve professional mathematicians in education and to bring modern concepts in mathematics to the attention of educators. To that end the summer institute features general sessions designed to encourage interaction among the various groups. In-year activities at the sites around the country form an integral part of the High School Teachers Program.

Each summer a different topic is chosen as the focus of the Research Program and Graduate Summer School. Activities in the Undergraduate Summer School deal with this topic as well. Lecture notes from the Graduate Summer School are being published each year in this series. The first twenty volumes are:

- Volume 1: *Geometry and Quantum Field Theory* (1991)
- Volume 2: *Nonlinear Partial Differential Equations in Differential Geometry* (1992)
- Volume 3: *Complex Algebraic Geometry* (1993)
- Volume 4: *Gauge Theory and the Topology of Four-Manifolds* (1994)
- Volume 5: *Hyperbolic Equations and Frequency Interactions* (1995)
- Volume 6: *Probability Theory and Applications* (1996)
- Volume 7: *Symplectic Geometry and Topology* (1997)
- Volume 8: *Representation Theory of Lie Groups* (1998)
- Volume 9: *Arithmetic Algebraic Geometry* (1999)
- Volume 10: *Computational Complexity Theory* (2000)
- Volume 11: *Quantum Field Theory, Supersymmetry, and Enumerative Geometry* (2001)
- Volume 12: *Automorphic Forms and their Applications* (2002)
- Volume 13: *Geometric Combinatorics* (2004)
- Volume 14: *Mathematical Biology* (2005)
- Volume 15: *Low Dimensional Topology* (2006)
- Volume 16: *Statistical Mechanics* (2007)
- Volume 17: *Analytic and Algebraic Geometry: Common Problems, Different Methods* (2008)
- Volume 18: *Arithmetic of L-functions* (2009)
- Volume 19: *Mathematics in Image Processing* (2010)
- Volume 20: *Moduli Spaces of Riemann Surfaces* (2011)

Volumes are in preparation for subsequent years.

Some material from the Undergraduate Summer School is published as part of the Student Mathematical Library series of the American Mathematical Society. We hope to publish material from other parts of the IAS/PCMI in the future. This will include material from the High School Teachers Program and publications documenting the interactive activities which are a primary focus of the PCMI. At the summer institute late afternoons are devoted to seminars of common interest to all participants. Many deal with current issues in education: others treat mathematical topics at a level which encourages broad participation. The PCMI has also spawned interactions between universities and high schools at a local level. We hope to share these activities with a wider audience in future volumes.

John C. Polking
Series Editor
Februrary 2013

Introduction

Benson Farb, Richard Hain, and Eduard Looijenga

IAS/Park City Mathematics Series
Volume 20, 2011

Introduction

Benson Farb, Richard Hain, and Eduard Looijenga

1. Introduction

This volume contains lecture notes from the nine lecture series on mapping class groups and moduli spaces of Riemann surfaces that were given at the Park City Mathematics Institute in July, 2011. These lecture series were directed at graduate students and researchers interested in these topics.

Let us recall how these two topics that were so central to this PCMI workshop are connected. The *mapping class group* of a closed, connected, oriented surface S is the group of isotopy classes of its orientation preserving diffeomorphisms. The *Teichmüller space* of that surface is the set of all isotopy classes of its conformal structures. If we assume the genus g of S to be positive, then this is a contractible complex manifold on which the mapping class group of S acts properly discontinuously, with a subgroup of finite index acting freely. The fact that the orbit space of this action may be understood as the moduli space $\mathcal{M}_g$ of Riemann surfaces of genus g accounts for the close relationship between the mapping class group and the topology of $\mathcal{M}_g$.

We can take this further: a compact Riemann surface is in a unique way a smooth complex-projective curve and so $\mathcal{M}_g$ is also the moduli space of smooth projective curves of genus g. With that interpretation it acquires a much finer structure, namely that of an algebraic variety (or more precisely, of a Deligne-Mumford stack, at least when $g \geq 2$) defined over the rational number field. As a result, moduli spaces of Riemann surfaces are fundamental objects of study in topology, complex analysis, and algebraic geometry; they can be studied from each of these points of view.

Each of the communities studying moduli spaces of curves — geometric topologists, algebraic topologists, complex analysts, algebraic and arithmetic geometers — generates its own set of problems and a set of techniques for resolving them. These problems and techniques are quite divergent. The goal of the PCMI summer school was cross-pollination: to educate students and researchers from these sometimes disparate communities in the basic problems and techniques of the others. Our premise was that interactions between these groups are fruitful in both the short and long term. This volume has the same goals; we believe that the authors have achieved this goal.

Benson Farb, Department of Mathematics, University of Chicago
E-mail address: farb@math.uchicago.edu
Richard Hain, Department of Mathematics, Duke University
E-mail address: hain@math.duke.edu
Eduard Looijenga, Mathematisch Instituut, Universiteit Utrecht
E-mail address: E.J.N.Looijenga@uu.nl

The lectures in this volume cover a broad range of topics, from introductory to advanced material. Several of the articles cross disciplinary boundaries. They are generally arranged from the more introductory to the more advanced. That said, such an ordering of the articles will depend upon one's background and taste. We encourage students to explore all of the lectures to develop a broad view of the subject. We encourage established researchers to get out of their comfort zone and explore the lectures in this volume that are not in their immediate areas of expertise. There they may find new ideas, problems and techniques.

A Brief Introduction to Mapping Class Groups

Yair N. Minsky

IAS/Park City Mathematics Series
Volume 20, 2011

A Brief Introduction to Mapping Class Groups

Yair N. Minsky

These notes summarize a short se tof lectures I gave at the PCMI program on mapping class groups and moduli spaces in the summer of 2011. My job was to introduce the mapping class group of a surface, discuss its basic features from a topologist's point of view, and give a description of the Nielsen-Thurston classification of mapping classes.

When writing out these notes I couldn't help wondering to myself what purpose they could possibly serve. All of this beautiful material is by now 30 years old or older, and has been described eloquently in a number of books and articles, notably (for the Thurston material) the Asterisque volume "Travaux de Thurston sur les surfaces" [**27**] and its new translation [**28**], Thurston's own famous article [**70**], Casson's lecture notes with Bleiler [**20**], and most recently Farb-Margalit's lovely text [**26**].

While giving the talks the most enjoyable part for me (and I hope for the audience) was drawing the many pictures one needs to really illustrate the ideas. Perhaps if nothing else, then, one goal for these notes is to recapture some of the feeling of an informal chalkboard lecture, with as many engaging pictures as possible.

In the last section, I have added a brief and highly biased meander through areas of the literature which I find particularly interesting. The field is much larger than I can really do it justice here, so this is by no means an exhaustive survey.

1. Definitions, examples, basic structure

Definitions. Throughout, $S = S_{g,n}$, the oriented surface of genus g with n punctures (Figure 1). Occasionally we will consider surfaces with boundary instead of punctures, but we will not emphasize this point.

The *mapping class group of* S, here denoted $\mathrm{Mod}(S)$, is the group

$$\mathrm{Mod}(S) = \mathrm{Homeo}^+(S)/\,\mathrm{Homeo}_0(S).$$

That is, the group of orientation-preserving homeomorphisms modulo the relation of isotopy. The group appears naturally in many contexts. It is relevant for describing gluings of 3-manifolds along boundary surfaces; for classifying bundles whose fibres are 2-dimensional; it is the natural symmetry group for the space of conformal or hyperbolic structures on a surface; it is important in the dynamical study of diffeomorphisms and flows.

Yale University, Department of Mathematics, New Haven, Connecticut 06520
E-mail address: yair.minsky@yale.edu

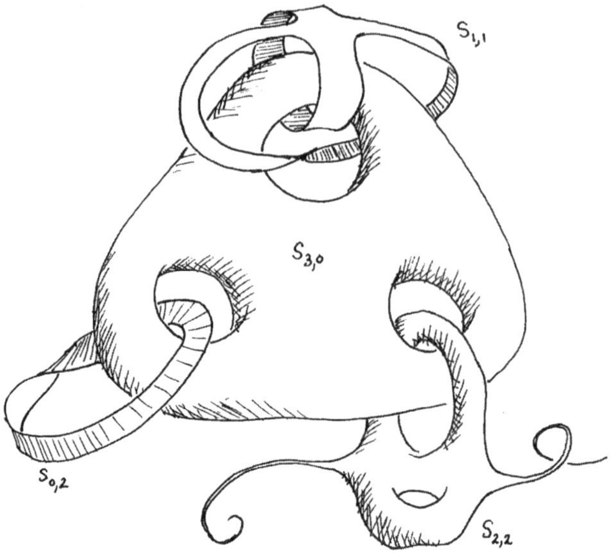

FIGURE 1. Some surfaces of finite type.

There are some choices implicit in this definition; let us remark on them briefly.

- *Orientation:* The restriction to oriented surfaces and orientation preserving homeomorphisms is mostly a matter of simplicity, and it is also natural when considering Riemann surfaces, which are naturally oriented. Allowing orientation-reversing maps of $S_{g,n}$ would, for example, produce an index 2 extension of $\mathrm{Mod}(S)$.
- *Homeo versus Diffeo*: We could endow S with smooth structure and consider $\mathrm{Diff}^+(S)/\mathrm{Diff}_0(S)$. This is the same group, but the proof is not trivial. We will be cavalier about this distinction, switching categories as convenient.
- *Homotopy versus Isotopy:* We could replace the relation of isotopy with homotopy, still yielding the same group. This is nontrivial, and not quite true in complete generality – in particular there exists an orientation-reversing homeomorphism of the disk or the annulus which is homotopic, but not isotopic, to the identity.
- *Homeomorphisms versus homotopy-equivalences:* The group of homotopy-equivalences, mod homotopy, is a completely algebraic object, namely

$$Out(\pi_1(S)) = Aut(\pi_1(S))/Inn(\pi_1(S)).$$

If S is closed ($n = 0$), then the natural map $\mathrm{Mod}(S) \to Out(\pi_1(S))$ is an isomorphism by the Dehn-Nielsen-Baer theorem. We will not prove this here. In general, one must restrict to automorphisms of $\pi_1(S)$ that "respect the punctures".

- *Punctures versus boundary:* If we give our surfaces boundary, we might consider the group of homeomorphisms that fix the boundary pointwise, modulo isotopies that do the same. This makes Dehn twists around the boundary nontrivial (see below for definitions), and so gives a group which is a extension of our group by this (abelian) group of twists.

It is not hard, with these remarks in mind, to show that $\text{Mod}(S)$ is a *countable* group – note that any map $S \to S$ *sufficiently close* to the identity is homotopic to the identity. However note also that this is false if we allow $g = \infty$ or $n = \infty$.

Simplest surfaces. For the sphere $S_{0,0}$ and the plane $S_{0,1}$, $\text{Mod}(S)$ is trivial. A key point here is to isotope the image of a circle back to itself, which is actually somewhat delicate (involving the Jordan curve theorem and the Alexander trick). This is the simplest case of the "homotopy implies isotopy" remark above.

For the annulus $S = S_{0,2}$, $\text{Mod}(S) = \mathbb{Z}/2\mathbb{Z}$. The nontrivial element is the 180 degree rotation illustrated in Figure 2. There is nothing else – an orientation-preserving homeomorphism fixing the punctures must be isotopic to the identity.

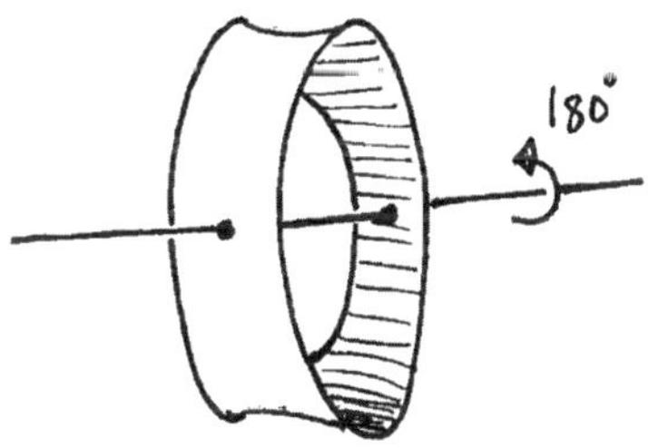

FIGURE 2. The rotation through the horizontal axis generates $\text{Mod}(S_{0,2})$.

Similarly, for the 3-holed sphere, $\text{Mod}(S_{0,3})$ is finite – it is identified with the group of permutations of the punctures. Generators are pictured in Figure 3.

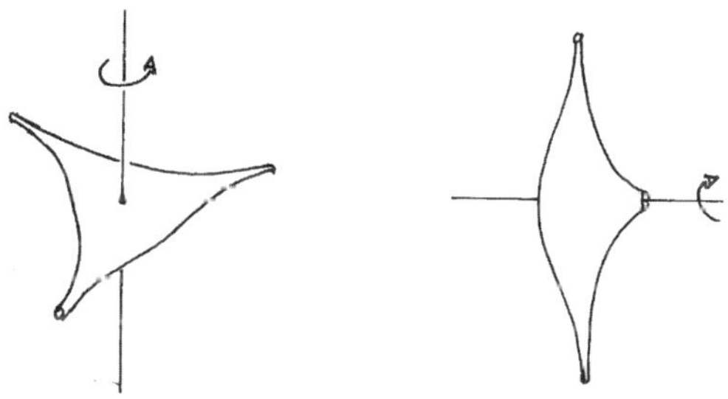

FIGURE 3. Order-2 and order-3 generators for $\text{Mod}(S_{0,3})$.

The Torus: $T = S_{1,0}$ is our first example of a surface with infinite mapping class group:

$$\text{Mod}(T) \cong \text{SL}(2,\mathbb{Z}) = \left\{ \left[\begin{smallmatrix} a & b \\ c & d \end{smallmatrix}\right] : a,b,c,d \in \mathbb{Z}, ad - bc = 1 \right\}.$$

To see this, note $\pi_1(T) = \mathbb{Z}^2$ and $Aut(\mathbb{Z}^2) = \text{GL}(2,\mathbb{Z})$, so there is a natural map $\text{Mod}(T) \to \text{GL}(2,\mathbb{Z})$. The map lands in $\text{SL}(2,\mathbb{Z})$ because of the orientation-preserving condition.

The map is surjective because any $A \in \mathrm{SL}(2, \mathbb{Z})$ acts linearly on $\mathbb{R}^2$ preserving $\mathbb{Z}^2$, hence the action gives a homeomorphism of $T = \mathbb{R}^2/\mathbb{Z}^2$.

It is injective because if a map is the identity on $\pi_1(T)$ it is *homotopic* to the identity. This can be done by hand in $\mathbb{R}^2$, or we can appeal to general notions – T is a $K(\pi, 1)$.

The nontrivial elements of $\mathrm{SL}(2, \mathbb{Z})$ fall into three distinct types:

- Finite order (elliptic): Consider $\left[\begin{smallmatrix} 0 & -1 \\ 1 & 0 \end{smallmatrix}\right]$, which has order 4. It acts by $\pi/2$ rotation on $\mathbb{R}^2$ (Figure 4).

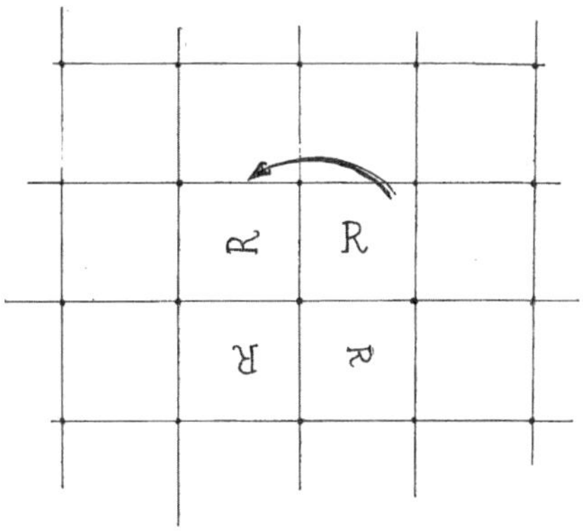

FIGURE 4

Similarly $\left[\begin{smallmatrix} 0 & -1 \\ 1 & 1 \end{smallmatrix}\right]$ has order 6. On the usual $\mathbb{Z}^2$ lattice it does not act as a rotation – it is better viewed on the *hexagonal* lattice $\mathbb{Z} + \mathbb{Z}e^{2\pi i/3}$ (identifying $\mathbb{R}^2$ with $\mathbb{C}$), where it is a rotation by $\pi/3$ (Figure 5). More precisely this rotation is obtained by conjugating $\left[\begin{smallmatrix} 0 & -1 \\ 1 & 1 \end{smallmatrix}\right]$ by the element of $\mathrm{SL}(2, \mathbb{R})$ that takes the square lattice to the hexagonal one.

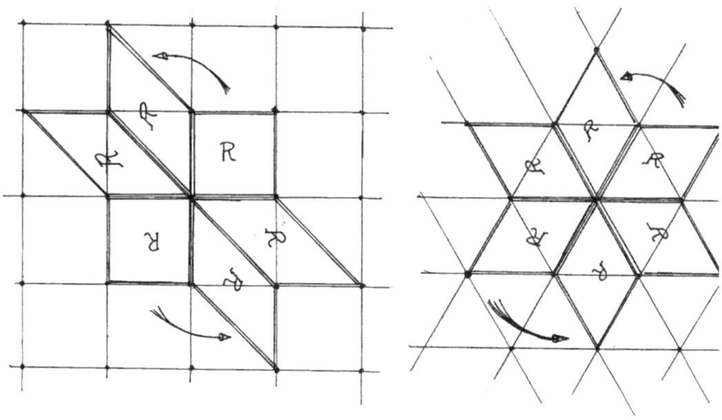

FIGURE 5. An order 6 mapping class preserves the geometry of the hexagonal lattice.

Finite order elements of $\mathrm{SL}(2, \mathbb{Z})$ have traces in $\{-1, 0, 1\}$. It is not hard to check that there are finitely many conjugacy classes like this, all corresponding to the examples we have given or to their powers.

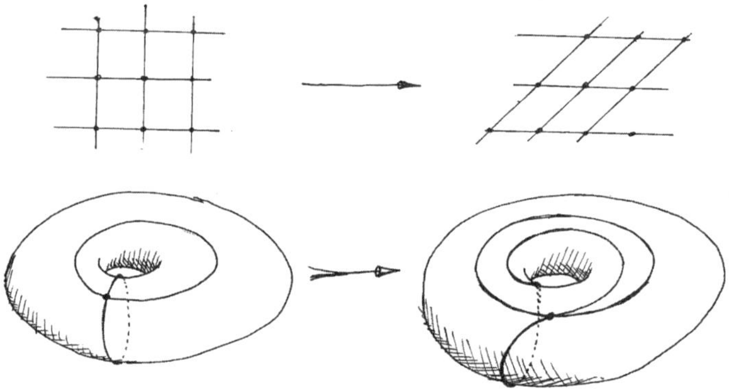

FIGURE 6. A parabolic transformation twists the torus around a simple loop.

- Twist maps (parabolic): Let $\tau = \left[\begin{smallmatrix} 1 & 1 \\ 0 & 1 \end{smallmatrix}\right]$. Then $\tau^n = \left[\begin{smallmatrix} 1 & n \\ 0 & 1 \end{smallmatrix}\right]$, so τ has infinite order. Note it preserves (up to isotopy) exactly one simple closed curve on the torus (Figure 6). Parabolic matrices have trace ± 2.

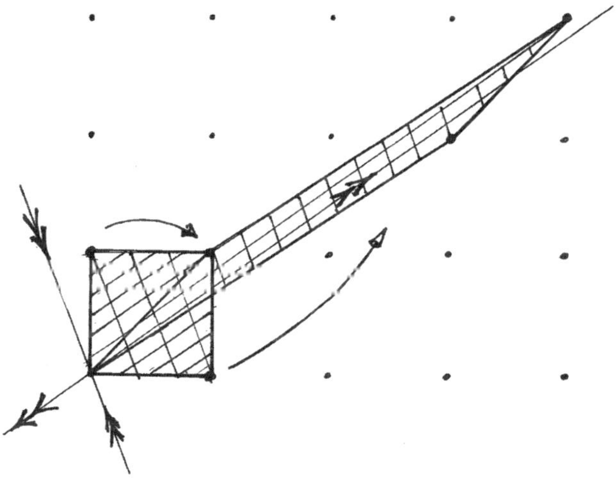

FIGURE 7. An Anosov map of the torus.

- Anosov (hyperbolic): Consider $A = \left[\begin{smallmatrix} 3 & 1 \\ 2 & 1 \end{smallmatrix}\right]$. It has two eigenvalues $\lambda^{\pm} > 0$ such that $\lambda^+\lambda^- = \det A = 1$. The eigenvectors give two *transverse invariant foliations* by straight lines, one expanded and one contracted by A (Figure 7). Anosov maps have traces outside $[-2, 2]$.

Our main goal, the Nielsen-Thurston classification, generalizes this picture to arbitrary surfaces.

Action on homology. A first attempt to generalize the torus can lead us to consider automorphisms of $H_1(S)$ which is $\mathbb{Z}^{2g}$ for $n = 0$, so we have

$$\Phi : \mathrm{Mod}(S) \to \mathrm{Aut}(H_1(S)).$$

We don't get $\mathrm{SL}(2g,\mathbb{Z})$ now: homeomorphisms preserve the *intersection form* which in a standard basis is the symplectic form J, a concatenation of $\left[\begin{smallmatrix} 0 & 1 \\ -1 & 0 \end{smallmatrix}\right]$'s along the diagonal. The subgroup of $\mathrm{GL}(2g,\mathbb{Z})$ preserving J is called the symplectic group $Sp(2g,\mathbb{Z})$. Hence we have a short exact sequence

$$1 \to \mathcal{I} \to \mathrm{Mod}(S) \to Sp(2g,\mathbb{Z}) \to 1.$$

The kernel, $\mathcal{I}$, is called the *Torelli group* and is an object of much interest. Surjectivity of the map can be obtained by identifying simple generators of $Sp(2g,\mathbb{Z})$ and showing each can be realized by a homeomorphism.

Dehn twists. Consider $\left[\begin{smallmatrix} 1 & 0 \\ 1 & 1 \end{smallmatrix}\right]$ applied to the band $B = [0,1] \times \mathbb{R}$. It descends to the annulus $A = B/\mathbb{Z} = [0,1] \times S^1$. Note that this is the *identity* on the boundary. Hence for any embedding of A in S we can apply this map on the image and extend by the identity. This is a (leftward) Dehn twist τ (see Figure 8). The isotopy class of τ depends only on the isotopy (in fact homotopy) class of the embedding, particularly of the core curve γ of the annulus. We write $\tau = \tau_\gamma$, or sometimes T_γ.

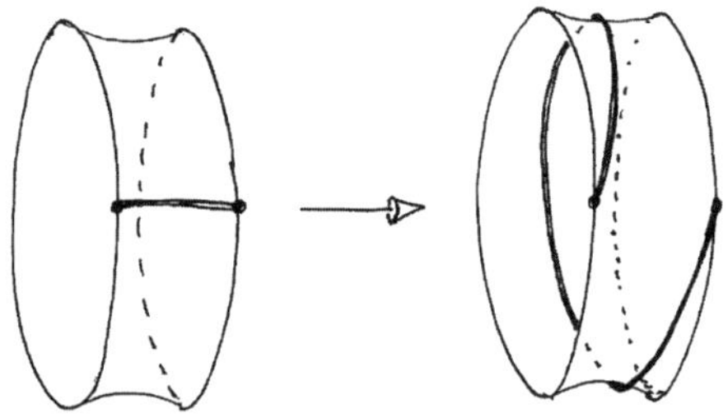

FIGURE 8. Dehn twist around an annulus.

If γ is trivial it bounds a disk (the Jordan curve theorem again) and τ_γ is trivial in $\mathrm{Mod}(S)$ by the Alexander trick.

If γ is nonseparating in S then $\Phi(\tau_\gamma)$ acts nontrivially in $H_1(S)$ and in fact has infinite order:

$$\Phi(\tau_\gamma)([\beta]) = [\beta] + i(\beta,\gamma)[\gamma]$$

where $i(\cdot,\cdot)$ is algebraic intersection number and $[\cdot]$ denotes homology class (see Figure 9). If γ is essential but separating then $[\gamma] = 0$ so $\Phi(\tau_\gamma)$ is the identity, but τ_γ still has infinite order in $\mathrm{Mod}(S)$. This requires more care; either some kind of careful discussion of intersection patterns of curves, or an appeal to the picture in the universal cover. We will return to this later. (Note "essential" means not bounding a disk *or* a once-punctured disk).

Pseudo-Anosov examples. In higher genus one can construct mapping classes similar to the Anosov case for the torus. For example, if an Anosov diffeomorphism lifts to a branched cover of the torus, we can lift the invariant foliations to obtain a pair of invariant foliations in the cover, defined everywhere except at the branch points. To make this explicit, let $T = \mathbb{R}^2/\mathbb{Z}^2$ be the square torus, and let X be obtained from two copies of T by slicing each along the image of the straight arc from $\mathbf{0} = (0,0)$ to $q = (\frac{1}{2},\frac{1}{2})$ and gluing crosswise. This gives a double branched cover $X \to T$ branched over $\mathbf{0}$ and q. The mapping $A = \left[\begin{smallmatrix} 4 & 1 \\ 7 & 2 \end{smallmatrix}\right]$ lifts to a map $\hat{A} : X \to X$, which preserves the lifted invariant foliations of A (Figure 10).

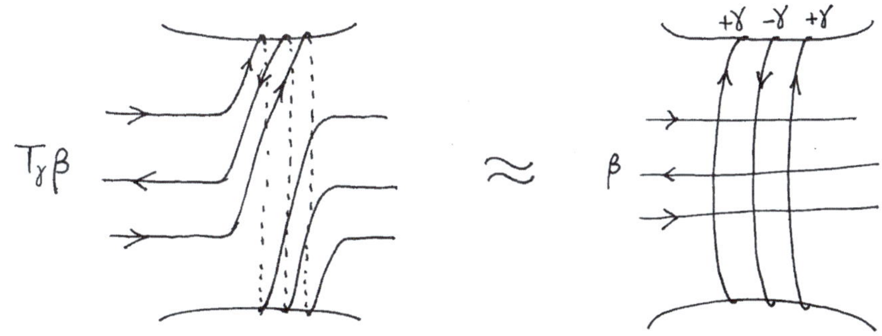

FIGURE 9. The effect of a Dehn twist on homology.

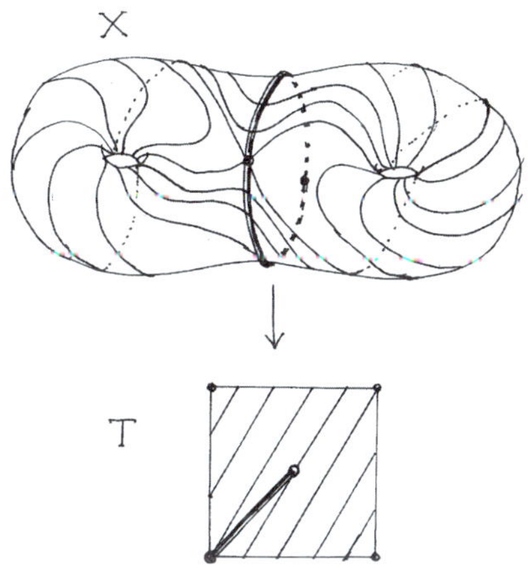

FIGURE 10. The double branched cover $X \to T$ and the expanding foliations of A and its lift.

The *L-shaped billiard table* is another example (Calta [**19**] and McMullen [**54**]). In Figure 11 we see the union of a 1×1 and $s \times s$ square, where $s = (1+\sqrt{5})/2$ is the golden ratio. Gluing the indicated opposite edges (A to A', B to B', etc) by Euclidean translations, we obtain a surface X of genus 2, with a Euclidean structure that has one singularity, the image of the eight indicated vertices.

There are two decompositions of X into a pair of annuli, as shown in Figure 12. The Dehn twists on the two vertical annuli can be simultaneously realized by an affine map, given in the Euclidean coordinates of the table as $\left[\begin{smallmatrix} 1 & 0 \\ 1 & 1 \end{smallmatrix}\right]$. The twists in the horizontal annuli can be similarly realized by $\left[\begin{smallmatrix} 1 & \lambda \\ 0 & 1 \end{smallmatrix}\right]$, where we leave it to the reader to deduce the value of λ. These two transformations generate a group of locally affine maps of X. In particular there are (many) compositions that result in a hyperbolic affine map, and hence admit the same eigendirection analysis as before.

Generators. It is a foundational fact that $\mathrm{Mod}(S)$ can be generated by finitely many elements. This allows the techniques of combinatorial group theory to be

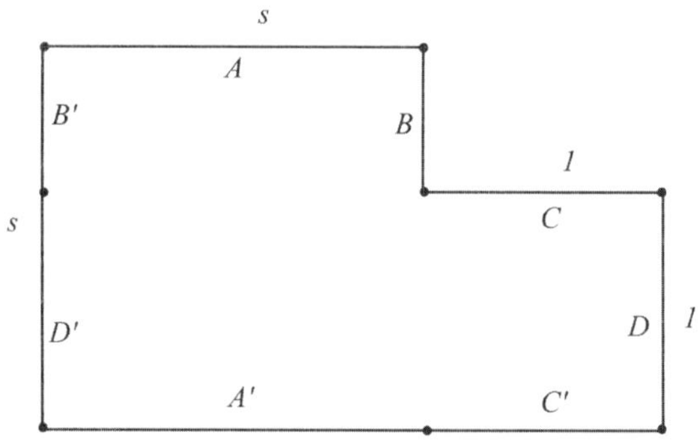

FIGURE 11. L-shaped table

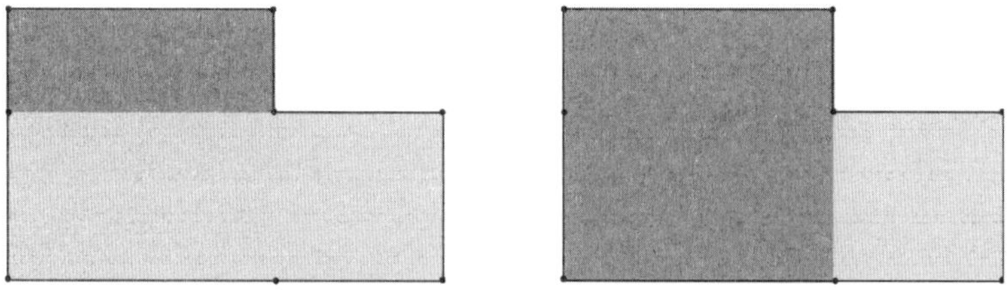

FIGURE 12. Two decompositions into pairs of annuli

applied to the study of Mod(S), as well as those of geometric group theory, as we will briefly discuss in the last section.

When S is closed the generators can be taken to be Dehn twists. If S has punctures we must be careful: A Dehn twist, being supported on a compact subset, cannot permute the punctures. Define the *pure mapping class group* PMod(S) to be the subgroup that fixes the punctures pointwise. We will sketch the proof of this theorem:

THEOREM 1.1. *For $g > 0$, $\mathrm{PMod}(S_{g,n})$ is generated by a finite collection of Dehn twists on nonseparating curves.*

Clearly $\mathrm{PMod}(S_{g,n}) = \mathrm{Mod}(S_{g,n})$ for $n \leq 1$. In general, it is easy to find a finite number of homeomorphisms that suffice to give all the permutations of the punctures, and append these to the twists to get a finite generating set for the whole group.

The proof we'll sketch, following Farb-Margalit [**26**], does not give a particularly efficient description of the generating set. In fact, quite explicit (and short) sets of twist generators exist – the first one was given by Dehn/Lickorish, and the shortest possible set ($2g + 1$ twists for $S_{g,0}$) was given by Humphries [**41**].

If one is not limited to Dehn twists, it is interesting to note that the number of generators need not grow with complexity of the surface. Indeed several authors, starting with Lickorish, have given generating sets as small as 2 for $S_{g,0}$ and $S_{g,1}$. See Wajnryb [**74**], Brendle-Farb [**16**] and Korkmaz [**47**] for recent results and discussions.

Sketch of the proof: Consider a graph $\mathcal{N}(S)$ whose vertices are (isotopy classes of) non-separating unoriented simple closed curves, and whose edges $[a,b]$ correspond to pairs of curves a and b that intersect exactly once. A configuration associated to an edge is indicated in Figure 13.

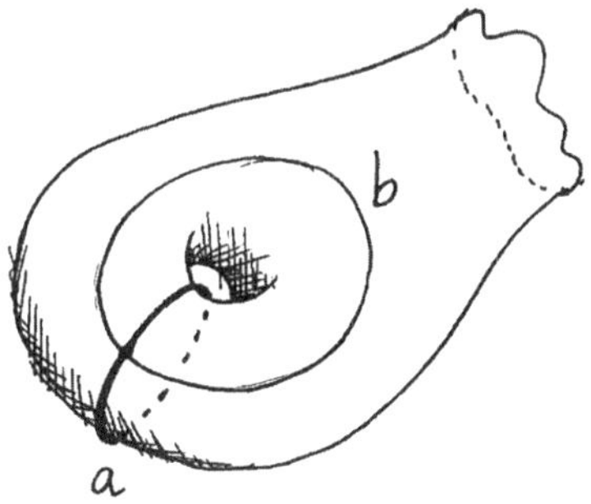

FIGURE 13. An edge $[a,b]$ in $\mathcal{N}(S)$.

Note that the natural action of PMod(S) on this graph is transitive on vertices, because the complement of a nonseparating curve in $S_{g,n}$ is always $S_{g-1,n+2}$. Thus if a and b are two such curves their complements are homeomorphic, and the homeomorphism can be chosen to "glue up" to a homeomorphism of S taking a to b. The action is transitive on edges because the complement of two curves that intersect once is always $S_{g-1,n+1}$. In fact, the action is transitive on *directed* edges: to find a homeomorphism that flips an edge $[a,b]$, observe that a regular neighborhood of $a \cup b$ is a torus with one hole. In the torus, the homeomorphism $\left[\begin{smallmatrix} 0 & -1 \\ 1 & 0 \end{smallmatrix}\right]$ interchanges the horizontal and vertical directions and fixes a point, so we can apply this to the regular neighborhood.

We will need this basic structural fact:

LEMMA 1.2. *$\mathcal{N}(S)$ is connected.*

as well as this inductive statement about stabilizers:

LEMMA 1.3. *The stabilizer $\Gamma_a <$ PMod(S) of a is contained in a subgroup generated by a finite collection of Dehn twists on nonseparating curves.*

(It is worth nothing that these nonseparating curves are not necessarily disjoint from a).

Assuming these lemmas for now, fix an edge $[a,b]$ of $\mathcal{N}(S)$. Consider any $f \in$ PMod(S). Lemma 1.2 implies that there is a sequence $a = a_0, a_1, \ldots, a_k = f(a)$ such that a_i and a_{i+1} intersect exactly once for each i.

The idea now, as indicated in Figure 14, is to successively "rotate" around the vertices a_i using the stabilizers Γ_{a_i}, carrying a to $f(a)$ only using a specified set of generators.

Consider the subgroup

$$H = \langle \Gamma_a, T_b \rangle.$$

Applying Lemma 1.3, we see that H is contained in a subgroup generated by a finite collection of twists on nonseparating curves.

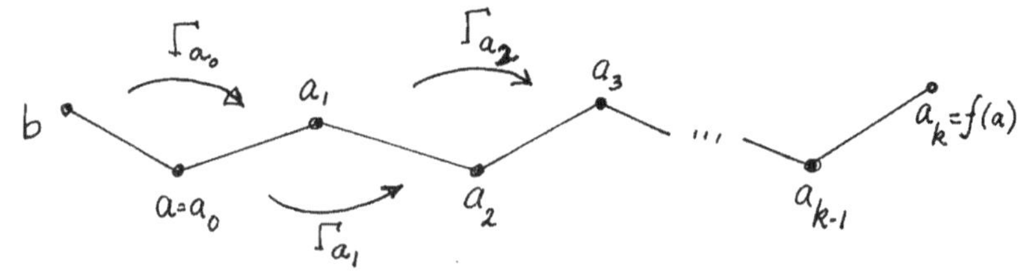

FIGURE 14. A path in $\mathcal{N}(S)$ connecting a to $f(a)$.

Since PMod(S) acts transitively on directed edges of $\mathcal{N}$, we see that Γ_a contains an element taking b to a_1, and hence (conjugating by this element) H contains the subgroup $H_0 = \langle \Gamma_a, T_{a_1} \rangle$.

Now note that $T_{a_i} T_{a_{i+1}}$ takes a_i to a_{i+1} (see the discussion on the braid relation, below). Hence since H_0 contains $T_{a_0} T_{a_1}$, conjugating by this element we see that H_0 contains Γ_{a_1}. Again by transitivity on directed edges, there is an element in Γ_{a_1} that takes a_0 to a_2. Conjugating by this element we find that H also contains T_{a_2}, and hence contains $H_1 = \langle \Gamma_{a_1}, T_{a_2} \rangle$.

This enables us to proceed inductively, concluding that H contains an element g taking a_0 to a_k. Thus $g^{-1} f \in \Gamma_a$, from which we can conclude that f itself is in H, and hence H is all of PMod(S). It remains to discuss the two lemmas.

Proof of Lemma 1.2: The first idea, given two curves α and ω that intersect many times, is to use surgery to replace ω by a curve ω' which is adjacent to it, and intersects α fewer times. Figure 15 shows the two cases that can arise when considering two intersections of ω with α that are successive along α. In the first, we indeed see that ω' intersects ω exactly once. However in the second case ω' and ω are disjoint, and ω' may be separating. This idea can easily be used to give a proof of connectivity of a related complex $X(S)$ whose vertices are all nontrivial isotopy classes and whose edges correspond to pairs of curves that intersect once or not at all.

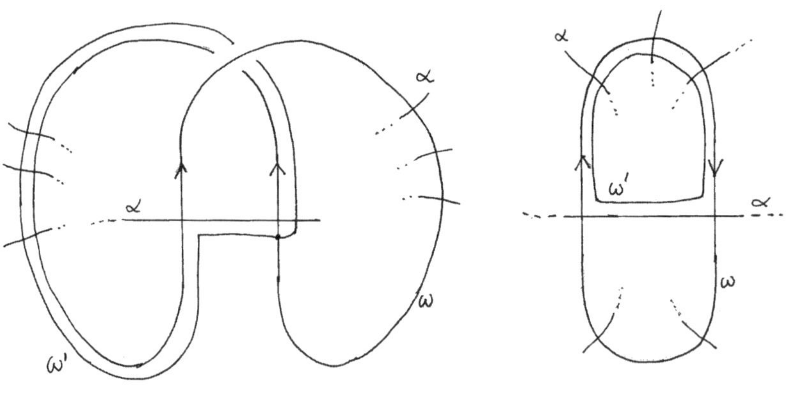

FIGURE 15. Surgery along the intersections of ω with α yields a curve ω' intersecting α fewer times.

Given a path (a_i) in $X(S)$, we can successively reduce the number of separating curves (keeping the same endpoints) with the following observation: if a_i separates S into two pieces then one, S', must have positive genus. If $a_{i\pm1}$ are in S' then, using induction (with base case $S_{1,1}$), they can be connected in $X(S')$ by a path with only nonseparating curves in its interior. If they are not in S' then a_i can be replaced by a nonseparating curve in S'.

Once we have a path with only nonseparating curves, we observe that for any two disjoint nonseparating curves there exists a third curve intersecting each of them exactly once. This gives us a path in $\mathcal{N}(S)$.

Proof of Lemma 1.3: The proof is by induction on complexity, with base case the torus (where the result is a standard fact about $\mathrm{SL}(2,\mathbb{Z})$).

Note first that we can generate the stabilizer Γ_a from $\mathrm{PMod}(S \setminus a)$ by adding a twist along a itself, and an element that interchanges the sides of a. A Dehn-twist construction for the latter is not hard to supply if $S \setminus a$ has positive genus: Let b be a curve intersecting a exactly once. A regular neighborhood of the union of a and b is a one-holed torus, and there a 180° rotation, which interchanges the sides of a, can be composed from T_a and T_b, as in $\mathrm{SL}(2,\mathbb{Z})$. Namely, T_aT_b is the order 6 map $\left[\begin{smallmatrix}0 & -1\\ 1 & 1\end{smallmatrix}\right]$ illustrated in Figure 5; hence we see that $(T_aT_b)^3$ corresponds in $\mathrm{SL}(2,\mathbb{Z})$ to $-I$, which interchanges the sides of a. (See however the end of the section for a bit more on this element).

Since $S \setminus a$ has strictly smaller genus than S, we would like to handle $\mathrm{PMod}(S \setminus a)$ inductively. However, note that the number of punctures has grown, so we also need an inductive step that reduces n. This can be done via the *Birman exact sequence,* which we will describe only briefly: The act of "forgetting" a puncture gives a map from $\mathrm{PMod}(S_{g,n+1})$ to $\mathrm{PMod}(S_{g,n})$, which fits into a short exact sequence like this:

$$1 \to \pi_1(S_{g,n}) \to \mathrm{PMod}(S_{g,n+1}) \to \mathrm{PMod}(S_{g,n}) \to 1.$$

The kernel is identified with $\pi_1(S_{g,n})$ because, given a map that is isotopic to the identity after forgetting the $n+1$-st puncture, following the puncture around the isotopy gives a loop in $S_{g,n}$. Conversely, given an element of π_1 we can "drag" the surface around the loop to get an element in the kernel of the forgetting map.

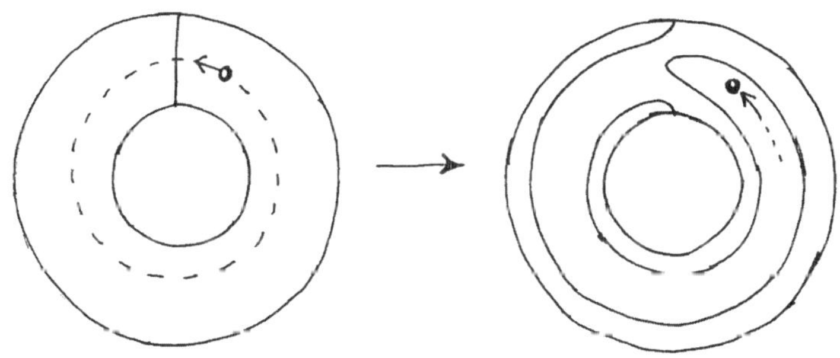

FIGURE 16. Dragging a puncture around a simple loop is a product of two Dehn twists in opposite directions.

Dragging around simple nonseparating loops in $S_{g,n}$ corresponds to products of Dehn twists on nonseparating loops in $S_{g,n+1}$ (see Figure 16, and Figure 17 for

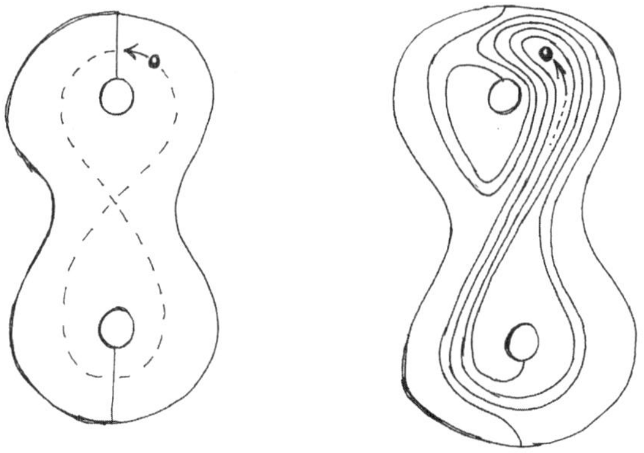

FIGURE 17. The dragging map corresponding to a self-intersecting loop.

an example with a nonsimple loop). Since $\pi_1(S_{g,n})$ is generated by such loops, we obtain enough twists in $S_{g,n+1}$ to enlarge a generating set for $\mathrm{PMod}(S_{g,n})$ to one for $\mathrm{PMod}(S_{g,n+1})$.

Relations. In fact $\mathrm{Mod}(S)$ is *finitely presented* – i.e. is completely determined by a finite list of generators and *relations*. We will not discuss the proof of this here, but will limit ourselves to pointing out some interesting relations:

Commuting elements: If two homeomorphisms are supported (i.e. different from the identity) on disjoint subsets, then they commute. This applies for instance to Dehn twists on disjoint curves.

Braid relation: If a and b intersect exactly once then one can easily check that $T_aT_b(a) = b$. This implies that T_aT_b conjugates T_a to T_b, or in other words

$$T_aT_bT_a = T_bT_aT_b.$$

This is known as the braid relation.

Root of a Dehn twist: Let c be a separating curve that cuts off a one-holed torus W, and let a and b be curves in W that intersect exactly once. Restricted to W, T_aT_b has order 6 (it corresponds to the elliptic element of $\mathrm{SL}(2,\mathbb{Z})$ we saw in the beginning of the section, and in the proof of Lemma 1.3). However, viewed as a homeomorphism of S as a whole, it satisfies

$$T_c = (T_aT_b)^6.$$

Figure 18 illustrates this. It indicates the lift of T_aT_b to the $\mathbb{Z}^2$-cover of the one-holed torus. This map is essentially the same as the $\pi/3$ rotation of Figure 5, except that it is required to move the circle boundaries by translations. It follows that after six iterations a full twist builds up around each circle.

Lantern relation: Let Y be the 4-holed sphere, with boundary components a, b, c, d and curves x, y, z labeled as in Figure 19. Imagine that Y is embedded in a surface S so each boundary component is essential. The reader is invited to check that

$$T_xT_yT_z = T_aT_bT_cT_d.$$

(See Farb-Margalit [**26**] for a more illuminating discussion of this).

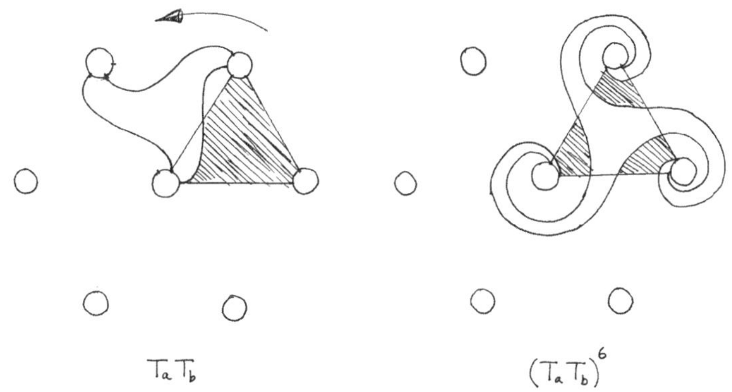

FIGURE 18. The map T_aT_b and its sixth power. The image of a selected triangle is shown for each.

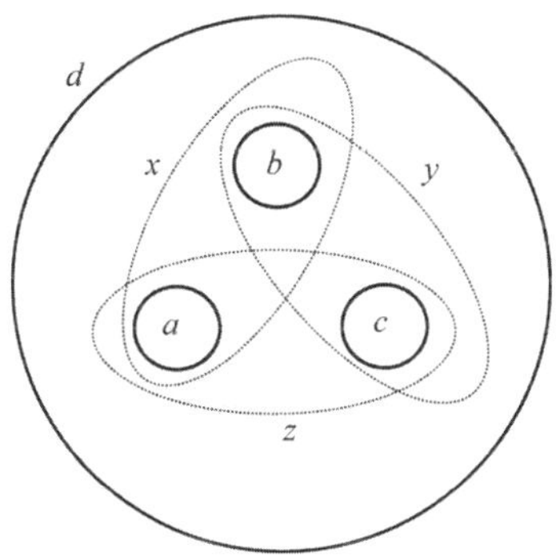

FIGURE 19. The lantern relation

2. Hyperbolic geometry, laminations and foliations

Hyperbolic plane and its boundary circle. We assume familiarity with the basic models of $\mathbb{H}^2$: First, the upper half plane

$$U = \{z \in \mathbb{C} : \operatorname{Im} z > 0\}$$

with the metric $ds^2/Im(z)^2$, where $ds^2 = dx^2 + dy^2$ is the Euclidean metric. The (orientation-preserving) isometry group is $\mathrm{PSL}(2,\mathbb{R})$, i.e. $\mathrm{SL}(2,\mathbb{R})/\pm 1$ acting by $z \mapsto (az+b)/(cz+d)$.

A suitable Möbius transformation M takes U to the unit disk, which gives another model of $\mathbb{H}^2$, where the metric is $4ds^2/(1-|z|^2)^2$. The isometry group is now the subgroup of $\mathrm{PSL}(2,\mathbb{C})$ conjugate to $\mathrm{PSL}(2,\mathbb{R})$ by M.

There is a natural circular boundary compactifying $\mathbb{H}^2$ to a disk on which isometries act homeomorphically. This boundary is identified with $\mathbb{R}\cup\{\infty\}$ for the upper half plane, and with the unit circle for the disk model.

Geodesics in both models are arcs of circles or lines that meet the boundary orthogonally. An important property (which is intrinsically a consequence of negative curvature) is that any two different geodesics are either disjoint or intersect exactly once. Geodesics are uniquely determined by their endpoints on the circle.

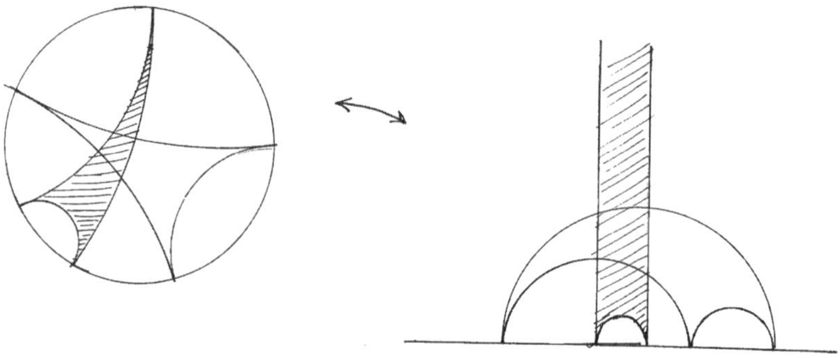

FIGURE 20. A configuration of ideal geodesic triangles in the hyperbolic plane, in both models.

Hyperbolic structures on surfaces. From now on we assume the Euler characteristic $\chi(S)$ is negative; that is, we exclude the sphere with 2 or fewer punctures, and the torus with no punctures. In this case the Uniformization Theorem gives (at least one) complete finite-area hyperbolic metric on S, or equivalently $S = \mathbb{H}^2/\Gamma$ for $\pi_1(S) \cong \Gamma < \mathrm{PSL}(2,\mathbb{R})$. Hence we can identify the universal cover of S with $\mathbb{H}^2$.

This picture yields some immediate consequences: Any nontrivial, nonperipheral (not homotopic to a puncture) homotopy class of curves has a *unique* geodesic representative. These representatives intersect themselves and each other in the minimal possible number of points for their homotopy classes (Figure 21).

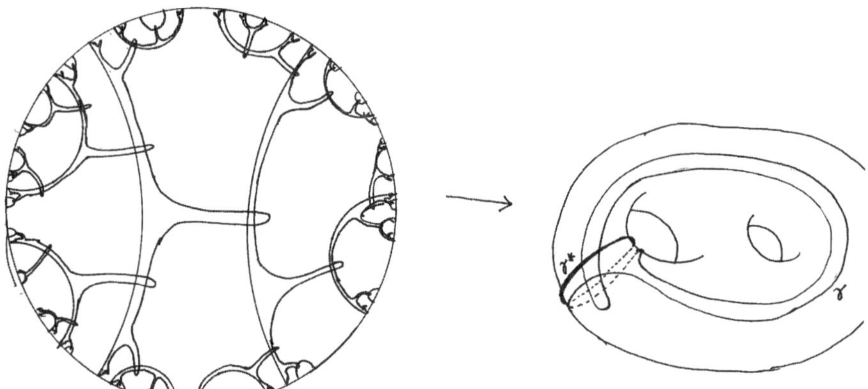

FIGURE 21. The curve γ has inessential self-intersections in S. Its geodesic representative γ^* does not. In the lift to $\mathbb{H}^2$ the untangling becomes visible.

Another consequence of uniqueness of geodesics is that an isometry isotopic to the identity must *be* the identity. We can see this by considering a system of geodesics that cuts S up into polygons. An isometry isotopic to the identity must fix all of these, hence their intersection points, and finally all the complementary polygons.

Hence the isometry group of S, which must be finite, injects in Mod(S). The Nielsen Realization Theorem gives a converse of this (see discussion in §4).

Lifting mapping classes to circle homeomorphisms. Any $[f] \in \text{Mod}(S)$ is realized by a bilipschitz map f, which can lift to a map $\widetilde{f} : \mathbb{H}^2 \to \mathbb{H}^2$ that conjugates $\Gamma \cong \pi_1(S)$ to itself. The lift is uniquely determined *up to the action of* Γ – that is, if $\beta \in \Gamma$ then $\beta \circ \widetilde{f}$ is also a lift, and conversely every lift has this form.

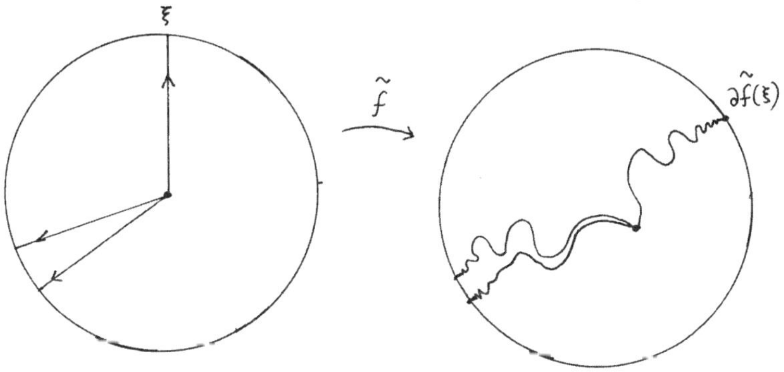

FIGURE 22. The image of a ray under a quasi-isometry has a well-defined endpoint.

Because $\widetilde{f}$ is bilipschitz, *it extends to the compactification* $\overline{\mathbb{H}^2}$, and in particular acts on the boundary circle $\partial\mathbb{H}^2$. To see this, consider a ray $r : [0, \infty) \to \mathbb{H}^2$ landing at a point $\xi \in \partial\mathbb{H}^2$. Assuming that r is parameterized by arclength, we see since $\widetilde{f}$ is bilipschitz that $\widetilde{f} \circ r$ escapes to infinity at a linear rate – that is, $d(\widetilde{f}(r(t)), 0) > Kt$. To simplify matters assume for a moment that f is differentiable. Then the Euclidean magnitude of the derivative of $\widetilde{f} \circ r$, in the disk model, is roughly equal to its hyperbolic magnitude multiplied by $\exp(-d(\widetilde{f}(r(t)), 0))$ (using the basic properties of the model). Integrating this we obtain the arclength of $\widetilde{f} \circ r$ in the model, and together with the linear escape rate we find this arclength is integrable and hence that the image path has a well-defined endpoint. The non-differentiable case follows from a similar coarse estimate. A slightly closer look at this discussion shows that the extension is continuous, and since the same analysis applies to $\widetilde{f}^{-1}$, the extension must be a homeomorphism.

We call the extension $\partial\widetilde{f}$. Moreover, note that if f and g are isotopic then this isotopy can be lifted to an isotopy between $\widetilde{f}$ and some lift $\widetilde{g}$ of g, which has *bounded-length tracks* (if there are punctures we need to take a bit more care). This implies that $\partial\widetilde{f} = \partial\widetilde{g}$ (Figure 23). Hence only the mapping class, and not its particular representative, matters when considering the circle maps.

The ambiguity in the lift $\widetilde{f} \mapsto \gamma\widetilde{f}$ for $\gamma \in \Gamma$ means that every element of Mod(S) determines a coset of Γ in $Homeo(S^1)$.

As an application, consider lifts of a Dehn twist on a (possibly separating) curve. An examination in the universal cover (Figure 24) shows that the extension to the circle is never a Möbius transformation, and hence the twist must be a nontrivial element of Mod(S) (and similarly for all its powers).

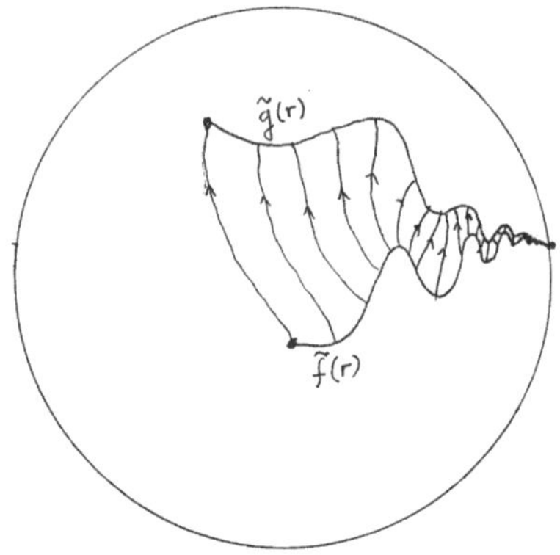

FIGURE 23. Lifts of homotopic homeomorphisms have the same extension to infinity.

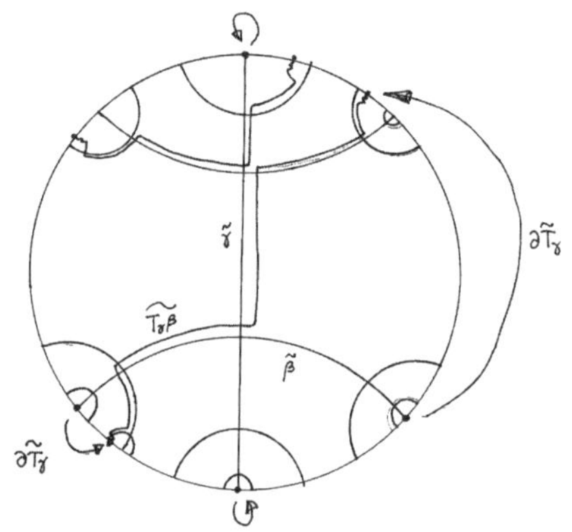

FIGURE 24. A lift of a Dehn twist T_γ to the universal cover is not a Möbius transformation. It fixes the endpoints of a lift $\tilde{\gamma}$, and moves the remaining two sides of the circle in opposite directions.

Geodesic laminations. If we consider a sequence of closed geodesics on a hyperbolic surface whose lengths go to infinity, we will notice that it tends to accumulate on Cantor-like patterns of infinite, disjoint geodesics. This phenomenon leads to the definition of geodesic laminations.

A *geodesic lamination* in a hyperbolic surface is a closed set λ foliated by geodesics. More precisely, for every point in λ there is a neighborhood U and a homeomorphism of pairs $(U, \lambda \cap U) \to ((0,1) \times (0,1), (0,1) \times K)$ where K is a compact subset of $(0,1)$. The preimages of the lines $(0,1) \times \{k\}$ for $k \in K$ are geodesic segments. (See Figure 25).

The simplest example is a simple closed geodesic; here K can be taken as a single point in a small enough neighborhood. For a "generic" lamination however, K will be a cantor set.

The complement $S \setminus \lambda$ is an open hyperbolic surface which is the interior of a complete surface with geodesic boundary (the completion is obtained by considering the *path metric* in $S \setminus \lambda$). The boundary naturally maps to certain leaves of λ,

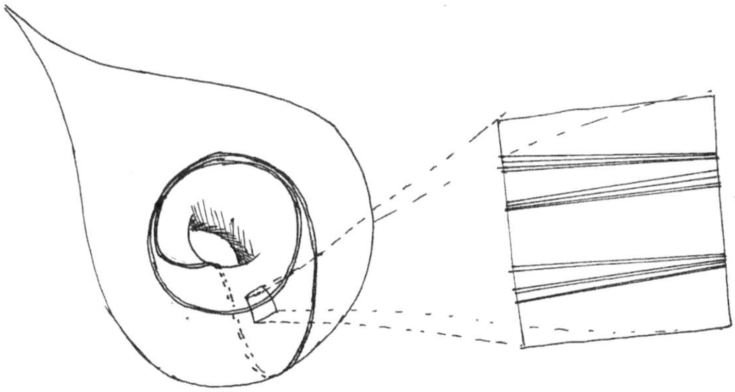

FIGURE 25. The local structure of a geodesic lamination.

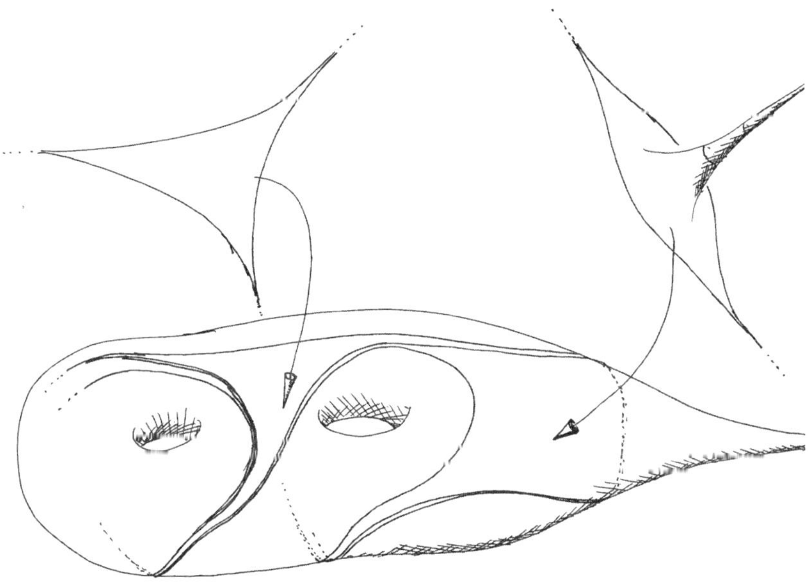

FIGURE 26. Two complementary components of this lamination are pictured; an ideal triangle and a one-cusped ideal bigon.

called boundary leaves. The simplest example of such a surface is an ideal polygon in $\mathbb{H}^2$ (Figure 26).

Assuming S has finite area, one can show that

$$area(S) = area(S \setminus \lambda);$$

and in particular λ *has zero area, and the transversal sets K have no interior.*

Here is a sketch of the argument: The Gauss-Bonnet theorem implies that the area of $S \setminus \lambda$ is $-2\pi\chi'(S \setminus \lambda)$, where χ' is the Euler characteristic $\chi(S \setminus \lambda)$ minus half the number of boundary cusps. By extending the tangent lines of $\partial(S \setminus \lambda)$ to a line field in the interior (see Figure 27) one sees that $\chi'(S \setminus \lambda)$ can be written as a Poincaré-Hopf sum of indices of singularities. By extending this line field continuously across λ one obtains $\chi'(S\smallsetminus\lambda) = \chi(S)$, and the Gauss-Bonnet Theorem applied to S completes the proof.

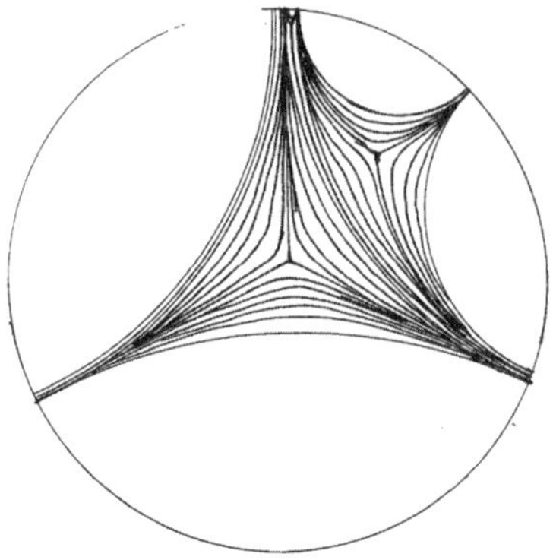

FIGURE 27. Extending a lamination to a singular line field.

(In fact, geodesic laminations are even thinner than that: A theorem of Birman-Series [**12**], see also Bonahon-Zhu [**75**], shows that a geodesic lamination has Hausdorff dimension 1, and more dramatically that the *union of all geodesic laminations in* S is a subset of Hausdorff dimension 1.)

The lift $\widetilde{\lambda}$ of λ to $\mathbb{H}^2$ is a lamination in $\mathbb{H}^2$, namely a Γ-invariant closed union of disjoint geodesics.

Note that $\widetilde{\lambda}$ is uniquely determined by the set of endpoint pairs of leaves, which gives a Γ-invariant closed set in $S^1 \times S^1 \smallsetminus$ *diagonal*. (Not every set will do: the endpoint pairs must satisfy a "non-linking" condition in the circle which corresponds to the leaves of the lamination not intersecting).

Let $\mathcal{GL}(S)$ be the set of all geodesic laminations on S. This set admits a natural $\mathrm{Mod}(S)$ action, via the lifts of mapping classes to the circle, and the description of laminations in terms of endpoint pairs. (Note that Γ-invariance of these sets implies the action of $f \in \mathrm{Mod}(S)$ is independent of choice of lift).

Topology of $\mathcal{GL}(S)$. Let X be a complete metric space and $C(X)$ the set of compact subsets of X. The *Hausdorff distance* between $A, B \in C(X)$ is the infimum of all $\epsilon > 0$ such that A is in the ϵ-neighborhood of B and vice versa. If X is compact it is not hard to see that $C(X)$ is compact.

We may consider $\mathcal{GL}(S)$ as a subset of $C(S)$, and claim:

LEMMA 2.1. *$\mathcal{GL}(S)$ is a closed subset of $C(S)$.*

When S is compact this implies that $\mathcal{GL}(S)$ is compact as well. In fact it is compact whenever S has finite area, but for simplicity we will focus from now on on the case that S is compact.

Let us sketch the proof. Given $\lambda_n \in \mathcal{GL}(S)$ converging to Z in $C(S)$ we must extract the leaf structure of Z. Note that Z must be a union of geodesics, because any limit in $C(S)$ of a sequence of geodesic segments of bounded length is a geodesic segment. However it is less obvious that these geodesics are all disjoint. A priori different subsequences can lead to intersecting geodesics, and in fact it is not clear that Z is not all of S!

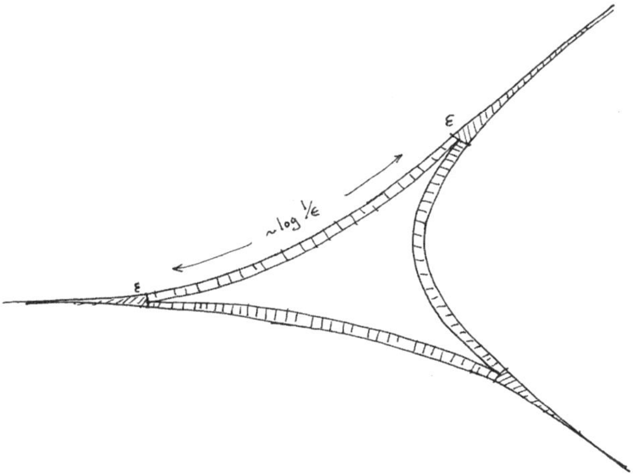

FIGURE 28. The area of an ϵ-neighborhood of the boundary of an ideal triangle.

The main enemy is the possibility that λ_n get "denser" in such a way that Z has nonempty interior. Hence, we need some type of "uniform sparseness". The main ingredient is to use the structure of $S \setminus \lambda$ to show that

$$area(\mathcal{N}_\epsilon(\lambda)) < O(\epsilon \log 1/\epsilon)$$

for any $\lambda \in \mathcal{GL}(S)$ (all that matters here is that the bound on the right is independent of λ, and goes to 0 with ϵ). In view of the fact that λ itself has area 0, it suffices to consider an ϵ-neighborhood of the boundary of a complementary component. To give the idea let us consider the case of an ideal triangle in $\mathbb{H}^2$ (Figure 28). Here, an ϵ neighborhood of the boundary is covered by three cusp neighborhoods bounded by horocycles of length roughly ϵ, each of area $O(\epsilon)$, and by three strips along the remaining segments of boundary, which have width ϵ and length $O(\log 1/\epsilon)$.

With this estimate in hand, consider a geodesic arc a of fixed length that makes an angle at least $\theta > 0$ with leaves of λ_n. The local product structure of λ_n, together with the estimate of area, implies that $a \cap \mathcal{N}_\epsilon(\lambda_n)$ has length going to 0 with ϵ (keeping θ fixed). Hence, for high n, a cannot be too close to a geodesic arc in Z. From this we can conclude that the geodesics in Z are simultaneously aligned with λ_n for all high enough n, and the result follows from this.

Measured laminations. A *transverse measure* on a lamination λ is a family of Borel measures on transversal arcs, invariant by holonomy (sliding transversals along λ, supported on the intersection with λ, and additive over concatenations of transversals.

The basic example is a simple closed geodesic with "counting measure" – the total measure of a transversal arc is the number of intersection points with λ.

Every lamination admits some measure: consider a leaf ℓ in λ and take the counting measure associated to a very long segment in ℓ. This is not a transverse measure – holonomy can slide transversals past the endpoint and the measure changes. However if we take a sequence of segments whose length goes to ∞, and then rescale the counting measures appropriately, we obtain a limit which

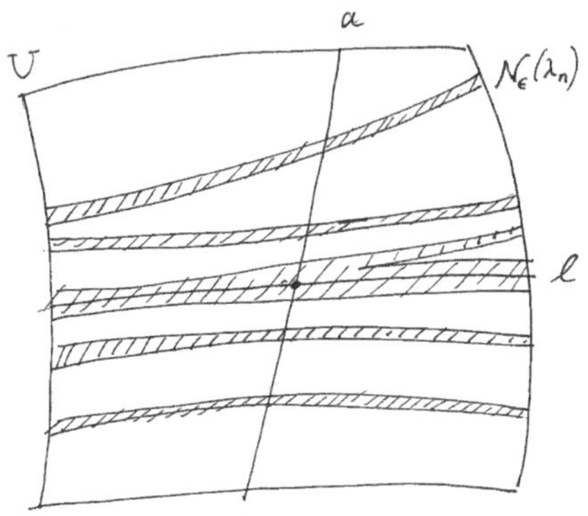

FIGURE 29. In a lamination chart for λ_n, the area estimate implies that a transversal making definite angle with λ_n has small intersection with an ϵ-neighborhood.

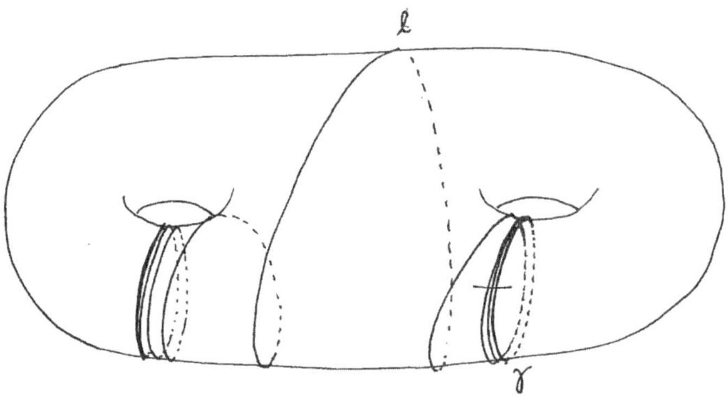

FIGURE 30. The leaf ℓ cannot support a positive measure because it intersects every transversal of γ infinitely many times.

does satisfy holonomy invariance. This is a weak-* limit, meaning that on any fixed transversal the measures converge, when integrated against any continuous function.

On the other hand the support of such a measure may not be all of λ. For example if γ is a closed leaf of λ and ℓ a leaf that spirals onto γ, ℓ cannot support a positive amount of measure, because then the measure on a transversal crossing γ would be infinite.

One can show that the number of such isolated leaves is uniformly bounded, and once they are removed the remainder falls into a bounded number of minimal components that are the supports of measures. We state this without proof:

LEMMA 2.2. *Each $\lambda \in \mathcal{GL}(S)$ can be written as a union*

$$\lambda = \delta \cup \lambda'$$

where λ' is a nonempty lamination admitting a transverse measure of full support, and δ is a finite (possibly empty) collection of isolated leaves that accumulate on λ'.

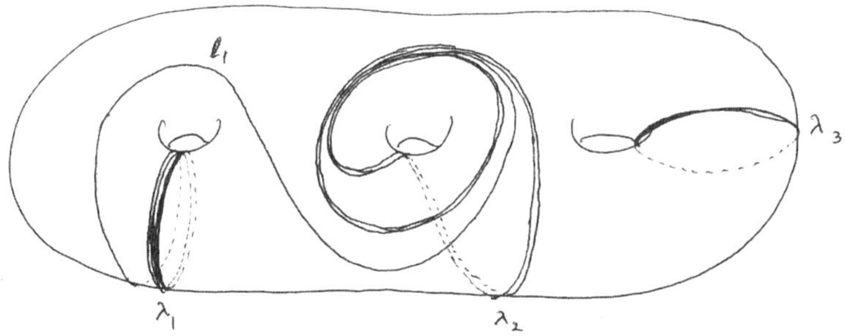

FIGURE 31. A lamination with three minimal pieces $\lambda' = \lambda_1 \cup \lambda_2 \cup \lambda_3$, and one isolated leaf $\delta = \ell_1$.

Each component of λ' is minimal, meaning that every leaf (and even half leaf) of it is dense in that component.

The number of components of λ' and leaves of δ is bounded in terms of the topological type of S.

The space $\mathcal{ML}(S)$ of all measured laminations admits a natural topology induced from weak-* convergence on transversals. One of Thurston's striking discoveries was that $\mathcal{ML}(S)$ is homeomorphic to Euclidean space, and that this has implications for Mod(S), for the structure of Teichmüller space, and for the theory of hyperbolic 3-manifolds. We will not develop this further here, however, because we will sketch a proof of the classification theorem that sidesteps this part of the theory. Let us also record the notation $\mathcal{PML}(S)$, or projectivized measured laminations, for the quotient of $\mathcal{ML}(S)$ minus the empty lamination by the natural equivalence of scaling of measures.

Measured foliations. For our description of pseudo-Anosov maps, it will be helpful to have the related and equivalent structure of *measured foliations*. A measured foliation is a foliation of all S minus a finite number of singular points, with a standard "pronged" local picture at each singularity (and puncture), together with a transverse measure defined as above, except that the holonomy maps cannot slide past the singular points. (See Figure 10 for an example).

Measured foliations do not have geodesic leaves, and we consider them up to an equivalence relation generated by isotopy and "Whitehead moves", which are collapses of singularities along compact leaves. With this we obtain a space $\mathcal{MF}(S)$ and $\mathcal{PMF}(S)$, admitting actions by Mod(S).

There is a natural homeomorphism $\mathcal{MF}(S) \to \mathcal{ML}(S)$, which respects all this structure. It is essentially the "straightening map": lifting a foliation to $\mathbb{H}^2$ one can show that each leaf is a "quasi-geodesic" and has distinct endpoints on the circle, which are then connected by a geodesic arc. Conversely the space between the leaves of a geodesic lamination can be "collapsed" to yield a foliation. (See Figure 32).

(We caution the reader that this is not the same as the extended foliation of Figure 27 – that foliation has many parallel leaves that do not support a non-zero transverse measure.)

Definition of pseudo-Anosov. The examples from Section 1 can now be described as part of a general class. A map $f : S \to S$ is *pseudo-Anosov* if, after

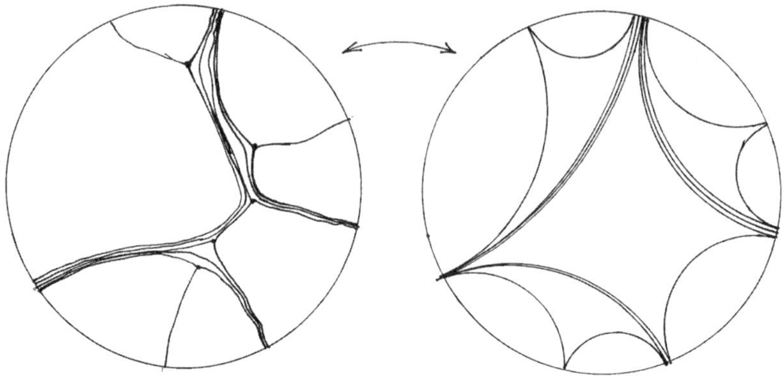

FIGURE 32. Straightening a foliation yields a lamination, and collapsing a lamination yields a foliation.

isotopy, there is a pair of measured foliations F^+, F^- that can be realized transversely and with the same singular points, so that f takes the leaves of F^+ to each other and those of F^- to each other, and there is an $m > 1$ so that f multiplies the transverse measure of F^+ by m and that of F^- by $1/m$.

Equivalently, the transverse measures of F^+ and F^- give *locally Euclidean charts* on S minus the singularities, such that leaves of F^+ and F^- map to vertical and horizontal lines, respectively, and the action of f is locally given, up to translation, by the matrix $\left[\begin{smallmatrix} m & 0 \\ 0 & 1/m \end{smallmatrix}\right]$.

One can of course describe this in terms of a transverse pair of measured geodesic laminations, which f takes to multiples of themselves by m and $1/m$. In our proof of the Nielsen-Thurston theorem much of the discussion will take place in the lamination setting.

3. The Nielsen-Thurston classification theorem

In Section 1 we observed that for nontrivial mapping classes of the torus there is a trichotomy: elliptic (rotation), parabolic (twist), and hyperbolic (Anosov). This classification is generalized by the following celebrated theorem:

THEOREM 3.1. *For every $f \in \mathrm{Mod}(S)$ one of the following holds:*

(1) (Elliptic) *f has finite order,*
(2) (Reducible) *There is a system C of disjoint essential simple closed curves such that $f(C) = C$ (up to isotopy),*
(3) *f has a pseudo-Anosov representative.*

Note that (1) and (2) are not mutually exclusive. On the other hand a pseudo-Anosov map cannot be elliptic or reducible, because every simple closed curve c crosses the foliations in an essential way, and the action of f stretches c in the direction of F^- and increases its length exponentially fast.

The reducible case generalizes the twist case for the torus, in which the core of the Dehn twist is the invariant curve system. In general once a reducing system C is found the classification can be applied inductively to its complementary surface. In the next section we will sketch a more detailed description of the reducible case, which involves identifying a *canonical* reducing system C.

The proof that we will sketch here closely follows the proof given by Casson in [**20**] (see also Gilman [**30**] for connections to Nielsen's work). It is different from Thurston's original proof, which depends more strongly on the topological structure of $\mathcal{PML}(S)$ and its use in the compactification of Teichmüller space, nor is it Bers' proof, which uses the metric structure of Teichmüller space.

Proof: (For simplicity we restrict to the case of a closed surface, of genus 2 or higher.) In this section we will actually prove a slightly weaker statement, namely that if f satisfies this condition:

(*) For every isotopy class of simple closed curves c in S, the orbit $\{f^n(c)\}$ is infinite, then f is pseudo-Anosov.

Note that (*) immediately implies that f is not finite-order or reducible, but the opposite implication requires a bit more work.

Given f satisfying (*), we need to find the invariant laminations/foliations. Since $\mathcal{GL}(S)$ is compact (Lemma 2.1), we can extract an accumulation point $\lambda \in \mathcal{GL}(S)$ for the infinite sequence $\{f^n(c)\}_{n>0}$. The rest of our argument can be summarized as follows:

Invariance: Lemma 2.2 on the structure of laminations gives us

$$\lambda = \delta \sqcup \lambda',$$

where λ' is the measurable part. Using condition (*) we prove that $f(\lambda') = \lambda'$, and that λ' *fills* S, i.e. components of $S \setminus \lambda$ are ideal polygons.

Dynamics: Study the actions of lifts $\widetilde{f}$ on $\partial\mathbb{H}^2$. Show that we get "north-south dynamics" where endpoints of leaves of $\widetilde{\lambda}'$ are attracting fixed points, and the repelling fixed points are endpoints of leaves of the corresponding lamination for f^{-1}.

This is enough to give the *topological* picture of two transverse invariant laminations (or, after collapse, foliations).

Measures: Build the invariant transverse measures, using a Markov partition and the Perron-Frobenius theorem.

We give more details below.

Invariance: We have $\lambda = \lim f^{n_i}(c)$ for some subsequence $n_i \to \infty$. It is not a priori clear from this that λ should be invariant. However, note that λ' has the following property:

(**) A leaf of λ' is not isolated in λ.

For if a leaf ℓ were isolated, it would be a closed leaf (each component of λ' is minimal). But then, being isolated means that eventually $f^{n_i}(c)$ just *equals* ℓ, since a neighborhood of ℓ meets no other curves in the limit. But this contradicts the irreducibility assumption.

Hence, there are infinitely many leaf segments of λ approximating any segment of λ'. This implies that

(***) λ' and $f^m(\lambda')$ have no transverse intersection for any $m \geq 0$.

For, suppose that they did: Fact (**) implies that λ and $f^m(\lambda)$ have infinitely many transverse intersections in a neighborhood of this point (Figure 33), and so $f^{n_i}(c)$ and $f^{n_i+m}(c)$ have a number of intersection points that goes to ∞. Since f^{n_i} is a homeomorphism this means that c and $f^m(c)$ have a number of intersections going to ∞. But m is fixed here! This is a contradiction.

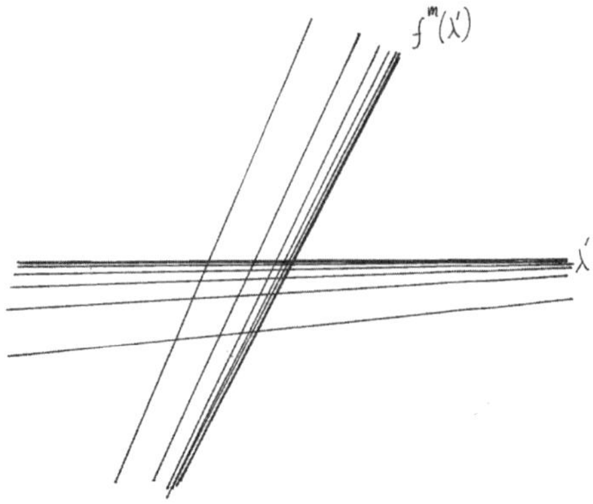

FIGURE 33. A transverse intersection of λ' and $f^m(\lambda')$ implies infinitely many intersections of λ and $f^m(\lambda)$.

We conclude that λ' and $f^m(\lambda')$ are either disjoint or their intersection is a sublamination of both. Hence their union is a lamination, and similarly $\lambda' \cup f(\lambda') \cup f^2(\lambda') \cup \cdots$ is a lamination. Lemma 2.2 gives a uniform upper bound to the number of components of a lamination, so this sequence eventually stabilizes. The union gives an invariant lamination μ.

However if μ is not filling, it is carried on a proper subsurface, and this gives a reducing system of curves for f.

We conclude that μ is filling, and hence minimal, and hence that all the pieces were already filling and equal to each other. So $\lambda' = f(\lambda')$, and λ' is filling.

North-South Dynamics: We now study the action on the circle of the lifts of f to the universal cover.

We can immediately learn something from the finitely many complementary regions of the lamination λ'. Each one is an ideal polygon, and f must permute them. Thus there is some power f^p which fixes each polygon, and moreover fixes each of its vertices. Lifting to the universal cover, we find that for each lift P of a complementary polygon there is a corresponding lift $\widetilde{f^p}$ which fixes its endpoints on the circle.

The following lemma gives more details about the dynamics of these fixed points, not just for these special lifts but for all lifts (see Figures 34 and 35).

LEMMA 3.2. *For any lift $\widetilde{f}$ of f, there is a power $\widetilde{f^p}$ which acts on the circle with an even number of fixed points, which are alternately attracting and repelling.*

When there are two or more attracting fixed points, they are the vertices of a leaf or a complementary region of $\widetilde{\lambda}'$. All other fixed points are not endpoints of leaves of $\widetilde{\lambda}'$.

The main step in the proof is an analysis of the following situation. Let ℓ be a leaf of $\widetilde{\lambda}'$. Its boundary divides the circle into two intervals I and J, and we say ℓ is *non-isolated in I* if there is a sequence of leaves of $\widetilde{\lambda}'$ converging to ℓ whose endpoints are in I (Figure 36).

The dynamics on the non-isolated side of a fixed leaf are determined by this lemma (see Figure 37):

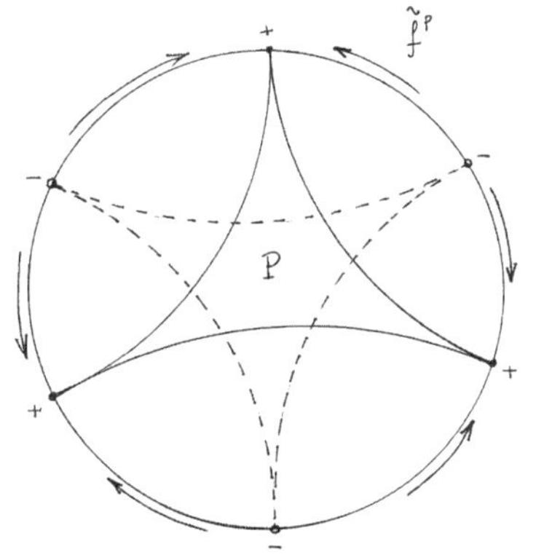

FIGURE 34. North-South dynamics of a lift of f^p which fixes the vertices of a complementary polygon.

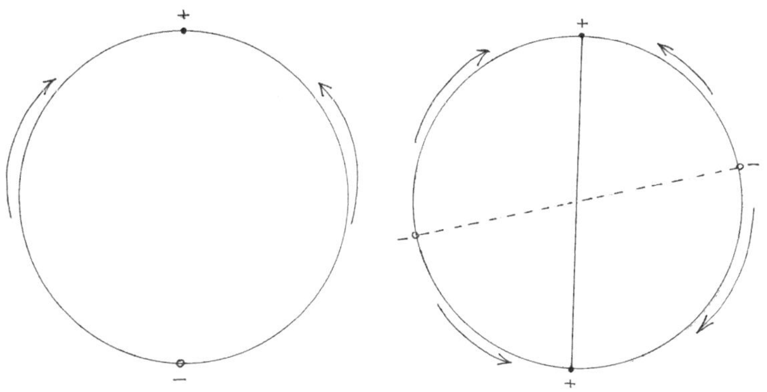

FIGURE 35. North-South dynamics in the remaining cases of 2 and 4 fixed points.

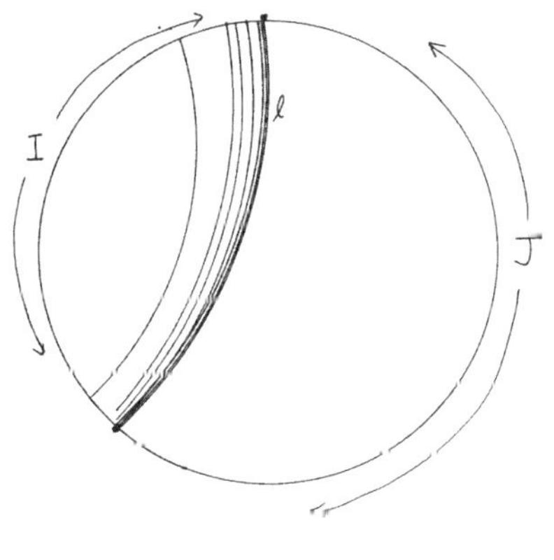

FIGURE 36. The leaf ℓ is non-isolated on the side I.

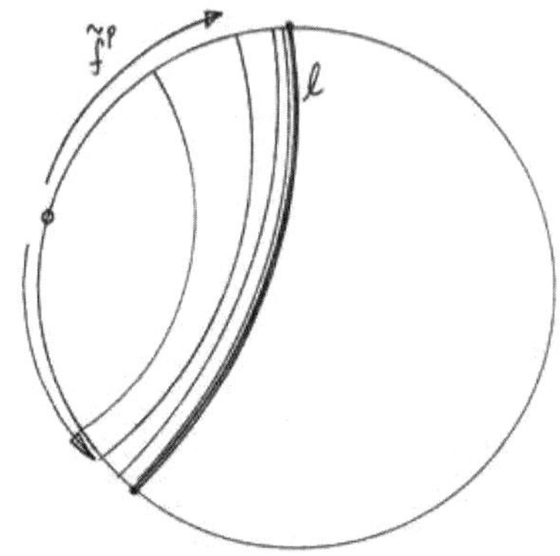

FIGURE 37. The dynamics of $\widetilde{f^p}$ on a non-isolated side of a fixed leaf.

LEMMA 3.3. *Let ℓ be a leaf of $\widetilde{\lambda}'$ whose endpoints are fixed by a power $\widetilde{f^p}$ of a lift of f. If ℓ is non-isolated on the side I then the endpoints of ℓ are attracting fixed points of $\widetilde{f^p}|_I$, and $\widetilde{f^p}|_I$ has a single repelling fixed point in int(I).*

For simplicity we will sketch the proof of this in the case where the power p is 1. Because every leaf of λ' is dense in λ' and λ' is filling, ℓ crosses a lift $\widetilde{c}$ of c. Since $f^{n_i}(c)$ converges to λ, the limit of $\widetilde{f^{n_i}}(\partial\widetilde{c})$ cannot have a point in the interior of I, because this would imply transversal self-intersections of λ (Figure 38).

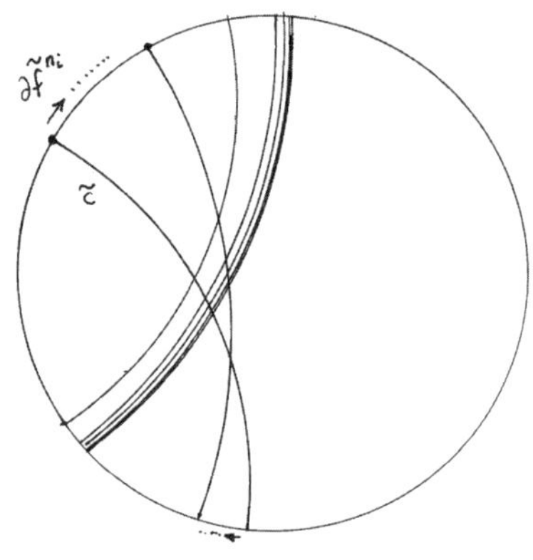

FIGURE 38. If the endpoint of $\widetilde{c}$ in I converges to the interior then since lim $f^{n_i}(c)$ is in λ, transversal self-intersections of λ occur.

Hence the endpoint of $\widetilde{c}$ in I converges to an endpoint of ℓ, so that at least that endpoint of I is attracting in a neighborhood in I.

A bit of care, again involving the avoidance of transverse self-intersections, shows that the other endpoint is attracting too. Now we claim that in I there is one other fixed point which is repelling, and that's the North-South dynamics we wanted. We show this by *backward*-iterating a leaf that lies near ℓ in I, and arguing that if it accumulates on a leaf (and not a single point) then that leaf has repelling behavior for $\widetilde{f}$, but that contradicts what we just proved applied to the new leaf (Figure 39).

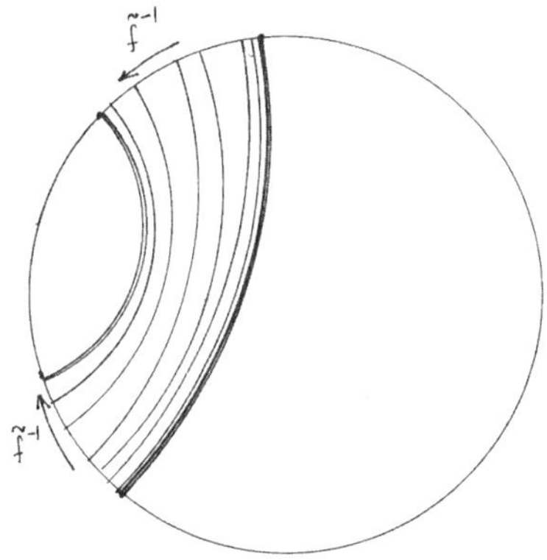

FIGURE 39. If the backward iterates of the leaf near ℓ converge to a leaf, we obtain a contradiction.

Now, given any polygon P of $\widetilde{\lambda}'$ which is fixed by a power of a lift $\widetilde{f}$, note that each side of P is non-isolated on the interval which is outside of P. Lemma 3.3 therefore gives us exactly the dynamical picture we are looking for.

A leaf of $\widetilde{\lambda}'$ which is not on the boundary of a polygon is non-isolated on both sides, and so if such a leaf is fixed by a power of a lift, we again get the desired picture.

Now considering an arbitrary lift $\widetilde{f}$, it remains to show that if no power fixes any leaves of $\widetilde{\lambda}'$ then a power must have exactly one attracting and one repelling fixed points. This can be done by similar methods, and we omit it.

Finally, this argument can also be applied to f^{-1}, and it follows immediately that the repelling fixed points of lifts of f must be the attracting fixed points for the corresponding lamination μ' for f^{-1}. Hence the two laminations are transverse and have the dynamics we expect.

Transverse measures: Let us collapse the laminations to transverse singular foliations L (for λ') and M (for μ'). (This isn't strictly necessary but improves the picture a bit.) Call the leaves of L "horizontal" and those of M "vertical". Note that a singular point appears for every transverse pair of polygons of λ' and μ' (Figure 40). Remember that we have replaced f by a power, so that it fixes each singular point as well as the leaves of L and M that meet it.

We find a decomposition of S into "rectangles" as follows: At each singular point choose horizontal initial segments of each leaf of L that meets it. Fixing these for the moment, extend from each singular point a vertical leaf that continues until it first meets one of the horizontal initial segments. This must occur because every leaf of L (and M) is dense, since the original laminations were filling.

Having fixed these vertical segments, continue the original horizontal segments until they hit the vertical segments (Figure 41).

The complement of this system of segments must be a finite union of disks, which we call "rectangles" because each one of them has two boundary segments along L and two along M, and is foliated on the interior by L and M, without singularities. Call the L and M boundaries of a rectangle R $\partial_h R$ and $\partial_v R$, respectively.

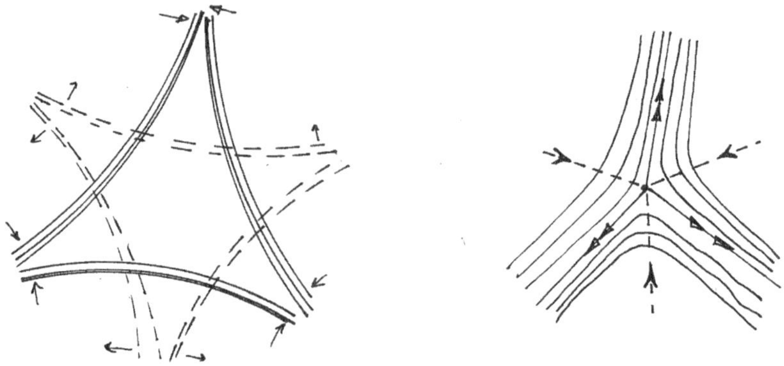

FIGURE 40. After collapsing, the boundary leaves of a polygon of λ' become expanding leaves meeting a singularity, and the boundary leaves of the corresponding polygon of μ' become contracting leaves.

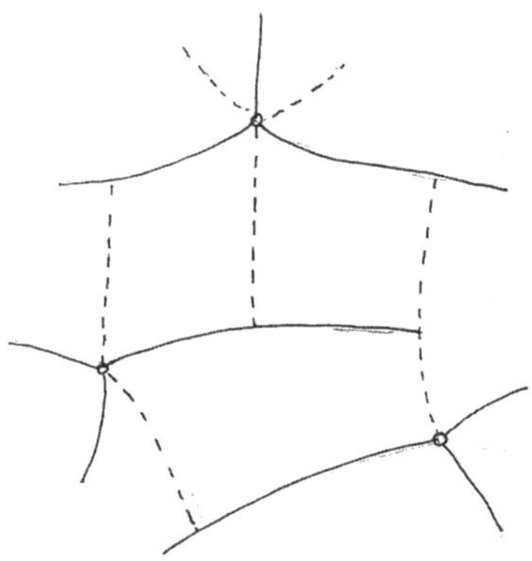

FIGURE 41. A configuration of rectangles bounded by singular leaves of L and M.

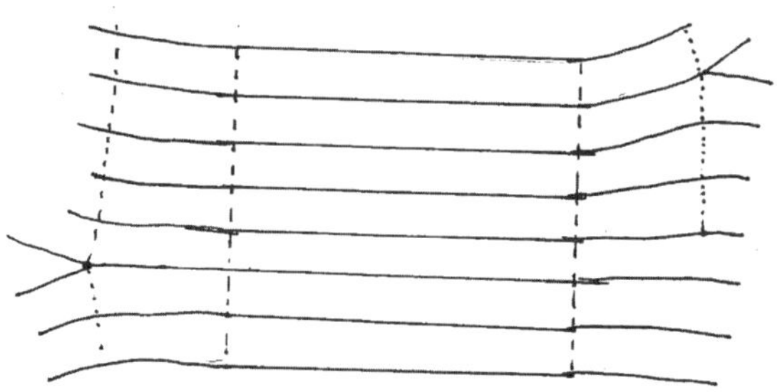

FIGURE 42. The images $f(R_i)$ cut each R_j into a vertical stack of subrectangles. Similarly R_j) is cut by $f^{-1}(R_i)$ into a horizontal sequence of subrectangles.

Number the rectangles $R_1, \dots, R_k$. Because f contracts leaves of M at the singularities and f^{-1} contracts leaves of L (this follows immediately from our north-south dynamics upstairs), we can see that the following holds, for each R_i:

$$f(\partial_v R_i) \subset \bigcup_j \partial_v R_j,$$

$$f^{-1}(\partial_h R_i) \subset \bigcup_j \partial_h R_j.$$

So $\{R_i\}$ is a Markov partition, and the images $f(R_i)$ cut each rectangle cleanly into subrectangles (Figure 42).

We are trying to find invariant transverse measures on λ' and μ', or what is equivalent, a (singular) Euclidean metric on the foliated surface such that L and M are horizontal and vertical straight lines, on which f acts like a hyperbolic linear map $\left[\begin{smallmatrix} m & 0 \\ 0 & 1/m \end{smallmatrix}\right]$, for some $m > 1$.

If we were to succeed, each R_i would inherit a height and a width. Moreover the vectors of heights and widths, $\mathbf{h} = (h(R_i))$ and $\mathbf{w} = (w(R_i))$, would satisfy an eigenvalue equation, as follows:

Because of the Markov condition, each R_i is decomposed by the images $f(R_j)$ into a vertical stack of subrectangles, and similarly each R_i is decomposed by the preimages $f^{-1}(R_j)$ into a horizontal stack of subrectangles. Let A be the incidence matrix, i.e. A_{ij} is the number of subrectangles in $R_i \cap f(R_j)$. Note that (applying f^{-1}) A_{ji} counts the subrectangles of $f^{-1}(R_j) \cap R_i$, so that A^t is the incidence matrix for f^{-1}.

If we have found the right metric then $h(R_i) = \sum_j A_{ij} 1/m h(R_j)$, or in other words $A\mathbf{h} = m\mathbf{h}$ where h is the vector of heights. Similarly for the widths we would have $A^t\mathbf{w} = m\mathbf{w}$.

Now A is a matrix of non-negative integers. Moreover for some power A^n, all of the entries are positive: this is because the forward image of each leaf segment starting at a singularity must eventually cover the entire leaf, so the image of each rectangle eventually meets every rectangle. A^n is of course the incidence matrix for f^n.

The Perron-Frobenius theorem tells us that A has a unique eigenvector with positive entries, and that its eigenvalue m is the unique largest positive eigenvalue of A. This gives us the vector of heights. Since A^t has the same set of eigenvalues, the same argument gives the desired vector of widths, with the same eigenvalue m.

4. Classification continued, and consequences

We are almost done with the proof of the classification theorem: we have shown that, given condition (*) (all curves have infinite orbits), the map f has a pair of invariant laminations (or foliations), with transverse measures that are expanded by f and its inverse, respectively, by the same factor.

It remains to check that condition (*) is equivalent to irreducibility. It is clear that (*) implies irreducibility, but we must also show that if (*) fails, i.e. if there is a curve with a finite orbit, then f admits an invariant system of disjoint curves, and hence is reducible.

To do this we introduce the idea of canonical reducing systems. This will also allow us to better understand the reducible mapping classes, and to obtain some consequences about the structure of subgroups of $\mathrm{Mod}(S)$.

DEFINITION 4.1. *The* canonical reducing system C_f *for* $f \in \mathrm{Mod}(S)$ *is the set of (isotopy classes of) simple closed curves* c *such that*

- *the orbit* $\{f^n(c)\}$ *is finite, and*
- *for every curve* b *crossing* c *essentially,* $\{f^n(b)\}$ *is infinite.*

It is immediately clear that:

- C_f is a system of *disjoint* curves (hence finite).
- C_f is empty if f is finite-order or if (*) holds.
- $C_{f^n} = C_f$ for any power $n \neq 0$.
- C_f is natural, i.e. $g(C_f) = C_{gfg^{-1}}$. In particular $f(C_f) = C_f$, and in fact $g(C_f) = C_f$ whenever g and f commute.

It takes a bit more care to check that

- C_f is non-empty if f does not satisfy condition (*) but is of infinite order.

This we do by induction. Given such an f, let c be a curve with finite orbit. For some power f^p, c is fixed and its complementary component(s) are each fixed.

Suppose in each component Y, f^p is finite order. Then a further power f^q is the identity on each component. Since f is infinite order, f^q must be a (power of a) Dehn twist on c, and then every curve that crosses c has infinite orbit, hence $c \in C_{f^q} = C_f$.

Suppose in some Y $f^p|_Y$ satisfies (*), and hence is pseudo-Anosov. Then (here we have to think a bit about the dynamical structure we have been discussing) every curve that crosses c essentially must intersect Y in some collection of essential arcs, and these cross the laminations of $f^p|_Y$, hence their images grow exponentially in length and in particular have infinite orbits. Again $c \in C_{f^p} = C_f$.

Finally we have the possibility that some $f^p|_Y$ fails to satisfy (*) but has infinite order. Then by induction $C_{f^p|_Y}$ is nonempty. It follows that C_f is nonempty, though one should take a bit of care showing that curves not contained in Y but crossing a curve of $C_{f^p|_Y}$ have infinite orbits.

This concludes the proof of the classification theorem, and we also have a more complete picture: Given an infinite-order reducible f, we have a canonical nonempty invariant curve system C_f. The complementary regions of C_f are permuted by f, and some power f^p fixes them. In each component, f^p is either pseudo-Anosov or finite order (else there would be even more curves of C_f inside). This gives us the beginnings of a discussion about basic types of subgroups of $\mathrm{Mod}(S)$.

Abelian subgroups. An obvious way to get an abelian subgroup of $\mathrm{Mod}(S)$ is to decompose the surface into pieces and choose homeomorphisms supported in each piece. Birman-Lubotzky-McCarthy [**11**] showed that this is essentially all that can happen; in fact they showed that these are the only possibilities for (virtually) solvable subgroups as well.

The canonical reduction system is the tool for doing this. If f and g commute then $g(C_f) = C_f$ and $f(C_g) = C_g$, by naturality. It follows immediately that C_f and C_g cannot intersect. Hence the union is an invariant curve system for the group generated by f and g. A finite-index subgroup must therefore fix every component of the resulting decomposition.

We will see below that two pseudo-Anosovs commute only if they are powers of a common element. An immediate consequence of all this is a bound $(3g-3+n)$ on the rank of any free abelian subgroup of $S_{g,n}$. The obvious maximal-rank groups are those generated by Dehn twists on the curves of a pants decomposition of S, but these are *not* the only possibilities (exercise).

Free groups, ping pong and the Tits alternative. One can find many free subgroups in $\mathrm{Mod}(S)$, once we have the classification theorem. Suppose we have elements f and g which are pseudo-Anosovs that do not have the same invariant laminations. That is, no two of the four laminations, which we denote $\lambda_f^\pm$ and $\lambda_g^\pm$, are equal.

To describe the proof we must appeal to a slightly finer description of the action of $\mathrm{Mod}(S)$ on laminations than we have had so far. We consider $\lambda_f^\pm$ and $\lambda_g^\pm$ as as points in $\mathcal{PML}(S)$. Once one forms a clearer picture of $\mathcal{PML}(S)$, it is not hard to see that in fact λ_f^+ is an *attracting* fixed point of f and λ_f^- is *repelling*. In fact f has *North-South dynamics* on $\mathcal{PML}(S)$: For every compact set K in $\mathcal{PML}(S) \setminus \{\lambda_f^-\}$, and any neighborhood U^+ of λ_f^+, there is a power f^p taking K into U; and similarly for f^{-1} with λ_f^+ and λ_f^- interchanged.

From this (and compactness of $\mathcal{PML}(S)$) we find that there is a pair of open neighborhoods $U^\pm$ (which can be taken as small as we like) and a power f^p such that f^p takes the complement of U^- into U^+, and f^{-p} takes the complement of U^+ to U^-.

We can do the same for g, obtaining $V^\pm$, and we can arrange for these open sets to all be disjoint, and for their complement to be nonempty. Denote $X = \mathcal{PML}(S)$, for brevity, and let $F = f^p$, $G = g^p$.

Now let

$$A = X \setminus (U^+ \cup U^-),$$

$$B = X \setminus (V^+ \cup V^-)$$

what we now have is, for any nonzero integer n,

$$F^n(A) \subset B \setminus A,$$

$$G^n(B) \subset A \setminus B.$$

We are ready to play ping-pong! Let w be a word composed of alternating non-zero powers of F and G, for example

$$w = F^{n_1} G^{n_2} \cdots F^{n_k}$$

where the n_i are nonzero. Starting with the (nonempty!) set $A \cap B$, we see that (e.g. if w has a power of F on the right) $A \cap B$ is taken to $B \setminus A$, then $A \setminus B$, and so on back and forth across the table. Hence in the end $w(A \cap B)$ must be in the complement of $A \cap B$, and we conclude that w cannot represent the identity.

In other words, the powers F and G generate a free subgroup of $\mathrm{Mod}(S)$.

A similar argument applies, with more care, to the case where f and g are reducible but have data that "cross" in some way – either their canonical reducing systems intersect, or in a common component they have pseudo-Anosov restrictions with distinct laminations.

If two pseudo-Anosovs f and g share at least one lamination, then in fact they share both, and they are commensurable, i.e. satisfy $f^p = g^q$ for some $p, q \in \mathbb{Z}$. Hence they generate a finite extension of $\mathbb{Z}$. This is not hard to prove but we will omit it.

A consequence of this discussion, together with the discussion of abelian groups, is a *Tits alternative* for Mod(S), analogous to Tits' foundational result for linear groups [**71**].

THEOREM 4.2. (McCarthy [**53**]) *A subgroup of* Mod(S) *is either virtually abelian or contains a free group of rank 2.*

Finite subgroups. Since it is easy to build symmetrical pictures of surfaces, Mod(S) has many finite subgroups. Indeed every finite group occurs for *some* S.

A seemingly special situation is a finite group of isometries of a hyperbolic surface, but in fact this turns out to be the general situation, by the *Nielsen realization theorem*:

THEOREM 4.3. *For every finite group $G <$ Mod(S) there is a hyperbolic metric on S so that G is realized as a group of isometries of S.*

This was proved by Steve Kerckhoff in the 1980's [**45**], after several decades of partial results by many people. The proof uses a "center of mass" argument in the Teichmüller space, and the geometry of Thurston's *earthquake paths*, which are generalizations of Dehn twists to the setting of laminations. Note that an apparently weaker consequence is that a finite group G lifts to the group of diffeomorphisms of S. It turns out that this is actually equivalent! If a finite G actually acts by diffeomorphisms (as opposed to diffeomorphisms modulo isotopy), then we can average any Riemannian metric to get an invariant one. Then this invariant metric is conformally equivalent to a hyperbolic metric by the Uniformization theorem, and conformal isomorphisms of a hyperbolic metric are automatically isometries by the Schwarz lemma.

Not a lot is known about the general problem of lifting a subgroup of Mod(S) to Diff(S). Morita proved that the entire group does not lift (i.e. the sequence $1 \to \mathrm{Diff}_0(S) \to \mathrm{Diff}(S) \to \mathrm{Mod}(S) \to 1$ does not split), and more recently Markovic-Saric [**50**] proved that Mod(S) does not even lift to $Homeo(S)$.

A consequence of the Nielsen realization theorem is a bound, in terms of genus, on the order of a finite subgroup $G <$ Mod(S). Since G acts by isometries the quotient S/G is a hyperbolic orbifold, or more mundanely a hyperbolic surface with (possibly) cone points with angles always of the form π/n. These can be classified, and in particular (this is just Gauss-Bonnet) there is a lower bound on the area of a hyperbolic orbifold. This gives an upper bound on the order of G in terms of the area of S, which depends only on g, again by Gauss-Bonnet.

There are, of course, many interesting subgroups of Mod(S) and we have only touched the tip of the iceberg. Examples of subgroups that arise naturally:

(1) Groups isomorphic to $\pi_1(S_g)$ inside Mod($S_{g,1}$), via the Birman exact sequence
(2) The Torelli group and its various relatives and descendants
(3) After identifying S_g with the boundary of a 3-manifold M, the group of mapping classes of S_g that come from restrictions of homeomorphisms of M
(4) various right-angled Artin groups – these are a natural category that combines the notion of free and free abelian

(5) stabilizers of certain "Teichmüller disks" in Teichmüller space (Veech groups). These are naturally lattices in $\mathrm{SL}(2,\mathbb{R})$ and are related to the dynamics of billiards.

5. Further reading and current events

We have only given a brief taste of a subject with classical roots that has had explosive growth over the past 30 years or so. Further information on the general theory of $\mathrm{Mod}(S)$ can be found in Farb-Margalit [**26**] and Ivanov's survey [**42**], as well as Birman's classic text [**10**].

One can read more about geodesic laminations in, for example, Bonahon [**13**], Levitt [**48**], as well as Penner-Harer [**65**].

We have only hinted at Teichmüller theory – the study of the space of hyperbolic (or conformal) structures on a surface, on which $\mathrm{Mod}(S)$ acts discretely with Riemann's moduli space as quotient. See for example Gardiner [**29**] or Hubbard [**39**], and Hamenstädt's notes in this volume.

There is a deep connection between mapping class groups and hyperbolic 3-manifolds, beginning of course with the mapping torus M_f of a mapping class f, a 3-manifold fibering over the circle given by $S \times \mathbb{R}$ modulo the action $(x,t) \mapsto (f(x), t+1)$. Thurston's celebrated hyperbolization theorem [**69, 68, 64**] states that M_f admits a hyperbolic structure if and only if f is pseudo-Anosov. More recently, the work in [**18**] gives explicit methods for connecting the finer structure of a mapping class to the details of the geometry of M_f. More generally, mapping classes are used to describe gluings of 3-manifolds along their boundaries. When these boundaries are incompressible (π_1-injective), this leads to a structure theory quite similar to that of surface bundles. Compressible boundaries present a number of formidable complications, as indicated for example by the rich but incomplete theory of Heegaard splittings of 3-manifolds.

The L-shaped table in Figure 11 gives an example of a subgroup of $\mathrm{Mod}(S)$ acting by affine automorphisms of a flat surface. This subgroup naturally embeds in $PSL(2,\mathbb{R})$, where in fact it is a lattice. Thurston originally gave constructions of this type in [**70**], and they were generalized and studied by Veech [**72**] in terms of their close connections to the dynamics of billiards and translation surfaces. This connection, pioneered also by Kerckhoff-Masur-Smillie [**43, 44**]), has given rise to a very active subject. See Hubert-Schmidt [**40**] for a good introduction.

One interesting point of view on the mapping class group is that of *coarse geometry*: As a finitely generated group, $\mathrm{Mod}(S)$ admits a *word metric*, or equivalently the metric induced from its Cayley graph, once a generating set has been fixed. The interaction between algebraic and geometric properties of groups can be very rich and its study goes back at least to the work of Milnor and Švarc on growth rate in groups [**57, 56, 67**] and to Gromov's polynomial growth theorem [**31**]. See also Gromov's 1983 ICM address [**32**].

An important geometric class of groups are the *word-hyperbolic* ones, whose geometry shares some of the coarse features of the classical hyperbolic spaces and of trees. $\mathrm{Mod}(S)$ is not hyperbolic, but it does admit an interesting action on a hyperbolic space, the *complex of curves* $\mathcal{C}(S)$, a close cousin of the complexes discussed earlier in the proof of the finite generation of $\mathrm{Mod}(S)$ (see Masur-Minsky [**51**] or Bowditch [**14**] for proofs of hyperbolicity). The action is not discrete –

reducible elements, while of infinite order, act with fixed points. However pseudo-Anosov elements act as translations, with quasi-geodesic axes, in analogy with translations in $\mathbb{H}^n$. Stabilizers of vertices of $\mathcal{C}(S)$ are closely related to mapping class groups of subsurfaces of S, and this leads to an inductive description that gives a detailed picture of the coarse-geometric structure of Mod(S) (see for example [**52**] and [**1**]). An expository account of this, including applications to hyperbolic 3-manifolds, can be found in [**58**].

One outcome of this type of analysis is a *quasi-isometric rigidity* theorem for Mod(S) (Behrstock-Kleiner-Minsky-Mosher [**3**] and Hamenstädt [**35**]), namely the statement that any group with the same coarse geometry as Mod(S) is related to it by simple operations such as restriction to finite-index subgroups or extension by finite groups.

A finitely-generated subgroup H of Mod(S) (or any finitely-generated group) is *undistorted* if the inclusion of H into Mod(S) is a quasi-isometry – that is, if distances suffer bounded additive and multiplicative distortion. There is a growing but still incomplete understanding of which subgroups of Mod(S) are undistorted. Work of Farb-Lubotzky-Minsky [**25**] and Masur-Minsky [**52**] tells us that abelian subgroups and more generally stabilizers of curve systems and subsurfaces are always undistorted. On the other hand, Broaddus-Farb-Putman [**17**] showed that a number of natural subgroups, such as the Torelli group and the kernel of the Birman exact sequence, are exponentially distorted. Similarly Hamenstädt-Hensel [**36**] showed that the handlebody group is exponentially distorted.

Coarse-geometric techniques also shed a little light on the question of *surface subgroups* of Mod(S) – that is, subgroups of the form $\pi_1(\Sigma) < \text{Mod}(S)$ where Σ is some closed surface. It is an open question whether such groups exist all of whose nontrivial elements are pseudo-Anosov, and this question has bearing on the existence of negatively-curved 4-manifolds which fibre over a surface. Bowditch [**15**] showed that there are at most finitely many conjugacy classes of such subgroups. Related restrictions on subgroups are obtained by Behrstock-Druţu-Sapir [**2**], through an examination of the limiting geometry of Mod(S) after unbounded rescaling. See also Mangahas [**49**] for applications to a uniform version of the Tits alternative, and Clay-Leininger-Mangahas [**21**] for applications to right-angled Artin subgroups.

The mapping class group also interacts with the rest of mathematics by way of *analogy*. Much of the study of Mod(S) is motivated by the classical theory of linear groups and lattices, and conversely the structural features of Mod(S) can inspire our study of other groups. Probably the richest single example of this is the study of $Out(F_n)$, the group of outer automorphisms of the free group on n letters. The analogy of course comes from the identification of Mod(S) with $Out(\pi_1(S))$ when S is closed.

Bestvina-Handel [**9**] and Bestvina-Feighn-Handel [**7, 8**] developed an analogue of the Nielsen-Thurston theory, and Thurston's work on laminations and train-tracks, for this setting. Culler-Vogtmann's *Outer Space* provides an analogy to Teichmüller space [**22, 73**]. The literature in this area is vast and we cannot do it justice here. It is interesting to note that quite recently an analogue of the hyperbolicity theorem for $\mathcal{C}(S)$ was proved for two natural $Out(F_n)$-complexes, by Bestvina-Feighn [**6**] and Handel-Mosher [**37**]. This promises to have interesting applications for the structure of this group.

The theory of laminations and foliations on surfaces has also inspired a significant generalization to the theory of *group actions on* $\mathbb{R}$*-trees*. A measured foliation on S lifts to a $\pi_1(S)$-invariant measured foliation on $\mathbb{H}^2$. Collapsing the leaves of this foliation to points yields a *tree*, called an $\mathbb{R}$-tree, which inherits a metric from the transverse measure, and an isometric action by $\pi_1(S)$. Such trees are typically non-simplicial, so this is a strict generalization of the theory of Bruhat-Tits. The study of group actions on $\mathbb{R}$-trees gives a powerful method in geometry for understanding degenerations of structures on manifolds. This point of view was introduced by Morgan-Shalen [**59, 60, 61**], with fundamental contributions from Rips (see Bestvina-Feighn [**5**]), and much subsequent development by many people. See [**4**] for a survey. This point of view also leads to a deep structure theory for groups in general, which echoes some of the features of the theory of 3-manifold groups, see for example Rips-Sela [**66**], Druţu-Sapir [**23**].

We should not conclude without a brief mention of the algorithmic aspects of Mod(S), notably the conjugacy problem [**38, 62**], automatic structure [**63**] and biautomatic structure [**34**]. One last interesting topic is that of the dilatation values of pseudo-Anosovs, which carry interesting geometric, dynamical and number-theoretic information. For a few recent articles in this direction see [**24, 55, 46, 33**].

Bibliography

1. J. Behrstock, *Asymptotic geometry of the mapping class group and Teichmüller space*, Geometry & Topology **10** (2006), 1523–1578. MR2255505 (2008f:20108)
2. J. Behrstock, C. Druţu, and M. Sapir, *Median structures on asymptotic cones and homomorphisms into mapping class groups*, Proc. Lond. Math. Soc. (3) **102** (2011), no. 3, 503–554. MR2783135 (2012c:20110)
3. J. Behrstock, B. Kleiner, Y. Minsky, and L. Mosher, *Geometry and rigidity of mapping class groups*, Geometry & Topology **16** (2012), no. 2, 781–888. MR2928983
4. M. Bestvina, *$\mathbb{R}$-trees in topology, geometry, and group theory*, Handbook of geometric topology, North-Holland, Amsterdam, 2002, pp. 55–91. MR1886668 (2003b:20040)
5. M. Bestvina and M. Feighn, *Stable actions of groups on real trees*, Invent. Math. **121** (1995), no. 2, 287–321. MR1346208 (96h:20056)
6. ———, *Hyperbolicity of the complex of free factors*, Preprint, 2011, arXiv:1107.3308.
7. M. Bestvina, M. Feighn, and M. Handel, *Laminations, trees, and irreducible automorphisms of free groups*, Geom. Funct. Anal. **7** (1997), no. 2, 215–244. MR1445386 (98c:20045)
8. ———, *The Tits alternative for* Out(F_n). *I. Dynamics of exponentially-growing automorphisms*, Ann. of Math. (2) **151** (2000), no. 2, 517–623. MR1765705 (2002a:20034)
9. M. Bestvina and M. Handel, *Train tracks and automorphisms of free groups*, Ann. of Math. (2) **135** (1992), no. 1, 1–51. MR1147956 (92m:20017)
10. J. Birman, *Braids, links, and mapping class groups*, Princeton University Press, Princeton, N.J., 1974, Annals of Mathematics Studies, No. 82. MR0375281 (51 #11477)
11. J. Birman, A. Lubotzky, and J. McCarthy, *Abelian and solvable subgroups of the mapping class groups*, Duke Math J. **50** (1983), 1107–1120. MR726319 (85k:20126)
12. J. S. Birman and C. Series, *Geodesics with bounded intersection number on surfaces are sparsely distributed*, Topology **24** (1985), no. 2, 217–225. MR793185 (87f:57012)
13. F. Bonahon, *Geodesic laminations on surfaces*, Laminations and Foliations in Dynamics, Geometry and Topology (Stony Brook, NY) (M. Lyubich, J. Milnor, and Y. Minsky, eds.), Contemporary Mathematics, vol. 269, AMS, 2001, pp. 1–37. MR1810534 (2001m:57023)
14. B. Bowditch, *Intersection numbers and the hyperbolicity of the curve complex*, J. Reine Angew. Math. **598** (2006), 105–129. MR2270568 (2009b:57034)
15. Brian H. Bowditch, *Atoroidal surface bundles over surfaces*, Geom. Funct. Anal. **19** (2009), no. 4, 943–988. MR2570310 (2011d:57045)
16. T. E. Brendle and B. Farb, *Every mapping class group is generated by 6 involutions*, J. Algebra **278** (2004), no. 1, 187–198. MR2068073 (2005c:57018)

17. N. Broaddus, B. Farb, and A. Putman, *Irreducible Sp-representations and subgroup distortion in the mapping class group*, Comment. Math. Helv. **86** (2011), no. 3, 537–556. MR2803852 (2012e:57002)
18. J. Brock, R. Canary, and Y. Minsky, *The classification of Kleinian surface groups II: the ending lamination conjecture*, Ann. of Math. (2) **176** (2012), no. 1, 1–149. MR2925381
19. K. Calta, *Veech surfaces and complete periodicity in genus two*, J. Amer. Math. Soc. **17** (2004), no. 4, 871–908. MR2083470 (2005j:37040)
20. A. J. Casson and S. A. Bleiler, *Automorphisms of surfaces after Nielsen and Thurston*, Cambridge University Press, 1988. MR964685 (89k:57025)
21. M. Clay, C. Leininger, and J. Mangahas, *The geometry of right angled Artin subgroups of mapping class groups*, Groups Geom. Dyn. **6** (2012), no. 2, 249–278. MR2914860
22. M. Culler and K. Vogtmann, *Moduli of graphs and automorphisms of free groups*, Invent. Math. **84** (1986), no. 1, 91–119. MR830040 (87f:20048)
23. C. Druţu and M. Sapir, *Tree-graded spaces and asymptotic cones of groups*, Topology **44** (2005), 959–1058, With an appendix by Denis Osin and Mark Sapir. MR2153979 (2006d:20078)
24. B. Farb, C. J. Leininger, and D. Margalit, *Small dilatation pseudo-Anosov homeomorphisms and 3-manifolds*, Adv. Math. **228** (2011), no. 3, 1466–1502. MR2824561 (2012f:37093)
25. B. Farb, A. Lubotzky, and Y. Minsky, *Rank-1 phenomena for mapping class groups*, Duke Math. J. **106** (2001), no. 3, 581–597. MR1813237 (2001k:20076)
26. B. Farb and D. Margalit, *A primer on mapping class groups*, Princeton U. Press, 2011. MR2850125 (2012h:57032)
27. A. Fathi, F. Laudenbach, and V. Poenaru, *Travaux de Thurston sur les surfaces*, vol. 66-67, Asterisque, 1979.
28. ———, *Thurston's work on surfaces*, Princeton U. Press, 2012, Translated by D. Kim and D. Margalit.
29. F. P. Gardiner, *Teichmüller theory and quadratic differentials*, Pure and Applied Mathematics (New York), John Wiley & Sons Inc., New York, 1987, A Wiley-Interscience Publication. MR903027 (88m:32044)
30. J. Gilman, *On the Nielsen type and the classification for the Mapping Class Group*, Advances in Mathematics **40** (1981), no. 1, 68–96. MR616161 (82i:57006)
31. M. Gromov, *Groups of polynomial growth and expanding maps*, Inst. Hautes Études Sci. Publ. Math. **53** (1981), 53–73. MR623534 (83b:53041)
32. ———, *Infinite groups as geometric objects*, Proceedings of the International Congress of Mathematicians, Warsaw, Amer. Math. Soc., 1983, pp. 385–392. MR804694 (87c:57033)
33. A. Hadari, *Algebraic entropy and the action of mapping class groups on character varieties*, Preprint, 2008, arXiv:0812.0853. MR2764888 (2012a:22040)
34. U. Hamenstädt, *Geometry of the mapping class groups II: A biautomatic structure*, arXiv:math.GR/0511349. MR2264542 (2007g:32008)
35. ———, *Geometry of the mapping class groups III: Quasi-isometric rigidity*, arXiv:math.GT/0512429.
36. U. Hamenstädt and S. Hensel, *The geometry of the handlebody groups I: Distortion*, J. Topol. Anal. **4** (2012), no. 1, 71–97. MR2914874
37. M. Handel and L. Mosher, *The free splitting complex of a free group I: Hyperbolicity*, Preprint, 2011, arXiv:1111.1994.
38. G. Hemion, *On the classification of homeomorphisms of* 2*-manifolds and the classification of* 3*-manifolds*, Acta Math. **142** (1979), no. 1-2, 123–155. MR512214 (80f:57003)
39. J. H. Hubbard, *Teichmüller theory and applications to geometry, topology, and dynamics. Vol. 1*, Matrix Editions, Ithaca, NY, 2006, Teichmüller theory, With contributions by Adrien Douady, William Dunbar, Roland Roeder, Sylvain Bonnot, David Brown, Allen Hatcher, Chris Hruska and Sudeb Mitra, With forewords by William Thurston and Clifford Earle. MR2245223 (2008k:30055)
40. P. Hubert and T. A. Schmidt, *An introduction to Veech surfaces*, Handbook of dynamical systems. Vol. 1B, Elsevier B. V., Amsterdam, 2006, pp. 501–526. MR2186246 (2006i:37099)
41. S. P. Humphries, *Generators for the mapping class group*, Topology of low-dimensional manifolds (Proc. Second Sussex Conf., Chelwood Gate, 1977), Lecture Notes in Math no. 722, Springer, 1979, pp. 44–47. MR547453 (80i:57010)

42. N. V. Ivanov, *Mapping class groups*, Handbook of geometric topology, North-Holland, Amsterdam, 2002, pp. 523–633. MR1886678 (2003h:57022)
43. S. Kerckhoff, H. Masur, and J. Smillie, *A rational billiard flow is uniquely ergodic in almost every direction*, Bull. Amer. Math. Soc. (N.S.) **13** (1985), no. 2, 141–142. MR799797 (86j:58080)
44. ______, *Ergodicity of billiard flows and quadratic differentials*, Ann. of Math. (2) **124** (1986), no. 2, 293–311. MR855297 (88f:58122)
45. S. P. Kerckhoff, *The Nielsen realization problem*, Ann. of Math. **117** (1983), 235–265. MR690845 (85e:32029)
46. T. Koberda, *Entropy of automorphisms, homology and the intrinsic polynomial structure of nilpotent groups*, Preprint, 2011, arXiv:1107.4129.
47. M. Korkmaz, *Generating the surface mapping class group by two elements*, Trans. Amer. Math. Soc. **357** (2005), no. 8, 3299–3310 (electronic). MR2135748 (2006c:57003)
48. G. Levitt, *Foliations and laminations on hyperbolic surfaces*, Topology **22** (1983), 119–135. MR683752 (84h:57015)
49. J. Mangahas, *Uniform uniform exponential growth of subgroups of the mapping class group*, Geom. Funct. Anal. **19** (2010), no. 5, 1468–1480. MR2585580 (2011d:57002)
50. V. Markovic and D. Saric, *The mapping class group cannot be realized by homeomorphisms*, Preprint, 2008, arXiv:0807.0182.
51. H. A. Masur and Y. Minsky, *Geometry of the complex of curves I: Hyperbolicity*, Invent. Math. **138** (1999), 103–149. MR1714338 (2000i:57027)
52. ______, *Geometry of the complex of curves II: Hierarchical structure*, Geom. Funct. Anal. **10** (2000), 902–974. MR1791145 (2001k:57020)
53. J. McCarthy, *A "Tits-alternative" for subgroups of surface mapping class groups*, Trans. Amer. Math. Soc. **291** (1985), 583–612. MR800253 (87f:57011)
54. C. T. McMullen, *Dynamics of* $\mathrm{SL}_2(\mathbb{R})$ *over moduli space in genus two*, Ann. of Math. (2) **165** (2007), no. 2, 397–456. MR2299738 (2008k:32035)
55. C. T. McMullen, *Entropy on Riemann surfaces and Jacobians of finite covers*, Preprint, 2010, http://www.math.harvard.edu/~ctm/papers.
56. J. Milnor, *Growth of finitely generated solvable groups*, J. Differential Geometry **2** (1968), 447–449. MR0244899 (39:6212)
57. ______, *A note on curvature and fundamental group*, J. Differential Geometry **2** (1968), 1–7. MR0232311 (38:636)
58. Y. N. Minsky, *Curve complexes, surfaces and 3-manifolds*, International Congress of Mathematicians. Vol. II, Eur. Math. Soc., Zürich, 2006, pp. 1001–1033. MR2275633 (2007m:57021)
59. J. W. Morgan and P. B. Shalen, *Valuations, trees, and degenerations of hyperbolic structures I*, Ann. of Math. (2) **120** (1984), no. 3, 401–476. MR769158 (86f:57011)
60. ______, *Degenerations of hyperbolic structures. II. Measured laminations in 3-manifolds*, Ann. of Math. (2) **127** (1988), no. 2, 403–456. MR932305 (89e:57010a)
61. ______, *Degenerations of hyperbolic structures. III. Actions of 3-manifold groups on trees and Thurston's compactness theorem*, Ann. of Math. (2) **127** (1988), no. 3, 457–519. MR942518 (89e:57010b)
62. L. Mosher, *The classification of pseudo-Anosovs*, Low-dimensional topology and Kleinian groups (Coventry/Durham, 1984), London Math. Soc. Lecture Note Ser., vol. 112, Cambridge Univ. Press, Cambridge, 1986, pp. 13–75. MR903858 (89f:57016)
63. L. Mosher, *Mapping class groups are automatic*, Ann. of Math. **142** (1995), 303–384. MR1343324 (96e:57002)
64. J.-P. Otal, *Le théorème d'hyperbolisation pour les variétés fibrées de dimension trois*, Astérisque No. 235, 1996. MR1402300 (97e:57013)
65. R. Penner and J. Harer, *Combinatorics of train tracks*, Annals of Math. Studies no. 125, Princeton University Press, 1992. MR1144770 (94b:57018)
66. E. Rips and Z. Sela, *Cyclic splittings of finitely presented groups and the canonical JSJ decomposition*, Ann. of Math. (2) **146** (1997), no. 1, 53–109. MR1469317 (98m:20044)
67. A. S. Švarc, *A volume invariant of coverings*, Dokl. Akad. Nauk SSSR (N.S.) **105** (1955), 32–34. MR0075634 (17:781d)
68. W. Thurston, *Hyperbolic structures on 3-manifolds, II: surface groups and manifolds which fiber over the circle*, preprint, arXiv:math.GT/9801045.

69. ———, *Three dimensional manifolds, Kleinian groups and hyperbolic geometry*, Bull. Amer. Math. Soc. **6** (1982), 357–381. MR648524 (83h:57019)
70. ———, *On the geometry and dynamics of diffeomorphisms of surfaces*, Bull. Amer. Math. Soc. **19** (1988), 417–431. MR956596 (89k:57023)
71. J. Tits, *Free subgroups in linear groups*, J. Algebra **20** (1972), 250–270. MR0286898 (44:4105)
72. W. A. Veech, *Teichmüller curves in moduli space, Eisenstein series and an application to triangular billiards*, Invent. Math. **97** (1989), no. 3, 553–583. MR1005006 (91h:58083a)
73. K. Vogtmann, *Automorphisms of free groups and outer space*, Proceedings of the Conference on Geometric and Combinatorial Group Theory, Part I (Haifa, 2000), vol. 94, 2002, pp. 1–31. MR1950871 (2004b:20060)
74. B. Wajnryb, *Mapping class group of a surface is generated by two elements*, Topology **35** (1996), no. 2, 377–383. MR1380505 (96m:57007)
75. X. Zhu and F. Bonahon, *The metric space of geodesic laminations on a surface. I*, Geom. Topol. **8** (2004), 539–564. MR2057773 (2005e:57052)

Teichmüller Theory

Ursula Hamenstädt

IAS/Park City Mathematics Series
Volume 20, 2011

Teichmüller Theory

Ursula Hamenstädt

Introduction

The goal of these notes is to give an introduction to geometric aspects of Teichmüller theory for a closed surface S of genus $g \geq 2$. There are very good recent books devoted to this and related subjects, for example [**B92, D11, FM11, FlM07, HM98, Hu06, IT89, Mi95, M07**], and we refer to these books throughout. There are moreover many survey articles on various aspects of the theory, notably the articles in the recent "Handbook of Teichmüller theory", Vol.I,II, edited by A. Papadopoulos. We do not aim at duplicating what can be found in these books and survey articles beyond what is needed to make these notes fairly self-contained. Rather, we give a subjective and selective summary of some of the recent developments which introduce new tools and explore connections to some areas of mathematics which had not been looked at closely before.

Throughout, we consider a closed oriented surface S of genus $g \geq 2$. A *marked complex structure* on S is a pair consisting of a Riemann surface X and an orientation preserving diffeomorphism $\varphi : S \to X$. Two such marked complex structures (X, φ) and (X', φ') are *equivalent* if there is a biholomorphic map $F : X \to X'$ such that $F \circ \varphi$ is isotopic to φ'. The *Teichmüller space* $\mathcal{T}(S)$ of S is the space of all equivalence classes of marked complex structures on S.

Teichmüller space is a complex manifold which is biholomorphic to a bounded domain in $\mathbb{C}^{3g-3}$ (see [**IT89**]). As any complex manifold, it can be equipped with the *Kobayashi pseudo-distance* $d_{\mathcal{T}}$ which is invariant under the group of biholomorphic automorphisms.

It turns out that $\mathcal{T}(S)$ is *complete Kobayashi hyperbolic* which means that the Kobayashi pseudo-distance is a complete distance. Then this distance is the intrinsic path metric of a Finsler metric. The Finsler metric can be described explicitly, and it has some nice properties. For example, any two points in $\mathcal{T}(S)$ can be connected by a unique geodesic. The metric $d_{\mathcal{T}}$ is more commonly called the *Teichmüller metric.*

The *mapping class group* $\text{Mod}(S)$ of all isotopy classes of orientation preserving diffeomorphisms of S acts on $\mathcal{T}(S)$ properly discontinuously by precomposition of marking. This action preserves the complex structure on $\mathcal{T}(S)$. In fact, if $g > 2$ then

Mathematisches Institut der Universität Bonn, Endenicher Allee 60, 53115 Bonn, Germany
E-mail address: ursula@math.uni-bonn.de

$\mathrm{Mod}(S)$ equals the group of biholomorphic automorphisms of $\mathcal{T}(S)$ (see **[IT89]**- in the case $g = 2$ one has to divide $\mathrm{Mod}(S)$ by the hyperelliptic involution which acts trivially on Teichmüller space). In particular, $\mathrm{Mod}(S)$ acts on $(\mathcal{T}(S), d_{\mathcal{T}})$ as a group of isometries. As a consequence, the Teichmüller metric descends to a distance on the *moduli space* $\mathcal{M}_g = \mathcal{T}(S)/\mathrm{Mod}(S)$ of Riemann surfaces. This moduli space is a complex orbifold. If $g > 2$ then its singular locus is the projection of the set of all Riemann surfaces which admit a non-trivial biholomorphic automorphism.

In Section 2 we construct the Teichmüller space as a real analytic manifold using the fact that by the uniformization theorem, a complex structure on S can be identified with an isotopy class of hyperbolic metrics. Section 3 contains an account of the basic properties of the Teichmüller metric with an emphasis on quasi-conformal analysis.

In Section 4 we have a closer look at Teichmüller space as a complex manifold. We adopt the differential-geometric viewpoint and state the basic algebraic-geometric facts about the moduli space and the Torelli map without proof. We introduce complex geodesics and a second natural $\mathrm{Mod}(S)$-invariant metric on $\mathcal{T}(S)$, the so-called *Weil-Petersson metric*.

The Weil-Petersson metric is not complete. The completion of $\mathcal{T}(S)$ with respect to this metric can be described explicitly, and this is explained in Section 5. The completion of moduli space can be identified with the *Deligne-Mumford compactification* of $\mathcal{M}_g$. We introduce the curve complex of S as a combinatorial tool and discuss some results which relate geometric properties of the curve complex to geometric properties of Teichmüller space with the Teichmüller metric.

In the final section of this note we look at the $SL(2, \mathbb{R})$-action on the moduli space of quadratic differentials and relate some of its properties to geometric properties of Teichmüller space and the mapping class group. As an application, we observe that the projection of a complex geodesic in $\mathcal{T}(S)$ intersects a fixed compact subset of $\mathcal{M}_g$ not depending on the geodesic, and it is unbounded.

There is nothing new contained in these notes. Everything presented is looked at with the eyes of a differential geometer. This leads to omissions of results of fundamental importance, of viewpoints and references in a beautiful and overwhelmingly rich theory. Most of the main contributors to this theory will not be given proper credit. My apologies to all of them.

LECTURE 1

Hyperbolic surfaces

In this section we introduce the Teichmüller space of a closed oriented surface S of genus $g \geq 2$ as the space of all marked hyperbolic structures on S. We discuss natural coordinates arising from hyperbolic geometry which equip Teichmüller space with a smooth (in fact real analytic) structure. Throughout, we use standard facts about the geometry of the hyperbolic plane, and we refer to the excellent treatment in [**B92**] for details.

The starting point is the observation that every closed surface of genus $g \geq 2$ admits a hyperbolic metric, i.e. a smooth Riemannian metric of constant Gauss curvature -1.

DEFINITION 1.1. Let S be a closed oriented surface of genus $g \geq 2$. A *marked hyperbolic surface* is a pair (X, φ) where X is a closed oriented hyperbolic surface of genus $g \geq 2$ and $\varphi : S \to X$ is an orientation preserving diffeomorphism. Two such marked hyperbolic surfaces $(X, \varphi), (X', \varphi')$ are *equivalent* if there exists an isometry $g : X \to X'$ such that φ' and $g \circ \varphi$ are isotopic. The space of equivalence classes is called the *Teichmüller space* $\mathcal{T}(S)$ of S.

In the sequel we often drop the diffeomorphism φ which defines the marking from our notation if no confusion is possible.

The *mapping class group*

$$\mathrm{Mod}(S) = \mathrm{Diff}(S)^+/\mathrm{Diff}_0^+(S)$$

of isotopy classes of orientation preserving diffeomorphisms of S acts on Teichmüller space by precomposition of marking: If ζ is a diffeomorphism of S then $\zeta(X, \varphi)$ is the point in $\mathcal{T}(S)$ which is given by the same hyperbolic surface X, but where the diffeomorphism φ has been replaced by $\varphi \circ \zeta^{-1}$. If η is isotopic to ζ then the marked hyperbolic structures $\eta(X, \varphi)$ and $\zeta(X, \varphi)$ are equivalent and hence this definition indeed defines an action of the mapping class group on $\mathcal{T}(S)$.

A closed curve α on a compact surface F, possibly with non-trivial boundary ∂F, is *essential* if α is not contractible and *non-peripheral*, i.e. not freely homotopic into the boundary. In the sequel we always assume that closed curves are essential.

Now let X be a marked hyperbolic surface. Let γ be an essential closed curve on S. Then the free homotopy class of γ can be represented by a unique closed geodesic on X. Here the identification of free homotopy classes on S with free homotopy classes on X is via the marking. The following observation is contained in Theorem 1.6.6 and Theorem 1.6.7 of [**B92**]. For its formulation, call a closed curve α on S *simple* if it is the image of an injective mapping $S^1 \to S$.

PROPOSITION 1.2. *Suppose that α is a simple closed curve on S.*

(1) *The closed geodesic on X freely homotopic to α is simple.*
(2) *If β is another simple closed curve on S which is disjoint from α and not freely homotopic to α then the closed geodesics on X representing the free homotopy classes of α, β are disjoint.*

By Proposition 1.2, if α is a simple closed curve on S then we can cut X open along the closed geodesic $\hat{\alpha}$ freely homotopic to α. The result is a hyperbolic surface with two geodesic boundary circles.

There are now two possibilities. The first case is that α is *non-separating*, i.e. $S - \alpha$ is connected. Then the genus of $S - \alpha$ equals $g - 1$. If α is *separating* then $S - \alpha$ is disconnected. Its two connected components S_1, S_2 are surfaces of genus $g_1 \geq 0, g_2 \geq 0$, respectively, with connected boundary. The Euler characteristic of S equals the sum of the Euler characteristics of the components. Since the Euler characteristic of each component is negative, the Euler characteristic of each component is strictly bigger than the Euler characteristic of S. This reasoning also applies to essential simple closed curves on surfaces with boundary.

By Proposition 1.2, if the simple closed curve β can be realized disjointly from α and is not freely homotopic to α then the geodesic $\hat{\beta}$ on X freely homotopic to β is disjoint from the geodesic $\hat{\alpha}$. Thus we can successively decompose X into $2g - 2$ hyperbolic *pairs of pants*, i.e. bordered hyperbolic surfaces with geodesic boundary which are homeomorphic to a sphere with 3 holes. Namely, the Euler characteristic of a sphere with 3 holes equals -1, moreover a sphere with 3 holes does not contain any non-peripheral simple closed curve.

DEFINITION 1.3. A *pants decomposition* $\mathcal{P}$ of S consists of $3g-3$ disjoint simple closed curves which decompose S into $2g - 2$ *pairs of pants.*

By the above discussion, for every pants decomposition $\mathcal{P}$ of S and every marked hyperbolic surface (X, φ), the hyperbolic geodesics representing the pants curves of $\mathcal{P}$ decompose X into $2g - 2$ *hyperbolic pairs of pants.*

Given a hyperbolic pair of pants P, for each pair of distinct boundary geodesics γ_1, γ_2 there is a unique embedded geodesic arc connecting γ_1 to γ_2 which meets γ_1, γ_2 orthogonally at its endpoints. We call such an arc a *seam*. Every boundary geodesic contains precisely two endpoints of seams which decompose the boundary circle into two arcs of equal length. Cutting P open along the three seams results in two isometric right angled convex hyperbolic hexagons (this is Proposition 3.1.5 of [**B92**]).

Now for an arbitrarily prescribed triple (a, b, c) of positive numbers there is up to isometry a unique right angled convex hyperbolic hexagon with three pairwise non-consecutive sides of length a, b, c (Theorem 2.4.1 of [**B92**]). In particular, glueing two such hexagons along the remaining sides yields a hyperbolic pair of

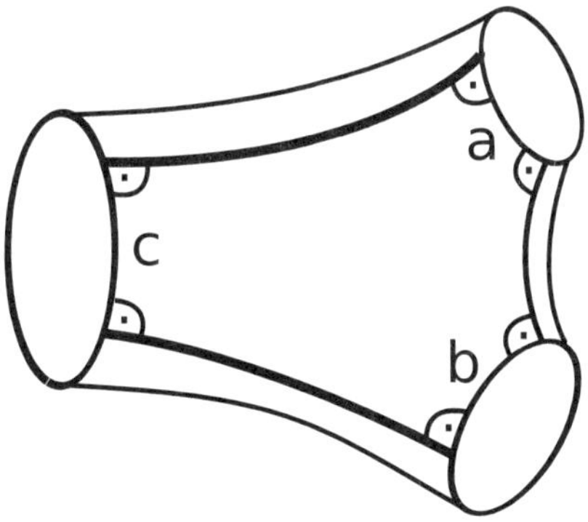

pants with geodesic boundary circles of length $2a, 2b, 2c$. Summarizing, we obtain Theorem 3.1.7 of [**B92**].

PROPOSITION 1.4. *For any triple (a, b, c) of positive numbers, there is up to isometry a unique hyperbolic pair of pants with boundary circles of length a, b, c.*

To reconstruct the hyperbolic surface X from the pairs of pants $X - \mathcal{P}$ we have to remember how the surface was glued from the pairs of pants P_i which are the components of $X - \mathcal{P}$. Boundary circles of pairs of pants are glued in pairs. In particular, since each pair of pants has precisely three boundary circles, the glueing pattern can be represented by a trivalent graph. Each vertex of this graph represents one of the pairs of pants, and each edge represents one of the simple closed curves of the pants decomposition $\mathcal{P}$. The vertices representing the pairs of pants P_i, P_j are connected by an edge if and only if P_i, P_j have a common boundary circle. Vice versa, every connected trivalent graph with $2g - 2$ vertices determines a glueing pattern for pairs of pants to a closed surface of genus g (see Section 3.5 of [**B92**]).

DEFINITION 1.5. A *combinatorial type of pants decomposition* for a closed surface S of genus g is a trivalent graph with $2g - 2$ vertices.

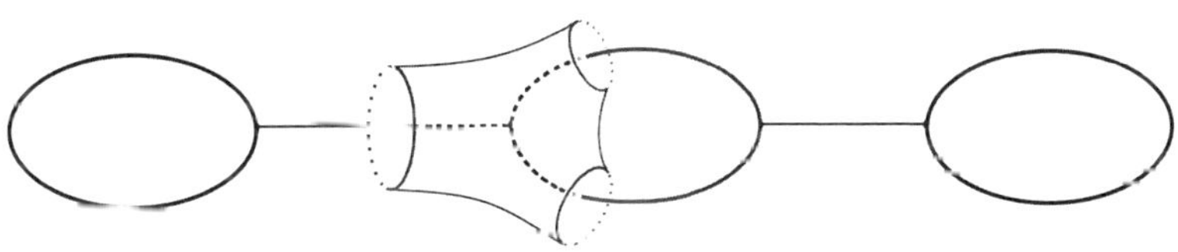

A diffeomorphism of S maps disjoint simple closed curves on S to disjoint simple closed curves. Therefore the mapping class group acts on free homotopy classes of simple closed curves preserving disjointness. As a consequence, it acts on isotopy classes of pants decompositions of S (this statement uses some subtle properties of surface topology which are discussed in the appendix of [**B92**]). In the sequel we always consider simple closed curves and pants decompositions up to isotopy. If $\mathcal{P}$ is a pants decomposition and if $\varphi \in \mathrm{Mod}(S)$ then $\varphi\mathcal{P}$ is a pants decomposition of the same combinatorial type. Vice versa, any two pants decompositions of the same combinatorial type can be transformed into each other by an element of $\mathrm{Mod}(S)$. This follows easily from the fact that any two pairs of pants are diffeomorphic.

To reconstruct the marked hyperbolic metric on X we have to glue the hyperbolic pairs of pants with isometries of their boundary circles. This requires that the lengths of these boundary circles coincide. Now given two pairs of pants P_1, P_2 with a boundary curve γ_1, γ_2 of the same length, the seams of P_1, P_2 which end on γ_1, γ_2 determine two preferred ways of glueing P_1 to P_2 along γ_1, γ_2. Namely, we require that the identification of γ_1 with γ_2 is by an orientation reversing isometry (where the orientation of γ_i is the boundary orientation of P_i) and that the two endpoints of the seams on γ_1 are glued to the endpoints of the seams on γ_2. This is possible because the seams decompose the boundary circles γ_1, γ_2 into two arcs of equal length. The surface Y_0 obtained by this identification is a hyperbolic sphere with four holes and geodesic boundary.

There are more possibilities for glueing P_1 to P_2 along γ_1, γ_2. Namely, we can start rotating P_2 along γ_1 with unit speed. Thus if $\gamma_1, \gamma_2 : \mathbb{R} \to P_1, P_2$ are unit speed parametrizations of the boundary geodesics of P_1, P_2 defining the boundary orientation and so that $\gamma_1(0)$ and $\gamma_2(0)$ is a point on a seam, then for each t we can glue P_2 to P_1 by identifying $\gamma_1(s)$ with $\gamma_2(t-s)$. Denote the resulting hyperbolic surface by Y_t. Note that if $r > 0$ is the length of γ_i then the surface Y_r obtained by rotating by the angle 2π is isometric to the surface Y_0, but any initial marking has been changed by a full *Dehn twist* about the image of γ_i. We refer to Chapter 3 of [**FM11**] for a detailed discussion of Dehn twists. The surface $Y_{r/2}$ obtained by rotating by the angle π corresponds to the second preferred way of glueing P_1 to P_2 along γ_1, γ_2 (i.e. endpoints of seams are glued to endpoints of seams).

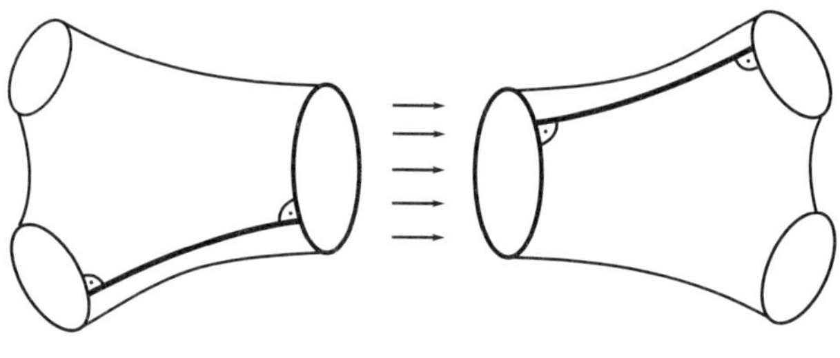

The following result is basic for Teichmüller theory. For its formulation, note first that every closed curve α on S defines a function ℓ_α on $\mathcal{T}(S)$ by associating to a marked hyperbolic surface X the length $\ell_\alpha(X)$ of the geodesic in the free homotopy class of α.

Let $\mathcal{P}$ be a fixed pants decomposition of S. Then the $3g-3$ pants curves $\alpha_1, \ldots, \alpha_{3g-3}$ of $\mathcal{P}$ define $3g-3$ length functions $\ell_i = \ell_{\alpha_i}$ on $\mathcal{T}(S)$ with values in $\mathbb{R}_+$. Moreover, for a fixed initial choice of glueing the pants in such a way that endpoints of seams are identified, there are $3g-3$ twist parameters $\tau_i \in \mathbb{R}$ for the glueing arising as in the previous paragraph by rotating the two pairs of pants adjacent to α_i with unit speed. Using a surface for which endpoints of seams are identified with endpoints of seams as a basepoint for the marking, this construction associates to a $(6g-6)$-tuple of lengths- and twist parameters a unique marked hyperbolic surface.

The following statement is an extension of the above discussion.

THEOREM 1.6. *For each pants decomposition $\mathcal{P}$ of S, the map which associates to a tuple*

$$(\ell_1, \ldots, \ell_{3g-3}, \tau_1, \ldots, \tau_{3g-3}) \in \mathbb{R}_+^{3g-3} \times \mathbb{R}^{3g-3}$$

of lengths and twist parameters the surface $(X, \varphi) \in \mathcal{T}(S)$ defined by these data is a bijection onto $\mathcal{T}(S)$.

The map $\mathbb{R}_+^{3g-3} \times \mathbb{R}^{3g-3} \to \mathcal{T}(S)$ is called a system of *Fenchel-Nielsen coordinates* based at $\mathcal{P}$. The above discussion shows that Fenchel Nielsen coordinates define a surjective map from $\mathbb{R}_+^{3g-3} \times \mathbb{R}^{3g-3}$ onto $\mathcal{T}(S)$. To show that this map is also injective we have to verify that distinct Fenchel Nielsen coordinates define surfaces X, X' which are not isometric with an isometry preserving the marking.

Clearly marked isometric surfaces have the same lengths parameters, so we have to study the effect of changing a twist parameter. To this end define the *twist vector field* $\mathbf{t}_\alpha$ of the component α of the pants decomposition $\mathcal{P}$ to be the vector field which integrates to the twist flow Ψ_t about α in Fenchel Nielsen coordinates. The image $\Psi_t X$ of X in Fenchel Nielsen coordinates equals the linear coordinate change obtained by adding t to the twist parameter of α.

Let β be any simple closed geodesic on X. If β does not intersect α then β is also a geodesic for $\Psi_t X$. In particular, in this case the length of β (i.e. the length of its geodesic representative) does not change along a flow line of the twist flow Ψ_t. In the case that β intersects α we can calculate the derivative of the length of β along the twist flow about α. The calculation is taken from [**K83**].

PROPOSITION 1.7.

$$\mathbf{t}_\alpha \ell_\beta = \sum_{p \in \alpha \cap \beta} \cos \theta_p$$

where for each point $p \in \alpha \cap \beta$ the angle θ_p is the angle between α and β measured counter-clockwise from β to α.

PROOF. We show the proposition in the case that β has a single point of intersection with α. Let α_1 be a lift of α to the hyperbolic plane $\mathbf{H}^2$ and let $p \in \alpha_1$ be a lift of the intersection point between α and β. There is a lift $\tilde{\beta}$ of β which meets p. For a fixed orientation of $\tilde{\beta}$ let q be the first intersection point distinct from p between $\tilde{\beta}$ and the preimage of α. Denote by α_2 the lift of α through q. The length of the subarc β_0 of $\tilde{\beta}$ connecting p to q is the length of β. Orient α_2 in such a way that the orientation of the basis β_0', α_2' of $T_q\mathbf{H}^2$ is positive.

After twisting along α the amount of t, the projection of β_0 to the twisted surface $\Psi_t X$ is not closed any more. To close it up, glide the endpoint of β_0 on α_2 the signed distance t. Let β_t be the resulting geodesic arc and let ℓ_{β_t} be its length. Then $t \to \beta_t$ is a variation of geodesic arcs with one fixed endpoint p and the second endpoint gliding along α_2. The first variational formula for geodesic lengths (see p.5 of [**CE75**]) states that the derivative at $t = 0$ of the lengths of these arcs equals the cosine of the oriented angle between the tangent of β_0 at q and the tangent of α_2 at q. Or, we have

$$\frac{d}{dt} \ell_{\beta_t}|_{t=0} = \cos \theta_p.$$

The geodesic arc β_t projects to a geodesic loop on the surface $\Psi_t X$. This loop has a breakpoint at its intersection point with α. The angle defect at this breakpoint (i.e. the angle between its incoming and outgoing angle) vanishes at $t = 0$.

By hyperbolic trigonometry [**B92**], the length of a closed geodesic γ' freely homotopic to a geodesic loop γ with an angle defect ζ at its breakpoint depends smoothly on the length of γ and the angle ζ, and the difference between the length of γ and γ' has a global minimum if $\zeta = 0$. Thus since the angle defect of β_t depends smoothly on t, the derivative at $t = 0$ of the length of the closed geodesic on $\Psi_t X$ freely homotopic to β coincides with the derivative at $t = 0$ of the function which associates to t the length of β_t. This completes the proof of the proposition. □

COROLLARY 1.8. (1) *For every simple closed curve β, the length function of β is continuously differentiable along Fenchel Nielsen twists.*
(2) $\mathbf{t}_\alpha \ell_\beta = -\mathbf{t}_\beta \ell_\alpha$ *for all* α, β.

PROOF. By Proposition 1.7, we have to show that for every simple closed curve β which intersects a pants curve α of $\mathcal{P}$, the intersection angle between α and the geodesic representative of β varies continuously with the twist parameter for α at each intersection point between α and β.

A piecewise geodesic γ on a hyperbolic surface X with breakpoints on a simple closed geodesic α is smooth if and only if at each of the intersection points between γ and α, the incoming and the outgoing intersection angles between γ and α coincide. This is a closed condition which can be checked on the surface cut open along α.

As a consequence, as the twist parameter for α varies, the geodesic arcs on the surface cut open along α which close up to the geodesic representatives of the curve β vary continuously up to and including their endpoints.

To be more precise, consider for the moment a simple closed geodesic β which intersects the pants curve α in a single point. Let β_0 be the geodesic arc with both endpoints on α obtained by cutting β at its intersection point with α. Gliding the two endpoints of β_0 along α defines a smooth two-parameter family of geodesic arcs $\beta_{s,u}$ through $\beta_0 = \beta_{0,0}$. For all small s, t the arc $\beta_{s,s+t}$ defines a loop on the surface $\Psi_t X$. Requiring that the incoming and outgoing angles of $\beta_{s,u}$ at the endpoints coincide defines a smooth one-parameter subfamily A of the two-parameter family $\beta_{s,u}$. For each t, A intersects the arc $s \to \beta_{s,s+t}$ in a single point which depends continuously on t.

Together we conclude that the intersection angles of the geodesic representatives of the curve β on the surface $\Psi_t X$ vary continuously with t. The first part of the corollary now follows from Proposition 1.7.

For the second part, simply note that $\cos\theta_p = -\cos(\pi - \theta_p)$. □

The Dehn twists about the components of $\mathcal{P}$ define a free abelian subgroup Γ of $\mathrm{Mod}(S)$ of rank $3g - 3$. The group Γ acts on Fenchel-Nielsen coordinates for $\mathcal{P}$ by preserving the length parameters and acting on the twist parameters as a cocompact group of translations. Thus to show that Fenchel-Nielsen coordinates are injective, it suffices to find for every fixed tuple of length parameters and for a fixed fundamental domain D for the action of Γ on twist parameters a collection $\beta_1, \ldots, \beta_{3g-3}$ of simple closed curves with the following properties.

(1) The curve β_i intersects α_i and no other curve from $\mathcal{P}$.
(2) The map which associates to $x \in D$ the tuple $(\ell_{\beta_1}(x), \ldots, \ell_{\beta_{3g-3}}(x))$ is injective.

Proposition 1.7 suggests how this can be accomplished: For a given hyperbolic surface X, find simple closed geodesics on X with property 1) above which intersect the geodesic representatives of the pants curves with a very small angle.

To this end we note

LEMMA 1.9. *Let α, β be simple closed geodesics on a hyperbolic surface X which intersect transversely. Let φ_α be the positive Dehn twist about α. Then as $k \to \infty$, the intersection angles between α and $\varphi_\alpha^k \beta$ measured counter-clockwise from $\varphi_\alpha^k \beta$ to α tend to zero.*

PROOF. Assume for simplicity that α, β intersect in a single point p (the argument we give is valid in general). Let $\tilde{\alpha}$, $\tilde{\beta}$ be lifts of α, β to the hyperbolic plane which intersect in a preimage $\tilde{p}$ of p. Let $\xi, \nu \in \mathrm{PSL}(2, \mathbb{R})$ be the hyperbolic isometries which preserve $\tilde{\alpha}, \tilde{\beta}$ and whose conjugacy classes define α, β.

For each k the hyperbolic isometry $\xi^k \circ \nu$ defines the conjugacy class of the simple closed curve $\varphi_\alpha^k \beta$ on S. This isometry fixes a unique pair of points (a_k, b_k) in the ideal boundary $\partial \mathbf{H}^2$ of $\mathbf{H}^2$ which lie in the two different components of $\partial \mathbf{H}^2$ cut out by the endpoints (a, b) of $\tilde{\alpha}$. Assume without loss of generality that a is the attracting fixed point of ξ. By hyperbolic trigonometry, it now suffices to show that as $k \to \infty$ the attracting fixed point a_k of $\xi^k \circ \nu$ converges to a and its repelling fixed point b_k converges to $\nu^{-1} b$.

To see that this is the case it suffices to show the following. For every neighborhood U of a and V of $\nu^{-1}(b)$ there is some $k_0 > 0$ such that $\xi^k \circ \nu(U) \subset U$ and $(\xi^k \circ \nu)^{-1}(V) \subset V$ for all $k \geq k_0$. Now observe that the points $\nu^{-1} a, \nu^{-1} b$ are the fixed points of the hyperbolic isometry $\sigma_k = \nu^{-1} \circ \xi^k \circ \nu$. These fixed points are distinct from the fixed points a, b of ξ. Moreover, the translation length of the isometry σ_k tends to infinity as $k \to \infty$. Thus if U is a small neighborhood of a not containing $\nu^{-1} b$ and W is a neighborhood of $\nu^{-1} a$ not containing $\nu^{-1} b$ which is mapped by ν into U then for sufficiently large k, the set U is mapped by σ_k into W. Or, $\xi^k \circ \nu$ maps U into itself. The same argument is also valid for a neighborhood V of $\nu^{-1}(b)$ and the maps $(\xi^k \circ \nu)^{-1}$. This is what we wanted to show.

Note that the above argument can be made uniform: For any compact set $A \subset \mathrm{PSL}(2, \mathbb{R})$ of hyperbolic isometries whose fixed points are contained in the different components of $\partial \mathbf{H}^2 - \{a, b\}$, for any open neighborhood U of a not containing any fixed point of an element $\varphi \in A$ and every compact neighborhood $K \subset \partial \mathbf{H}^2 - \{a, b\}$ of the repelling fixed points of the elements in A, there is some $k_0 > 0$ so that for each $\varphi \in A$ and every $k \geq k_0$ the attracting fixed point of $\xi^k \circ \varphi$ is contained in U

and the repelling fixed point of $\xi^k \circ \varphi$ is contained in K. In particular, as $k \to \infty$ the angle of intersection with $\tilde{\alpha}$ of an axis of $\xi^k \circ \varphi$ is arbitrarily close to 0, uniformly in φ. □

Since intersection angles change continuously with twist parameters, the last paragraph in the proof of Lemma 1.9 shows that the estimate of intersection angles in Lemma 1.9 can be made uniform whenever the hyperbolic structure on S is modified by changing the twist parameter for α in Fenchel Nielsen coordinates by a bounded amount. From this observation, injectivity of the Fenchel-Nielsen coordinate map follows.

To be more precise, let $\mathcal{P}$ be a pants decomposition of S and let x, x' be two tuples of Fenchel Nielsen coordinates defining hyperbolic structures X, X' which are marked isometric. Then the length parameters of x, x' coincide. Let $\alpha \in \mathcal{P}$ and let τ, τ' be the twist parameters of x, x' for α. By Lemma 1.9, there is a simple closed curve β which is disjoint from $\mathcal{P} - \alpha$ and such that for each hyperbolic structure Y obtained from X by varying the twist parameter for α within the line segment connecting τ to τ', the geodesic representative of β on Y intersects α with an oriented angle smaller than $\pi/4$. Proposition 1.7 shows that the length of β is strictly increasing as the twist parameter is increasing. Since the length of β on X, X' coincides, the twist parameters τ, τ' coincide.

To summarize, Fenchel Nielsen coordinates for a pants decomposition $\mathcal{P}$ of S parametrize Teichmüller space. In particular, they define a topology on $\mathcal{T}(S)$. Our next goal is to check that this topology does not depend on the choice of the pants decomposition.

To this end let again $\mathcal{P}$ be a pants decomposition and let β be a simple closed curve which intersects one of the pants curves α but does not intersect any other pants curve. Let φ_α be the Dehn twist about α. Lemma 1.9 shows that the angle of intersection between α and the curve $\varphi_\alpha^k \beta$ tends to zero as $k \to \infty$.

By hyperbolic trigonometry [**B92**], two geodesics in the hyperbolic plane which intersect with a very small intersection angle remain uniformly close for a very large time. Or, the curve $\varphi_\alpha^k \beta$ remains in a small tubular neighborhood of α for a long time (i.e. it wraps around α many times) and therefore its length tends to infinity. This growth in length is uniform as we vary the Fenchel Nielsen coordinates within a fixed compact set. Since the length of $\varphi_\alpha^k \beta$ on the Riemann surface $X \in \mathcal{T}(S)$ equals the length of β on the surface $\varphi_\alpha^{-k}(X)$, the length of β is arbitrarily large on surfaces whose twist parameter for α in Fenchel Nielsen coordinates is large in absolute value provided that the length parameters are contained in a compact set (in fact, this is even the case if the length parameter of α is arbitrarily large or small).

For a marked hyperbolic surface X and $\epsilon > 0$ define a set $U(X, \epsilon) \subset \mathcal{T}(S)$ as follows. $X' \in U(X, \epsilon)$ if and only if $|\log \ell_c(X) - \log \ell_c(X')| < \epsilon$ for every simple closed curve c on S.

LEMMA 1.10. *The sets $U(X, \epsilon)$ $(\epsilon > 0)$ define a neighborhood basis of X in Fenchel Nielsen coordinates.*

PROOF. (Sketch) Let $\mathcal{P}$ be a pants decomposition of S which defines Fenchel Nielsen coordinates. We first show that for every $X \in \mathcal{T}(S)$ and for every $\epsilon > 0$ there is a neighborhood of X in Fenchel Nielsen coordinates which is contained in $U(X, \epsilon)$.

Since a closed geodesic is the shortest curve in its free homotopy class, for this it suffices to show the following. Let $L > 1$ be arbitrary; then there is a neighborhood V of X in Fenchel Nielsen coordinates such that for every $Y \in V$, there is an L-bilipschitz diffeomorphism $F : X \to Y$ compatible with the markings. Namely, let γ be any closed geodesic on X of length $\ell_\gamma(X)$. Then $F(\gamma)$ is a closed curve on Y of length at most $L\ell_\gamma(X)$. Since F is compatible with the marking, $\ell_\gamma(Y)$ is the shortest length of a simple closed curve on Y freely homotopic to $F(\gamma)$ and hence $\ell_\gamma(Y) \leq L\ell_\gamma(X)$. By symmetry, we also have $\ell_\gamma(X) \leq L\ell_\gamma(Y)$.

Such a neighborhood can be obtained as follows. For a given number $L > 1$ and for every hyperbolic pair of pants P with boundary geodesics α_i of lengths a_i $(i = 1, 2, 3)$, find a number $\delta < \min\{a_i \mid i\}/2$ with the following property. Let $\tilde{a}_i \in (a_i - \delta, a_i + \delta)$ $(i = 1, 2, 3)$ and let $\tilde{P}$ be a hyperbolic pair of pants with boundary geodesics $\tilde{\alpha}_i$ of lengths $\tilde{a}_i$. Let moreover $s_i \in (-\delta, \delta)$; then there is an L-bilipschitz map $F : P \to \tilde{P}$ which maps a boundary geodesic of P parametrized by arc length to a boundary geodesic of $\tilde{P}$ parametrized proportional to arc length. Moreover, the endpoints of the seams on γ_i are mapped to points of oriented distance s_i to the endpoints of the seams on $\tilde{\gamma}_i$.

Cut X open along the geodesic representatives of the pants curves of $\mathcal{P}$. The maps described in the previous paragraph on the components of the cut open surface can be glued to an L-bilipschitz map respecting the markings of X onto any surface whose Fenchel Nielsen coordinates differ from the Fenchel Nielsen coordinates of X componentwise by at most δ.

As a consequence of this discussion, length functions are continuous in the topology defined by Fenchel Nielsen coordinates. Moreover, for every compact set K in Fenchel Nielsen coordinates, there is a number $L > 1$ and for any $X, Y \in K$ there is an L-bilipschitz map $X \to Y$ respecting the marking.

We are left with showing that any neighborhood V of X in Fenchel Nielsen coordinates contains a set of the form $U(X, \epsilon)$ for some $\epsilon > 0$. For this we argue by contradiction and we assume that there is a neighborhood V of X in Fenchel Nielsen coordinates and for each $i > 0$ there is some $X_i \in U(X, \frac{1}{i}) - V$. Then as $i \to \infty$, the length parameters in Fenchel Nielsen coordinates for the surfaces X_i converge to the lengths parameters for X. By the discussion preceding this lemma, the twist parameters of the surfaces X_i are bounded independent of i. As a consequence, after passing to a subsequence we may assume that the Fenchel Nielsen coordinates for X_i converge to the Fenchel Nielsen coordinates of some surface $Y \neq X$. By continuity of length functions in Fenchel Nielsen coordinates, we have $\ell_c(X) = \ell_c(Y)$ for every simple closed curve c. However, we observed in the proof of injectivity of Fenchel Nielsen coordinates that this implies that $Y = X$. This contradiction completes the proof of the lemma. □

Since the sets $U(X,\epsilon)$ are defined independently of the choice of a pants decomposition, we conclude that transition maps for Fenchel Nielsen coordinates defined by distinct pants decompositions are homeomorphisms.

Now any $\varphi \in \mathrm{Mod}(S)$ maps Fenchel Nielsen coordinates for a pants decomposition $\mathcal{P}$ to Fenchel Nielsen coordinates for $\varphi(\mathcal{P})$ and therefore

PROPOSITION 1.11. *$\mathcal{T}(S)$ has a $\mathrm{Mod}(S)$-invariant topology which is homeomorphic to $\mathbb{R}^{6g-6}$.*

The above discussion also indicates that Teichmüller space can be parametrized by suitably chosen length functions. We will not discuss this fact and rather state a useful easy property of length functions which can largely be generalized and analyzed quantitatively. In fact, for hyperbolic surfaces, the lengths of the closed geodesics are related to many other geometric invariants. We refer to [**B92**] for more information.

LEMMA 1.12. *For every $X \in \mathcal{T}(S)$ and every $\ell > 0$ there are only finitely many closed geodesics on X of length at most ℓ.*

PROOF. Let $X \in \mathcal{T}(S)$. Choose any finite collection $\mathcal{C} = \{\gamma_1, \dots, \gamma_k\}$ of simple closed geodesics on X whose union is a graph $G \subset X$ which decomposes X into simply connected regions. Then each free homotopy class on X can be represented by a closed edge-path on G. There are only finitely many homotopy classes of edge-paths of uniformly bounded combinatorial length, where the combinatorial length is the number of edges crossed through by the path.

Let γ be a closed geodesic on X which is distinct from a curve from $\mathcal{C}$. Then γ intersects each geodesic $\gamma_i \in \mathcal{C}$ transversely. It now suffices to show that the length of γ is bounded from below by a constant multiple of the number $\iota(\gamma, \mathcal{C})$ of these intersection points. Namely, since the components of $X - G$ are simply connected, there is an edge path on G which is homotopic to γ and whose combinatorial length is bounded from above by a constant multiple of the number $\iota(\gamma, \mathcal{C})$.

For this note that each curve from the collection $\mathcal{C}$ has a tubular neighborhood in X. Since the number of curves in $\mathcal{C}$ is finite, the width of each such neighborhood is bounded from below by a fixed number $\kappa > 0$. Then each essential intersection of γ with a component of $\mathcal{C}$ contributes at least 2κ to the length of γ. Now $\iota(\gamma, \mathcal{C}) \leq k \max_i \iota(\gamma, \gamma_i)$ and hence

$$\ell_\gamma(X) \geq 2\kappa\iota(\gamma, \mathcal{C})/k$$

which completes the proof of the lemma. □

For more details about the proof of the following proposition we refer to Section 6.3 of [**IT89**].

PROPOSITION 1.13. *The action of $\mathrm{Mod}(S)$ on $\mathcal{T}(S)$ is properly discontinuous.*

PROOF. It suffices to show the following. For every $X \in \mathcal{T}(S)$ and every pants decomposition $\mathcal{P}$ for S there is a compact neighborhood K of X in Fenchel Nielsen coordinates for $\mathcal{P}$ so that $\varphi K \cap K \neq \emptyset$ only for finitely many $\varphi \in \mathrm{Mod}(S)$.

For this we claim that there is a compact neighborhood K of X in Fenchel Nielsen coordinates and there are finitely many pants decompositions $\mathcal{P}_1, \dots \mathcal{P}_k$ so that $\varphi\mathcal{P} \in \{\mathcal{P}_1, \dots \mathcal{P}_k\}$ for all $\varphi \in \mathrm{Mod}(S)$ with $\varphi(K) \cap K \neq \emptyset$.

Let $\ell = \max\{\ell_\alpha(X) \mid \alpha \in \mathcal{P}\}$. Let K be a compact neighborhood of X so that for every $Y \in K$ there is a marked 2-bilipschitz map $F : Y \to X$. If $\varphi \in \mathrm{Mod}(S)$ is such that $\varphi(Y) \in K$ for some $Y \in K$ then $\varphi\mathcal{P}$ is a pants decomposition whose pants curves have geodesic representatives on X of length at most 2ℓ (since φ just changes the marking). Thus the claim is an immediate consequence of Lemma 1.12.

Now observe that if φ_1, φ_2 map $\mathcal{P}$ to a fixed pants decomposition $\mathcal{P}_i$ then $\varphi_2^{-1} \circ \varphi_1$ preserves $\mathcal{P}$. Since every diffeomorphism of a pair of pants which preserves each of the boundary circles is isotopic to the identity (see[**B92**]), the stabilizer of $\mathcal{P}$ in $\mathrm{Mod}(S)$ equals the free abelian group of Dehn twists about the pants curves (we refer once more to [**FM11**] for a detailed account on Dehn twists). However, by construction of the Fenchel Nielsen coordinates, the action of the stabilizer of $\mathcal{P}$ in $\mathrm{Mod}(S)$ is properly discontinuous. From this the proposition follows. ☐

Example: Let P be a pair of pants all of whose boundary curves have the same length. Glue two boundary curves together so that endpoints of seams are glued to endpoints of seams, and glue two copies of the resulting one-holed torus identifying endpoints of seams with endpoints of seams. The resulting hyperbolic surface X of genus 2 has an isometric involution φ exchanging the two pairs of pants which form X so that X/φ is a torus. Thus $\varphi \in \mathrm{Mod}(S)$ is a non-trivial element which fixes X. Similar constructions can be carried out for surfaces of any genus $g \geq 2$ and show that the action of $\mathrm{Mod}(S)$ on $\mathcal{T}(S)$ is not free. For more details and bounds on the order of the isometry group of a hyperbolic surface we refer to [**B92**].

Fenchel Nielsen coordinates also define a real analytic structure on $\mathcal{T}(S)$, but this is more difficult to see. In the remainder of this section, we discuss the special case of the Teichmüller space $\mathcal{T}(S_0)$ of a four-holed sphere S_0. This is the space of marked hyperbolic structures on S_0 so that the boundary consists of four closed geodesics. Any simple closed curve α on S_0 defines a pants decomposition, and hence there are Fenchel Nielsen coordinates consisting of five length functions and one twist parameter about α.

PROPOSITION 1.14. *Let S_0 be a four-holed sphere. Then Teichmüller space for S_0 has a structure of a real analytic manifold so that all length functions of simple closed geodesics are real analytic.*

PROOF. Let Y be a hyperbolic four-holed sphere. The universal covering of Y is a convex subset of the hyperbolic plane $\mathbf{H}^2$ with geodesic boundary. The fundamental group $\pi_1(Y)$ of Y is a free subgroup of $\mathrm{PSL}(2, \mathbb{R})$ which is generated by three hyperbolic isometries $\varphi_1, \varphi_2, \varphi_3 \in \mathrm{PSL}(2, \mathbb{R})$. The fixed points of these elements on the boundary $\partial\mathbf{H}^2$ of the hyperbolic plane are pairwise distinct.

The group $\mathrm{PSL}(2,\mathbb{R})$ acts triply transitive on the ideal boundary $\partial\mathbf{H}^2$ of the hyperbolic plane. If γ is an invariant geodesic for a hyperbolic isometry g and if $\varphi \in \mathrm{PSL}(2,\mathbb{R})$ then $\varphi\gamma$ is invariant for $\varphi \circ g \circ \varphi^{-1}$. On the other hand, two conjugate discrete torsion free subgroups of $\mathrm{PSL}(2,\mathbb{R})$ define the same marked hyperbolic surface. Therefore we may assume that in the upper half-plane model for $\mathbf{H}^2$, the fixed points of φ_1 on $\partial\mathbf{H}^2$ are the points $0,\infty$ and that 1 is a fixed point for φ_2. Since the group $\pi_1(Y)$ is generated by $\varphi_1, \varphi_2, \varphi_3$ and each element in $\mathrm{PSL}(2,\mathbb{R})$ is determined by three real parameters (=matrix entries), there are 6 real parameters which determine the conjugacy class of $\pi_1(Y) < \mathrm{PSL}(2,\mathbb{R})$.

In order to verify that we obtain 6 free parameters (i.e. there are no additional constraints) it suffices to show the following. Let $\tilde{\varphi}_i \in \mathrm{PSL}(2,\mathbb{R})$ be three elements which are close to φ_i with respect to the topology of $\mathrm{PSL}(2,\mathbb{R})$ as a quotient of $\mathrm{SL}(2,\mathbb{R})$. Then the elements $\tilde{\varphi}_i$ are hyperbolic and generate a free group which is the fundamental group of a hyperbolic four holed sphere with geodesic boundary.

An element in $\mathrm{PSL}(2,\mathbb{R})$ is hyperbolic if and only if it has a preimage in $\mathrm{SL}(2,\mathbb{R})$ whose trace is bigger than two. This clearly is an open condition.

Since Y is a four-holed sphere, there are six pairwise disjoint closed intervals $I_1,\dots,I_6 \subset \partial\mathbf{H}^2$ so that φ_i maps the exterior of I_i homeomorphically onto the interior of I_{i+3}. The intervals I_i, I_{i+3} contain the fixed points of φ_i. (This numbering of the intervals is not the numbering obtained by their counter-clockwise order on $\partial\mathbf{H}^2$). A nearby triple of hyperbolic elements $\tilde{\varphi}_i$ determines six new closed intervals $\tilde{I}_i$. The intervals $\tilde{I}_i$ can be chosen to depend continuously on the elements $\tilde{\varphi}_i$. In particular, for $\tilde{\varphi}_i$ sufficiently close to φ_i these intervals are pairwise disjoint.

A *ping pong argument* shows that any subgroup Γ of $\mathrm{PSL}(2,\mathbb{R})$ generated by three elements with this property is free and consists of hyperbolic elements. For this let $u = u_1 \cdots u_s$ be a nontrivial reduced word in the generators $\tilde{\varphi}_i$. Let $x \in \partial\mathbf{H}^2$ be a point in the complement of all intervals $\tilde{I}_j$. If $u_s = \tilde{\varphi}_j$ then $u_s x \in \tilde{I}_{j+3}$ and inductively, if $u_1 = \tilde{\varphi}_p$ then $u(x) \in \tilde{I}_{p+3}$ (indices are taken modulo six and $\tilde{\varphi}_{j+3} = \tilde{\varphi}_j^{-1}$). In particular, $u(x) \neq x$ and hence the group generated by $\tilde{\varphi}_1, \tilde{\varphi}_2, \tilde{\varphi}_3$ is free (compare [**M88**]).

The same argument also shows that the group Γ generated by these elements is discrete. Namely, otherwise for the point $x \in \partial\mathbf{H}^2$ chosen as in the previous paragraph there is a sequence $v_j \subset \Gamma$ so that $v_j \neq e$ and that $v_j x \to x$. However, the discussion in the previous argument shows that for a compact neighborhood U of x in $\partial\mathbf{H}^2 - \cup_i \tilde{I}_i$ we have $vx \notin U$ for $v \neq e \in \Gamma$.

Now the length of a closed geodesic on the hyperbolic surface Y can be calculated from the trace of a lift to $\mathrm{SL}(2,\mathbb{R})$ of the corresponding element of $\mathrm{PSL}(2,\mathbb{R})$. Namely, if this trace equals $\lambda + \lambda^{-1}$ for some $\lambda > 1$ then the length is $2\log\lambda$ (we refer to [**B92**] for this fact). As a consequence, length functions are real analytic with respect to the analytic structure on $\mathcal{T}(S_0)$ defined by the matrix components. □

Remark: Proposition 1.14 and its proof are equally valid for the Teichmüller space of a sphere with an arbitrary number $\ell \geq 4$ of geodesic boundary components. It can be extended to define coordinates for the Teichmüller space of closed surfaces as well. These so-called *Fricke coordinates* are important for many aspects of Teichmüller theory. For our purpose, however, they play no further role. We refer the reader to [**IT89**] for details and more.

COROLLARY 1.15. *Fenchel Nielsen coordinates are real analytic coordinates for the Teichmüller space of* S_0. *In particular, length functions are real analytic in Fenchel Nielsen coordinates.*

PROOF. Let α be a simple closed curve in S_0 and let $(\ell_1, \dots, \ell_4, \ell_5, \tau)$ be coordinates for $\mathcal{T}(S_0)$ defined by the 4 lengths ℓ_i of the boundary circles of S_0, the length ℓ_5 of α and the twist parameter τ. We claim that the map which associates to $X \in \mathcal{T}(S_0)$ its Fenchel Nielsen coordinates is real analytic.

By Proposition 1.14, for this it suffices to show that the twist parameter is an analytic function on $\mathcal{T}(S_0)$. This can be seen as follows.

Let $\tilde{\alpha}$ be a lift of α to $\mathbf{H}^2$. We may assume that in the upper half-plane model its endpoints are $0, \infty$. There are lifts β_1, β_2 of seams of the two components P_1, P_2 of $S_0 - \alpha$ with one endpoint on $\tilde{\alpha}$ and the second endpoint on a lift $\tilde{\beta}_i$ of a boundary curve of P_i distinct from α $(i = 1, 2)$. The geodesics $\tilde{\beta}_1, \tilde{\beta}_2$ are axes of hyperbolic isometries $\varphi_1, \varphi_2 \in \pi_1(S_0)$, and the segments β_1, β_2 realize the distance between the geodesic lines $\tilde{\alpha}$ and $\tilde{\beta}_1, \tilde{\beta}_2$.

As the isometries φ_1, φ_2 vary in $\mathrm{PSL}(2, \mathbb{R})$ in an analytic family, their axes vary in an analytic family as well. As a consequence, the shortest distance projections of these axes to the fixed geodesic $\tilde{\alpha}$ depend in a real analytic fashion on the elements. However, the signed distance between these projections is just the twist parameter τ of the Fenchel Nielsen coordinates in the analytic family. As a consequence, Fenchel Nielsen coordinates are real analytic.

To show that Fenchel Nielsen coordinates are real analytic coordinates for $\mathcal{T}(S_0)$ it now suffices to show that the map which associates to $X \in \mathcal{T}(S_0)$ its Fenchel Nielsen coordinates is of maximal rank differentiable at every point. This can easily be seen explicitly. We omit the proof and refer to [**A80, IT89**]. □

We formulate without proof the corresponding result for the Teichmüller space of a closed surface S. We refer to [**A80**] and to [**IT89**] for a more comprehensive discussion and a proof.

THEOREM 1.16. *Fenchel Nielsen coordinates define a* $\mathrm{Mod}(S)$*-invariant real analytic structure on* $\mathcal{T}(S)$.

Remark: Everything in this section is equally valid for marked hyperbolic metrics of finite area on a surface with punctures.

LECTURE 2

Quasiconformal maps

In this section we discuss some analytic aspects of Teichmüller theory. We introduce quasiconformal maps and abelian and quadratic differentials, and we discuss Teichmüller's existence and uniqueness theorem.

Recall the upper half-plane model for the hyperbolic plane $\mathbf{H}^2$. An oriented hyperbolic surface can be defined as a surface S together with a covering of S by orientation preserving charts $\varphi : U \subset S \to \varphi(U) \subset \mathbf{H}^2$ so that chart transitions are orientation preserving isometries. Since an orientation preserving isometry of $\mathbf{H}^2$ is in particular a biholomorphic map, these charts define a complex structure on S. Thus we have

PROPOSITION 2.1. *An oriented hyperbolic surface is a Riemann surface.*

By the uniformization theorem, the converse is also true. Namely, the *conformal class* of a complex structure on S is the space of all Riemannian metrics g on S with the property that the fibrewise multiplication with i in the tangent bundle of S preserves g. Since $i^2 = -1$, this means that multiplication by i in a fibre is just rotation by the angle $\pi/2$.

PROPOSITION 2.2. *The conformal class of a complex structure on S contains a unique hyperbolic metric.*

Namely, the universal covering of a closed Riemann surface X of genus $g \geq 2$ is the unit disc $D \subset \mathbb{C}$ which is a model for the hyperbolic plane $\mathbf{H}^2$ (see [**FK80**]). Then X is the quotient of D by a discrete group of biholomorphic automorphisms of D. But the group of all biholomorphic automorphisms of $D = \mathbf{H}^2$ is just the group $\mathrm{PSL}(2,\mathbb{R})$ of hyperbolic isometries. In particular, there is a hyperbolic metric on X so that the orientation preserving charts determined by this metric are holomorphic for the complex structure. Or, the complex structure coincides with the structure constructed from the hyperbolic metric in the first paragraph of this section.

In Section 2 we saw that length functions of simple closed curves can be used to define a $\mathrm{Mod}(S)$-invariant topology on $\mathcal{T}(S)$. This topology is described by measuring the deviation from the existence of a marked isometry between two marked hyperbolic surfaces. (We remark here that Thurston initiated a systematic study of Teichmüller theory via (bi)-Lipschitz maps with optimal Lipschitz constants, and there is a substantial current activity in this line of investigation. We refer to [**PT07**] for a recent survey and references on this subject).

We can also define the topology on $\mathcal{T}(S)$ by measuring the deviation from the existence of a marked biholomorphic map between two marked Riemann surfaces. Recall that a map F between two Riemann surfaces is holomorphic if in complex charts we have $\bar{\partial} F = 0$.

DEFINITION 2.3. Let X, X' be Riemann surfaces. For a number $K \geq 1$, a map $F : X \to X'$ is *K-quasiconformal* if F is a homeomorphism which is continuously differentiable, with differential of maximal rank, outside a finite set Σ of points and if

$$|\bar{\partial} F| \leq k|\partial F|$$

outside Σ where $0 \leq k < 1$ and $K = \frac{1+k}{1-k}$.

The geometric significance is as follows. Let g, g' be any Riemannian metric in the conformal class defined by the complex structure on X, X'. Then with respect to these metrics, the differential of F at $x \in X - \Sigma$ maps a round circle in the tangent space $T_x X$ of X at x to an ellipse in the tangent space $T_{F(x)} X'$ of X' whose axes have length ratio bounded by K. Note that this is independent of the choice of the metrics g, g' in the conformal class defined by the complex structures. (In Lemma 4.8 of [**IT89**], this easy calculation is carried out in detail). As a consequence, the Jacobian of F at x is bounded from below by K^{-1} times the square of the operator norm of the differential of F at x.

Example: Since S is compact by assumption, every orientation preserving diffeomorphism $F : X \to X'$ is quasiconformal, i.e. there is a number $K > 0$ so that F is K-quasiconformal.

It turns out that particularly nice quasiconformal maps can be constructed using objects directly defined by the complex structure of the Riemann surface X. We next introduce these structures and these maps and establish some of their most important properties.

DEFINITION 2.4. An *abelian differential* on a Riemann surface X is a nontrivial holomorphic one-form ω on X. Such a one-form is a holomorphic section of the *holomorphic cotangent bundle* $T'X$ of X.

In a holomorphic local coordinate z, a holomorphic one-form ω can be represented as $f(z)dz$ with a holomorphic function f. Now the exterior differential d for complex valued one-forms on $\mathbb{C}$ can be decomposed as $d = \partial + \bar{\partial}$ with $\partial\partial = \bar{\partial}\bar{\partial} = 0 = \partial\bar{\partial} + \bar{\partial}\partial$ (see pp. 77 and 78 of [**D11**] for a calculation in local coordinates) and hence $d\omega = df \wedge dz = 0$ since $\bar{\partial} f = 0$. As a consequence, ω defines a complex valued de Rham cohomology class on S.

Thus there is a natural homomorphism from the vector space $\Omega(X)$ of holomorphic one-forms on X into the first cohomology group $H^1(X, \mathbb{C})$ of X with complex coefficients which associates to $\omega \in \Omega(X)$ its de Rham cohomology class. The complex vector space $H^1(X, \mathbb{C})$ is isomorphic to $\mathbb{C}^{2g}$.

PROPOSITION 2.5. *The homomorphism $\Omega(X) \to H^1(X, \mathbb{C})$ is an embedding onto a complex subspace of dimension g.*

PROOF. (Incomplete sketch) The complex Laplacian for smooth functions $f : X \to \mathbb{C}$ is defined by

$$\Delta = 2i\bar{\partial}\partial.$$

A function f is *harmonic* if $\Delta f = 0$. Similarly, a one-form α on X is *harmonic* if locally α is the differential of a harmonic function. Since $d = \partial + \bar{\partial}$ in a complex coordinate, a harmonic one-form is closed. In fact, a one-form α on X is harmonic if and only if it is both closed and coclosed (see p.27 of [**FK80**]).

Now a non-zero harmonic one-form α is not exact. Namely, otherwise $\alpha = df$ for a function f. Then f is harmonic and non-constant. By Stokes theorem, we have $\int d(f\bar{\partial}f) = 0 = \int \partial(f\bar{\partial}f)$, moreover $\bar{\partial}\partial = -\partial\bar{\partial}$ and therefore

$$\int f\Delta f = 0 = 2i\int \partial(f\bar{\partial}f) - f\partial\bar{\partial}f = 2i\int \partial f \wedge \bar{\partial}f = i\int df \wedge \overline{df}$$

which is possible only if df vanishes identically: In a local complex coordinate z, a complex valued one-form α can be written in the form $\alpha = adz + bd\bar{z}$ and hence $i\alpha \wedge \bar{\alpha} = i(a\bar{a} + b\bar{b})dz \wedge d\bar{z}$ which is a positive multiple of the euclidean area form at any point where α does not vanish (we refer to [**D11**] for more details).

By the Cauchy Riemann equations, a real valued function f on a domain in $\mathbb{C}$ is harmonic if and only if it is the real part of a holomorphic function. Thus a real one-form α on X is harmonic if and only if it is the real part of a holomorphic one-form. Since by the Hodge de Rham theorem every de Rham cohomology class of degree one can be represented by a unique harmonic one-form (see Theorem 6 of [**D11**]), this implies that the homomorphism $\Omega(X) \to H^1(X, \mathbb{C})$ is injective. Moreover, the $\mathbb{R}$-linear map which associates to a holomorphic one-form ω the de Rham cohomology class of its real part maps $\Omega(X)$ onto $H^1(X, \mathbb{R})$ and hence the complex dimension of the image $\Omega(X)$ in $H^1(X, \mathbb{C})$ equals g. □

We also need related but more general objects which are defined as follows.

DEFINITION 2.6. A *quadratic differential* on X is a holomorphic section of $T'X \otimes T'X$.

In local coordinates, q can be represented in the form $q = f(z)dz^2$ where f is a nontrivial holomorphic function. In particular, if $q \neq 0$ then q has only finitely many zeros.

Example: The square of every abelian differential is a quadratic differential.

The Riemann Roch formula (Theorem 7 of [**D11**]) implies

THEOREM 2.7. *The complex vector space $\mathcal{Q}(X)$ of quadratic differentials on X is of dimension $3g - 3$.*

Quadratic differentials can be described as follows. Let $x \in X$ be a point so that $q \in \mathcal{Q}(X)$ does not vanish at x. Then with respect to a holomorphic coordinate u near x, the differential q can be represented in the form fdu^2 with a non-vanishing holomorphic function f. Use a branch of the square root to write $q = g^2du^2$ for a non-vanishing holomorphic function g (perhaps defined on a smaller neighborhood of x). Then locally near x, gdu is the differential of a holomorphic function z defined by

$$z(y) = \int_x^y g(u)du$$

so that

$$q = dz^2.$$

The coordinate z is unique up to translation and perhaps multiplication with -1, i.e. up to an euclidean isometry. In particular, every quadratic differential q defines an euclidean metric outside its zero set Σ. We refer to [**S84**] for a detailed account on this construction.

The line fields $\{v \in T(X - \Sigma) \mid q(v) \geq 0\}$ and $\{v \in T(X - \Sigma) \mid q(v) \leq 0\}$ define local foliations of $X - \Sigma$ called the *horizontal* and *vertical* foliation, respectively. In the distinguished coordinate z for q, these foliations are precisely the foliations into lines parallel to the real and imaginary axis. Rectangles for these transverse foliations are rectangles for the euclidean metric defined by q.

At a zero x we can write $q = g^p dz^2$ for some $p > 0$ and some holomorphic function g which has a simple zero at x. Taking appropriate roots shows that the metric has a standard $p+2$-pronged singularity at x (we refer to the standard reference [**S84**] for quadratic differentials for details). A neighborhood of such a singular point is a neighborhood of 0 in the following space. Take $p+2$ copies H_i ($i = 1, \dots, p+2$) of the upper half-plane and glue these half-planes along their boundaries in cyclic order so that the ray $\{\mathrm{Im} = 0, \mathrm{Re} \leq 0\}$ in the boundary of the half-plane H_i is glued to the corresponding ray in the boundary of the half-plane H_{i-1} with an orientation reversing isometry, and the ray $\{\mathrm{Im} = 0, \mathrm{Re} \geq 0\}$ in the boundary of H_i is glued to the corresponding ray in the boundary of H_{i+1}. In particular, the area of the singular euclidean metric defined by q is finite.

As a consequence, we can define a norm $\| \ \|$ on the space $\mathcal{Q}(X)$ by

$$\|q\| = \mathrm{area}(q) = \int_X |q|.$$

For a finite subset Σ of S, a foliation of $S - \Sigma$ is *orientable* if its tangent bundle admits a smooth nowhere vanishing section.

LEMMA 2.8. *A quadratic differential is the square of a holomorphic one-form if and only if its horizontal and vertical foliations are orientable.*

PROOF. If $q = \omega^2$ then $\omega > 0$ defines an orientation of the horizontal foliation, and $i\omega > 0$ defines an orientation of the vertical foliation.

On the other hand, if q has orientable foliations then away from the zeros of q we can take a square root of q defining this orientation, and these square roots are consistent under change of charts. □

If the underlying complex structure of the quadratic differential is allowed to be arbitrary then it is very easy to construct quadratic differentials explicitly. For this purpose we define

DEFINITION 2.9. Two simple closed curves α, β *bind* the surface S if α, β are in minimal position (i.e. they intersect in the minimal number of points in their free homotopy classes) and decompose S into simply connected regions.

The following construction is due to Thurston. We refer to Section 9 of [**V89**] for a slightly more general account.

PROPOSITION 2.10. *A pair of simple closed curves on S which bind S defines a quadratic differential.*

PROOF. We construct explicitly a quadratic differential from two simple closed curves α, β which bind S as follows.

Note first that α, β decompose S into polygons with an even number of sides. The sides alternate between subarcs of α and subarcs of β. Place an euclidean square of side length one over each intersection point between α and β. This can be done in such a way that the squares only meet along their sides or at their vertices, and that an intersection of sides corresponds to a component of $\alpha - \beta$ or a component of $\beta - \alpha$. This means that each side contains precisely one point of $\alpha - \beta$ or of $\beta - \alpha$, and each component of $\alpha - \beta$ or of $\beta - \alpha$ intersects the sides in precisely one point. Each component of $S - \{\alpha \cup \beta\}$ contains exactly one vertex. If this component is a $2p$-gon for some $p \geq 2$ then there are $2p$ squares coming together at that vertex.

The euclidean structures on the squares define a singular euclidean metric on S with one p-pronged singularity for each complementary polygon with $2p \geq 6$ sides. Away from the singularities, this structure is just given by the standard euclidean charts. It can be extended across the singular points since each such point is a standard p-pronged singularity.

Up to maps of the form $z \to \pm z + c$ for some $c \subset \mathbb{C}$, each square has a preferred isometric embedding into $\mathbb{C}$ which maps the sides parallel to α to horizontal straight segments parallel to the real axis, and which maps the sides parallel to β to vertical straight segments parallel to the imaginary axis. Then transitions for these rectangles on their overlaps are of the form $z \to \pm z + c$ $(c \in \mathbb{C})$, and these transitions preserve the quadratic differential dz^2. As a consequence, these differentials define a quadratic differential q on the complement of the singular points which naturally extends to the singular points. □

We can use quadratic differentials to construct quasiconformal maps as follows. Let q be a holomorphic quadratic differential with zero set Σ on a marked Riemann surface X. Then on $S - \Sigma$, the differential q defines an euclidean metric q and horizontal and vertical foliations. For $t > 0$ let g_t be the singular euclidean metric on $S - \Sigma$ obtained by stretching the horizontal direction by the factor $e^{t/2}$ and contracting the vertical direction by the factor $e^{-t/2}$. The metric extends to a singular euclidean metric on S with singular points at the zeros of q. This metric then defines a marked complex structure X_t on S and a holomorphic quadratic differential q_t for X_t. The marking of X_t is the composition of the marking of X with the stretch map. The stretch map $X \to X_t$ is quasi-conformal with constant e^t.

It is called a *Teichmüller map* with initial differential q and stretch factor $e^{t/2}$. The area of q_t equals the area of q.

As is customary in Riemannian geometry, for $t > 0$ and a quadratic differential $q \in \mathcal{Q}(X)$ of area one (= norm one) we associate to the quadratic differential tq of norm t the image $\Psi(tq)$ of the Teichmüller map with initial differential q and stretch factor $e^{t/2}$. This convention defines a map $\Psi : \mathcal{Q}(X) \to \mathcal{T}(S)$ with $\Psi(0) = X$. We call the image under Ψ of a ray in $\mathcal{Q}(X)$ a *Teichmüller geodesic* in $\mathcal{T}(S)$.

For two Riemann surfaces X, X' define

$$d_{\mathcal{T}}(X, X') = \frac{1}{2} \inf\{\log K \mid$$
$$\text{there is a } K\text{-quasiconformal map } f : X \to X' \text{ respecting the marking}\}.$$

The next result implies that Teichmüller maps are optimal for this invariant. For its formulation, recall that a quadratic differential q determines an area element dA on S. Its total area is the norm $\|q\|$ of q. A *separatrix* for q is a maximal segment or ray (i.e. a geodesic for the singular euclidean metric) which begins at a singular point and does not contain any singular point in its interior. The proof of the following theorem is taken from [**FM11**].

THEOREM 2.11. *Let $f : X \to X'$ be a quasiconformal map which is homotopic to a Teichmüller mapping with initial differential q and stretch factor L. Then*

$$\int_X |f_x| dA \geq L\|q\|$$

where f_x is the derivative of f in the horizontal direction and the norm is taken with respect to the singular euclidean metric of q.

PROOF. For $p \in X$ and $T > 0$ let

$$\delta(p, T) = \int_{-T}^{T} |f_x| dx$$

where the integration is over the horizontal arc α of length $2T$ centered at p. If such a horizontal arc passes through a singular point then $\delta(p, T)$ is undefined. Since there are only finitely many horizontal separatrices and hence the union of these separatrices has area zero, the value $\delta(p, T)$ is defined almost everywhere with respect to the area element dA of the quadratic differential q.

By Fubini's theorem,

$$\int_X \left(\int_{-T}^{T} |f_x| dx \right) dA = 2T \int_X |f_x| dA.$$

However, $\delta(p, T)$ is just the length of $f(\alpha)$ with respect to the singular euclidean metric defined by the terminal differential q' for the Teichmüller mapping. Thus

$$\int_X \left(\int_{-T}^{T} |f_x| dx \right) dA \geq \int_X (2LT - M) dA = (2LT - M)\|q\|$$

where $M > 0$ is a fixed constant which is obtained as follows.

The diameter of the singular euclidean metric defined by q' is finite. By assumption, there is a homotopy connecting f to the Teichmüller mapping with initial differential q and stretch factor L. For each horizontal arc α, such a homotopy determines a homotopy class of arcs connecting the endpoints of $f(\alpha)$ to the endpoints of the stretched arc. The q'-length of any q'-geodesic representative of such an arc is uniformly bounded, independent of α. If $M/2$ is an upper bound for these lengths then the estimate holds true for M.

As a consequence, we have

$$\int_X |f_x| dA \geq (L - \frac{M}{2T}) \|q\|.$$

As $T \to \infty$ the theorem follows. □

COROLLARY 2.12. *A Teichmüller map between two marked Riemann surfaces X, X' is the unique $e^{2d_{\mathcal{T}}(X,X')}$-quasi-conformal map in its homotopy class mapping X to X'.*

PROOF. Let q be an area one quadratic differential on a Riemann surface X and let X' be the image of the Teichmüller mapping with initial differential q and stretch factor L. If $f : X \to X'$ is K-quasi-conformal then with respect to the singular euclidean metrics on X, X', the Jacobian $\mathrm{Jac}(f)$ of f at any point which is both regular for the map f and the quadratic differential q is not smaller than K^{-1} times the square of the operator norm of its differential df.

Now the K-quasi-conformal map $f : X \to X'$ is a homeomorphism and therefore using the Cauchy Schwarz inequality, Theorem 2.11 and the fact that $|df| \geq |f_x|$, with the notations from Theorem 2.11 we have

$$1 = \int_X \mathrm{Jac}(f) dA \geq \int_X |df|^2 dA/K \geq \left(\int |f_x| dA \right)^2 /K \geq L^2/K.$$

This shows that the quasi-conformal constant K of f is at least L^2.

Equality only holds if equality holds true in the Cauchy-Schwarz inequality and if moreover $|f_x| = |df|$ almost everywhere. This implies that $|df| = |f_x| = L$ almost everywhere.

The same argument applies to the restriction of f to the vertical foliation. Thus since f is continuously differentiable on the complement of a finite set of points, if $K = L^2$ then the composition of f with the inverse of the Teichmüller map is an isometry for the singular euclidean metric defined by q which is continuously differentiable on the complement of a finite set of points. This isometry is moreover isotopic to the identity.

However, an isometry Φ of the singular euclidean metric defined by q which is isotopic to the identity is the identity.

Namely, since the metric defined by q is singular euclidean with singular points of cone angle bigger than 2π, it is a metric of non-positive curvature. An isotopy between an isometry Φ and the identity determines for each point $x \in S$ a homotopy class of an arc connecting x to $\Phi(x)$. The homotopy class can be represented by a unique geodesic $\alpha_x : [0,1] \to S$ depending continuously on x. By convexity, the map $\Phi_0 : x \to \alpha_x(\frac{1}{2})$ is distance non-increasing. If $\Phi \neq \mathrm{Id}$ then it is not isometric. Then Φ_0 is a map of degree one whose Jacobian is bounded from above by one and which is different from one on some nontrivial open set. This is impossible. □

Theorem 2.11 and its proof can also be used to classify annuli up to biholomorphic equivalence. Here an annulus is a compact Riemann surface which is biholomorphic to $A_a = [0,1] \times [0,a]/\sim$ where $(0,t) \sim (1,t)$ for all t. Define the *modulus* of A_a to be a. The reasoning in the proof of Theorem 2.11 shows that the modulus is an invariant of the complex structure of such annuli.

COROLLARY 2.13. *For $b \geq a$ the smallest quasiconformal dilatation of a quasiconformal map $A_a \to A_b$ equals b/a.*

If $A \subset \mathbb{C}$ is any annulus then uniformization implies that A is biholomorphic to A_a for some $a > 0$, and a is called the *modulus* of A. Corollary 2.13 immediately shows

COROLLARY 2.14. *The modulus classifies annuli up to biholomorphic equivalence.*

We can use Corollary 2.14 to relate lengths of geodesics for hyperbolic metrics directly to invariants of complex structures. As in Section 2, denote by ℓ_c the hyperbolic length function on $\mathcal{T}(S)$ of a simple closed curve c. Wolpert showed (see [**W10**] for references and for more details)

PROPOSITION 2.15. *Let X, X' be hyperbolic surfaces and let $f : X \to X'$ be a K-quasiconformal map. Then for any simple closed curve c we have*

$$\frac{\ell_c(X)}{K} \leq \ell_c(X') \leq K\ell_c(X).$$

PROOF. Let γ_1, γ_2 be the closed geodesics on X, X' in the free homotopy class of c. There is a covering A_1, A_2 of X, X' whose fundamental group is generated by γ_1, γ_2. If a_1, a_2 is the length of c on X, X' then we may assume that $A_i = \mathbf{H}^2/ < b_i >$ where $b_i(z) = e^{a_i} z$ (here as before, we use the upper half-plane $\{\mathrm{Im} > 0\}$ as a model for the hyperbolic plane).

There is a branch of the logarithm which maps the upper half-plane biholomorphically onto the infinite strip

$$\mathrm{Im} \in (0, \pi).$$

Under this identification, the group $< b_i >$ corresponds to the infinite cyclic group of translations generated by $z \to z + a_i$. As a consequence, the annulus A_i is biholomorphic to a standard flat cylinder of circumference a_i and height π. In other words, the modulus of A_i equals $m_i = \pi/a_i$.

A K-quasiconformal map $X \to X'$ lifts to a K-quasiconformal map of the cylinder A_1 onto the cylinder A_2. By Corollary 2.13, this just means that $m_2/K \leq m_1 \leq Km_2$ as claimed. □

As an immediate consequence we obtain

PROPOSITION 2.16. *$d_{\mathcal{T}}$ is a* Mod(S)*-invariant metric on $\mathcal{T}(S)$ which defines the topology given by Fenchel Nielsen coordinates.*

PROOF. Clearly $d_{\mathcal{T}}$ is a Mod(S)-invariant pseudo-metric on $\mathcal{T}(S)$. Proposition 2.15 and Lemma 1.10 show that $d_{\mathcal{T}}$ is in fact a metric (i.e. we have $d_{\mathcal{T}}(X,X')>0$ if $X\neq X'$) defining a topology which is finer than the topology induced by Fenchel Nielsen coordinates. This means that the identity $(\mathcal{T}(S),d_{\mathcal{T}})\to\mathcal{T}(S)$ is continuous.

Thus for the proof of the proposition we are left with showing that the identity map $\mathcal{T}(X)\to(\mathcal{T}(X),d_{\mathcal{T}})$ is continuous as well. For this it suffices to show that in Fenchel Nielsen coordinates, if coordinate functions converge then the minimal dilatation of a quasiconformal map between the two surfaces defined by these coordinates converges to one. However, quasi-conformal maps with small quasiconformal dilatation between hyperbolic surfaces with nearby Fenchel Nielsen coordinates can be constructed explicitly using elementary hyperbolic geometry exactly as in the proof of Lemma 1.10. Namely, note that for some $L>1$, an L-bilipschitz diffeomorphism F between two hyperbolic surfaces is L^2-quasi-conformal. □

As before, let $\mathcal{Q}(X)$ be the vector space of quadratic differentials on the Riemann surface X. Recall the definition of the map $\Psi:\mathcal{Q}(X)\to\mathcal{T}(S)$. The following important result is much more difficult than the rather elementary facts discussed so far. We are not going to give a proof but rather indicate an elementary strategy to this end which we hope gives some geometric intuition why the result holds true. For this and later purpose, we define

DEFINITION 2.17. A *saddle connection* of a quadratic differential q is a straight line segment for the singular euclidean metric defined by q which connects two singular points and does not have a singular point in its interior.

Thus a saddle connection is a compact separatrix.

In a sequel we mean by a triangulation of a surface S a decomposition of S into triangles whose vertices are not necessarily distinct. The following simple observation will be used several times.

LEMMA 2.18. *Let q be a quadratic differential on S. Then there exists a triangulation of S consisting of saddle connections whose vertex set is the set of singular points of S.*

PROOF. Throughout this proof, distances are taken with respect to the singular euclidean metric.

Let $B\subset S$ be any embedded graph whose vertices are singular points for q and whose edges are saddle connections. If there is a component C of $S-B$ which is not simply connected then choose an arc $\alpha\subset\overline{C}$ with endpoints at singular points which is not homotopic into B with fixed endpoints, which does not cross through an edge of B and which is shortest with this property. Then α is a piecewise geodesic for the singular euclidean metric. Since the length of α is minimal, it does not contain an edge from the graph B. Moreover, it does not meet the interior of an edge of B since otherwise α has a breakpoint at such an interior point with nontrivial angle defect and hence it can be shortened by a local homotopy. As a consequence, α is a saddle connection whose interior is disjoint from the interiors of the edges of B. In other words, $B \cup \alpha$ is an embedded graph whose vertices are singular points and whose edges are saddle connections.

By successively adding edges to B with the above procedure, one constructs an embedded graph B' in S whose edges are saddle connections and which decomposes S into simply connected regions. Any such region can be subdivided into a union of triangles by first connecting any singular point in the interior to a boundary vertex with a saddle connection of minimal length and then subdividing the resulting euclidean polygons. □

THEOREM 2.19. *The map Ψ is continuous.*

PROOF. The standard proof evokes the measurable Riemann mapping theorem which is beyond the scope of these notes. We refer to Section 11 of [**FM11**] for a nice account. Another reference is [**IT89**].

There is a partial result which can be proven with elementary methods. Namely, let $q_i, q \in \mathcal{Q}(X)$ and assume that $q_i \to q$. Since q_i, q are holomorphic sections of the same line bundle over X, the zeros of q_i converge to the zeros of q. Assume that the zeros of q are *simple*, i.e. all singular points for the singular euclidean metric defined by q are 3-pronged singularities. Then for large i, the zeros of q_i are simple as well.

By Lemma 2.18 we can choose a triangulation T of X by saddle connections for q which contains all singular points as vertices.

Every triangle of the triangulation T is isometric to an euclidean triangle. For large enough i there is a corresponding triangulation T_i of the surface S whose edges consist of saddle connections for q_i. This triangulation is constructed by choosing a diffeomorphism F of S isotopic to the identity which maps the singular points of q to the singular points of q_i and replacing the image under F of each edge from T by the unique shortest arc for q_i in the same homotopy class with fixed endpoints. Since such an arc is a concatenation of saddle connections, for large enough i these arcs are saddle connections.

Let $\tilde{q}, \tilde{q}_i$ be the terminal quadratic differentials of the Teichmüller maps $\Psi(q)$, $\Psi(q_i)$. Since the stretch maps are affine in euclidean coordinates, the triangulations of S constructed in the previous paragraph define triangulations of S whose edges are saddle connections for $\tilde{q}, \tilde{q}_i$. As $i \to \infty$, the side lengths of the triangles for $\tilde{q}_i$ converge to the side lengths of the corresponding triangles for $\tilde{q}$.

Choose a number $r > 0$ so that the circular discs of radius $2r$ about the singular points of $\tilde{q}$ are pairwise disjoint. Note that such a disc is up to isometry determined by its radius and the order of the singular point. The intersection of each triangle with the complement of the discs of radius r about the singular points is the complement of discs of radius r about the vertices of the triangles.

Since the side lengths of the triangles for $\tilde{q}_i$ converge to the side lengths of the triangles for $\tilde{q}$ as $i \to \infty$, for large i one can construct explicitly a bilipschitz diffeomorphism between the truncated triangles whose bilipschitz constants tend to one as $i \to \infty$. This can be done in such a way that the restrictions of these maps to the truncated sides match up. Moreover, we can require that they match up with suitably constructed bilipschitz maps between the circles of radius r which are isometries near the singular point and whose bilipschitz constant tends to one as $i \to \infty$. Then these maps can be glued to a quasi-conformal map between the Riemann surfaces $\Psi(\tilde{q}_i)$, $\Psi(\tilde{q})$, with constant tending to one as $i \to \infty$ (compare the proof of Lemma 1.10).

A variation of this elementary argument can also be used in the general case, but apart from its intuitive appeal, it does not have any advantage over the elegant proof using the important measurable Riemann mapping theorem. □

LEMMA 2.20. *Ψ is proper.*

PROOF. Since Ψ is continuous, it suffices to show that the preimage under Ψ of a compact set in $\mathcal{T}(S)$ is bounded in $\mathcal{Q}(X)$.

To this end let $\kappa(Y) = d_{\mathcal{T}}(X, Y)$. By Proposition 2.16, κ is a continuous function on $\mathcal{T}(S)$. In particular, if $K \subset \mathcal{T}(S)$ is compact then κ assumes a maximum $M > 0$ on K. By Corollary 2.12, this implies that $\|q\| \leq M$ for every $q \in \mathcal{Q}(X)$ with $\Psi(q) \in K$. The lemma follows. □

Now Ψ is an injective continuous proper map between topological manifolds homeomorphic to $\mathbb{R}^{6g-6}$. By invariance of domain, the map Ψ is in fact a homeomorphism. As a consequence, we obtain

COROLLARY 2.21. *Any two points in $\mathcal{T}(S)$ can be connected by a unique Teichmüller geodesic.*

LECTURE 3

Complex structures, Jacobians and the Weil Petersson form

The goal of this section is a differential geometric look at the complex geometry of Teichmüller space and moduli space. We also evoke without proof some connections to the algebraic geometry of the moduli space of curves.

We begin with having a closer look at the vector space $\Omega(X)$ of holomorphic one-forms on a Riemann surface X. By integration along loops, any 1-form $0 \neq \omega \in \Omega(X)$ defines a homomorphism $\pi_1(S) \to \mathbb{C}$ called the *period map*.

Choose a basis $a_1, \ldots, a_g, b_1, \ldots, b_g$ for $H_1(S, \mathbb{Z})$ so that $a_i \cdot b_j = \delta_{ij}$ where $\cdot$ is the homology intersection form. Such a basis is called a *symplectic basis.* An example can be constructed as follows. Let Y be a $4g$-gon with a counter-clockwise numbering of the edges. For each $0 \leq i \leq g-1$ identify the edge $a_i = 4i+1$ with the edge $a_i^{-1} = 4i+3$ with an orientation reversing homeomorphism, and identify the edge $b_i = 4i+2$ with the edge $b_i^{-1} = 4i+4$ with an orientation reversing homeomorphism. The result is a closed surface S of genus g. The image in S of each oriented edge of Y is a simple closed curve which is identified with its edge label, and these curves intersect in a single point. The curves $a_1, \ldots, a_g, b_1, \ldots, b_g$ define a symplectic basis of $H_1(S, \mathbb{Z})$.

Now assume that the above polygon Y is obtained by cutting the Riemann surface X open along smooth simple closed loops intersecting in a single point. Let α_i, β_i be the periods of ω for the corresponding symplectic basis of $H_1(S, \mathbb{Z})$. On the polygon Y, ω is the differential df of a holomorphic function f. By Stokes' theorem,

$$\int_X \omega \wedge \bar{\omega} = \int_Y df \wedge \bar{\omega} = \int_{\partial Y} f \wedge \bar{\omega}.$$

Observe that if p, p' are two points which are identified under the glueing of the sides of Y and if p projects to an interior points of a_i and p' projects to an interior point of a_i^{-1} then

$$f(p') - f(p) = \int_{b_i} \omega = \beta_i.$$

Similarly, if u, u' are identified under the glueing and if u projects to an interior point of b_i and u' projects to an interior point of b_i^{-1} then $f(u') - f(u) = -\alpha_i$. We therefore have

$$\int_{a_i} (f(p) - f(p')) \wedge \bar{\omega} = -\beta_i \bar{\alpha}_i \quad \text{and}$$
$$\int_{b_i} (f(u) - f(u')) \wedge \bar{\omega} = \alpha_i \bar{\beta}_i.$$

Since the curve a_i occurs precisely twice along ∂Y, together this yields the formula

$$\int_S \omega \wedge \bar{\omega} = \sum \alpha_i \bar{\beta}_i - \beta_i \bar{\alpha}_i.$$

The formula implies that if the a-periods of ω vanish then so does ω. But there are exactly g a-periods and hence the linear map $\omega \to (\int_{a_1} \omega, \dots, \int_{a_g} \omega)$ is a linear isomorphism $\Omega(X) \to \mathbb{C}^g$ (recall from Proposition 2.5 that $\Omega(X)$ is isomorphic to $\mathbb{C}^g$; in fact, the above discussion coincides with the discussion in the proof of Proposition 2.5, only the viewpoint taken here is slightly different). Or, for a given symplectic basis (a_i, b_j) of $H_1(S, \mathbb{Z})$, there is a canonical basis of $\Omega(X)$ such that $\omega_i(a_j) = \delta_{ij}$. Then the *Riemann period matrix* of X is given by

$$Z_{i,j} = \omega_i(b_j).$$

Thus if $\omega \in \Omega(X)$ has a-periods α_i, the b-periods are $\beta_i = \sum_j Z_{ij} \alpha_j$, or, equivalently, $\beta = Z\alpha$. Together we get

$$|\omega|^2 = \frac{i}{2} \int \omega \wedge \bar{\omega} = \frac{i}{2}(\alpha \bar{Z} \bar{\alpha} - \bar{\alpha} Z \alpha) = \mathrm{Im}(\bar{\alpha} Z \alpha).$$

As a consequence, the matrix Z is symmetric, and its imaginary part is positive definite. This is one half of what is called *Riemann binary relations* in the literature (see e.g p.262 of [**Mi95**] for more information).

For a given symplectic basis (a_i, b_j) of $H_1(S, \mathbb{Z})$ and every $X \in \mathcal{T}(S)$ let $B(X)$ be the basis of $\Omega(X)$ defined by $\omega_i(a_j) = \delta_{ij}$. This defines a trivialization of the space

$$\mathcal{H} = \cup_{X \in \mathcal{T}(S)} \Omega(X)$$

which equips $\mathcal{H}$ with the structure of a complex vector bundle on $\mathcal{T}(S)$. Any other symplectic basis of $H_1(S, \mathbb{Z})$ defines a new trivialization which differs from the initial one by a bundle isomorphism.

Indeed, that this is pointwise the case is immediate from the above discussion. That these pointwise isomorphisms depend smoothly on the basepoint requires an argument. One fairly easy way to see that this is the case is to observe that the Laplacian depends smoothly on the marked hyperbolic structure in Fenchel Nielsen coordinates, compare Proposition 2.5.

A mapping class defines a symplectic automorphism of $H_1(S, \mathbb{Z})$ and therefore the group $\mathrm{Mod}(S)$ acts on this bundle by bundle automorphisms. In fact, the above trivializations define on $\mathcal{H}$ a $\mathrm{Mod}(S)$-invariant real analytic structure as well. The bundle $\mathcal{H} \to \mathcal{T}(S)$ is called the *Hodge bundle*.

Since the imaginary part of the period matrix Z is positive definite the map

$$H_1(S, \mathbb{Z}) \to \Omega(X)^*$$

defined by the periods is non-degenerate (here $\Omega(X)^*$ is the dual of $\Omega(X)$). In particular, its image is a lattice in $\Omega(X)^*$, and $\Omega(X)^*/H_1(S,\mathbb{Z})$ is a complex torus.

To summarize, the period matrix determines a marked complex torus where the marking is defined by the choice of a symplectic basis of $H_1(S,\mathbb{Z})$. The (real) symplectic structure on $H_1(S,\mathbb{Z})$ defined by the intersection form $\cdot$ is called the *principal polarization.* Recall that the intersection form is given by $a\wedge b=(a\cdot b)[S]$ where $[S]$ denotes the fundamental cycle of S.

For each $X\in\mathcal{T}(S)$ there is an anti-symmetric real bilinear form $\langle,\rangle$ on $\Omega(X)$ defined by

$$\langle\omega,\zeta\rangle=\mathrm{Im}\frac{i}{2}\int\omega\wedge\bar{\zeta}.$$

This form clearly is invariant under the complex multiplication J (here J is just standard multiplication with the imaginary i), i.e. we have $\langle J\omega,J\zeta\rangle=\langle\omega,\zeta\rangle$ for all ω,ζ. Moreover, $(u,v)\to\langle Ju,v\rangle$ is an inner product defining an hermitian metric with imaginary part $\langle,\rangle$. A complex multiplication J with these two properties is called *compatible* with the symplectic structure $\langle,\rangle$.

The Hermitian metric determined by the inner product $\langle J\cdot,\cdot\rangle$ defines a complex linear isomorphism $\Omega^*(X)\to\Omega(X)$ which maps the principal polarization on $H_1(S,\mathbb{Z})\subset\Omega(X)^*$ to the real bilinear form $\langle,\rangle$ on $\Omega(X)$. Namely, separating into real and imaginary part and identifying $\Omega(X)^*$ with a subspace of $H_1(S,\mathbb{C})$, the isomorphism $\Omega^*(X)\to\Omega(X)$ is just the isomorphism given by Poincaré duality. In particular, the complex torus defined by the period matrix can be viewed as a quotient of $\Omega(X)$.

As the Hermitian metric on $\Omega(X)$ is determined by the complex structure on $\Omega(X)$ and the principal polarization, the unitary group $U(g)$ for this metric does not depend on any choices made. Moreover, it preserves the complex torus defined by the period matrix.

The symplectic group $\mathrm{Sp}(2g,\mathbb{R})$ acts simply transitively on the space of symplectic bases of $\mathbb{R}^{2g}$. It also acts transitively on the space of complex structures compatible with the symplectic structure. The standard identification of $\mathbb{R}^{2g}$ with $\mathbb{C}^g$ determines a particular symplectic basis, namely the standard basis e_i,f_j $(i,j=1,\dots,g)$ of $\mathbb{R}^{2g}$ where $e_1,\dots,e_g$ is a complex basis of $\mathbb{C}^g$ and where the symplectic structure $\langle,\rangle$ satisfies $\langle e_i,e_j\rangle=\langle f_i,f_j\rangle=0$ and $\langle e_i,f_j\rangle=\delta_{ij}$.

The stabilizer in $\mathrm{Sp}(2g,\mathbb{R})$ of the standard complex structure on $\mathbb{C}^g$ is just the unitary group $\mathrm{U}(g)$. Thus the *Siegel upper half-space* $\mathrm{Sp}(2g,\mathbb{R})/\mathrm{U}(g)$ is the space of complex tori which are quotients of $\mathbb{R}^{2g}$ equipped with a compatible complex structure by a lattice generated by a symplectic basis. With this interpretation, the map which associates to a marked Riemann surface X the complex torus defined by the period matrix for a given symplectic basis of $H_1(S,\mathbb{Z})$ can be viewed as a map from Teichmüller space $\mathcal{T}(S)$ into the Siegel upper half-space.

By Proposition 1.13 the mapping class group $\mathrm{Mod}(S)$ acts properly discontinuously on Teichmüller space. The quotient orbifold $\mathcal{M}_g=\mathcal{T}(S)/\mathrm{Mod}(S)$ is called the *moduli space.*

The automorphism group $\mathrm{Sp}(2g, \mathbb{Z})$ of the integral homology $H_1(S, \mathbb{Z})$ of S equipped with the intersection form changes the marking of the complex torus defined by the period matrix of $\Omega(X)$. The action of a mapping class on the first homology $H_1(S, \mathbb{Z})$ defines a surjective homomorphism $\mathrm{Mod}(S) \to \mathrm{Sp}(2g, \mathbb{Z})$ (we refer to [**FM11**] for surjectivity) and hence the map which associates to a marked Riemann surface the marked complex torus defined by the period matrix descends to a map $\mathcal{M}_g \to \mathrm{Sp}(2g, \mathbb{Z})\backslash\mathrm{Sp}(2g, \mathbb{R})/\mathrm{U}(g)$ which is called the *Torelli map*. The Torelli map does not depend on any choices made. It associates to an unmarked Riemann surface X its *Jacobian* $\mathrm{Jac}(X)$ which is the unmarked complex torus defined by a period matrix and equipped with its principal polarization.

The following is due to Torelli. A proof can be found in Section III of [**FK80**].

THEOREM 3.1. *The Torelli map* $\mathcal{M}_g \to \mathrm{Sp}(2g, \mathbb{Z})\backslash\mathrm{Sp}(2g, \mathbb{R})/\mathrm{U}(g)$ *is injective.*

The Siegel upper half-space is a Hermitean symmetric space of non-compact type. In particular, it has an $\mathrm{Sp}(2g, \mathbb{R})$-invariant complex structure given by a system of charts with values in $\mathbb{C}^m$. Its complex dimension equals $\frac{g(g+1)}{2}$ and hence for $g \geq 4$, this dimension exceeds half of the real dimension $3g-3$ of the Teichmüller space. Moreover, for $g \geq 3$ the Torelli map is *not* an embedding in the sense of differentiable orbifolds. Namely, it is branched at the *hyperelliptic locus* which is the set of *hyperelliptic curves*, i.e. Riemann surfaces X with a biholomorphic involution τ so that $X/\tau = \mathbb{C}P^1$. In particular, its differential fails to be of maximal rank at the hyperelliptic locus. The well known *Schottky problem* asks for determining the image of the Torelli map (see [**HM98**] for an account of what was known about this problem 15 years ago).

The usual way to analyze this and related questions is via complex algebraic geometry. This requires the existence of a $\mathrm{Mod}(S)$-invariant complex structure on $\mathcal{T}(S)$. Such a structure indeed exists (see [**IT89, Hu06**]), but it is difficult to construct.

Instead we will adopt a differential geometric viewpoint and explore a (much weaker) structure.

DEFINITION 3.2. An *almost complex structure* on $\mathcal{T}(S)$ is a continuous $(1,1)$-tensor field J on the tangent bundle of $\mathcal{T}(S)$ with $J^2 = -\mathrm{Id}$.

It follows from the discussion in Section 3 that for every $X \in \mathcal{T}(S)$ the complex vector space $\mathcal{Q}(X)$ can be identified with the cotangent space of $\mathcal{T}(S)$ at X (why this is the cotangent space rather than the tangent space is not apparent from our discussion. Instead we refer to [**Hu06**]). Moreover,

$$\mathcal{Q}(S) = \cup_X \mathcal{Q}(X)$$

is a smooth (in fact real analytic) vector bundle over $\mathcal{T}(S)$ on which $\mathrm{Mod}(S)$ acts as a group of bundle automorphisms (again this requires an argument which we do not give here. Again we refer to [**IT89, Hu06**]).

For $X \in \mathcal{T}(S)$, the complex structure on $\mathcal{Q}(X)$ is given by the usual multiplication of a quadratic differential with a complex number. These fibrewise complex structures define a complex structure on the bundle $\mathcal{Q}(S)$.

If $q \in \mathcal{Q}(X)$ then locally $q = f dz^2$ for a holomorphic function f. The complex conjugate of q then equals $\bar{q} = \bar{f} d\bar{z}^2$. In particular, if u is another quadratic differential then $q\bar{u}$ is locally of the form $\alpha dz^2 d\bar{z}^2$ for a complex valued function α. Now the hyperbolic metric h on X is a tensor field of the form $\beta dz d\bar{z}$ and hence we can define

$$g(q_1, q_2) = \int_S \frac{q_1 \bar{q}_2}{h}.$$

This construction defines an Hermitian inner product on the cotangent bundle $\mathcal{Q}(S)$ of $\mathcal{T}(S)$ which is called the *Weil-Petersson metric*. The Weil-Petersson metric is invariant under the action of the mapping class group. It determines a Mod(S)-equivariant bundle isomorphism $T\mathcal{T}(S) \to \mathcal{Q}(S)$ which maps the complex structure on $\mathcal{Q}(S)$ to an almost complex structure J on $\mathcal{T}(S)$.

An almost complex structure J on a manifold M is called *integrable* if it can be defined by charts with values in $\mathbb{C}^m$ for some $m > 0$ in the sense that in these charts, the almost complex structure is just multiplication with i in the tangent bundle. The following fundamental result is quite involved and will not be proven in these notes. Proofs can be found in [**Hu06, IT89**].

THEOREM 3.3.

(1) *The almost complex structure J on $\mathcal{T}(S)$ is integrable.*
(2) *The Torelli map is holomorphic.*

Since the almost complex structure J is Mod(S)-invariant, it follows that Mod(S) acts on $\mathcal{T}(S)$ properly discontinuously as a group of biholomorphic automorphisms.

The Hermitian metric g on $\mathcal{T}(S)$ defines a two-form

$$\omega(q, z) = \operatorname{Im} g(q, z)$$

which is called the *Weil-Petersson form*. The most important properties of the Weil-Petersson metric and the form ω are summarized in the following theorem. We refer to [**IT89, W10, DW07**] for more details and for references.

THEOREM 3.4.

(1) *ω is closed, i.e. the Weil-Petersson metric is Kähler.*
(2) *The Weil-Petersson metric has negative sectional curvature which is neither bounded from below nor bounded from above by a negative constant.*
(3) *The Weil-Petersson metric is incomplete.*
(4) *Any two points in $\mathcal{T}(S)$ can be connected by a unique Weil-Petersson geodesic*
(5) *Lengths functions are convex along Weil-Petersson geodesics.*
(6) *The group of orientation preserving isometries for the Weil-Petersson metric is the mapping class group.*

The strength of this result is illustrated by the following short proof of the *Nielsen realization theorem*. This result was first established by Kerckhoff [**K83**]. The proof below is due to Wolpert (see [**W10**]).

THEOREM 3.5. *Let $\Gamma <$ Mod(S) be a finite subgroup. Then there is $X \in \mathcal{T}(S)$ such that $\Gamma(X) = X$.*

PROOF. Choose a Γ-invariant finite set $\mathcal{C}$ of simple closed curves on S so that $\sum_{c\in\mathcal{C}} \ell_c$ is a proper function on $\mathcal{T}(S)$ where as before, $\ell_c(X)$ is the length of the simple closed curve c on the hyperbolic surface X. Such a set of curves can easily be found as follows. Let $\mathcal{P}$ be a pants decomposition of S. For each pants curve $\alpha \in \mathcal{P}$ choose a simple closed curve β_α which is disjoint from $\mathcal{P} - \alpha$ and which intersects α in one or two points depending on whether the component containing α of $S - (\mathcal{P} - \alpha)$ is a bordered torus or a four-holed sphere. Let $\mathcal{C}_0$ be the union of these curves. It follows from the discussion in Section 2 that $\sum_{c\in\mathcal{C}_0} \ell_c$ is a proper function on $\mathcal{T}(S)$. Let $\mathcal{C} = \Gamma\mathcal{C}_0$.

For sufficiently large $b > 0$ the set $K = \{\sum_{c\in\mathcal{C}} \ell_c \leq b\} \subset \mathcal{T}(S)$ is compact and non-empty. Since length functions are convex along Weil-Petersson geodesics, this set is moreover convex for the Weil-Petersson metric (i.e. the geodesic segment connecting any two points in K is contained in K), and by Γ-invariance of $\mathcal{C}$, it is invariant under Γ.

The restriction to K of the Weil-Petersson metric is a smooth Riemannian metric of negative sectional curvature. Thus K has a unique center for this metric which can be found as follows. For each $p \in K$ let $\kappa(p) = \max\{d_{WP}(p,u) \mid u \in K\}$ where d_{WP} denotes the distance defined by the Weil-Petersson metric. The function $p \to \kappa(p)$ is continuous. The center of K then is the unique point $p \in K$ for which $\kappa(p)$ is minimal. Here uniqueness follows from convexity of K and the fact that the Weil-Petersson metric is negatively curved and therefore distance functions are strictly convex along Weil-Petersson geodesics. Since K is Γ-invariant and Γ preserves d_{WP}, this center is Γ-invariant. □

Recall from Corollary 1.8 the definition of the Fenchel-Nielsen twist vector field $\mathbf{t}_\alpha$ for a simple closed curve α on S. The following important result is due to Wolpert and can be found in [**W10**]. Its proof uses harmonic Beltrami differentials and exceeds the scope of these notes.

THEOREM 3.6. *Let α be a simple closed curve. Then*

$$2\mathbf{t}_\alpha = J\mathrm{grad}\ell_\alpha$$

and $2\omega(\cdot, \mathbf{t}_\alpha) = d\ell_\alpha$.

COROLLARY 3.7. *ω is invariant under the Fenchel Nielsen twist flow.*

PROOF. Since ω is closed, the Lie derivative $L_{\mathbf{t}_\alpha}\omega$ of ω in direction of $\mathbf{t}_\alpha$ equals

$$2L_{\mathbf{t}_\alpha}\omega = 2d(\mathbf{t}_\alpha \lrcorner \omega) = -d(d\ell_\alpha) = 0.$$

□

Wolpert uses this to show (Theorem 3.14 of [**W10**])

THEOREM 3.8. *In Fenchel Nielsen coordinates,*

$$\omega = \frac{1}{2}\sum_i d\ell_i \wedge d\tau_i.$$

PROOF. Let $\mathcal{P}$ be a pants decomposition of S defining the Fenchel Nielsen coordinates. For $i \neq j$ we have

$$2\omega(\text{grad}\ell_j, \mathbf{t}_i) = d\ell_i(\text{grad}\ell_j) = \delta_{ij}.$$

Our goal is to express ω in the length-twist parameters at a given surface X. Since ω is invariant under the Fenchel Nielsen twist flow, we can modify the surface X along the flow lines of the (commuting) twist flows about the pants curves of $\mathcal{P}$ to a surface X' with vanishing twist coordinates without changing the expression for ω in length-twist parameters. Then for X', the seams of a pairs of pants from $S - \mathcal{P}$ are glued to the seams of the neighboring pants.

A pair of pants has an orientation reversing isometric reflection which preserves all seams pointwise and exchanges the two right angled hexagons which together form the pair of pants. By the choice of X', the involution of each individual pair of pants extends to an isometric reflection ρ of X'. The reflection reverses the orientation of the twist vector fields and preserves the length coordinates.

The reflection ρ reverses the complex structure on $\mathcal{T}(S)$ and preserves the real part of the Weil-Petersson metric and hence ω is anti-invariant under ρ. But X' is a fixed point for ρ, moreover its differential at X' preserves the tangents of the length parameter and reverses the tangents of the twist parameters. In particular, the ρ-anti-invariant subspace of the second exterior power of the tangent space of $\mathcal{T}(S)$ at X' is spanned by the two-forms $d\ell_i \wedge d\tau_j$. As a consequence, in the expression for ω the coefficients of the two-forms $d\tau_i \wedge d\tau_j$ and $d\ell_i \wedge d\ell_j$ vanish. The theorem now follows from the second paragraph of this proof. □

As $\mathcal{T}(S)$ is a complex manifold, we can ask for *holomorphic discs* in $\mathcal{T}(S)$. Such a disc is a holomorphic map $D \to \mathcal{T}(S)$ where $D \subset \mathbb{C}$ is the standard unit disc or, equivalently, the hyperbolic plane $\mathbf{H}^2$. Thus we are looking for holomorphic maps $\mathbf{H}^2 \to \mathcal{T}(S)$. Particular such maps can easily be constructed.

Let $q \in \mathcal{Q}(S)$ be arbitrary. Outside its set Σ of zeros, the quadratic differential q defines a singular euclidean metric. This metric determines a family of charts $U_j \subset S - \Sigma \to \varphi_j U_j \subset \mathbb{C}$, unique up to translation and perhaps reflection (multiplication by -1). Chart transitions are translations, perhaps composed with multiplication by -1. The charts are holomorphic for the complex structure X defined by q. For every $A \subset \mathrm{SL}(2, \mathbb{R})$ we can postcompose these charts with A. The resulting family $A \circ \varphi_i$ of charts has chart transitions $(A \circ \varphi_i) \circ (A \circ \varphi_j)^{-1} = A(\varphi_i \circ \varphi_j^{-1})A^{-1}$ which are translations, perhaps composed with -1. As a consequence, these charts define a new complex structure and a new quadratic differential. If A is a diagonal matrix, then the new Riemann surface is just the image of X under a Teichmüller map defined by q. This construction defines an action of $\mathrm{SL}(2, \mathbb{R})$ on $\mathcal{Q}(S)$ preserving the norm (=area) of the differential.

If $A \in \mathrm{SO}(2)$ then the new charts for S differ from the old ones by a rotation by some angle θ. This amounts to replacing q by $e^{i\theta}q$. This multiplication does not change the complex structure on the underlying Riemann surface and therefore the map $\mathrm{SL}(2, \mathbb{R}) \to \mathcal{Q}(S)$ projects to a map $\mathrm{SL}(2, \mathbb{R})/\mathrm{SO}(2) = \mathbf{H}^2 \to \mathcal{T}(S)$. We call such a map a *complex geodesic*. Note that the orbits of the action of the diagonal subgroup of $\mathrm{SL}(2,\mathbb{R})$ project to Teichmüller geodesics on $\mathcal{T}(S)$. In particular, a complex geodesic is totally geodesic for the Teichmüller metric. This means that any geodesic segment connecting two points in the disc is entirely contained the disc.

LEMMA 3.9. *A complex geodesic is holomorphic for the almost complex structure J and totally geodesic for the Teichmüller metric.*

PROOF. Let $P : \mathcal{Q}(S) \to \mathcal{T}(S)$ be the canonical projection. By construction, the differential at $z = Pq$ of a complex geodesic $\sigma : \mathbf{H}^2 \to \mathcal{T}(S)$ associates to $e^{i\theta} \in S^1 \subset \mathbb{C} = T_i\mathbf{H}^2$ the cotangent at z of the Teichmüller geodesic with initial velocity $e^{i\theta}q$. But this just means that this differential satisfies $J \circ d\sigma = d\sigma \circ i$, i.e. it is complex linear. □

The mapping class group acts on $\mathcal{T}(S)$ preserving the complex geodesics. Every real Teichmüller geodesic is contained in a unique complex geodesic.

The *Kobayashi pseudo-metric* of a complex manifold M is defined as follows. For any two points $x, y \in M$ let $\rho_1(x,y)$ be the infimum of the hyperbolic distances in $\mathbf{H}^2$ between two points x_0, y_0 so that there is a holomorphic map $f : \mathbf{H}^2 \to M$ with $f(x_0) = x, f(y_0) = y$. For any positive integer n we put

$$\rho_n(x,y) = \inf \sum_{i=0}^{n-1} \rho_1(x_i, x_{i+1})$$

where the infimum is over all n-chains with $x_0 = x, x_n = y$. Finally put $\rho(x,y) = \inf_n \rho_n(x,y)$. Clearly ρ is symmetric and satisfies the triangle inequality, so it is a pseudo-metric.

The significance of this construction is as follows. If $\varphi : M \to N$ is any holomorphic map then φ is distance non-increasing for the Kobayashi pseudo-metric. In particular, the Kobayashi pseudo-metric is invariant under all biholomorphic automorphisms of M.

The above discussion immediately implies one part of the following theorem due to Royden (see [**IT89**]).

THEOREM 3.10. *The Kobayashi pseudo-metric on $\mathcal{T}(S)$ equals the Teichmüller metric.*

PROOF. Any two points $x, y \in \mathcal{T}(S)$ are contained in a complex geodesic. This means that there is a holomorphic map $\mathbf{H}^2 \to \mathcal{T}(S)$ which is an isometric embedding for the hyperbolic metric on $\mathbf{H}^2$ and the Teichmüller metric on $\mathcal{T}(S)$. Thus by definition, the Kobayashi pseudo-metric is not bigger than the Teichmüller metric.

The other inequality is much harder. We refer to [**IT89**] for a proof. □

A complex geodesic (or *Teichmüller disc*) is a special case of the following

DEFINITION 3.11. A *holomorphic family of Riemann surfaces* is a holomorphic surjection $P : U \to T$ of complex manifolds such that each fibre is a Riemann surface and that P admits horizontally holomorphic trivializations.

Example: For any Riemann surface S and every complex manifold M, $S \times M$ is a holomorphic family of Riemann surfaces.

Note that if $P \to T$ is a holomorphic family of Riemann surfaces, then the holomorphic cotangent bundle of the fibre is a holomorphic line bundle over P. Moreover, the Hodge bundle induces a holomorphic vector bundle over P etc. We refer to the book [**HM98**] for a careful discussion of this and related results.

THEOREM 3.12. *The functor which associates to a complex manifold T the set of isomorphism classes of holomorphic families of Riemann surfaces $P : Y \to T$ with an equivalence class of markings is equivalent to the functor which associates to T the set of holomorphic maps $T \to \mathcal{T}(S)$.*

Remark: The above theorem is false without the requirement that the marking is being remembered. The difficulty is the existence of surfaces with non-trivial automorphisms. This results in the non-existence of a fine moduli space of Riemann surfaces which causes substantial difficulties for an algebraic geometric approach. We refer once more to [**HM98**] for a comprehensive discussion of this problem and the various ways to overcome it.

LECTURE 4

The curve graph and the augmented Teichmüller space

In Section 4 we saw that there is a natural complex structure on Teichmüller space which descends to a complex structure on moduli space. It turns out that moduli space admits a natural compactification, the so-called Deligne-Mumford compactification. The goal of this section is to give a differential geometric description of this compactification and relate its structure to the geometry of Teichmüller space.

The idea is as follows. Consider a set of Fenchel Nielsen coordinates for a pants decomposition $\mathcal{P}$ of S. For a given pants curve α, we can shrink the length of α to zero and obtain a surface where the curve α has been degenerated to a pair of cusps. The next observation is an immediate consequence of the classical collar lemma in hyperbolic geometry (see [**B92**]). It is used to understand this degeneration.

LEMMA 4.1. *For every $\epsilon > 0$ there is a number $M(\epsilon) > 0$ with the following property. Let α be a simple closed curve on the hyperbolic surface X of length at most ϵ. Then α is the core curve of an embedded annulus of width and modulus at least $M(\epsilon)$. Moreover, $M(\epsilon) \to \infty$ as $\epsilon \to 0$.*

PROOF. Let γ be a simple closed geodesic of length ϵ in a hyperbolic surface X. Let A be the cover of X with fundamental group $< \gamma >$. In the proof of Proposition 2.15 we saw that A is an annulus of modulus π/ϵ.

Let $\tilde{\gamma}$ be the lift of γ to A. Define $w(\epsilon)$ by $\sinh w(\epsilon) \sinh(\epsilon/2) = 1$. An explicit calculation shows that the cylinder about $\tilde{\gamma}$ of width $w(\epsilon)$ embeds into X (we refer to [**B92**] for details) Its modulus can be calculated as in the proof of Proposition 2.15 using a branch of the logarithm. It goes to infinity as $\epsilon \to 0$. ☐

As a consequence, the lengths of two simple closed curves on X can both be small only if the curves can be realized disjointly. This suggests to look more closely at the combinatorial structure of the set of all simple closed curves on S.

DEFINITION 4.2. The *curve complex* $\mathcal{C}(S)$ of S is the simplicial complex whose vertices are isotopy classes of simple closed curves on S and where a k-tuple of such curves spans a simplex if and only if they can be realized disjointly.

Example: 1) Let α be a simple closed non-separating curve. Then $S - \alpha$ is a surface of genus $g - 1$ with two boundary components. In particular, there are infinitely many distinct free homotopy classes of simple closed curves in $S - \alpha$. As a consequence, $\mathcal{C}(S)$ is locally infinite.

2) Any k-tuple of mutually disjoint simple closed curves can be completed to a pants decomposition of S. Thus any k-simplex in $\mathcal{C}(S)$ is a face of a simplex of maximal dimension $3g - 4$.

The mapping class group $\text{Mod}(S)$ acts on free homotopy classes of simple closed curves preserving disjointness and hence it acts on $\mathcal{C}(S)$ as a group of simplicial automorphisms. The quotient $\mathcal{C}(S)/\text{Mod}(S)$ is a finite simplicial complex. Namely, for any two non-separating simple closed curves α, β there is a mapping class which maps α to β. If α, β are separating then there is a mapping class which maps α to β if and only if the disconnected surfaces $S - \alpha, S - \beta$ are homeomorphic. Since these surfaces are determined up to homeomorphism by the minimum of the genus of one of their components, the number of $\text{Mod}(S)$-orbits of separating curves equals $[g/2]$. As a consequence, $\mathcal{C}(S)/\text{Mod}(S)$ has only $[g/2] + 1$ vertices. Finiteness of the number of simplices of dimension bigger than zero is seen in the same way.

The following important result is due to Ivanov (see [**I02**] for more and for references) with some cases due to Korkmaz and Luo.

THEOREM 4.3. *For $g \geq 3$ the automorphism group of the curve complex is the extended mapping class group of S.*

Here the extended mapping class group is the group of isotopy classes of all diffeomorphisms of S including orientation reversing diffeomorphisms.

Consider again a pants decomposition $\mathcal{P}$ and its associated system of Fenchel Nielsen coordinates $(\ell_1, \dots, \ell_{3g-3}, \tau_1, \dots, \tau_{3g-3})$. Let α_i be the pants curve corresponding to the length parameter ℓ_i. Fixing all coordinates but the length function ℓ_1 and letting ℓ_1 go to zero defines a smooth curve of hyperbolic surfaces which degenerate to a surface with two cusps:

The tuple $(\ell_2, \dots, \ell_{3g-3}, \tau_2, \dots, \tau_{3g-3})$ defines as before Fenchel Nielsen coordinates for a surface with two cusps (which may be disconnected if the pants curve α_1 is separating). The twist parameter τ_1 is not defined any more.

Replacing simple closed curves by punctures can simultaneously be done with every tuple of disjoint simple closed curves, i.e. with all vertices of a simplex σ in $\mathcal{C}(S)$. We denote by $\mathcal{S}(\sigma)$ the corresponding Teichmüller space of punctured Riemann surfaces. This Teichmüller space is defined exactly in the same way as the Teichmüller space of a closed surface. In particular, Fenchel Nielsen coordinates define a real analytic structure which is invariant under the mapping class group. Note that if σ is a maximal simplex whose vertices define a pants decomposition then $\mathcal{S}(\sigma)$ consists of a unique point.

DEFINITION 4.4. *Augmented Teichmüller space* is $\overline{\mathcal{T}}(S) = \mathcal{T}(S) \cup_{\sigma \in \mathcal{C}(S)} \mathcal{S}(\sigma)$.

We can define a topology on $\overline{\mathcal{T}}(S)$ as follows. Let $\sigma \in \mathcal{C}(S)$ be a simplex of dimension $k - 1$. Let $\mathcal{P}$ be a pants decomposition of S with pants curves α_i $(1 \leq i \leq 3g - 3)$ with the property that σ is spanned by the curves $\alpha_1, \dots, \alpha_k$. Let

$(\ell_1,\dots,\ell_{3g-3},\tau_1,\dots,\tau_{3g-3})$ be Fenchel Nielsen coordinates for the pants decomposition $\mathcal{P}$. Then every $X\in\mathcal{S}(\sigma)$ determines a tuple of Fenchel Nielsen coordinates

$$(\ell_{k+1},\dots,\ell_{3g-3},\tau_{k+1},\dots,\tau_{3g-3}).$$

Define a sequence $(X_i)\subset\overline{\mathcal{T}}(S)$ to converge to X if the following holds. Let $(\ell_1^i,\dots,\ell_{3g-3}^i,\tau_1^i,\dots,\tau_{3g-3}^i)$ be the Fenchel Nielsen coordinates of X_i with respect to the pants decomposition $\mathcal{P}$. We require that $\ell_j^i\to 0$ $(1\leq j\leq k)$ and $\ell_j^i\to\ell_j,\tau_j^i\to\tau_j$ $(j\geq k+1)$. Since Fenchel Nielsen coordinates parametrize Teichmüller space for all hyperbolic surfaces, this definition does not depend on the choice of the pants decomposition $\mathcal{P}$ extending σ. It defines a topology on $\overline{\mathcal{T}}(S)$ which is invariant under the action of the mapping class group $\mathrm{Mod}(S)$.

PROPOSITION 4.5. *$\overline{\mathcal{T}}(S)$ is a stratified non-locally compact* $\mathrm{Mod}(S)$*-space.*

Here the stratification is the decomposition of $\overline{\mathcal{T}}(S)$ into subsets $\mathcal{S}_j$ called *strata*. Each stratum $\mathcal{S}_j$ is a (topological) manifold of dimension n_j. The closure of a stratum is the union of the stratum with a collection of strata of smaller dimension. A stratum in $\overline{\mathcal{T}}(S)$ is defined by a simplex $\mathcal{S}(\sigma)\subset\mathcal{C}(S)$ (which is allowed to be empty). Its closure is the union of $\mathcal{S}(\sigma)$ with the sets $\mathcal{S}(\sigma')$ where σ' runs through all simplices in $\mathcal{C}(S)$ which contain σ as a face. $\overline{\mathcal{T}}(S)$ is not locally compact since a neighborhood basis of a point $X\in\mathcal{S}(\sigma)$ consists of sets with the property that for a pants decomposition $\mathcal{P}$ extending σ, the Fenchel Nielsen length parameters for the curves in σ are small but the twist parameters are arbitrary.

In spite of the fact that $\overline{\mathcal{T}}(S)$ is not locally compact we have

THEOREM 4.6. *$\overline{\mathcal{T}}(S)/\mathrm{Mod}(S)$ is compact. It is called the* Deligne Mumford compactification *of moduli space.*

For the proof of compactness one uses an observation of independent interest.

LEMMA 4.7. *There is a number $\chi=\chi(g)>0$ so that every closed hyperbolic surface of genus g admits a pants decomposition with pants curves of length at most χ.*

PROOF. (Sketch, a detailed proof can be found in Chapter 5 of [**B92**] which also gives explicit bounds.)

Let X be a hyperbolic surface. By Lemma 4.1, any very short simple closed geodesic γ on X is contained in an embedded annulus A of very large width. Moreover, if γ' is any very short simple closed curve which is freely homotopic to γ then γ' is contained in A. In other words, in the complement C of the standard annuli about the very short simple closed geodesics, the injectivity radius of X is bounded from below by a universal constant $12r_0>0$. We may assume that these annuli are convex with smooth boundary of constant curvature and uniformly bounded length. Then C is a possibly disconnected surface with smooth boundary which is diffeomorphic to the surface obtained from X by cutting X open along the very short simple closed geodesics. In particular, C has at most $2g-2$ components.

By the Gauss Bonnet theorem, the area of a hyperbolic surface X of genus $g \geq 2$ equals $2\pi(2g-2)$. The area of a hyperbolic disc of radius $r > 0$ is $2\pi \sinh r$ and hence the cardinality of a maximal collection of disjoint embedded discs of radius r_0 in X with center in C is at most $(2g-2)/\sinh r_0$.

Let $x_1, \dots, x_k$ be the set of centers of these discs. The open discs D_i of radius $3r_0$ and center x_i cover C. Two such discs intersect only if the distance between their centers is at most $6r_0$. Since the injectivity radius at a center point is at least $12r_0$, if two such discs intersect then their union is contained in a contractible disc. In particular, their centers can be connected by a unique minimal geodesic of length at most $6r_0$.

The union of all geodesic arcs connecting centers of intersecting discs from the collection $\{D_i\}$ is a graph $G \subset X$ of uniformly bounded length. The inclusion $G \to X$ maps the fundamental group of G onto the fundamental group of C. Namely, every closed curve ρ in C travels successively through a chain of discs among the discs D_i so that any two adjacent discs intersect. Let $D_{i_1}, \dots, D_{i_k}$ be this chain of discs. For each $j \leq k$ choose some t_j so that $t_j < t_{j+1}$, that $\rho(t_j) \in D_{i_j}$ and such that $\rho[t_j, t_{j+1}] \subset D_{i_j} \cup D_{i_{j+1}}$. Connect $\rho(t_j)$ by a path α_j entirely contained in D_{i_j} to the center x_{i_j} of D_{i_j}. Then $\alpha_{j+1} \circ \rho[t_j, t_{j+1}] \circ \alpha_j^{-1}$ is a path connecting x_{i_j} to $x_{i_{j+1}}$ which is entirely contained in $D_{i_j} \cup D_{i_{j+1}}$. Therefore this path is homotopic with fixed endpoints to an edge of G. Now ρ is homotopic to the concatenation of these paths and hence ρ is homotopic to an edge path in G.

Since the number of edges of G is uniformly bounded, the fundamental group of C can be generated by simple edge loops in G whose lengths are bounded from above by a universal constant, and these loops pairwise intersect in uniformly few points. Then there also is a pants decomposition consisting of curves of uniformly bounded length. Such a pants decomposition contains every very short simple closed curve on X. The lemma follows. $\square$

A pants decomposition of a hyperbolic surface X with the properties stated in Lemma 4.7 is called a *Bers decomposition* of X. The number χ is called a *Bers constant.*

Let as before $\mathcal{M}_g = \mathcal{T}(S)/\mathrm{Mod}(S)$ be the moduli space of S. Bers decompositions are used in the following

PROPOSITION 4.8. *Let $(x_i) \subset \mathcal{M}_g$ be a sequence which exits every compact set. Then the shortest length of a simple closed curve on x_i tends to zero as $i \to \infty$.*

PROOF. We show that a sequence of surfaces $(x_i) \subset \mathcal{M}_g$ so that the shortest length ρ_i of a simple closed curve on x_i is bounded away from zero is relative compact in moduli space.

For this let $\tilde{X}_i \in \mathcal{T}(S)$ be a preimage of x_i and let $\mathcal{P}_i$ be a Bers decomposition for $\tilde{X}_i$. Since there are only finitely many combinatorial types of pants decompositions for S and any two pants decompositions of the same combinatorial type can be mapped to each other with an element of Mod(S), up to passing to a subsequence and changing the lift $\tilde{X}_i$ we may assume that these pants decompositions all coincide. Let $\mathcal{P}$ be this fixed pants decomposition. Let ℓ_j ($j = 1, \ldots, 3g-3$) be the length functions of the pants curves of $\mathcal{P}$. The lengths $\ell_j(\tilde{X}_i)$ are bounded from above and below by a positive constant independent of i.

Dehn twists about the pants curves act with compact quotient on the twist parameters. In other words, up to modifying the preimages of the points x_i by suitably chosen Dehn multitwists about the pants curves of $\mathcal{P}$, these preimages are contained in a compact subset of $\mathcal{T}(S)$. Therefore the sequence $(x_i) \subset \mathcal{M}_g$ is relative compact. □

To complete the proof that $\overline{\mathcal{T}}(S)/\mathrm{Mod}(S)$ is compact, observe that if the sequence $(x_i) \subset \mathcal{M}_g$ exits every compact set then by Proposition 4.8, up to passing to a subsequence we may assume that for each i there is at least one simple closed curve on x_i whose length is at most $1/i$.

Assume as in the proof of Proposition 4.8 that there are lifts $\tilde{X}_i \in \mathcal{T}(S)$ of x_i which admit a fixed pants decomposition $\mathcal{P}$ as a Bers decomposition. Let $\epsilon_0 > 0$ be sufficiently small that for any hyperbolic surface X, any two distinct simple closed geodesics on X of length at most ϵ_0 are disjoint and moreover every simple closed geodesic of length at most ϵ_0 is contained in a Bers decomposition of X. Such a number exists since a very short simple closed geodesic is the core curve of an annulus of very large width. For sufficiently small $\epsilon < \epsilon_0$ and sufficiently large i let σ_i be the simplex in the curve complex corresponding to the curves of $\tilde{X}_i$-length at most ϵ. Then σ_i is a face of the maximal simplex defined by $\mathcal{P}$ and hence by passing to a subsequence, we may assume that σ_i does not depend on i.

Let σ be this fixed simplex. The above argument implies that up to passing to another subsequence, $(\tilde{X}_i)$ converges to a point in the stratum $\mathcal{S}(\sigma)$ and hence (x_i) converges in $\overline{\mathcal{T}}(S)/\mathrm{Mod}(S)$. This completes the proof that $\overline{\mathcal{T}}(S)/\mathrm{Mod}(S)$ is compact.

We also have (see [**W10**] for references)

THEOREM 4.9. *Augmented Teichmüller space is the completion of $\mathcal{T}(S)$ with respect to the Weil Petersson metric. Strata are convex for the completed Weil-Petersson metric.*

The *curve graph* $\mathcal{CG}(S)$ is the one-skeleton of the curve complex. It is a locally infinite graph. Providing each edge with a metric of length one defines the structure of a metric space on $\mathcal{CG}(S)$. The mapping class group acts on this metric graph as a group of simplicial isometries.

The curve graph can be used to obtain information on the geometry of Teichmüller space and the mapping class group. To this end we look more closely at the geometry of $\mathcal{CG}(S)$.

Let $\iota(c,d)$ be the geometric intersection number between two simple closed curves c,d. By definition, $\iota(c,d)$ is the minimal number of intersection points between any two curves c',d' which are freely homotopic to c,d. We call curves which realize this intersection number *in minimal position.*

LEMMA 4.10. *The distance in $\mathcal{CG}(S)$ between two simple closed curves c,d does not exceed $\iota(c,d)+1$.*

PROOF. Let c,d be simple closed curves in minimal position. We may assume that c and d intersect. If c and d intersect in a single point then the boundary of a tubular neighborhood of $c\cup d$ in S is a simple closed curve which is disjoint from both c and d. Thus the distance between c and d is at most two.

If $\iota(c,d)\geq 2$ then let β be a component of $d-c$. If β leaves and returns to the same side of c then the concatenation of β with a subarc of c connecting the endpoints of β is an essential simple closed curve disjoint from c whose intersection number with d is strictly smaller than the intersection number between c and d.

If β leaves and returns to different sides of c and if $\beta\neq d$ then a similar argument produces a simple closed curve c' whose intersection number with d does not exceed $\iota(c,d)-2$ and which intersects c in at most one point. We saw above that the distance between c and c' does not exceed 2. The claim now follows by induction on $\iota(c,d)$. □

Choose a map

$$\Upsilon:\mathcal{T}(S)\to\mathcal{CG}(S)$$

which associates to a hyperbolic metric a simple closed curve of length at most χ where as before, χ is a Bers constant for S.

Define a map Ψ between metric spaces to be *coarsely Lipschitz* if there is some $L>0$ so that $d(\Psi(x),\Psi(y))\leq Ld(x,y)+L$. Proposition 2.15 and Lemma 4.10 are used to show

PROPOSITION 4.11. *The map Υ is coarsely Lipschitz.*

PROOF. Let γ be a simple closed curve on X of length at most χ. By Lemma 4.1, there is a universal number $c>0$ such that the length of every simple closed curve β on X is at least $c\iota(\gamma,\beta)$.

By Proposition 2.15, if $d_{\mathcal{T}}(X,X')\leq 1$ then $\ell_\beta(X)\leq e^4\chi$ for every curve β with $\ell_\beta(X')\leq\chi$. Then the intersection number between γ,β does not exceed $e^4\chi/c$ and therefore by Lemma 4.10, the distance between γ and β in $\mathcal{CG}(S)$ is at most $e^4\chi/c+1$. □

In fact, much more is true. The following fundamental result is due to Masur and Minsky [**MM99**]. For its formulation, we say that a map $\gamma:\mathbb{R}\to\mathcal{CG}(S)$ is an *L-quasi-geodesic* if

$$|s-t|/L-L\leq d(\gamma(s),\gamma(t))\leq L|s-t|+L.$$

The map is an *unparametrized L-quasi-geodesic* if there is a homeomorphism $\rho : \mathbb{R} \to \mathbb{R}$ so that $\gamma \circ \rho$ is an L-quasigeodesic.

A geodesic metric space X is *hyperbolic* if there exists a constant $\delta > 0$ with the following property. Let Δ be a geodesic triangle in X with sides a, b, c. Then the side c is contained in the δ-neighborhood of $a \cup b$.

THEOREM 4.12. *The curve graph is hyperbolic. The image under Υ of a Teichmüller geodesic is a uniform unparametrized quasi-geodesic.*

This result turned out to be very important not only for an understanding of the geometry of the mapping class group, but also for an understanding of the Teichmüller metric and the behavior of Teichmüller geodesics. To give a glimpse of what has been recently achieved in this direction, denote for $\epsilon > 0$ by $\mathcal{T}(S)_\epsilon$ the *ϵ-thick part* of Teichmüller space. This set consists of all hyperbolic metrics on S for which the shortest length of a closed geodesic is at least ϵ. Clearly $\mathcal{T}(S)_\epsilon$ is invariant under $\mathrm{Mod}(S)$, moreover by Proposition 4.8, the action of $\mathrm{Mod}(S)$ on $\mathcal{T}(S)_\epsilon$ is cocompact.

A closed unbounded subset B of $\mathcal{T}(S)$ is called *coarsely convex* if there is a number $r > 0$ such that for any two points $X, Y \in B$ the Teichmüller geodesic connecting X to Y is contained in the r-neighborhood of B (note that this definition is meaningless if B is bounded). With this terminology we have [**H10**]

PROPOSITION 4.13. *The restriction of the map Υ to a coarsely convex subset B of $\mathcal{T}(S)_\epsilon$ is a quasi-isometry onto a coarsely convex subset of $\mathcal{CG}(S)$.*

As a consequence, coarsely convex subsets of $\mathcal{T}(S)_\epsilon$ are hyperbolic for the Teichmüller metric.

There is a particularly well known consequence of this statement. Namely, a *pseudo-Anosov mapping class* is an element $g \in \mathrm{Mod}(S)$ which preserves a Teichmüller geodesic γ and acts on it as a non-trivial translation. The geodesic γ is called the *axis* of the pseudo-Anosov element. Since g acts cocompactly on its axis, for some $\epsilon > 0$ the geodesic is entirely contained in $\mathcal{T}(S)_\epsilon$. Now a Teichmüller geodesic is coarsely convex and hence this shows the "only if" part of the following statement.

COROLLARY 4.14. *A mapping class is pseudo-Anosov if and only if it acts on $\mathcal{CG}(S)$ with unbounded orbits.*

The "if" part of the corollary is a consequence of the fact that for a mapping class $g \in \mathrm{Mod}(S)$ which is not pseudo-Anosov there is some $k \geq 1$ so that g^k fixes a simple closed curve (see [**FM11**]) In particular, every g-orbit on $\mathcal{CG}(S)$ is bounded.

In the thin part of Teichmüller space, the Teichmüller metric is very far from being hyperbolic. Indeed, Minsky's product region theorem [**M96**] shows that "coarse positive curvature" prevails in the thin part of Teichmüller space.

For the formulation of his result, let again $\mathcal{P}$ be a pants decomposition of S and let $\sigma \subset \mathcal{P}$ be a simplex in $\mathcal{C}(S)$ with $u \geq 1$ components $\gamma_1, \dots, \gamma_u$. For some sufficiently small $\delta > 0$ let $V \subset \mathcal{T}(S)$ be the region where the length of each of the components γ_i is at most δ. Let S_0 be the (possibly disconnected) surface obtained by replacing each of the components γ_i of γ by a pair of punctures. The $(6g-6-2u)$-tuple of Fenchel Nielsen coordinates at pants curves in $\mathcal{P}-\sigma$ defines Fenchel Nielsen coordinates for S_0. The natural coordinate forgetful map determines a projection $\Pi_0 : V \to \mathcal{T}(S_0)$.

For each component γ_i of γ let $\mathbf{H}_i$ be a copy of the upper half-plane equipped with the hyperbolic metric. Define a map $\Pi_i : V \to \mathbf{H}_i$ by $\Pi_i(x) = (\tau_i, 1/\ell_i)$ where (τ_i, ℓ_i) are the twist and length parameters for γ_i. In Fenchel Nielsen coordinates, the positive Dehn twist about the curve γ_i preserves the length and twist parameter of the pants curves $\zeta \in \mathcal{P}-\gamma_i$ and projects to the transformation $(\tau_i, \ell_i) \to (\tau_i + \ell_i, \ell_i)$. For two points $x, y \in V$ define

$$d_P(x,y) = \max\{d_{\mathcal{T}(S_0)}(\Pi_0(x), \Pi_0(y)), d_{\mathbf{H}_i}(\Pi_i(x), \Pi_i(y))\}$$

where $d_{\mathcal{T}(S_0)}$ denotes the Teichmüller distance on $\mathcal{T}(S_0)$ and where $d_{\mathbf{H}_i}$ is the hyperbolic metric on $\mathbf{H}_i$. The following is Theorem 6.1 of [**M96**].

THEOREM 4.15. *There is a constant $a > 0$ only depending on δ such that for all $x, y \in V$,*

$$|d_{\mathcal{T}(S)}(x,y) - d_P(x,y)| \leq a.$$

More recently, this result has been used by Rafi [**R07b**] to establish a coarse distance formula for the Teichmüller metric.

The product region theorem is silent about the precise behavior of Teichmüller geodesics connecting any two given points in $\mathcal{T}(S)$- the latter are unique while geodesics for an L^∞-metric are not. However, coarsely such geodesics can be understood explicitly [**R10**]. An example of possible behavior is illustrated in the following [**H12**]

THEOREM 4.16. *For every $\epsilon > 0$ there is a pseudo-Anosov element $g \in \mathrm{Mod}(S)$ whose axis is entirely contained in $\mathcal{T}(S) - \mathcal{T}(S)_\epsilon$.*

In fact, in imperfect analogy to Weil-Petersson geodesics (we refer once more to [**W10**] for references), Teichmüller geodesics in the thin part or Teichmüller space approach Teichmüller geodesics in boundary strata [**H11, R10**].

LECTURE 5

Geometry and dynamics of moduli space

In this final section we relate some geometric properties of moduli space to dynamics of the Teichmüller flow.

A basic question concerning the geometry of moduli space is as follows: To what extend does $\mathcal{M}_g$ equipped with the Teichmüller metric resemble a locally symmetric space of finite volume?

The Torelli map embeds the moduli space $\mathcal{M}_g$ into the finite volume locally symmetric space $\mathrm{Sp}(2g,\mathbb{Z})\backslash\mathrm{Sp}(2g,\mathbb{R})/\mathrm{U}(g)$. However, the dimension of this locally symmetric space is $\frac{g(g+1)}{2}$ and hence for g large, the image of $\mathcal{M}_g$ is of high codimension. Moreover, the kernel of the induced map on orbifold fundamental groups is the *Torelli group*, a group which is large and complicated (in fact, for $g = 2$ the Torelli group is an infinitely generated free group, see [**I02**] for this result of Mess).

The following result is due to Kaimanovich-Masur and Masur-Farb. Another proof was given later by Bestivina and Fujiwara. We refer to [**I02**] for a history and references.

THEOREM 5.1. *Let Γ be an irreducible lattice in a semisimple Lie group of non-compact type and rank at least two. Then every homomorphism $\Gamma \to \mathrm{Mod}(S)$ has finite image.*

An important geometric property of a locally symmetric space Y of non-compact type and finite volume is the following. Let $r \geq 1$ be the rank of Y and let $T \subset Y$ be an embedded flat torus of rank r. Then there is a compact set $K \subset Y$ such that T can not be homotoped outside K [**PS09**]. If $r = 1$ then this means that a closed geodesic can not be homotoped outside K.

The mapping class group $\mathrm{Mod}(S)$ is finitely presented and hence it can be equipped with a word metric defined by a finite generating set. Such a word metric is unique up to bilipschitz equivalence. The *geometric rank* of the mapping class group is defined to be the maximal dimension of a quasi-isometrically embedded euclidean space. Since any abelian subgroup of $\mathrm{Mod}(S)$ is quasi-isometrically embedded [**FLM91**], the geometric rank is not smaller than the number $3g-3$ of components of a pants decomposition of S. Namely, the group of Dehn twists about the components of a pants decomposition is free abelian of rank $3g-3$. In fact, the geometric rank of $\mathrm{Mod}(S)$ equals $3g-3$ [**BM08**]. There should be a similar statement for Teichmüller space with the Teichmüller metric as indicated by the product region theorem, but we are not aware of such a statement in the literature.

In contrast to locally symmetric spaces, using the Deligne Mumford compactification of moduli space it is easy to show

PROPOSITION 5.2. *For every compact set $K \subset \mathcal{M}_g$, an embedded torus in $\mathcal{M}_g$ whose fundamental group is the abelian group of Dehn twists about the components of a pants decomposition can be homotoped into $\mathcal{M}_g - K$.*

PROOF. Let $\mathcal{P}$ be a pants decomposition of S. For any $\epsilon > 0$, a torus as in the proposition can be represented by the projection to $\mathcal{M}_g$ of the set of all points in $\mathcal{T}(S)$ whose Fenchel Nielsen coordinates with respect to $\mathcal{P}$ have lengths parameters equal to ϵ. □

It turns out, however, that *complex* geodesics in $\mathcal{M}_g$ (i.e. projections of complex geodesics in $\mathcal{T}(S)$) are much more rigid. The remainder of this section is devoted to the proof of

THEOREM 5.3.

(1) *There is a compact set $K \subset \mathcal{M}_g$ so that every complex geodesic intersects K.*
(2) *A complex geodesic is unbounded in moduli space.*

The (easier) second part of the theorem is due to Masur [**M86**]. The strategy is to use the geometry of the singular euclidean metrics on S of the quadratic differentials defining the complex geodesic. For this we have to understand a bit more explicitly the topology of $\mathcal{Q}(S)$. We begin with organizing quadratic differentials as follows.

DEFINITION 5.4. A *stratum* of quadratic or abelian differentials consists of differentials with the same number and orders of zeros.

Each stratum is a complex orbifold. *Relative periods* of *abelian* differentials can be used to define coordinates on strata. By this we mean the following.

Let ω be any abelian differential. Choose a triangulation T of S whose set of vertices is the set of singular points of ω and whose edges are saddle connections for ω. The existence of such a triangulation was shown in Lemma 2.18. The tangent vector of each edge from the triangulation is a vector in $\mathbb{C}$. These vectors determine the singular euclidean metric defined by ω as well as the horizontal and vertical line fields for ω and hence they determine ω.

A nearby abelian differential ζ in the stratum has nearby singular points of the same orders and hence the triangulation T of S is isotopic to a triangulation T' whose vertices are singular points for ζ and whose edges are saddle connections. Thus we obtain a new set of vectors from ζ by integrating the differential over the edges of T'. As a consequence, integration over the edges of T defines an embedding of a neighborhood of ω in its stratum into $\mathbb{C}^r$ for some large $r > 0$, and this embedding defines a topology on strata which is independent of any choices made. Indeed, changing the triangulation amounts to an affine change of coordinates. As another consequence, strata of abelian differentials are complex manifolds. Note that this structure of a complex manifold is compatible with the structure on the

Hodge bundle $\mathcal{H}$ defined by period coordinates as discussed in Section 4. In particular, the closure of a stratum of abelian differentials is a union of strata in $\mathcal{H}$ of smaller dimension obtained by merging two or more zeros of the differentials. (Using the not very difficult fact that every Riemann surface admits an abelian differential with simple zeros (see p.98 of [**FK80**]), it seems possible to use the natural complex structure on the maximal stratum of abelian differentials to give a direct and elementary proof that Teichmüller space admits a Mod(S)-invariant integrable complex structure).

Strata of quadratic differentials can be treated in the same way. Namely, given a quadratic differential q which is not the square of a holomorphic one-form, there is a two-sheeted cover $\hat{S}$ of S ramified at some of the zeros of q such that q lifts to an abelian differential $\hat{q}$ on $\hat{S}$. Namely, for a simple closed loop α on S not passing through a singular point, define the *holonomy* by parallel transport of a non-zero vector v along α with respect to the singular euclidean metric defined by q. Then the image of v either equals v or $-v$, and this association of sign defines a homology class in $H_1(S - \Sigma, \mathbb{Z}/2\mathbb{Z})$ where Σ is the set of singular points of q. The double cover of $S - \Sigma$ associated to this class determines a branched cover of S with the desired properties.

Any nearby quadratic differential in the stratum of q has the same orientation cover and hence a neighborhood of q in its stratum injects into a neighborhood of $\hat{q}$ in its stratum. The closure of a stratum in $\mathcal{Q}(S)$ is a union of strata of smaller dimension obtained by merging two or more zeros of the differentials.

Strata are invariant under scaling. Their intersections with the space $\mathcal{Q}^1(S)$ of area one quadratic differentials will be called strata in $\mathcal{Q}^1(S)$. Such strata in $\mathcal{Q}^1(S)$ are invariant under the natural SL$(2, \mathbb{R})$-action on $\mathcal{Q}^1(S)$ and under the action of the mapping class group. Therefore strata in $\mathcal{Q}^1(S)$ project to strata in the moduli space $\mathcal{Q}^1(S)/\text{Mod}(S) = \mathcal{V}$ of area one quadratic differentials.

Compact subsets of strata in $\mathcal{V}$ are easy to understand. The next definition is used to this end.

DEFINITION 5.5. The *extremal length* of a simple closed curve α on a Riemann surface X is defined to be

$$E(\alpha) = \sup_{\rho} \frac{\ell_\rho(\alpha)^2}{\text{area}_\rho(S)}$$

where the supremum is taken over all Borel metrics ρ in the conformal class of X and where $\ell_\rho(\alpha)$ is a shortest ρ length of a curve in the free homotopy class of α.

If S contains a cylinder of modulus $a > 0$ then the extremal length of the core curve of the cylinder is at most $1/a$ (we refer to [**IT89**] for references).

PROPOSITION 5.6. *A closed subset K of a stratum in $\mathcal{V}$ is compact if and only if there is a number $\epsilon > 0$ so that no $q \in K$ has a saddle connection of length smaller than ϵ.*

PROOF. Let C be a stratum in $\mathcal{V}$ and let $K \subset C$ be compact. Assume that there is a sequence $(q_i) \subset K$ so that for each i there is a saddle connection γ_i for q_i of length less than $1/i$. By passing to a subsequence we may assume that $q_i \to q \in K$. But then two zeros of q_i have collided to a single zero of q and hence $q \notin C$, a contradiction.

For the reverse direction, let K be a closed subset of a stratum $C \subset \mathcal{V}$ so that there is no saddle connection of length smaller than ϵ for any point in K. We have to show that K is compact. For this note that for $q \in K$ there is no essential simple closed curve of q-length at most ϵ. Namely, by the Arzela Ascoli theorem, any closed curve on S is freely homotopic to a closed geodesic for the singular euclidean metric defined by q whose length is minimal in its free homotopy class. Such a closed geodesic either is a concatenation of saddle connections and hence its length is bigger than ϵ by assumption, or it does not pass through a singular point. Since the metric q is euclidean, in the latter case the geodesic is the core curve of a flat cylinder foliated by closed geodesics of the same length. A boundary circle of a maximal such cylinder contains a singular point and hence it is a path composed of saddle connections (we refer to [**S84**] for more details). Once again, the length of the curve is at least ϵ.

Now the q-length of any simple closed curve on S is at least ϵ, and the area of q equals one. This implies that the extremal length of any simple closed curve on the Riemann surface X defined by q is bounded from below by ϵ^2. As a consequence, there is no embedded cylinder in X of modulus bigger than $1/\epsilon^2$. On the other hand, a short hyperbolic geodesic on a surface S is the core curve of an embedded cylinder whose modulus tends to infinity as the length of the curve tends to zero. This implies that K projects to a compact subset of $\mathcal{M}_g$ and hence it is relative compact in $\mathcal{V}$.

If there is an accumulation point of K in $\mathcal{V}$ which is not contained in C then the shortest length of a saddle connection for points in K is not bounded from below by a positive constant. The proposition follows. □

The *Teichmüller flow* Φ^t on $\mathcal{Q}^1(S)$ is the flow defined by the action of the diagonal subgroup of $\mathrm{SL}(2,\mathbb{R})$. Thus

$$\Phi^t q = \begin{pmatrix} e^{t/2} & \\ & e^{-t/2} \end{pmatrix} q.$$

This flow commutes with the action of the mapping class group and hence it descends to a flow on $\mathcal{V}$ which is called again the Teichmüller flow.

COROLLARY 5.7. *Let q be a quadratic differential which admits a vertical saddle connection. Then the image of q under the Teichmüller flow does not have an accumulation point in the interior of the stratum containing q.*

PROOF. Let q be a quadratic differential with a vertical saddle connection of length s. Then for every $t > 0$, the differential $\Phi^t q$ has a vertical saddle connection of length $e^{-t/2}s$. Now use Proposition 5.6. □

If q is any quadratic differential and if v is any saddle connection for q then there is some $\theta \in [0, 2\pi)$ so that v is vertical for $e^{i\theta}q$. Thus Corollary 5.7 shows that an $\mathrm{SL}(2,\mathbb{R})$-orbit is not relative compact in any stratum. However, there are quadratic differentials with arbitrarily short saddle connections which project to a fixed compact subset of moduli space and hence this does not imply that a complex geodesic intersects the complement of every compact subset of moduli space. That this nevertheless holds true is due to Masur [**M86**].

PROPOSITION 5.8. *Let $H \subset \mathcal{M}_g$ be a complex geodesic and let $K \subset \mathcal{M}_g$ be any compact set. Then $H \cap (\mathcal{M}_g - K) \neq \emptyset$.*

PROOF. Let $q \in \mathcal{V}$ be an area one quadratic differential. Let $E \subset \mathcal{V}$ be the set of accumulation points of the $\mathrm{SL}(2,\mathbb{R})$-orbit of q, i.e. the set of all points z so that there is an unbounded sequence $(g_i) \subset \mathrm{SL}(2,\mathbb{R})$ such that $g_i q \to z$. Then E is a closed $\mathrm{SL}(2,\mathbb{R})$-invariant subset of $\mathcal{V}$. We may assume that $E \neq \emptyset$.

For $u \in \mathcal{V}$ with zeros of order k_i define $\mathcal{O}(u) = \sum_i (k_i - 1)$. Let $E_{\max} = \{u \in E \mid \mathcal{O}(u) \geq \mathcal{O}(z) \text{ for all } z \in E\}$. Since E is closed, the set $E_{\max}$ is closed as well. Namely, if q is a quadratic differential with s zeros of order k_i, then any nearby differential has at least s zeros. By the Gauss Bonnet theorem, the sum of the orders of the zeros is constant which implies that $\mathcal{O}(z) < \mathcal{O}(q)$ if z has more than s zeros. It now suffices to show that $E_{\max}$ is not compact.

For this we argue by contradiction and we assume that $E_{\max}$ is compact. For a quadratic differential u let $d(u)$ be the shortest length of a saddle connection for the singular euclidean metric defined by u. Let $d_0 = \inf_{u \in E_{\max}} d(u)$.

If $d_0 = 0$ then choose a sequence $u_i \subset E_{\max}$ so that $d(u_i) \to 0$. Since we assume that $E_{\max}$ is compact, by passing to a subsequence we may assume that $u_i \to u \subset E_{\max}$. But then two zeros of u_i connected by a short saddle connection collide to a single zero in u. As a consequence, we have $\mathcal{O}(u) > \mathcal{O}(u_i)$ (see the discussion above) which contradicts the definition of $E_{\max}$.

If $d_0 > 0$ then by the same reasoning, there is some $u \in E_{\max}$ with $d(u) = d_0$. Let γ be a saddle connection of length d_0 on u. There is some $\theta_0 \in [0, 2\pi]$ so that γ is vertical for $e^{i\theta_0}u$. Since the Teichmüller flow contracts vertical distances, there is some $s > 0$ so that $\Phi^s e^{i\theta_0} u$ has a saddle connection of length $d_0/2$.

The set E is $\mathrm{SL}(2,\mathbb{R})$-invariant and therefore $\Phi^s e^{i\theta_0} u \in E$. However, the $\mathrm{SL}(2,\mathbb{R})$-action preserves the strata of $\mathcal{V}$ and hence we have $\Phi^s e^{i\theta_0} u \in E_{\max}$. This is a contradiction to the definition of d_0. The proposition follows. □

A degeneration of quadratic differentials which is particularly easy to understand can be described as follows. A *metric cylinder* for a singular euclidean metric q is an embedded cylinder in S which is foliated by closed geodesics for q. If these geodesics are vertical then their lengths decrease and the height of the cylinder increases under the Teichmüller flow Φ^t. Lemma 4.1 then shows that the hyperbolic length of the core curve of the cylinder tends to zero as $t \to \infty$. In particular, the underlying family of Riemann surfaces leaves every compact subset of moduli space.

Masur [**M86**] showed the following result which is much stronger than Proposition 5.8 (see also the survey [**MT02**]).

THEOREM 5.9. *For every quadratic differential q, the set of directions θ so that $e^{i\theta}q$ has a vertical metric cylinder is dense in the unit circle.*

The *horocyle flow* h_t $(t \in \mathbb{R})$ is the flow on $\mathcal{V}$ defined by the action of the unipotent subgroup

$$\begin{pmatrix} 1 & t \\ 0 & 1 \end{pmatrix} \tag{1}$$

of $\mathrm{SL}(2, \mathbb{R})$.

The following is a non-quantitative version of a result of Minsky and Weiss [**MW02**]. Together with Proposition 5.8 it implies Theorem 5.3.

THEOREM 5.10. *For every stratum $C \subset \mathcal{V}$ there is a compact subset K of C which is intersected by the orbit of the horocycle flow through every quadratic differential $q \in C$ without horizontal saddle connection.*

Example: Let q be a quadratic differential which defines a flat metric consisting of a single cylinder which is foliated by vertical simple closed geodesics. This cylinder is bounded by vertical saddle connections (see [**S84**] for examples). Assume moreover that there is a closed horizontal curve. Then there is a closed orbit of the horocycle flow through q. Choosing such quadratic differentials carefully results in closed orbits for the horocycle flow which are disjoint from any given compact subset K of the stratum containing q. Thus the statement of Theorem 5.10 does not hold for all quadratic differentials in C. This behavior is familiar for other unipotent flows. An example of such a flow is the horocycle flow on the unit tangent bundle of a non-compact hyperbolic surface of finite volume (which is defined to be the action of same unipotent subgroup of $\mathrm{PSL}(2, \mathbb{R})$). For this flow every orbit is closed or equidistributed for the Haar measure (see [**MW02**] for an overview and references), and closed orbits exist arbitrarily far out in the cusps.

Slightly stronger versions of Theorem 5.10 can be found in the appendix of [**LM08**] and of [**H09**]. Beyond this, not much is known about the horocycle flow on strata (see however [**SW04**]).

The remainder of these notes is devoted to the proof of Theorem 5.10. We follow [**MW02**]. The proof uses an idea from the theory of unipotent flows: make effective use of slow divergence of orbits to find a point on the orbit which does not have a short saddle connection.

We begin with having a closer look at an orbit of the horocycle flow through an area one quadratic differential q without horizontal saddle connection. Let δ be a saddle connection for q. Then δ is a straight line segment for the singular euclidean metric defined by q. Its tangent can be decomposed into its horizontal part $x = x(\delta, q)$ and its vertical part $y = y(\delta, q)$. Since q does not have a horizontal saddle connection, we have $y \neq 0$ and hence $h_t(x, y) = (x + ty, y) \neq (x, y)$ for $t \neq 0$

(note that the image under h_t of a saddle connection for q is a saddle connection for $h_t q$). Write

$$\ell(\delta, q) = \max\{|x(\delta, q)|, |y(\delta, q)|\}$$

and $\ell_{q,\delta}(t) = \ell(\delta, h_t q)$. The next easy lemma describes these functions explicitly and gives a quantitative version of the idea of slow divergence of orbits of the horocycle flow.

LEMMA 5.11. (1) *There are $t_0 \in \mathbb{R}, c > 0$ so that*

$$\ell_{q,\delta}(t) = \max\{c, c|t - t_0|\}.$$

(2) *Let $f, \tilde{f}$ be two functions of the form $t \to \max\{c, c|t - t_0|\}$ for some $c > 0$. Suppose that for some $b > 0$ and $s \in \mathbb{R}$ we have $f(s) < b, \tilde{f}(s) < b$. Then possibly after exchanging f and $\tilde{f}$, $f(t) < 3b$ whenever $\tilde{f}(t) < b$.*

PROOF. Put $c = |y| > 0$ and $t_0 = -x/y$. The first part of the lemma follows.

For the second part, let c, t_0 and $\tilde{c}, \tilde{t}_0$ be the constants for $f, \tilde{f}$. Assume that $0 < c \leq \tilde{c}$.

By the hypothesis on s, t we have $\tilde{c}|s - \tilde{t}_0| < b, \tilde{c}|t - \tilde{t}_0| < b$ and hence

$$\tilde{c}|t - s| \leq \tilde{c}|t - \tilde{t}_0| + \tilde{c}|\tilde{t}_0 - s| < 2b.$$

Now $f(s) < b$ and therefore $c < b$. If t_0 is between s and t then

$$c|t - t_0| \leq \tilde{c}|t - t_0| \leq \tilde{c}|s - t| < 2b.$$

If t_0 is not between s and t then

$$c|t - t_0| = c|s - t_0| + c|t - s| < f(s) + 2b < 3b.$$

In either case we get $f(t) < 3b$. □

Denote by $|A|$ the Lebesgue measure of a subset of the real line. From the first part of Lemma 5.11 one obtains the following. Let δ be any saddle connection, let $\theta > 0$, $I \subset \mathbb{R}$ be an interval and let

$$I_{\delta,\theta} = \{s \in I \mid \ell_{q,\delta}(s) < \theta\}$$

and

$$||\ell_{q,\delta}||_I = \sup_{s \in I} \ell_{q,\delta}(s).$$

Then

$$\frac{|I_{\delta,\theta}|}{|I|} \leq 2\left(\frac{\theta}{||\ell_{q,\delta}||_I}\right).$$

Namely, using the above notation, if $\theta \leq c$ then $I_{\delta,\theta} = \emptyset$. If $\theta > c$ then either $I \cap \{t > t_0\} \subset I_{\delta,\theta}$ or for $b = \sup\{s \in I\}$ we have $\frac{|I_{\delta,\theta} \cap \{t \geq t_0\}|}{|I \cap \{t \geq t_0\}|} = \frac{\theta}{c|b - t_0|} \leq \frac{\theta}{||\ell_{q,\delta}||_I}$.

From this we deduce

LEMMA 5.12. *Assume that there is some $M > 0$, there is an interval I, a number $\rho > 0$ and a collection Δ of saddle connections for q with the following properties.*

(1) *For any* $t \in I$, $\ell_{q,\delta}(t) < \rho$ *for at most* M *saddle connections* $\delta \in \Delta$.
(2) $\|\ell_{q,\delta}\|_I \geq \rho$ *for every* $\delta \in \Delta$.

Then

$$|\{s \in I \mid \ell_{q,\delta}(s) < \rho/4M^2 \text{ for some } \delta \in \Delta\}| < |I|/2M.$$

PROOF. Let

$$J = \int_I \sharp\{\delta \in \Delta \mid \ell_{q,\delta}(s) < \rho\} ds.$$

The integrand of J is bounded from above by M and therefore $J \leq M|I|$. Let $\beta \leq \rho/4M^2$. By the discussion after Lemma 5.11, for every $\delta \in \Delta$ we have $|I_{\delta,\beta}| \leq 2|I_{\delta,\rho}|\frac{\beta}{\rho}$. Since the set of all saddle connections is countable, the sum

$$\sum_{\delta \in \Delta} |I_{\delta,\rho}|$$

is well defined and equals J. Together we obtain

$$J \geq [2\frac{\beta}{\rho}]^{-1} \sum_{\delta \in \Delta} |I_{\delta,\beta}|.$$

Since $\beta \leq \rho/4M^2$ we conclude that $\sum_{\delta \in \Delta} |I_{\delta,\beta}| \leq |I|/2M$. □

While for every hyperbolic metric on S there are at most $3g - 3$ short simple closed curves, there is no uniform bound on the number of short saddle connections for a quadratic differential. Examples can be constructed as follows.

Let T_1, T_2 be two flat tori of area one. For a small number $\delta > 0$, cut T_2 open along an embedded line segment of length δ. For each i let T_2^i be the flat torus obtained from T_2 by scaling the flat metric by the constant $1/i$. Cut T_1 open along a line segment of length δ/i and glue the tori T_1 and T_2^i along the two slits of length δ/i to obtain a surface T^i of genus 2. The euclidean metrics on the tori T_1 and T_2^i define a singular euclidean metric on T^i which is the metric of an abelian differential ω^i. The images in T^i of the endpoints of the slits in T_1, T_2^i are singular points for ω^i, and the slits are saddle connections. As $i \to \infty$, the length of any saddle connection of ω^i which is entirely contained in T_2^i tends zero (with the obvious interpretation). Now observe that there are countably many such saddle connections.

To overcome this difficulty one looks at "isolating" saddle connections which play the role of the slits in the above example. One verifies that the number of isolating saddle connections is uniformly bounded and that moreover in the absence of isolating saddle connections, there are no short saddle connections at all.

To define such isolating saddle connection, we observe

LEMMA 5.13. *There is a number* $M > 0$ *so that the cardinality of any set of saddle connections with mutually disjoint interior is at most* M.

PROOF. Assume that q has k singular points. Let $\alpha_1, \dots, \alpha_s$ be saddle connections with disjoint interior. As in the proof of Lemma 2.18, by adding a finite additional collection of saddle connections we may assume that the arcs α_i decompose S into triangles in such a way that all singular points of q occur as vertices. We do not require that the vertices of these triangles are distinct.

The number of edges of this triangulation of S equals s. Any edge is contained in the boundary of 2 triangles, and every triangle has 3 sides. Euler's formula shows that $\frac{2}{3}s - s + k = 2 - 2g$ and hence $s = 3(2g - 2 + k)$. □

Let $\mathcal{L}_q$ be the set of all saddle connections for q. For $r \leq M$ define

$$\mathcal{E}_r = \{E \subset \mathcal{L}_q \mid E \text{ consists of } r \text{ segments with disjoint interior}\}.$$

For $E \in \mathcal{E}_r$ define $S(E)$ to be the closure of the simply connected components of $S - \cup_{\delta \in E}\delta$. Then $S(E)$ is a possibly empty subsurface of S whose boundary $\partial S(E)$ is contained in E. Let $W(E)$ be the union of $\partial S(E)$ with those saddle connections in E whose interiors are disjoint from $S(E)$. Then $W(E)$ is a closed subset of S which is empty only if $S(E) = S$.

The next lemma indicates how to find "isolating" saddle connections: As saddle connections on the boundary of regions of S which can be decomposed by short saddle connections into simply connected components. Note that the constant θ in the assumptions of the lemma is a gap for lengths of saddle connections. The main remaining task for the completion of the proof of Theorem 5.10 will then be to show the existence of such a gap number to which the lemma can be applied.

LEMMA 5.14. *Suppose that $E \in \mathcal{E}_r$ consists of segments of length $\leq \theta$. Suppose furthermore that there is no saddle connection of length $\leq 4\theta$ whose interior is disjoint from E. Then any saddle connection δ' which intersects the interior of a saddle connection $\delta \subset W(E)$ has length at least 2θ.*

PROOF. Let δ be a saddle connection in $W(E)$ and assume that the saddle connection δ' intersects δ in an interior point p. Let $\omega \subset \delta'$ be a subsegment of δ' in $S - (S(E) \cup E)$ with one endpoint p and such that the second endpoint of ω either is an interior point of a saddle connection in E or is a singular point. Then ω is contained in a component C of $S - E$ which is not simply connected.

The component C is an oriented surface with singular euclidean metric with cone points of cone angles bigger than 2π and finite area. Its metric completion $\overline{C}$ is a compact surface with piecewise geodesic boundary and finitely many singular points. Each boundary arc with the boundary orientation is an oriented saddle connection in S. The closure of C in S can be obtained from $\overline{C}$ by identifying some of the boundary arcs with an orientation reversing isometry. The boundary of C contains the saddle connection $\delta \in W(E)$.

Our goal is to show that the length of ω is at least 2θ. For this we argue by contradiction and we assume that the length of ω is smaller than 2θ.

Assume furthermore for the moment that the second endpoint of ω is a singular point. Let ρ_1, ρ_2 be the two subarcs of the saddle connection $\delta \subset \overline{C}$ connecting p

to the two endpoints x_1, x_2 of δ. The concatenation of ω with ρ_1, ρ_2 is a broken geodesic in $\overline{C}$ with endpoints at singular points and length smaller than 3θ.

The shortest arc ρ_i' in $\overline{C}$ which is homotopic with fixed endpoints to the concatenation of ω with ρ_i is composed of saddle connections all of whose lengths are smaller than 3θ. The saddle connections either are contained in E or their interiors are disjoint from E. By the assumption in the lemma, these saddle connections are all contained in E.

Now ω is a smooth geodesic arc in $\overline{C}$ and hence it is the unique arc of minimal length in its homotopy class with fixed endpoints. The concatenation $\rho_i^{-1} \circ \rho_i'$ is homotopic to ω with fixed endpoints. In particular, ρ_i', ρ_i, ω bound an embedded disc Ω_i in $\overline{C}$. Since ρ_i' is minimal, the angle at each singular point of $\partial\Omega_i$ which is distinct from x_i and the endpoints of ω is at least π. Since the cone angle at each interior singular point of Ω_i is bigger than 2π, by the Gauss Bonnet theorem the angle at x_i is smaller than π.

As a consequence, the concatenation $\delta \circ \rho_1'$ is not length minimizing. By construction, ρ_2' is a shortest arc in $\overline{C}$ with the same endpoints which is homotopic to $\rho_1' \circ \delta$. A homotopy between these two arcs with fixed endpoints covers a disc which necessarily contains ω. By the definition of $S(E)$, this region is contained in $S(E)$ which is a contradiction to the choice of ω.

If both endpoints of ω are interior points of saddle connections in E then we argue in exactly the same way. In this case we construct from ω and the subarcs of the saddle connections containing the endpoints of ω a simply connected quadrangle $Q \subset S(E)$. Two opposite sides of Q are the saddle connections $\delta, \delta' \in W(E)$ containing the endpoints of ω. The other two sides are geodesic arcs in $S - (S(E) \cup E)$ homotopic to the concatenation of ω with the two subsegments of δ, δ' which are to the left (or right) of ω for some choice of orientation. □

Lemma 5.14 shows that under the assumption of the existence of a "gap" the surface S can be decomposed into subsurfaces bounded by paths which are composed of short saddle connections, and these subsurfaces either are small in size or the short saddle connections they contain is a forest, i.e. a finite union of trees.

Our next goal is to find points on the orbit $h_t q$ which satisfy the assumptions in Lemma 5.14 for a suitably chosen gap constant. Let $M > 0$ be as in Lemma 5.13 and for $r \in \{1, \dots, M\}$ define

$$\alpha_r(t) = \min_{E \in \mathcal{E}_r} \max\{\ell_{q,\delta}(t) \mid \delta \in E\}.$$

Note that $\alpha_1(t)$ is just the shortest length of a saddle connection for $h_t q$.

LEMMA 5.15. *There is a number $\theta_0 > 0$ with the following property. Let $E \in \mathcal{E}_r$ be such that $S(E) = S$; then for each t, E contains at least one saddle connection of $h_t q$-length at least θ_0. In particular, we have $\alpha_M(t) \geq \theta_0$ for all t.*

PROOF. Let E be a collection of saddle connections with pairwise disjoint interiors which decompose S into simply connected regions. Each of these regions is a singular euclidean polygon. The number of these polygons is uniformly bounded. Since the area of the singular euclidean metric defined by $h_t q$ equals one, there is at least one side of these polygons whose length is bounded from below by a constant $\theta_0 > 0$ only depending on S.

This shows the first part of the lemma, and the second part follows from the fact that by Lemma 5.13, any collection of M saddle connections with pairwise disjoint interior decomposes S into simply connected components. □

In the following lemma, the number $T > 0$ depends in an essential way on the initial quadratic differential q. Note that this corresponds to the fact that the amount of time the h_t-orbit through q spends before entering the fixed compact set whose existence is stated in Theorem 5.10 can be arbitrarily large.

LEMMA 5.16. *There is a number $T > 0$ such that $\|\ell_{q,\delta}\|_{[0,T]} \geq \theta_0$ for every saddle connection δ.*

PROOF. Since q is an arbitrary but fixed quadratic differential without horizontal saddle connection there are only finitely many saddle connections for q whose length is smaller than θ_0. By Lemma 5.11, for each of these saddle connections δ there is a number $\tau(\delta) > 0$ so that $\ell_{q,\delta}(\tau(\delta)) \geq \theta_0$. The maximum T of these numbers $\tau(\delta)$ satisfies the requirement in the lemma. □

Write $I = [0, T]$. For $k \leq M - 1$ define

$$L_k = \theta_0/(48M^2)^{M-k}$$

and for each t let

$$r(t) = \max\{k \mid \alpha_k(t) < L_k\}.$$

If $E \in \mathcal{E}_{r(t)}$ is any collection of $r(t)$ disjoint saddle connections for $h_t q$ of length at most L_k then any saddle connection whose interior is disjoint from E has length at least $48M^2 L_k$. The sets

$$V_k = \{t \in I \mid r(t) = k\} \; (k = 1, \ldots, M-1)$$

are disjoint, and their union equals the set V of all $t \in I$ so that $h_t q$ has a saddle connection of length smaller than L_1. In particular, there is some $k \leq M - 1$ such that

$$|V_k| \geq \frac{|V|}{M-1}.$$

For $\delta \in \mathcal{L}_q$, let $H(\delta)$ be the set of all $t \in I$ so that $\ell_{q,t}(t) < L_k$, and whenever $\delta \cap \delta' \neq \emptyset$ for some $\delta' \in \mathcal{L}_q$ then

$$\ell_{q,\delta'}(t) \geq 24M^2 L_k.$$

Thus if $t \in H(\delta)$ then δ satisfies the conclusion of Lemma 5.14 for $h_t q$ and hence it is "isolating". Define

$$\mathcal{F}_0 = \{\delta \in \mathcal{L}_q \mid V_k \cap H(\delta) \neq \emptyset\}.$$

In other words, $\mathcal{F}_0$ is the set of all saddle connections which are "isolating" for some time in V_k. The following statement is the main remaining step towards Theorem 5.10.

LEMMA 5.17. *For all* $t \in I$,

$$\sharp\{\delta \in \mathcal{F}_0 \mid \ell_{q,\delta}(t) \leq 4M^2 L_k\} \leq M.$$

PROOF. All we have to show is the following. If δ, δ' are any two saddle connections with $H(\delta) \cap V_k \neq \emptyset, H(\delta') \cap V_k \neq \emptyset$ and if $\ell_{q,\delta}(t) \leq 4M^2 L_k, \ell_{q,\delta'}(t) \leq 4M^2 L_k$ then δ, δ' are disjoint.

Now choose $s \in V_k \cap H(\delta)$ so that $\ell_{q,\delta}(s) < L_k < 4M^2 L_k$; such an s exists by the assumption on δ. After perhaps exchanging δ and δ', the second part of Lemma 5.11 shows that $\ell_{q,\delta'}(s) < 24M^2 L_k$ (the factor 2 in this estimate comes from the fact that the function f measures the sup of the vertical and horizontal component rather than the euclidean metric) and hence the claim follows from the definition of the set $\mathcal{F}_0$. □

Now

$$V_k \subset \cup_{\delta \in \mathcal{L}_q} H(\delta).$$

Namely, if $t \in V_k$ then there is a collection $E \subset \mathcal{E}_k$ of k saddle connections of length at most L_k with disjoint interior so that any additional saddle connection whose interior is disjoint from E has length at least $48M^2 L_k$. By Lemma 5.15 and the choice of L_k, the union $S(E)$ of the closures of the simply connected components of $S - E$ is not all of S. Then $W(E) \neq \emptyset$ and by Lemma 5.14, we have $t \in H(\delta)$ for every $\delta \in W(E)$.

Apply Lemma 5.12 to the interval $I = [0, T]$ and to $\rho = 4M^2 L_k < \theta_0$. It follows that $|V_k| \leq |I|/2M$. Since $|V_k| \geq |V|/M - 1$ this shows that $I - V$ is non-empty. This completes the proof of Theorem 5.10.

Particularly nice complex geodesics are complex geodesics whose stabilizers in Mod(S) act with cofinite volume on the geodesic. We refer to [**HS06**] for an introduction to this fascinating subject.

The proof of Theorem 5.10 relies on a careful analysis of the geometry of a singular euclidean metric defined by a quadratic differential and its change under the horocycle flow.

Rafi [**R07a**] obtained a fairly precise understanding of such singular euclidean metrics. To explain his result, define for a number $\epsilon > 0$ an *ϵ-thick piece* of a hyperbolic surface X to be a component of $X - \cup\alpha$ where α runs through the simple closed geodesics of length at most ϵ.

Provided that $\epsilon > 0$ is sufficiently small, if $q \in \mathcal{Q}(X)$ then for every ϵ-thick piece Y there is an associated (perhaps degenerate) subset Y_q of S. This set is bounded by q-geodesics freely homotopic to the boundary circles of Y. Moreover, Y_q is disjoint from the interior of any flat cylinder whose core curve is freely homotopic to a boundary circle of Y. Rafi showed [**R07a**]

THEOREM 5.18. *For every ϵ-thick piece Y of X there is a number $\lambda(Y) > 0$ so that the q-length of every essential closed curve $\alpha \subset Y$ equals $\lambda(Y)\ell_\alpha(X)$ up to a multiplicative constant which is bounded independent of X.*

In other words, maximal subsets of X which can be decomposed into simply connected regions by short saddle connections correspond to ϵ-thick pieces Y of X for which the scaling constant $\lambda(Y)$ is small.

Bibliography

[A80] W. Abikoff, *The real analytic theory of Teichmüller space*, Lecture Notes in Math. 820, Springer, Berlin New York 1980. MR590044 (82a:32028)

[BM08] J. Behrstock, Y. Minsky, *Dimension and rank for mapping class groups*, Ann. Math. 167 (2008), 1055–1077. MR2415393 (2009d:57031)

[BF02] M. Bestvina, K. Fujiwara, *Bounded cohomology of subgroups of mapping class groups*, Geom. Top. 6 (2002), 69–89. MR1914565 (2003f:57003)

[B92] P. Buser, *Geometry and spectra of compact Riemann surfaces*, Progress in Mathematics, Birkhäuser 1992. MR1183224 (93g:58149)

[CE75] J. Cheeger, D. Ebin, *Comparison theorems in Riemannian geometry*, North Holland/American Elsevier 1975. MR0458335 (56:16538)

[DW07] G. Daskalopoulos, R. Wentworth, *Harmonic maps and Teichmüller theory*, Handbook of Teichmüller theory I, 34–109, A. Papadopoulos, Ed., European Math. Soc. 2007. MR2349668 (2010c:58021)

[FM11] B. Farb, D. Margalit, *A primer on mapping class groups*, Princeton University Press, Princeton 2011. MR2850125 (2012h:57032)

[FlM07] A. Fletcher, V. Markovic, *Quasiconformal maps and Teichmüller theory*, Oxford Graduate Texts in Math. 11, Oxford 2007. MR2269887 (2007g:30001)

[D11] S. Donaldson, *Riemann Surfaces*, Oxford Graduate Text in Mathematics 22, Oxford University Press 2011. MR2856237

[FLM91] B. Farb, A. Lubotzky, Y. Minsky, *Rank one phenomena in mapping class groups*, Duke Math. J. 106 (2001), 581–597. MR1813237 (2001k:20076)

[FK80] H. Farkas, I. Kra, *Riemann surfaces*, Springer Graduate Texts in Math. 71, Springer New York 1980. MR583745 (82c:30067)

[H09] U. Hamenstädt, *Invariant Radon measures on measured lamination space*, Invent. Math. 176 (2009), 223–272. MR2495764 (2010c:37047)

[H10] U. Hamenstädt, *Stability of quasi-geodesics in Teichmüller space*, Geom. Dedicata 146 (2010), 101–116. MR2644273 (2011d:32021)

[H11] U. Hamenstädt, *Symbolic dynamics of the Teichmüller flow*, arXiv:1112.6107.

[H12] U. Hamenstädt, *Counting periodic orbits in strata: Precise bounds*, to appear.

[HM98] J. Harris, I. Morrison, *Moduli of curves*, Graduate Texts in Math. 187, Springer, New York 1998. MR1631825 (99g:14031)

[Hu06] J. Hubbard, *Teichmüller theory*, Matrix editions 2006. MR2245223 (2008k:30055)

[HS06] P. Hubert, T. Schmidt, *An introduction to Veech surfaces*, in "Handbook of dynamical systems", Vol. 1B, 501–526, Elsevier B.V., Amsterdam 2006. MR2186246 (2006i:37000)

[IT89] Y. Imayoshi, M. Taniguchi, *An introduction to Teichmüller spaces*, Springer 1989. MR1215481 (94b:32031)

[I02] N. Ivanov, *Mapping class groups*, Handbook of geometric topology, 523–633, North Holland, Amsterdam 2002. MR1886678 (2003h:57022)

[K83] S. Kerckhoff, *The Nielsen realization problem*, Ann. Math. 117 (1983), 235–265. MR690845 (85e:32029)

[M07] A. Marden, *Outer circles*, Cambridge University Press, Cambridge 2007. MR2355387 (2008i:57001)

[LM08] E. Lindenstrauss, M. Mirzakhani, *Ergodic theory of the space of measured laminations*, Int. Math. Res. Not. IMRN 2008, no. 4. MR2424174 (2009j:37003)

[M88] B. Maskit, *Kleinian groups*, Grundlehren der mathematischen Wissenschaften 187, Springer, Berlin 1988. MR959135 (90a:30132)

[M86] H. Masur, *Closed trajectories for quadratic differentials with an application to billiards*, Duke Math. J. 53 (1986), 307–314. MR850537 (87j:30107)

[MM99] H. Masur, Y. Minsky, *Geometry of the complex of curves I: Hyperbolicity*, Invent. Math. 138 (1999), 103–149. MR1714338 (2000i:57027)

[MT02] H. Masur, S. Tabachnikov, *Rational billiards and flat structures*, in "Handbook of dynamical systems", Vol 1.A, 1015–1098, B. Hasselblatt and A. Katok, Edts., Elsevier Science 2002. MR1928530 (2003j:37002)

[M96] Y. Minsky, *Extremal length and product regions in Teichmüller space*, Duke Math. J. 83 (1996), 249–268. MR1390649 (97b:32019)

[MW02] Y. Minsky, B. Weiss, *Nondivergence of horocyclic flows on moduli space*, J. Reine Angew. Math. 552 (2002), 131–177. MR1940435 (2004d:32012)

[Mi95] R. Miranda, *Algebraic curves and Riemann surfaces*, Graduate texts in Math. 5, AMS 1995. MR1326604 (96f:14029)

[PT07] A. Papadopoulos, G. Theret, *On Teichmüller's metric and Thurston's asymmetric metric on Teichmüller space*, in Handbook of Teichmüller theory, Vol.I, 111–204, IRMA Lect. Math. Theor. Phys., 11, Eur. Math. Soc., Zürich, 2007. MR2349669 (2008k:32037)

[PS09] A. Pettet, J. Souto, *Periodic maximal flats are not peripheral*, arXiv:0909.2899.

[R07a] K. Rafi, *Thick-thin decompositions for quadratic differentials*, Math. Res. Lett. 14 (2007), 333–341. MR2318629 (2008g:30035)

[R07b] K. Rafi, *A combinatorial model for the Teichmüller metric*, Geom. Funct. Anal. 17 (2007), 936–959. MR2346280 (2008j:30063)

[R10] K. Rafi, *Hyperbolicity in Teichmüller space*, arXiv:1011.6004.

[SW04] J. Smillie, B. Weiss, *Minimal sets for flows on moduli space*, Israel J. Math. 142 (2004), 249–260. MR2085718 (2005g:37067)

[S84] K. Strebel, *Quadratic differentials*, Ergebnisse der Mathematik 5, Springer, Berlin 1984. MR743423 (86a:30072)

[V89] W. Veech, *Teichmüller curves in moduli space, Eisenstein series and an application to triangular billiards*, Invent. Math. 97 (1989), 553–583. MR1005006 (91h:58083a)

[W10] S. Wolpert, *Families of Riemann surfaces and Weil-Petersson geometry*, CBMS Regional Conference Series 113, AMS, Providence 2010. MR2641916 (2011c:32020)

The Mumford Conjecture, Madsen-Weiss and Homological Stability for Mapping Class Groups of Surfaces

Nathalie Wahl

IAS/Park City Mathematics Series
Volume 20, 2011

The Mumford Conjecture, Madsen-Weiss and Homological Stability for Mapping Class Groups of Surfaces

Nathalie Wahl

Introduction

The Mumford conjecture [**26**] says that the rational cohomology ring of the moduli space of Riemann surfaces is a polynomial algebra on the so-called Mumford-Morita-Miller classes, in a range of degrees increasing with the genus of the surface. This conjecture is now known to be true, using the following two theorems as main ingredients: *Harer's stability theorem* [**16**] which tells us that the rational cohomology of the moduli space is independent of the genus in a range of dimensions, and *Madsen-Weiss' theorem* [**23**] which identifies the stable cohomology with that polynomial algebra.

Harer's and Madsen-Weiss' theorems are both statements about the integral homology of the mapping class groups of surfaces, using, in the Madsen-Weiss' case, Earle and Eells' theorem [**9**] relating the diffeomorphism groups to the mapping class groups of surfaces. In Lecture 1, we describe the relationship between the moduli space of Riemann surfaces, the mapping class groups and the diffeomorphism groups of surfaces. We then give a definition of the Mumford-Morita-Miller classes, state the main part of Harer's stability theorem, and give a first statement of Madsen-Weiss' theorem.

The last three lectures are devoted to a sketch proof of Harer stability theorem, using improvements by Ivanov [**19, 20**], Hatcher [**17**], Boldsen [**4**] and Randal-Williams [**32**]. This proof uses two spectral sequences associated to the action of the mapping class groups on certain simplicial complexes of arcs in the surfaces. In Lecture 2, we give the general strategy, define the relevant arc complexes and study the properties of the action of the mapping class groups on these complexes. In Lecture 3, we give the spectral sequence argument (following Randal-Williams). Harer's stability theorem is proved by studying the spectral sequences carefully, using the properties of the action given in Lecture 2, as well as a connectivity property of the arc complexes, whose proof is sketched in Lecture 4. The last three are mostly based on the survey [**37**].

This series of lectures is supplemented by Galatius' lectures in this volume [**12**], which present a sketch proof of the Madsen-Weiss theorem.

University of Copenhagen
E-mail address: wahl@math.ku.dk
Supported by the Danish National Sciences Research Council (DNSRC) and the European Research Council (ERC), as well as by the Danish National Research Foundation (DNRF) through the Centre for Symmetry and Deformation

LECTURE 1

The Mumford conjecture and the Madsen-Weiss theorem

In this lecture, we give a brief introduction to important players in the proof of the Mumford conjecture by Madsen and Weiss. We introduce the moduli space of Riemann surface, the Teichmüller space and describe their relationship to diffeomorphism groups and mapping class groups of surfaces. We state the Mumford conjecture, the Madsen-Weiss theorem and Harer's stability theorem.

1. The Mumford conjecture

Let S_g be a closed, smooth, oriented surface of genus g and let $\mathrm{Diff}(S_g)$ denote the topological group of orientation preserving diffeomorphisms of S_g. The moduli space $\mathcal{M}_g$ can be defined in many ways:

$$\begin{aligned}
\mathcal{M}_g &= \textit{Moduli space of Riemann surfaces}\\
&= \text{Space of conformal classes of Riemannian metrics on } S_g\\
&= \{\text{Riemannian metrics on } S_g\}/\mathrm{Diff}(S_g)\\
&= \text{Isometry classes of hyperbolic structures on } S_g\\
&= \text{Biholomorphy classes of complex structures on } S_g\\
&= \text{Isomorphy classes of smooth algebraic curves homeomorphic to } S_g
\end{aligned}$$

We would like to describe $\mathcal{M}_g$, and, for example, compute its (co)homology. The present lectures, together with Galatius' lectures in the same volume [**12**], are centered around the following result about $\mathcal{M}_g$:

THEOREM 1.1 (Mumford conjecture, proved by Madsen-Weiss [**23, 26**]).

$$H^*(\mathcal{M}_g;\mathbb{Q}) \cong_{(*)} \mathbb{Q}[\kappa_1,\kappa_2,\dots] \quad \textit{with} \quad |\kappa_i| = 2i$$

where the isomorphism $(*)$ *is in a range of dimension growing with* g.

We will reformulate this theorem in terms of diffeomorphism groups and mapping class groups of surfaces. The classes κ_i are called the *Mumford-Morita-Miller classes* and are defined below, and the range for the isomorphism is the homological stability range of the mapping class group of surfaces, also given explicitly below.

2. Moduli space, mapping class groups and diffeomorphism groups

We will in this section give a direct relationship between the moduli space described above and the mapping class group and diffeomorphism group of the corresponding surface, when the surface has genus at least two.

Define the *Teichmüller space*

$$\mathcal{T}_g = \{\text{Riemannian metrics on } S_g\}/\,\mathrm{Diff}_0(S_g)$$

with $\mathrm{Diff}_0(S_g)$ the topological group of diffeomorphisms of S_g isotopic to the identity, i.e. the component of the identity in $\mathrm{Diff}(S_g)$, acting on the space of metrics by pull-back (see [**10**, 10.1]).

Let

$$\Gamma_g = \Gamma(S_g) := \mathrm{Diff}(S_g)/\,\mathrm{Diff}_0(S_g) = \pi_0\,\mathrm{Diff}(S_g)$$

denote the *mapping class group* of S_g. (We note that Γ_g is also often denoted $\mathrm{Mod}(S_g)$, and sometimes referred to as the *modular group*.) The group Γ_g acts on $\mathcal{T}_g$, via the full action of $\mathrm{Diff}(S_g)$ on the space of metrics (see [**10**, 12.1]), and

$$\mathcal{M}_g = \mathcal{T}_g/\Gamma_g$$

As $\mathcal{T}_g \simeq *$ (in fact $\mathcal{T}_g \cong \mathbb{R}^{6g-6}$ when $g \geq 2$, see [**10**, 10.6]) and Γ_g acts properly discontinuously on $\mathcal{T}_g$ with finite stabilizers [**10**, 12.1,12.3], we have

$$H^*(\mathcal{M}_g;\mathbb{Q}) \;\cong\; H^*(B\Gamma_g;\mathbb{Q})$$

where $B\Gamma_g$ is a *classifying space* for Γ_g, i.e. $B\Gamma_g = E\Gamma_g/\Gamma_g$, for $E\Gamma_g \simeq *$ with free properly discontinuous Γ_g–action. [See exercises after Lecture 3.]

Recall moreover that $H^*(B\Gamma_g;\mathbb{Z}) = H^*(\Gamma_g;\mathbb{Z})$ is the group cohomology of Γ_g (see [**5**, I.4]).

To relate these homology groups to the diffeomorphism group, we need the following

THEOREM 1.2 (Earle-Eells [**9**]). *For $g \geq 2$, $\mathrm{Diff}(S_g)$ has contractible components.*

In other words, the theorem says that the homomorphism

$$\mathrm{Diff}(S_g) \to \pi_0\,\mathrm{Diff}(S_g) = \Gamma_g$$

is a homotopy equivalence. As one can build compatible models of $E\,\mathrm{Diff}(S_g)$ and $E\Gamma_g$ (via the *standard resolution* and a topological version of it [**5**, I.5]), it follows that, when $g \geq 2$,

$$B\,\mathrm{Diff}(S_g) = E\,\mathrm{Diff}(S_g)/\,\mathrm{Diff}(S_g) \xrightarrow{\simeq} E\Gamma_g/\Gamma_g = B\Gamma_g$$

and thus

$$H^*(\mathcal{M}_g;\mathbb{Q}) \;\cong\; H^*(\Gamma_g;\mathbb{Q}) \;\cong\; H^*(B\Gamma_g;\mathbb{Q}) \;\cong\; H^*(B\,\mathrm{Diff}(S_g);\mathbb{Q})$$

giving us many formulations of the Mumford conjecture.

The space $B\,\mathrm{Diff}(S_g)$ is a *classifying space for S_g–bundles*: there is a 1-1 correspondence

$$\{S_g \to E \xrightarrow{\pi} X\}/_{\cong} \quad\longleftrightarrow\quad \mathrm{Maps}(X, B\,\mathrm{Diff}(S_g))/_{\simeq}$$

between isomorphism classes of bundles and homotopy classes of maps. Moreover, elements of $H^*(B\,\mathrm{Diff}(S_g))$ are *characteristic classes* for S_g–bundles: they give an assignment of a cohomology class $c(E,\pi) \in H^*(X)$ to any bundle $S_g \to E \xrightarrow{\pi} X$, which is natural in the sense that $c(g^*(E,\pi)) = g^*c(E,\pi)$ for any map $g : Y \to X$. Given a class $c \in H^*(B\,\mathrm{Diff}(S_g))$, the associated characteristic class is defined by $c(E,\pi) = f^*(c)$ for $f : X \to B\,\mathrm{Diff}(S_g)$ classifying (E,π) via the above correspondence. (See [**12**, Cor 1.5,1.6] for more details, using an embedding model for $B\,\mathrm{Diff}(S_g)$.)

3. The Mumford-Morita-Miller classes

The Mumford-Morita-Miller-classes κ_i, $i = 1, 2, \ldots$, are characteristic classes for surface bundles defined as follows: to the bundle of oriented surfaces

$$S_g \to E \xrightarrow{\pi} X,$$

one associates the *vertical tangent bundle*

$$\mathbb{R}^2 \to T_v E \xrightarrow{\pi} E$$

with fiber at $e \in E$ the tangent plane to the surface $\pi^{-1} \circ \pi(e) = F_{\pi(e)}$, the fiber over $\pi(e)$. (See Figure 1.)

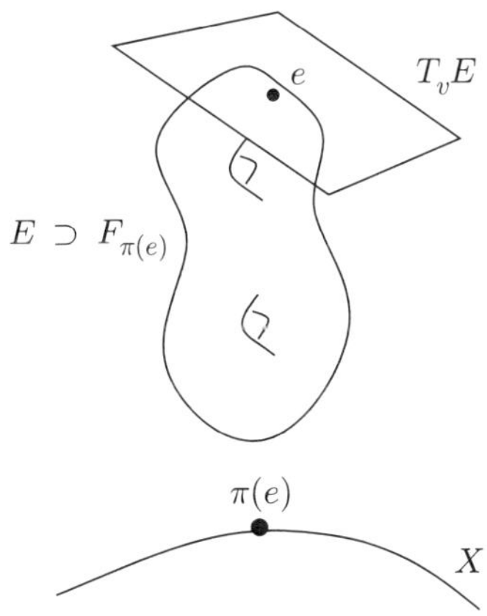

FIGURE 1. The vertical tangent bundle

This plane bundle is canonically oriented, as the surfaces in the original bundle were oriented. Hence it has a first Chern class $c_1(T_v E) \in H^2(E)$. Then

$$\kappa_i := (-1)^{i+1} \pi_*(c_1(T_v E)^{i+1}) \in H^{2i}(X)$$

were $\pi_* : H^{2i+2}(E) \to H^{2i}(X)$ is the *Gysin homomorphism* or *integration along the fibers* (see e.g. [**28**, 4.2.1,4.2.3]). (Note that Mumford originally defined κ_i using the cotangent bundle.)

The cohomology class $\kappa_i \in H^{2i}(B\operatorname{Diff}(S_g))$ (or $H^{2i}(\mathcal{M}_g; \mathbb{Q})$) corresponding to this characteristic class is obtained by doing the same construction on the universal bundle

$$S_g \longrightarrow E\operatorname{Diff}(S_g) \times_{\operatorname{Diff}(S_g)} S_g \longrightarrow B\operatorname{Diff}(S_g)$$

or on the universal curve

$$S_g \longrightarrow \mathcal{M}_{g,1} \longrightarrow \mathcal{M}_g$$

where $\mathcal{M}_{g,1}$ is the moduli space of Riemann surfaces with one marked point.

4. Homological stability

In the previous section, we have defined a class $\kappa_i \in H^{2i}(B\operatorname{Diff}(S_g))$ for each $i = 1, 2, \ldots$, and for each genus g, that is we have define a map of graded rings

$$\mathbb{Q}[\kappa_1, \kappa_2, \ldots] \longrightarrow H^*(B\operatorname{Diff}(S_g)).$$

The Mumford conjecture says that this map is an isomorphism in a range of dimensions grodiffeomorphismswing with g. Note that $\mathbb{Q}[\kappa_1, \kappa_2, \dots]$ is independent of g, so part of the Mumford conjecture is a stability statement, which says that the cohomology of $\mathcal{M}_g$ in any given degree is independent of g if g is sufficiently large. This is known as Harer's stability theorem, which we state in this section.

A family of groups

$$G_1 \hookrightarrow G_2 \hookrightarrow \dots \hookrightarrow G_n \hookrightarrow \dots$$

satisfies *homological stability* if the induced maps

$$H_i(G_n) \longrightarrow H_i(G_{n+1})$$

are isomorphisms in a range $i \ll n$, where $H_*(G_n)$ denotes the group homology of G_n.

Examples: Families of groups satisfying homological stability are $G_n =$ the symmetric group Σ_n [**29**], the braid group β_n [**1**], the linear group $GL_n(\mathbb{Z})$ [**7**].

Define $G_\infty = \bigcup_{n\geq 1} G_n$ to be the "*stable group*". If $\{G_n\}_{n\geq 1}$ satisfies homological stability, then

$$H_i(G_n) \cong H_i(G_\infty) \quad \text{in the range } i \ll n$$

and $H_*(G_\infty)$ is the "*stable homology*".

Let $S_{g,1}$ be a surface of genus g with one boundary component, and let $\Gamma_{g,1} = \pi_0 \operatorname{Diff}(S_{g,1} \text{ rel } \partial)$ be the group of components of the diffeomorphisms restricting to the identity on the boundary. Consider the family of groups

$$\Gamma_{1,1} \hookrightarrow \Gamma_{2,1} \hookrightarrow \dots \hookrightarrow \Gamma_{g,1} \hookrightarrow \dots$$

where the map $\Gamma_{g,1} \hookrightarrow \Gamma_{g+1,1}$ is induced by including $S_{g,1}$ inside $S_{g+1,1}$ as in Figure 2 and extending the diffeomorphisms by the identity on $S_{g+1,1} \backslash S_{g,1}$.

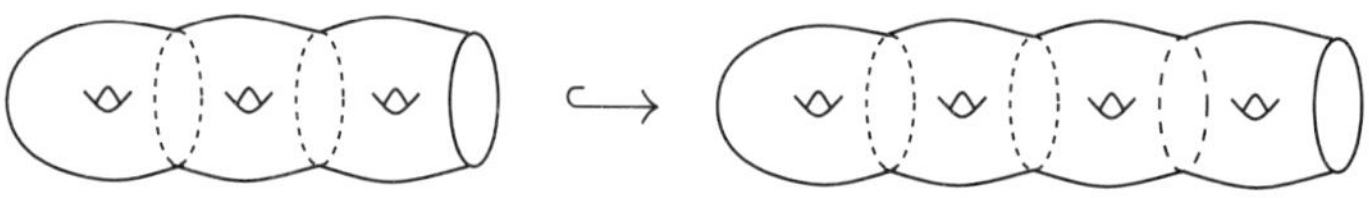

FIGURE 2. Inclusion $S_{3,1} \hookrightarrow S_{4,1}$

As we are interested in mapping class groups of closed surfaces, we also consider the map $\Gamma_{g,1} \to \Gamma_g$ induced by gluing a disc on the boundary component of $S_{g,1}$.

THEOREM 1.3 (Harer's stability theorem, improved by Ivanov, Boldsen, and Randal-Williams [**16, 19, 4, 32**]).

$$H_i(\Gamma_{g,1}; \mathbb{Z}) \xrightarrow{\cong} H_i(\Gamma_{g+1,1}; \mathbb{Z}) \quad \textit{for } i \leq \frac{2}{3}(g-1)$$

and

$$H_i(\Gamma_{g,1}; \mathbb{Z}) \xrightarrow{\cong} H_i(\Gamma_g; \mathbb{Z}) \quad \textit{for } i \leq \frac{2}{3}g$$

The range $i \leq \frac{2}{3}(g-1)$ is the range of degrees i in which the isomorphism in the Mumford conjecture holds. We will give a sketch proof of the stability theorem in the next three lectures.

5. The Madsen-Weiss theorem

The Madsen-Weiss theorem gives a computation of the stable (co)homology of mapping class groups, i.e. the group (co)homology of

$$\Gamma_\infty = \bigcup_{g \geq 1} \Gamma_{g,1}$$

or singular (co)homology of its classifying space $B\Gamma_\infty$. We give now a first formulation of this theorem, without defining all the players yet:

THEOREM 1.4 (Madsen-Weiss [**23**]). *There is a homology isomorphism*

$$B\Gamma_\infty \longrightarrow \Omega_0^\infty MTSO(2)$$

where the target is the 0th component of the infinite loop space of the spectrum $MTSO(2)$.

The spectrum $MTSO(2)$, defined in Galatius' lectures [**12**], is build out of Grassmannians of 2–planes in $\mathbb{R}^n$, in the limit as $n \to \infty$, and the map uses a vertical tangent bundle type construction. (See Definition 1.7, Theorem 1.8 and Corollary 1.10 in [**12**].)

Using homological stability, this can be restated as saying that

$$H_i(\Gamma_g; \mathbb{Z}) \cong H_i(\Omega^\infty MTSO(2); \mathbb{Z}) \quad \text{for } i \leq \frac{2}{3}(g-1)$$

(and the same for cohomology using the universal coefficient theorem).

The target space is computable and $H^*(\Omega_0^\infty MTSO(2); \mathbb{Q}) = \mathbb{Q}[\kappa_1, \kappa_2, \ldots]$ with κ_i in degree $2i$ corresponding to the Mumford-Morita-Miller class of the same name (see [**12**, 2.1]). Combining these two facts gives the Mumford conjecture, namely that

$$H^*(\mathcal{M}_g; \mathbb{Q}) \cong H^*(\Gamma_g, \mathbb{Q}) \cong_{(*)} \mathbb{Q}[\kappa_1, \kappa_2, \ldots]$$

where the isomorphism $(*)$ is up to degree $\frac{2}{3}(g-1)$.

This type of theorem for the symmetric groups and braid groups were already proved in the early 70's:

THEOREM 1.5 (Symmetric groups). $H_*(\Sigma_\infty) \cong H_*(\Omega_0^\infty S^\infty)$

THEOREM 1.6 (Braid groups). $H_*(\beta_\infty) \cong H_*(\Omega_0^2 S^2)$

The first theorem is known as the *Barratt-Priddy theorem* [**31**]. Both theorems can be seen as special cases of the *approximation theorem*, which says that the map $C_n X \to \Omega^n \Sigma^n X$ is a group completion for C_n the little n–cubes monad. Here one needs to take $X = S^0$ and $n = \infty$ in the first case, and the same X but $n = 2$ in the second case. (See May [**24**, Thm 2.7] and [**8**, p.486 (15)], or Segal [**35**, Thm 1], for the approximation theorem—see also the work of Boardman-Vogt [**3**] and Barratt-Eccles [**2**]. See [**12**, Lec 4] or [**25**] for the group completion theorem.) The proof of the Madsen-Weiss Theorem presented in Galatius' lecture series [**12**] follows Galatius–Randal-Williams [**13**], which can be seen as a generalization of Segal's proof of the approximation theorem.

The Madsen-Weiss Theorem was generalized to other types of mapping class groups of surfaces (Non-orientable [**36**], framed, Spin and Pin mapping class groups [**33**]). Other examples are:

THEOREM 1.7 (Automorphisms of free groups [**11**]). $H_*(\mathrm{Aut}_\infty) \cong H_*(\Omega_0^\infty S^\infty)$

Here $\mathrm{Aut}_\infty = \bigcup_{n\geq 1} \mathrm{Aut}(F_n)$ with F_n the free group on n letters.

THEOREM 1.8 (Handlebody groups [**18**]). $H_*(\mathcal{H}_\infty) \cong H_*(\Omega_0^\infty \Sigma^\infty(BSO(3)_+))$

Here $\mathcal{H}_\infty = \bigcup_{g\geq 1} \mathcal{H}_{g,1}$ with $\mathcal{H}_{g,1} = \pi_0 \,\mathrm{Diff}(H_g \text{ rel } D^2)$ the mapping class group of a handlebody H_g of genus g fixing a disc in the boundary of H_g.

All the above examples are computations of the homology of the stable group of a family of groups satisfying homological stability, so they all can be restated as computations of the homology of the unstable groups in a range of degrees. Stability is however not a necessary ingredient of a "Madsen-Weiss theorem"—it is rather an interpretational tool.

6. Exercises

1) Show that S^2 and T^2 do not satisfy the Earle-Eells theorem, i.e. that $\mathrm{Diff}(S^2)$ and $\mathrm{Diff}(T^2)$ do not have contractible components.
2) Give a definition of the group $\Gamma_\infty = \bigcup_{g\geq 1} \Gamma_{g,1}$ in terms of an infinite genus surface S_∞.
3) Let $\Gamma_g^1 = \pi_0 \,\mathrm{Diff}(S_g^1)$ denote the mapping class group of a once punctured genus g surface S_g^1. Use homological stability and a factorization of the map $\Gamma_{g,1} \to \Gamma_{g,0}$ to show injectivity of the map $H_i(\Gamma_{g,1}) \to H_i(\Gamma_g^1)$ in a range.

LECTURE 2

Homological stability: geometric ingredients

In this lecture, we briefly describe a general strategy for proving homological stability for families of groups and then give the main geometric ingredients needed for the case of the mapping class group of surfaces, with an emphasis on the case of surfaces with boundaries. We follow Randal-Williams [**32**] and the survey [**37**], which contains further details.

1. General strategy of proof

A *simplicial complex* $X = (X_0, \mathcal{F})$ is a set of vertices X_0 together with a collection $\mathcal{F}$ of subsets of X_0 closed under taking subsets and containing all the singletons. The subsets of cardinality $p+1$ are called the p–*simplices* of X.

To a simplicial complex X, one can associate its *realization* $|X|$ which has a topological p–simplex Δ^p for each p–simplex of X.

A space or simplicial complex X is called n–*connected* if $\pi_i(X) = 0$ for all $i \leq n$ (where $\pi_i(X) := \pi_i(|X|)$ if X is a simplicial complex). Note that, by Hurewicz theorem, a simply connected space X is n–connected, $n \geq 2$, if and only if $H_*(X) = 0$ for $2 \leq * \leq n$.

Given a family of groups

$$G_1 \hookrightarrow G_2 \hookrightarrow \dots \hookrightarrow G_n \hookrightarrow \dots$$

we want to find a simplicial complex (or simplicial set) X_n for each n such that

- G_n acts on X_n,
- the stabilizer $\mathrm{Stab}(\sigma_p) \cong G_{n-p-1}$ for any p–simplex σ_p,
- the action is as transitive as possible,
- X_n is highly connected.

There is then a spectral sequence for the action of G_n on X_n which decomposes the homology of G_n in terms of the homology of the stabilizers. As these are assumed to be previous groups in the sequence, this spectral sequence allows an inductive argument. (See Lecture 3 for more details.)

2. The case of the mapping class group of surfaces

To prove homological stability for the groups Γ_g, we will need to consider surfaces with any number of boundary components. Let $S = S_{g,r}$ be a surface of genus g with r boundary components. We will consider 3 maps

$$\begin{aligned}
\alpha_g : \Gamma(S_{g,r}) &\to \Gamma(S_{g+1,r-1}) \\
\beta_g : \Gamma(S_{g,r}) &\to \Gamma(S_{g,r+1}) \\
\delta_g : \Gamma(S_{g,r}) &\to \Gamma(S_{g,r-1})
\end{aligned}$$

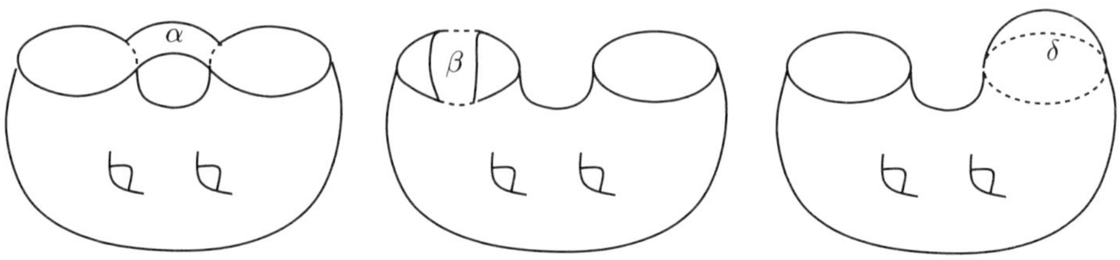

FIGURE 1. The maps α, β and δ

induced by gluing a strip which identifies arcs lying in different (for α) or the same (for β) boundary component(s) of S, and by gluing a disc (for δ). (See Figure 1.)

The proof of homological stability for mapping class groups presented here involves three simplicial spaces: an arc complex for each of α and β, and a disc space for δ. In the remainder of the lecture, we will define the two arc complexes and study their properties. These are the complexes $\mathcal{O}^2$ and $\mathcal{O}^1$ defined below.

3. The ordered arc complex

We will work here with collections of disjointly embedded arcs in a surface. We say that a collection of arcs $\langle a_0, \ldots, a_p \rangle$ is *non-separating* if its complement $S \backslash (a_0 \cup \cdots \cup a_p)$ is connected.

Given a surface S with points b_0, b_1 in its boundary, define $\mathcal{O}(S, b_0, b_1)$ to be the simplicial complex with

vertices =: isotopy classes of non-separating arcs with boundary $\{b_0, b_1\}$
p–simplices =: non-separating collections of $p+1$ distinct isotopy classes of arcs $\langle a_0, \ldots, a_p \rangle$ disjointly embeddable (away from b_0, b_1) in such a way that the anticlockwise ordering of $a_0, \ldots, a_p$ at b_0 agrees with the clockwise ordering at b_1.

Up to isomorphism, there are two such complexes:

$\mathcal{O}^1(S)$ =: $\mathcal{O}(S, b_0, b_1)$ with b_0, b_1 on the same boundary component,
$\mathcal{O}^2(S)$ =: $\mathcal{O}(S, b_0, b_1)$ with b_0, b_1 on different boundary components.

The mapping class group $\Gamma(S_{g,r}) = \Gamma_{g,r}$ acts on $\mathcal{O}^1(S_{g,r})$ and $\mathcal{O}^2(S_{g,r})$. We give now the properties of this action that are key for us.

PROPERTY 1. *$\Gamma(S)$ acts transitively on p–simplices of $\mathcal{O}^i(S)$ for each p.*

PROOF SKETCH. (See the proof of [**37**, Prop 2.2 (1)] for more details.) Consider the case $\mathcal{O}^1(S)$ and let $\sigma_p = \langle a_0, \ldots, a_p \rangle$ be a p–simplex. Then $S \backslash \sigma_p = S \backslash \{a_0, \ldots, a_p\}$ is a connected surface with $r+p+1$ boundary components (see Figure 2) and Euler characteristic $\chi(S)+p+1$. Thus it must have genus $g-p-1$ and any two $S \backslash \sigma_p$ and $S \backslash \sigma'_p$ are diffeomorphic. Moreover, a diffeomorphism $S \backslash \sigma_p \to S \backslash \sigma'_p$ can be chosen so that it identifies the arcs of σ_p with those of σ'_p and thus glues back to a diffeomorphism of S mapping σ_p to σ'_p. □

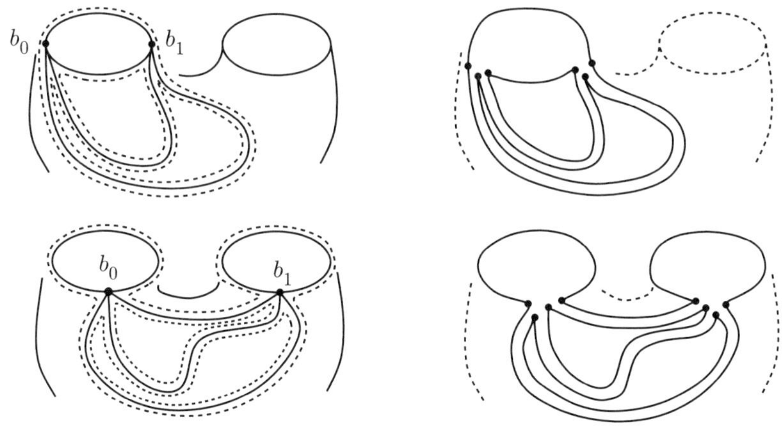

FIGURE 2. Cutting along simplices of $\mathcal{O}^1$ and $\mathcal{O}^2$

PROPERTY 2. *There is an isomorphism $Stab_{\mathcal{O}}(\sigma_p) \cong \Gamma(S\backslash\sigma_p)$, i.e.*

$$Stab_{\mathcal{O}^1(S_{g,r})}(\sigma_p) \cong \Gamma_{g-p-1,r+p+1} \quad \text{and} \quad Stab_{\mathcal{O}^2(S_{g,r})}(\sigma_p) \cong \Gamma_{g-p,r+p-1}.$$

The proof of this second property will rely on two useful results of manifold topology which we first state.

For M, N two manifolds, let $\mathrm{Emb}(N, M)$ denote the space of embeddings of M into N (with the C^∞ topology).

THEOREM 2.1 (*Fibering Theorem*, Palais and Cerf [**30, 6**]). *Let M, N be manifolds and V a compact submanifold of M. Then the restriction map*

$$\mathrm{Emb}(M, N) \longrightarrow \mathrm{Emb}(V, N)$$

is a locally trivial fibration.

For $M = N$, one gets a fibration

$$\mathrm{Diff}(M \operatorname{rel} V) \to \mathrm{Diff}(M) \to \mathrm{Emb}(V, M)$$

See [**6**, Chap. II 3.4.2], or the exercises for a sketch of proof.

This theorem has the following very useful corollary (see [**30**, Sec. 5]):

THEOREM 2.2 (*Isotopy extension theorem*). *For a compact submanifold $V \subset M$, and any path $\gamma : I \to \mathrm{Emb}(V, M)$ with $\gamma(0)$ the inclusion, there exits a path $\hat{\gamma} : I \to \mathrm{Diff}(M)$ with $\hat{\gamma}(0) = id$ and $\hat{\gamma}(t)|_V = \gamma(t)$.*

Note in particular that the theorem produces a diffeomorphism of M (namely $\hat{\gamma}(1)$) which takes V to its isotoped image $\gamma(1)(V)$ in M.

SKETCH OF PROOF OF PROPERTY 2. (See the proof of [**37**, Prop 2.2 (2)] for more details.) Suppose $\sigma_p = \langle a_0, \dots, a_p \rangle$. There is a map

$$\Gamma(S\backslash\sigma_p) \to \mathrm{Stab}_{\mathcal{O}}(\sigma_p)$$

induced by gluing the surface back together along the arcs.

Surjectivity: Consider $\phi \in \mathrm{Stab}_{\mathcal{O}}(\sigma_p)$. So $\phi(a_i)$ is isotopic to $a_{\theta(i)}$ for all i for some permutation $\theta \in \Sigma_{n+1}$. Applying the isotopy extension theorem to the isotopy

$\phi(a_0) \simeq a_{\theta(0)}$ gives a diffeomorphism ψ_0 of S with $\psi_0 \simeq id$ and $\psi_0(\phi(a_0)) = a_{\theta(0)}$. Hence we can replace ϕ by the isotopic diffeomorphism $\phi_0 = \psi_0 \circ \phi$ which satisfies that $\phi_0(a_0) = a_{\theta(0)}$. Proceed in the same way with the other arcs, one by one, away from the arcs already dealt with. Hence we can replace ϕ with a diffeomorphism that fixes the arcs, possibly up to a permutation. But the permutation must be trivial as ϕ fixes the boundary. Then ϕ can be reinterpreted as a diffeomorphism of $S\backslash\sigma_p$.

Injectivity: Suppose $p = 0$ for simplicity. We would like to show that the map $\Gamma(S\backslash I) \to \Gamma(S)$ is injective for any (non-separating) arc I. A relative version of Theorem 2.1 above gives a fibration

$$\mathrm{Diff}(S \text{ rel } \partial S \cup I) \to \mathrm{Diff}(S \text{ rel } \partial S) \to \mathrm{Emb}^{\partial}(I, S)$$

where $\mathrm{Emb}^{\partial}(I, S)$ denotes the space of embeddings of an arc I in S with ∂I mapping to chosen points $A, B \in \partial S$. By [**15**, Thm 5], each component of $\mathrm{Emb}^{\partial}(I, S)$ is contractible. The result then follows from looking at the long exact sequence of homotopy groups of the fibration for the component of the non-separating arcs. □

Note that gluing strips identifying arcs of the boundary induce maps (still denoted α and β)

$$\alpha : \mathcal{O}^2(S_{g,r}) \to \mathcal{O}^1(S_{g+1,r-1}) \quad \text{and} \quad \beta : \mathcal{O}^1(S_{g,r}) \to \mathcal{O}^2(S_{g,r+1})$$

(see Figure 3).

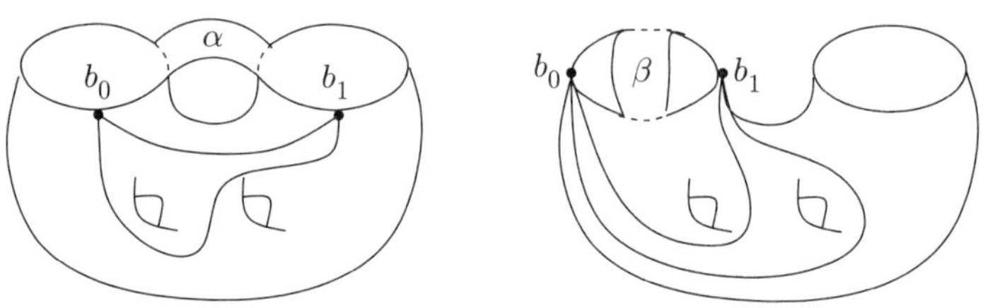

FIGURE 3. The maps α, β on the complexes of arcs

PROPERTY 3. *The map α on the complexes induces β on stabilizers and vice versa.*

(See [**37**, Prop 2.3] for a more detailed formulation.)

PROOF SKETCH. (See the proof of [**37**, Prop 2.3] for more details.) Patching together Figure 2 and Figure 3 shows how the strips defining α and β glue on the boundary components of the surface cut along the arcs of a simplex σ: in the case of α, the strip is glued on the cut surface to a unique boundary component of $S\backslash\sigma$, that is it induces β, while for β it is glued to two different boundaries of $S\backslash\sigma$, that is it induces α. □

PROPERTY 4. *Let $S = S_{g,r}$ and let $S_\alpha \cong S_{g+1,r-1}$ and $S_\beta \cong S_{g,r+1}$ denote S union a strip glued via α and β respectively as in Figure 3. Then for any vertex σ_0 of $\mathcal{O}^i(S)$, there are curves c_α, c_β (given in Figure 4) in S_α and S_β such that* *conjugation by Dehn twists t_{c_α} and t_{c_β} along these curves fits into commutative diagrams*

$$\begin{array}{ccc} St_{\mathcal{O}^2}(\sigma_0) & \longrightarrow & St_{\mathcal{O}^1}(\alpha(\sigma_0)) \\ \downarrow & \overset{t_{c_\alpha}}{\swarrow} & \downarrow \\ \Gamma(S) & \xrightarrow{\ \alpha\ } & \Gamma(S_\alpha) \end{array} \qquad \begin{array}{ccc} St_{\mathcal{O}^1}(\sigma_0) & \longrightarrow & St_{\mathcal{O}^2}(\beta(\sigma_0)) \\ \downarrow & \overset{t_{c_\beta}}{\swarrow} & \downarrow \\ \Gamma(S) & \xrightarrow{\ \beta\ } & \Gamma(S_\beta) \end{array}$$

i.e. there are conjugations $St_{\mathcal{O}^1}(\alpha(\sigma_0)) \sim_{t_{c_\alpha}} \alpha(\Gamma(S))$ in $\Gamma(S_\alpha)$ relative to $\alpha(St_{\mathcal{O}^2}(\sigma_0))$, and $St_{\mathcal{O}^2}(\beta(\sigma_0)) \sim_{t_{c_\beta}} \beta(\Gamma(S))$ in $\Gamma(S_\beta)$ relative to $\beta(St_{\mathcal{O}^1}(\sigma_0))$.

Note that it follows from the existence of these conjugation that the maps $\alpha\colon \Gamma(S) \to \Gamma(S_\alpha)$ and $\beta\colon \Gamma(S) \to \Gamma(S_\beta)$ are injective, as we already know by Property 2 that the stabilizers of vertices are abstractly isomorphic to $\Gamma(S)$.

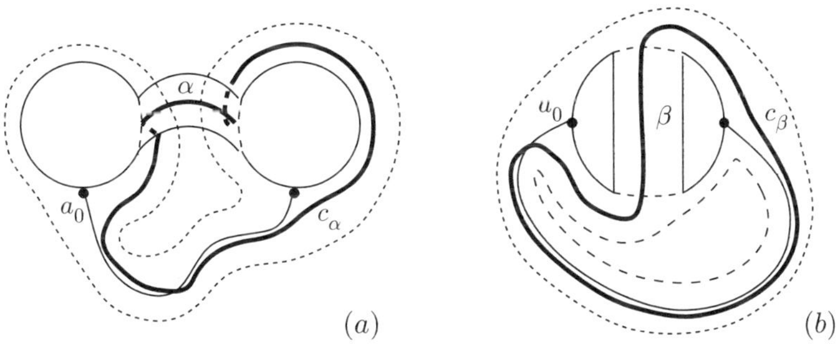

FIGURE 4. The curves c_α and c_β of Property 4 for $\sigma_0 = \langle a_0 \rangle$

PROOF SKETCH. (See the proof of [**37**, Prop 2.4] for more details.) Let $\sigma_0 = \langle a_0 \rangle$ be as in Figure 4. Check that the left Dehn twist along c_α (resp. c_β) takes a_0 to an arc whose stabilizer is $\Gamma(S)$. (Hint: to identify the stabilizer with $\Gamma(S)$, thicken a neighborhood of ∂S union the arc.) □

PROPERTY 5. *$\mathcal{O}^i(S_{g,r})$ is $(g-2)$-connected.*

The proof of this last property is the topic of Lecture 4.

4. Curve complexes and disc spaces

When the surface has no boundary components, one cannot work with arcs. Harer originally worked with embedded curves instead of arcs. These are not quite as well-behaved as arcs (see the exercises below). Randal-Williams introduced instead a space of embedded discs which gives a slightly better stability range for closing the last boundary component, but does require working with a semi-simplicial space build out of spaces of embedded discs, instead of a simplicial complex as we have done so far, because isotopy classes of embedded discs are not so interesting... (See [**32**, Sect. 10] or [**37**, Sect. 5].)

5. Exercises

1) Define the n–simplex Δ^n and its boundary $\partial\Delta^n$ as simplicial complexes.
2) Show that $\mathcal{O}(S, b_0, b_1)$ is indeed a simplicial complex.
3) Let S_g be a closed surface of genus g and let $\mathcal{C}_0(S_g)$ be the simplicial complex whose p–simplices are non-separating collections $\{[C_0], \dots, [C_p]\}$ of isotopy classes of disjoint embedded *curves* in S_g, i.e. embeddings $S^1 \hookrightarrow S_g$, for which there exist representatives whose images $\{C_1, \dots, C_p\}$ satisfy
   - $C_i \cap C_j = \emptyset$ whenever $i \neq j$,
   - the complement $S \setminus (C_0 \cup \dots \cup C_p)$ is connected.

   (a) Construct a natural action of $\Gamma_g = \Gamma(S_g)$ on $\mathcal{C}_0(S_g)$.
   (b) Show that this action is transitive on p–simplices for each $p \geq 0$. (Hint: as in the proof of Property 1 above, first use the classification of surfaces to prove that the complement of two such collections of $p+1$ circles are diffeomorphic.)
   (c) Construct a map from $\Gamma_{g-1,2}$ to the stabilizer of a vertex of $\mathcal{C}_0(S_g)$ and prove that it is surjective using the isotopy extension theorem as in the proof of Property 2. Is it injective?
4) Complete the proof of Property 4.
5) Let M and N be smooth manifolds. Denote by $C^0(M, N)$ the set of continuous maps from M to N, and denote by $\mathrm{Emb}^0(M, N) \subset C^0(M, N)$ the subset consisting of topological embeddings. Inductively, for $k > 0$ denote by $C^k(M, N) \subset C^{k-1}(M, N)$ the subset consisting of differentiable maps $f : M \to N$ for which the induced map on tangent spaces Tf is in $C^{k-1}(TM, TN)$. We topologize the set $C^k(M, N)$ inductively: we use the compact-open topology on $C^0(M, N)$, and we note that $D : C^k(M, N) \to C^{k-1}(TM, TN)$ is an inclusion, so we give $C^k(M, N)$ the subspace topology. We let $C^\infty(M, N)$ denote the inverse limit of the $C^k(M, N)$.

   Denote by $\mathrm{Emb}^k(M, N) \subset C^k(M, N)$ the subspace consisting of topological embeddings e for which Te is in $\mathrm{Emb}^{k-1}(TM, TN)$, and write the inverse limit as $\mathrm{Emb}(M, N)$. Denote by $\mathrm{Diff}(M) \subset \mathrm{Emb}(M, M)$ the subspace consisting of those invertible maps ϕ for which $\phi^{-1} \in \mathrm{Diff}(M)$.
   (a) Prove that a sequence of maps $f_n \in C^1(\mathbb{R}, \mathbb{R})$ converges if and only if the sequences $f_n \in C^0(\mathbb{R}, \mathbb{R})$ and $f'_n \in C^0(\mathbb{R}, \mathbb{R})$ converge.
   (b) Prove that this inclusion $\mathrm{Emb}^k([0,1], \mathbb{R}) \subset C^k([0,1], \mathbb{R})$ is open when $k = 1$ but not when $k = 0$.
   (c) (difficult) In this exercise we will prove that if N is a compact submanifold of M, then the restriction map

   $$j : \mathrm{Emb}(M, \mathbb{R}^n) \longrightarrow \mathrm{Emb}(N, \mathbb{R}^n)$$

   is locally trivial fibration (i.e. a fibre bundle with structure group the full homeomorphism group of the fibre). This was first proved by Palais and Cerf, but we follow Lima [**22**].
      (i) Given a $f \in \mathrm{Emb}(N, \mathbb{R}^n)$, show there is a neighbourhood U of f and a map $\xi : U \to \mathrm{Diff}(\mathbb{R}^n)$ such that $\xi(g) \circ f = g$. [Hint: note that $\mathrm{Diff}(\mathbb{R}^n)$ is an open subset of $C^\infty(\mathbb{R}^n, \mathbb{R}^n)$.]
      (ii) Hence construct a trivialisation of j over U.

REMARK 2.3. Note that this implies it is in particular a Serre fibration, and hence the existence of a long exact sequence on homotopy groups

$$\cdots \to \pi_{n+1}F \to \pi_n \operatorname{Emb}(M, \mathbb{R}^n) \to \pi_n \operatorname{Emb}(N, \mathbb{R}^n) \to \pi_n F \to \cdots,$$

where F is the fiber of the restriction map over some base point $f \in \operatorname{Emb}(N, \mathbb{R}^n)$.

LECTURE 3

Homological stability: the spectral sequence argument

In this lecture, we present a spectral sequence argument (the one used by Randal-Williams in [**32**]) which allows us to prove homological stability for the mapping class groups of surfaces in the range stated in Lecture 1. We will consider the two spectral sequences associated to a double chain complex built from a pair of groups acting on a pair of spaces. We will then use all the geometric properties presented in Lecture 2 to analyze these spectral sequences.

1. Double complexes associated to actions on simplicial complexes

To X a simplicial complex, one can associate a chain complex $(\tilde{C}_*, \partial)$ (its augmented cellular complex) which computes the reduce homology of its realization $|X|$. It has

- $\tilde{C}_p(X) = \mathbb{Z}X_p$, the free module on the set of p–simplices
- $\tilde{C}_{-1}(X) = \mathbb{Z}$

with boundary maps coming from the face maps and the augmentation.

We are interested here in simplicial complexes X admitting a simplicial G–action for some group G. For such, one can construct a double complex

$$E_*G \otimes_G \tilde{C}_*(X)$$

where $\cdots \to E_qG \to \cdots \to E_0G \to \mathbb{Z} \to 0$ is a free resolution of $\mathbb{Z}$ over $\mathbb{Z}G$. This is the basic double complex commonly used to prove homological stability results. We will use here a relative version of it, which we construct now.

Suppose Y in addition is a simplicial complex with a simplicial H–action and $f : X \to Y$ a simplicial map equivariant with respect to a map $G \to H$. Then we get a map of double complexes

$$F : E_*G \otimes_G \tilde{C}_*(X) \longrightarrow E_*H \otimes_H \tilde{C}_*(Y)$$

(The two examples of interest to us are the maps $\alpha : \mathcal{O}^2 \to \mathcal{O}^1$ and $\beta : \mathcal{O}^1 \to \mathcal{O}^2$ of Lecture 2 and their companion maps on the mapping class groups.) We will use the double complex

$$C_{p,q} = (E_{q-1}G \otimes_G \tilde{C}_p(X)) \oplus (E_qH \otimes_H \tilde{C}_p(Y))$$

with horizontal differential $(a\otimes b, a'\otimes b') \mapsto (a\otimes \partial b, a'\otimes \partial b')$ and vertical differential $(a \otimes b, a' \otimes b') \mapsto (da \otimes b, da' \otimes b' + F(a \otimes b))$, i.e. for each p we take the mapping cone of F in the q–direction.

The horizontal and vertical filtrations of such a double complex give two spectral sequences, both converging to the homology of the total complex. We now analyze these two spectral sequences.

2. The spectral sequence associated to the horizontal filtration

As $E_{q-1}G$ and E_qH are free G– (resp. H–)modules and the horizontal differential is that of $\tilde{C}(X)$ and $\tilde{C}(Y)$, taking first the homology in the p–direction computes copies of the reduced homology of X plus that of Y.

In particular, if X is $(c-1)$–connected and Y is c–connected, the E^1–term of the horizontal spectral sequence, which is the homology of $C_{p,q}$ with respect to the horizontal differential, is 0 in the range $p+q \leq c$ (noting that $\tilde{C}_p(X)$ only contributes to $C_{p,q}$ when $q>0$).

It follows that the other spectral sequence, obtained using the vertical filtration instead, also converges to 0 in the range $p+q \leq c$.

3. The spectral sequence associated to the vertical filtration

For each p, the module $\tilde{C}_p(X) = \mathbb{Z}X_p$ is a G–module, which decomposes as a sum of modules corresponding to the orbits of the G–action on X_p. (We define here $X_{-1} = \{*\}$ with the trivial action.) Given an orbit $o \in O(X_p)$, the set of orbits of X_p, we let $\mathrm{Stab}(o) \leq G$ denote the stabilizer subgroup of some chosen simplex σ_p in the orbit o. Assuming that the stabilizer of a simplex fixes the simplex pointwise, we can rewrite the G–module $\tilde{C}_p(X)$ as

$$\tilde{C}_p(X) \cong \bigoplus_{o\in O(X_p)} G \otimes_{\mathrm{Stab}(o)} \mathbb{Z}$$

(See e.g. [**4**, Sect. 1] for a description of the case when the stabilizers do not fix the simplices pointwise.) The chain complex $E_*(G) \otimes_G \tilde{C}_p(X)$, where p is now fixed, computes the homology of G with coefficients in that module. (This is the definition of the homology of a group with twisted coefficients.)

We will use a relative version (left as an exercise) of the following well-known lemma (see e.g. [**5**, III 6.2]):

LEMMA 3.1 (*Shapiro's lemma*). *Let $H < G$ be groups and M be an H–module, with $G \otimes_H M$ the induced G–module. Then*

$$H_*(G, \mathbb{Z}G \otimes_H M) \cong H_*(H, M)$$

Hence for any G–simplicial complex X as above, we have, for each p, that

$$H_*(E_*G \otimes \tilde{C}_p(X)) \cong \bigoplus_{o\in O(X_p)} H_*(\mathrm{Stab}(o)).$$

The E^1–term of the vertical spectral sequence is the homology of the double complex $C_{p,q} = (E_{q-1}G \otimes_G \tilde{C}_p(X)) \oplus (E_qH \otimes_H \tilde{C}_p(Y))$ with respect to the vertical differential. This is the relative homology group

$$E^1_{p,q} = H_q\big(E_*H \otimes_H \tilde{C}_p(Y), E_*G \otimes_G \tilde{C}_p(X)\big)$$

as the columns of $C_{p,q}$ are the mapping cones of the map F (with p fixed).

Now if the actions of G and H are transitive on X and Y (which is the case we are interested in), a relative version of Shapiro's lemma identifies the E^1–term of the vertical spectral sequence with

$$E^1_{p,q} = H_q(\mathrm{Stab}_Y(\sigma_p), \mathrm{Stab}_X(\sigma_p))$$

where $\mathrm{Stab}_X(\sigma_p)$ and $\mathrm{Stab}_Y(\sigma_p)$ are the stabilizers in X and Y of some chosen p–simplex σ_p of X and its image in Y. Note that this formulation in the case $p = -1$ gives

$$E^1_{-1,q} = H_q(H, G)$$

4. The proof of stability for surfaces with boundaries

Recall from the previous lecture the maps

$$\alpha_g \colon \Gamma(S_{g,r+1}) \to \Gamma(S_{g+1,r}) \quad \text{and} \quad \beta_g \colon \Gamma(S_{g,r}) \to \Gamma(S_{g,r+1}).$$

Denote by $H_*(\alpha_g)$ the relative homology group $H_*(\Gamma_{g+1,r}, \Gamma_{g,r+1}; \mathbb{Z})$ corresponding to the map α_g, and $H_*(\beta_g)$ the relative homology group $H_*(\Gamma_{g,r+1}, \Gamma_{g,r}; \mathbb{Z})$ corresponding to β_g. (The number of boundaries r here will not play a role here.) Harer's improved stability theorem (Theorem 1.3) can be restated as follows:

THEOREM 3.2. *(1) $H_i(\alpha_g) = 0$ for $i \leq \frac{2g+1}{3}$ and (2) $H_i(\beta_g) = 0$ for $i \leq \frac{2g}{3}$.*

The proof of this theorem uses the spectral sequences just described in the case of the maps α and β of Lecture 2. The argument will need Properties 1–5 of Lecture 2.

PROOF. We prove the theorem by induction on g. To start the induction, note that statements (1) for genus 0 and (2) for genus 0,1 are trivially true as they are just concerned with H_0. Let (1_g) and (2_g) denote the truth of (1) and (2) in the theorem for genus g. The induction will go in two steps:

Step 1: For $g \geq 1$, $(2_{\leq g})$ implies (1_g).
Step 2: For $g \geq 2$, $(1_{<g})$ and (2_{g-1}) imply (2_g).

Step 1: We consider the spectral sequence described above for the actions of $G = \Gamma_{g,r+1}$ on $X = \mathcal{O}^2(S_{g,r+1})$ and of $H = \Gamma_{g+1,r}$ on $Y = \mathcal{O}^1(S_{g+1,r})$ with the homomorphism $\phi\colon G \to H$ and the map $f\colon X \to Y$ both induced by the map $\alpha\colon S_{g,r+1} \to S_{g+1,r}$ of Figure 3. These actions satisfy that the stabilizer of a simplex fixes the simplex pointwise as an orientation preserving diffeomorphism of the surface will preserve the order of the arcs defining the simplex as their endpoints. As the action is transitive in both cases (Property 1), we get, as explained above, that the vertical spectral sequence has the form $E^1_{p,q} = H_q(\mathrm{Stab}_Y(\alpha(\sigma_p)), \mathrm{Stab}_X(\sigma_p))$, with σ_p a chosen p–simplex of X and $\alpha(\sigma_p)$ its image in Y. That is, when $p = -1$, we have

$$E^1_{-1,q} = H_q(\Gamma_{g+1,r}, \Gamma_{g,r+1}) = H_q(\alpha_g)$$

which are the groups we are interested in. By Properties 2 and 3, the other groups are identified with

$$E^1_{p,q} = H_q(\beta_{g-p}) \quad \text{for} \ \ p \geq 0.$$

Hence we will be able to apply induction to these terms of the spectral sequence. We want to deduce that $E^1_{-1,q} = 0$ for $q \leq \frac{2g+1}{3}$. This will follow from the following three claims:

Claim 1: $E^\infty_{-1,q} = 0$ for $q \leq \frac{2g+1}{3}$.

Claim 2: The E^1–term is as in Figure 1, i.e. there are no possible sources of differentials to kill classes in $E^1_{-1,q}$ with $q \leq \frac{2g+1}{3}$, except possibly for $d^1\colon E^1_{0,q} \to E^1_{-1,q}$ when $q = \frac{2g+1}{3}$ (i.e. when the fraction is an integer).

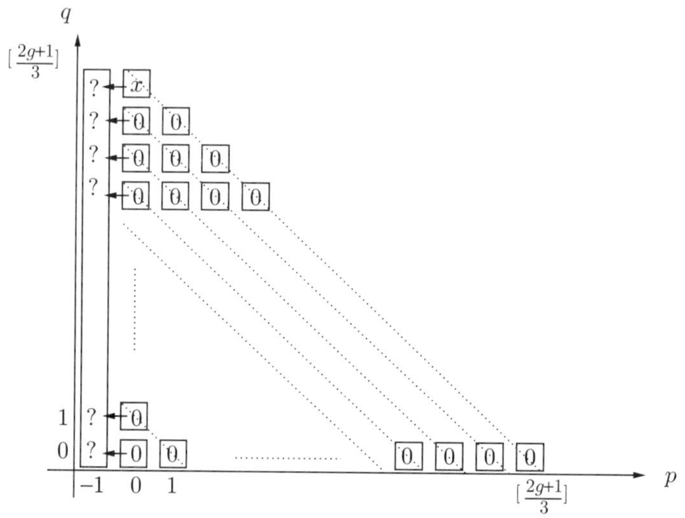

FIGURE 1. Spectral sequence for Step 1. The possible sources of differentials for the "?" are along the dotted diagonals.

Claim 3: The map $d^1\colon E^1_{0,q} \to E^1_{-1,q}$ is the 0-map.

Claims 1 and 2 imply immediately that $E^1_{-1,q} = 0$ for $q < \frac{2g+1}{3}$ as "it must die by E^∞" (Claim 1) and "nobody can kill it" (Claim 2). Claim 3 gives that this also holds when $q = \frac{2g+1}{3}$ as the only differential with a possibly non-trivial source is the zero map, and hence won't kill anything in the target.

Proof of Claim 1: By Property 5, X is $(g-2)$–connected and Y is $(g-1)$–connected. By the analysis of the horizontal spectral sequence above, and the fact that both spectral sequences converge to the same target, we get that $E^\infty_{p,q} = 0$ for $p+q \le g-1$. In particular, $E^\infty_{-1,q} = 0$ for $q \le g$, and $\frac{2g+1}{3} \le g$ when $g \ge 1$.

Proof of Claim 2: The sources of differentials to $E^1_{-1,q}$ are the terms $E^{p+1}_{p,q-p}$ for $p \ge 0$. As $E^1_{p,q} = H_q(\beta_{g-p})$ when $p \ge 0$, by induction we know that $E^1_{p,q} = 0$ when $q \le \frac{2(g-p)}{3} = \frac{2g-2p}{3}$ and $p \ge 0$. Hence $E^1_{p,q-p} = 0$ for $q \le \frac{2g+p}{3}$ and $p \ge 0$, i.e. they are all 0 for any $p \ge 0$ if $q \le \frac{2g}{3}$ or for $p \ge 1$ if $q = \frac{2g+1}{3}$.

Proof of Claim 3: The map $d^1\colon E^1_{0,q} \to E^1_{-1,q}$ is the map

$$H_q(\mathrm{Stab}_{\mathcal{O}^1}(\alpha(\sigma_0)), \mathrm{Stab}_{\mathcal{O}^2}(\sigma_0)) \to H_q(\Gamma(S_\alpha), \Gamma(S))$$

for σ_0 a vertex of $\mathcal{O}^2(S)$ with $S = S_{g,r+1}$ and $S_\alpha = S_{g+1,r}$ the source and targets of α. This map comes precisely from mapping the top row to the bottom row of the first square in Property 4. It is tempting to deduce immediately from Property 4 that the map is zero but it does require a little extra argument: From the diagram, it follows that the map factors as

$$H_q(\mathrm{Stab}_{\mathcal{O}^1}(\alpha(\sigma_0)), \mathrm{Stab}_{\mathcal{O}^2}(\sigma_0)) \xrightarrow{\partial} H_{q-1}(\mathrm{Stab}_{\mathcal{O}^2}(\sigma_0)) \to H_q(\Gamma(S_\alpha), \Gamma(S))$$

where the second map "crosses" with the Dehn twist t_{c_α}. (See Lemma 2.5 in [**37**].) The triviality of the map then follows from the fact that c_α can be conjugated to a curve in S fixing the support of $\mathrm{Stab}_{\mathcal{O}^2}(\sigma_0)$. (This uses that c_α is non-separating in the neighborhood pictured in Figure 4.)

For Step 2, the argument is essentially the same, though proving Claim 3 is a little more subtle because c_β is separating in the corresponding neighborhood in Figure 4. It is necessary to use induction an extra time here to finish the argument. (See the proof of [**37**, Thm 3.1 (2)] for the details.) □

5. Closing the boundaries

To prove that the map $\delta : \Gamma_{g,1} \to \Gamma_{g,0}$ also induces a homology isomorphism in a range, one uses a similar spectral sequence for the action of the mapping class groups on the disc semi-simplicial space of [**32**] (or the curve complex as in [**16**]) and compare the spectral sequence for each case (a comparison argument going back to [**20**]). See [**37**, Sect 5] for details.

6. Exercises

1) State and prove the relative version of Shapiro's lemma needed in the analysis of the vertical spectral sequence above.
2) Show that Theorem 3.2 implies the first part of Theorem 1.3 of Lecture 1.
3) This exercise sets up a way of approaching the proof of homological stability for the symmetric groups.
   (a) Let Δ^n denote the n–simplex. It can be thought of as a simplicial complex with $n+1$ vertices $\{0,\dots,n\}$, the set of p–simplices being the set of subsets of $\{0,\dots,n\}$ of cardinality $p+1$. Consider the action of the symmetric group Σ_{n+1} on Δ^n induced by permuting the vertices. Show that the action is transitive on the set of p–simplices for each p. What is the stabilizer of a vertex? of a p–simplex in general? (Note that the stabilizer of a simplex is the subgroup of symmetries that map the simplex to itself, as a set of vertices.)
   (b) Replace Δ^n in the above exercise by the semi-simplicial set (=simplicial set without degeneracies) X_{n+1} whose p–simplices are injective maps $\sigma : \{0,1,\dots,p\} \to \{0,\dots,n\}$. Is the action of Σ_{n+1} still transitive on the set of p–simplices for any p? What is the stabilizer of a p–simplex in this case?
   (c) For a semi-simplicial set $Y = Y_*$, let $||Y|| = \coprod_{p>0} \Delta^p \times Y_p/\sim$ denote its realization, where the equivalence relation $\sim$ is induced by the face relations. For X_1, X_2, X_3 as in (b), show that $||X_1|| = *$, $||X_2|| \cong S^1$ and that $||X_3||$ is simply-connected.
   (d) Consider the cellular chain complex $C_*(X_{n+1})$ with its induced action of Σ_{n+1}. Pick a free resolution $E_*\Sigma_{n+1}$ of $\mathbb{Z}$ considered as a $\mathbb{Z}[\Sigma_{n+1}]$–module with a trivial action of Σ_{n+1}. Now consider the double complex $C_{p,q} = C_p(X_{n+1}) \otimes_{\Sigma_{n+1}} E_q\Sigma_{n+1}$. Write down the E^1 terms of the associated spectral sequences in both filtrations.

REMARK 3.3. Given that X_n is $(n-2)$–connected (see [**34**, Prop 3.2] or [**21**]), one can use the above spectral sequences to prove Nakaoka's stability theorem: the map $H_q(\Sigma_n) \to H_q(\Sigma_{n+1})$ induced by the inclusion of groups is an isomorphism for $q \leq n/2$.

4) The action of the mapping class group Γ_g on the Teichmüller space $\mathcal{T}_g$ satisfies that the stabilizers of points are finite (most often trivial) groups (see e.g. [**10**, 12.1]). Let $C_*(\mathcal{T}_g, \mathbb{Q})$ denote the singular chain complex of $\mathcal{T}_g$ with rational coefficients. The action of the mapping class group gives rise, as above, to a double complex $C_*(\mathcal{T}_g, \mathbb{Q}) \otimes_{\Gamma_g} E_*\Gamma_g$, with $E_*\Gamma_g$ now a free resolution of $\mathbb{Q}$ as a trivial $\mathbb{Q}[\Gamma_g]$–module. Using the spectral sequences associated to the double complex, show that the coarse moduli space $\mathcal{T}_g/\Gamma_g$ is rationally a classifying space for Γ_g, i.e. that $H_*(\mathcal{T}_g/\Gamma_g, \mathbb{Q}) \cong H_*(B\Gamma_g, \mathbb{Q})$.

LECTURE 4

Homological stability: the connectivity argument

In this lecture, we give a sketch proof of the last ingredient of the proof of homological stability for mapping class groups of surfaces with boundaries, namely the fact that the complex $\mathcal{O}(S; b_0, b_1)$ of Lecture 2 is $(g-2)$–connected for any surface S of genus g. To close the surfaces, an analogous statement is needed for the disc space or curve complex, which is not presented here. We refer instead [**32**], [**37**], or [**20**] for that case.

1. Strategy for computing the connectivity of the ordered arc complex

We will prove that $\mathcal{O}(S; b_0, b_1)$ is highly connected by working our way through the following sequence of smaller and smaller simplicial complexes:

$$\mathcal{A}(S,\Delta) \overset{i_1}{\hookleftarrow} \mathcal{B}(S,\Delta_0,\Delta_1) \overset{i_2}{\hookleftarrow} \mathcal{B}_0(S,\Delta_0,\Delta_1) \overset{i_3}{\hookleftarrow} \mathcal{O}(S,b_0,b_1)$$

where $\Delta, \Delta_0, \Delta_1$ are sets of points in ∂S and

$\mathcal{A}(S,\Delta)$ is the simplicial complex whose vertices are isotopy classes of all non-trivial arcs in S with boundary in Δ. A p–simplex of $\mathcal{A}(S,\Delta)$ is a collection of $p+1$ distinct isotopy classes of arcs $\langle a_0, \dots, a_p \rangle$ representable by arcs with disjoint interiors.

$\mathcal{B}(S,\Delta_0,\Delta_1) \subset \mathcal{A}(S,\Delta_0 \cup \Delta_1)$ is the subcomplex of arcs having each one boundary point in Δ_0 and one in Δ_1.

$\mathcal{B}_0(S,\Delta_0,\Delta_1) \subset \mathcal{B}(S,\Delta_0,\Delta_1)$ is the subcomplex of non-separating collections.

$\mathcal{O}(S,b_0,b_1) \subset \mathcal{B}_0(S,\{b_0\},\{b_1\}$ is the ordered subcomplex (defined in Lecture 2, Section 3).

The largest complex $\mathcal{A}(S,\Delta)$ is contractible in most cases, and one slowly looses connectivity as one goes down to smaller and smaller subcomplexes.

The connectivity arguments used are of three types:

(1) *direct calculation* showing contractibility,
(2) exhibition of a complex as a *suspension* (or wedge of such) of a "previous" complex,
(3) *inductive deduction* from the connectivity of a larger complex.

The argument for the connectivity of $\mathcal{A}(S,\Delta)$ is a mix of type (1) and (2), the deduction along i_1 in the sequence is the most intricate argument and is a mix of the three types of arguments, while deduction along i_2 and i_3 are purely (and simpler) type (3) arguments.

We do here, to exemplify, a type (1) and a type (3) argument and refer to [**37**] for the complete proofs.

2. Contractibility of the full arc complex

THEOREM 4.1. [**16**, Thm 1.5] *Suppose S is not a disc or a cylinder with Δ included in a single boundary of S. Then $\mathcal{A}(S, \Delta)$ is contractible.*

SKETCH OF PROOF OF THE MAIN CASE. (Following Hatcher [**17**]. See also the proof of [**37**, Thm 4.1] for a complete proof.) We will prove the theorem under the extra assumption that S has at most one point of Δ in each boundary component. Reducing to that case requires an extra (type (2)) argument.

Non-emptiness: Clear if $|\Delta| > 1$ by choosing any arc in the surface between two distinct points of Δ–such an arc is always non-trivial as the points lie on different boundary components. For $|\Delta| = 1$, we have assumed that S is not a disc or a cylinder, i.e. S has non-zero genus or at least three boundary components. In both cases, there are non-trivial arcs.

In a simplicial complex X, the *star* of a simplex σ is the union of all the simplices τ of X such that $\sigma \cup \tau$ is again a simplex of X. This is a contractible subcomplex.

Contraction: As $\mathcal{A}(S, \Delta)$ is non-empty, we can choose an arc $a \in \mathcal{A}(S, \Delta)$. We will contract the complex to the star of a.

Let σ be a simplex of $\mathcal{A}(S, \Delta)$. It is in the star of a precisely if the interior of the arcs of σ do not intersect a. If they do intersect a, the idea is to modify σ by cutting the arcs one by one at the intersection points with a, replacing each time an arc by one or two arcs with fewer intersection points with a. This is illustrated by the following figure:

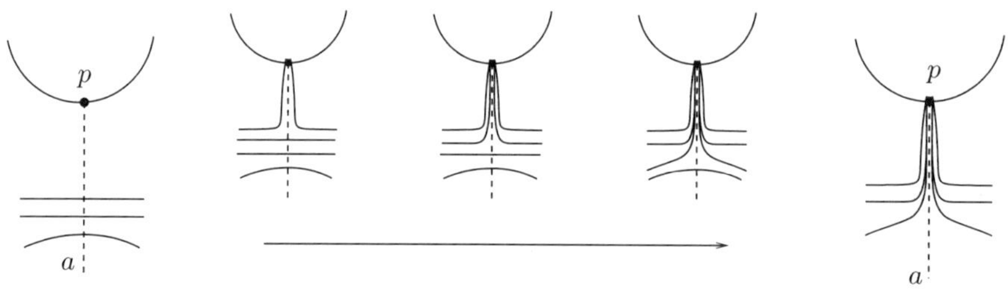

FIGURE 1. Retraction of $\mathcal{A}(S, \Delta)$ in the case of 3 crossings with a

More precisely, if σ has k intersection points with a, we produce a sequence of consecutive simplices $r_1(\sigma), \ldots, r_k(\sigma)$ such that σ is a face of $r_1(\sigma)$ while a face of $r_k(\sigma)$ lies in the star of a. A continuous retraction of X can be defined from there using the barycentric coordinates (thought of as weights on the arcs of σ) to go through this sequence of simplices at a speed and with a weight on the arcs depending on those coordinates.

Note that, to be well-defined, the above argument requires that

- the intersection $\sigma \cap a$ is suitably independent of the chosen representative of σ
- the new arcs created during the deformation contain each time at least one non-trivial arc.

The first issue is addressed by choosing representative with minimal intersection with a, and the second is addressed by the additional assumption we worked with, namely that there is at most one point of Δ in each boundary component. □

3. Deducing connectivity of smaller complexes

The connectivity of the subcomplex $\mathcal{B}(S, \Delta_0, \Delta_1)$ of arcs between two subsets Δ_0 and Δ_1 of Δ is deduced from that of $\mathcal{A}(S, \Delta)$ by a long argument... To be able to state the connectivity bound, we need a couple of definitions: Disjoint sets $\Delta_0, \Delta_1 \subset \partial S$ define a decomposition of ∂S into vertices (the points of $\Delta_0 \cup \Delta_1$), edges between the vertices, and circles without vertices. We say that an edge is *pure* if both its endpoints are in the same set, Δ_0 or Δ_1. We say that an edge is *impure* otherwise. Note that the number of impure edges is always even.

THEOREM 4.2 ([**16**, Thm 1.6]). *The complex $\mathcal{B}(S, \Delta_0, \Delta_1)$ is $(4g+r+r'+l+m-6)$–connected, where g is the genus of S, r its number of boundary components, r' the number of components of ∂S containing points of $\Delta_0 \cup \Delta_1$, l is half the number of impure edges and m is the number of pure edges.*

See [**37**, Thm 4.3] for a detailed proof.

We will now use the connectivity of $\mathcal{B}(S, \Delta_0, \Delta_1)$ to deduce that of the subcomplex $\mathcal{B}_0(S, \Delta_0, \Delta_1)$ of non-separating simplices.

The *join* $X * Y$ of two simplicial complexes X and Y is the simplicial complex with vertices $X_0 \sqcup Y_0$ and a $(p+q+1)$–simplex $\sigma_X * \sigma_Y = \langle x_0, \ldots, x_p, y_0, \ldots, y_q \rangle$ for each p–simplex $\sigma_X = \langle x_0, \ldots, x_p \rangle$ of X and q–simplex $\sigma_Y = \langle y_0, \ldots, y_q \rangle$ of Y. Note that $|X * Y| = |X| * |Y|$, i.e. the realization of the join complex is the (topological) join of the realization of the two complexes. This follows from the fact that it is true for each pair of simplices.

Recall that a space (or simplicial complex) X is called *n–connected* if $\pi_i(X) = 0$ for all $i \leq n$. For $n = -1$, we use the convention that (-1)–*connected* means non-empty. (For $n \leq -2$, n–connected is a void property.)

The following proposition tells us how to compute the connectivity of a join in terms of the connectivity of the pieces.

PROPOSITION 4.3 ([**27**, Lem 2.3]). *Consider the join $X = X_1 * \cdots * X_k$ of k non-empty spaces. If each X_i is n_i–connected, then X is $\left(\left(\sum_{i=1}^{k}(n_i+2)\right)-2\right)$–connected.*

THEOREM 4.4 ([**16**, Thm 1.4]). *If Δ_0, Δ_1 are two disjoint non-empty sets of points in ∂S, then the complex $\mathcal{B}_0(S, \Delta_0, \Delta_1)$ is $(2g+r'-3)$–connected, for g and r' as above.*

PROOF. We prove the theorem by induction on the lexicographically ordered triple (g, r, q), where $r \geq r'$ is the number of components of ∂S and $q = |\Delta_0 \cup \Delta_1| \geq 2$. To start the induction, note that the theorem is true when $g = 0$ and $r' < 2$ for any $r > r'$ and any q, and more generally that the complex is non-empty whenever $r' \geq 2$ or $g \geq 1$.

So fix (S, Δ_0, Δ_1) satisfying $(g, r, q) \geq (0, 3, 2)$. Then $2g+r'-3 \leq 4g+r+r'+l+m-6$. Indeed, $r \geq 1$ and $l+m \geq 1$. Moreover we assumed that either $r \geq 3$ or $g \geq 1$.

Let $k \leq 2g + r' - 3$ and consider a map $f\colon S^k \to \mathcal{B}_0(S, \Delta_0, \Delta_1)$, which we may assume to be simplicial for some PL triangulation of S^k (see [**37**, Sect 6]). This map can be extended to a simplicial map $\hat{f}\colon D^{k+1} \to \mathcal{B}(S, \Delta_0, \Delta_1)$ by Theorem 4.2 and the above calculation, for a PL triangulation of D^{k+1} extending that of S^k. We call a simplex σ of D^{k+1} *regular bad* if $\hat{f}(\sigma) = \langle a_0, \ldots, a_p\rangle$ and each a_j separates $S\backslash(a_0 \cup \ldots \widehat{a_j} \cdots \cup a_p)$. Let σ be a regular bad simplex of maximal dimension p. Write $S\backslash\hat{f}(\sigma) = X_1 \sqcup \cdots \sqcup X_c$ with each X_i connected. By maximality of σ, $\hat{f}$ restricts to a map

$$\mathrm{Link}(\sigma) \longrightarrow J_\sigma = \mathcal{B}_0(X_1, \Delta_0^1, \Delta_1^1) * \cdots * \mathcal{B}_0(X_c, \Delta_0^c, \Delta_1^c)$$

where each Δ_ϵ^i is inherited from Δ_ϵ and is non-empty as the arcs of $\hat{f}(\sigma)$ are impure. Each X_i has $(g_i, r_i, q_i) < (g, r, q)$, so by induction $\mathcal{B}_0(X_i, \Delta_0^i, \Delta_1^i)$ is $(2g_i + r_i' - 3)$–connected. The Euler characteristic gives $\sum_i(2 - 2g_i - r_i) = 2 - 2g - r + p' + 1$, where $p'+1 \leq p+1$ is the number of arcs in $\hat{f}(\sigma)$. We also have $\sum_i(r_i - r_i') = r - r'$, so $\sum(2g_i + r_i') = 2g + r' - p' + 2c - 3$. Now J_σ is $(\sum_i(2g_i + r_i' - 1) - 2)$–connected (using Proposition 4.3), that is $(2g + r' - p' + c - 5)$–connected. As $c \geq 2$ and $p' \leq p$, we can extend the restriction of $\hat{f}$ to $\mathrm{Link}(\sigma) \simeq S^{k-p}$ to a map $F\colon K \to J_\sigma$ with K a $(k - p + 1)$–disc with boundary the link of σ. We modify $\hat{f}$ on the interior of the star of σ using $\hat{f} * F$ on $\partial\sigma * K \simeq \mathrm{Star}(\sigma)$. If a simplex $\alpha * \beta$ in $\partial\sigma * K$ is regular bad, β must be trivial since β does not separate $S\backslash\hat{f}(\alpha)$, so that $\alpha * \beta = \alpha$ is a face of σ. We have thus reduced the number of regular bad simplices of maximal dimension and the result follows by induction. □

From there, one can prove by a similar type of argument that the ordered subcomplex is also highly connected:

THEOREM 4.5 (Property 5). *$\mathcal{O}(S, b_0, b_1)$ is $(g - 2)$–connected.*

See [**37**, Thm 4.9] for a detailed proof.

4. Exercises

1) The complex $\mathcal{C}_0(S)$ of the exercises of Lecture 2 is a subcomplex of the complex $\mathcal{C}(S)$ with vertices all isotopy classes of non-trivial circles in S and p–simplices all disjointly embeddable collections of $p + 1$ distinct isotopy classes. Assuming that $\mathcal{C}(S)$ is $(2g + r - 4)$–connected if S has genus g and r boundary components, show that $\mathcal{C}_0(S)$ is $(g - 2)$–connected.

Bibliography

[1] Arnol'd, V. I. Certain topological invariants of algebraic functions. (Russian) Trudy Moskov. Mat. Obsc. 21 1970 27–46. English transl. in Trans. Moscow Math. Soc. 21 (1970), 30–52. MR0274462 (43:225)

[2] Barratt, M. G.; Eccles, Peter J. Γ^+–structures. I. A free group functor for stable homotopy theory. Topology 13 (1974), 25–45. MR0348737 (50:1234a)

[3] Boardman, J. M.; Vogt, R. M. Homotopy invariant algebraic structures on topological spaces. Lecture Notes in Mathematics, Vol. 347. MR0420609 (54:8623a)

[4] Boldsen, Søren K. Improved homological stability for the mapping class group with integral or twisted coefficients, Math. Z. 270 (2012), no. 1-2, 297–329. MR2875835 (2012k:57026)

[5] Brown, Kenneth S. Cohomology of groups. Corrected reprint of the 1982 original. Graduate Texts in Mathematics, 87. Springer-Verlag, New York, 1994. MR1324339 (96a:20072)

[6] Cerf, Jean. Topologie de certains espaces de plongements. (French) Bull. Soc. Math. France 89 1961 227–380. MR0140120 (25:3543)

[7] Charney, Ruth M. Homology stability for GL_n of a Dedekind domain. Invent. Math. 56 (1980), no. 1, 1–17. MR557579 (81h:18010)

[8] Cohen, Frederick R.; Lada, Thomas J.; May, J. Peter. The homology of iterated loop spaces. Lecture Notes in Mathematics, Vol. 533. Springer-Verlag, Berlin-New York, 1976. MR0436146 (55:9096)

[9] Earle, Clifford J.; Eells, James. A fibre bundle description of Teichmüller theory. J. Differential Geometry 3 1969 19–43. MR0276999 (43:2737a)

[10] Farb, Benson; Margalit, Dan. A primer on mapping class groups. Princeton Mathematical Series, 49. Princeton University Press, Princeton, NJ, 2012. MR2850125 (2012h:57032)

[11] Galatius, Søren. Stable homology of automorphism groups of free groups. Ann. of Math. (2) 173 (2011), no. 2, 705–768. MR2784914 (2012c:20149)

[12] Galatius, Søren. Lectures on the Madsen-Weiss theorem (PCMI lecture, same volume).

[13] Soren Galatius, Oscar Randal-Williams. Monoids of moduli spaces of manifolds. Geom. Topol. 14 (2010), no. 3, 1243-1302. MR2653727 (2011j:57047)

[14] Galatius, Søren; Tillmann, Ulrike; Madsen, Ib; Weiss, Michael. The homotopy type of the cobordism category. Acta Math. 202 (2009), no. 2, 195–239. MR2506750 (2011c:55022)

[15] Gramain, André. Le type d'homotopie du groupe des difféomorphismes d'une surface compacte. (French) Ann. Sci. École Norm. Sup. (4) 6 (1973), 53–66. MR0326773 (48:5116)

[16] Harer, John L. Stability of the homology of the mapping class groups of orientable surfaces. Ann. of Math. (2) 121 (1985), no. 2, 215–249. MR786348 (87f:57009)

[17] Hatcher, Allen. On triangulations of surfaces. Topology Appl. 40 (1991), no. 2, 189–194. (Updated version available at: http://www.math.cornell.edu/~hatcher/Papers/TriangSurf.pdf.) MR1123262 (92f:57020)

[18] Hatcher Allen. In preparation.
(See http://www.math.cornell.edu/~hatcher/Papers/luminytalk.pdf.)

[19] Ivanov, Nikolai V. Stabilization of the homology of Teichmüller modular groups. (Russian) Algebra i Analiz 1 (1989), no. 3, 110–126; translation in Leningrad Math. J. 1 (1990), no. 3, 675–691. MR1015128 (91g:57010)

[20] Ivanov, Nikolai V. On the homology stability for Teichmüller modular groups: closed surfaces and twisted coefficients. Mappping Class Groups and Moduli Spaces of Riemann Surfaces, Contemporary Mathematics, V. 150, American Mathematical Society, 1993, 149–194. MR1234264 (94h:57022)

[21] Kerz, Moritz. The complex of words and Nakaoka stability. Homology Homotopy Appl. 7 (2005), no. 1, 77–85. MR2155519 (2006d:18005)

[22] Lima, Elon L. On the local triviality of the restriction map for embeddings. Comment. Math. Helv. 38 (1964), 163–164. MR0161343 (28:4551)

[23] Madsen, Ib; Weiss, Michael. The stable moduli space of Riemann surfaces: Mumford's conjecture. Ann. of Math. (2) 165 (2007), no. 3, 843–941. MR2335797 (2009b:14051)

[24] May, J. P. The geometry of iterated loop spaces. Lectures Notes in Mathematics, Vol. 271. Springer-Verlag, Berlin-New York, 1972. MR0420610 (54:8623b)

[25] McDuff, D.; Segal, G. Homology fibrations and the "group-completion" theorem. Invent. Math. 31 (1975/76), no. 3, 279–284. MR0402733 (53:6547)

[26] Mumford, David. Towards an enumerative geometry of the moduli space of curves. Arithmetic and geometry, Vol. II, Progr. Math. 36 (1983), 271–328. MR717614 (85j:14046)

[27] Milnor, John. Construction of universal bundles. II. Ann. of Math. (2) 63 (1956), 430–436. MR0077932 (17:1120a)

[28] Morita, Shigeyuki. Geometry of characteristic classes. Translated from the 1999 Japanese original. Translations of Mathematical Monographs, 199. Iwanami Series in Modern Mathematics. American Mathematical Society, Providence, RI, 2001. MR1826571 (2002d:57019)

[29] Nakaoka, Minoru. Decomposition theorem for homology groups of symmetric groups. Ann. of Math. (2) 71 (1960), 16–42. MR0112134 (22:2989)

[30] Palais, Richard S. Local triviality of the restriction map for embeddings. Comment. Math. Helv. 34 1960 305–312. MR0123338 (23:A666)

[31] Priddy, Stewart B. On $\Omega^\infty S^\infty$ and the infinite symmetric group. Algebraic topology (Proc. Sympos. Pure Math., Vol. XXII, Univ. Wisconsin, Madison, Wis., 1970), pp. 217–220. Amer. Math. Soc., Providence, R.I., 1971. MR0358767 (50:11226)

[32] Randal-Williams, Oscar. Resolutions of moduli spaces and homological stability. Preprint, 2009 (arXiv:0909.4278).

[33] Randal-Williams, Oscar. Homology of the moduli spaces and mapping class groups of framed and Pin surfaces. Preprint (arXiv:1001.5366).

[34] Randal-Williams, Oscar. Homological stability for unordered configuration spaces. Preprint, 2011 (arXiv:1105.5257). MR2764873 (2012d:57044)

[35] Segal, Graeme Configuration-spaces and iterated loop-spaces. Invent. Math. 21 (1973), 213–221. MR0331377 (48:9710)

[36] Wahl, Nathalie. Homological stability for the mapping class groups of non-orientable surfaces. Invent. Math. 171 (2008), no. 2, 389–424. MR2367024 (2008m:57047)

[37] Wahl, Nathalie. Homological stability for mapping class groups of surfaces. Preprint (2011). To appear in the Handbook of Moduli.

Lectures on the Madsen–Weiss Theorem

Søren Galatius

IAS/Park City Mathematics Series
Volume 20, 2011

Lectures on the Madsen–Weiss Theorem

Søren Galatius

These are the revised lecture notes for my four lectures at the Park City Math Institute, Utah 2011. The lectures form a continuation of Nathalie Wahl's lectures, and I refer to her lectures for context and motivation.

My goals for the lectures were to give a precise statement of the Madsen–Weiss theorem and give (most of) a proof. The theorem was first proved by Madsen and Weiss in [**5**]. The lectures will concern the proof given in [**1**] (first outlined in [**2**]), and the reader is referred there for more details. The exposition is influenced by Hatcher's survey [**3**].

Stanford University, Department of Mathematics, Building 380, Stanford, California 94305
E-mail address: galatius@stanford.edu

LECTURE 1

Spaces of submanifolds and the Madsen–Weiss Theorem

The main goal of this lecture is to give a precise statement of the Madsen–Weiss theorem ([5]). In the process, we will introduce several objects which will be useful in the proof.

1.1. Spaces of manifolds

We begin with a definition of a certain topological space which will be used throughout the lectures, first defining the underlying set, and then describing a topology. The relation to Madsen–Weiss' theorem will become clear during this lecture. Before defining the underlying set, let us point out a potential cause of confusion about words: For a submanifold $W \subseteq \mathbb{R}^n$, being "closed" could have two very different meanings. It could mean either that W is a closed manifold, i.e. it is compact and has no boundary (this property does not depend on how W sits in $\mathbb{R}^n$), or it could mean that W is a closed subset of $\mathbb{R}^n$. To avoid this confusion, we shall say that $W \subseteq \mathbb{R}^n$ is *topologically closed* to mean that W is merely a closed subset.

DEFINITION 1.1. Let $\Psi(\mathbb{R}^n)$ be the set of pairs (W, ω), where $W \subseteq \mathbb{R}^n$ is a smooth 2-manifold without boundary which is topologically closed in $\mathbb{R}^n$, and ω is an orientation of W.

We emphasize that there are no further conditions required, and in particular W is allowed to be empty, and is not required to be path connected. In general $H_i(W)$ could have infinite rank for all or some $i \leq 2$. This set has something akin to a smooth structure (although it is not in any good sense an infinite dimensional manifold).

DEFINITION 1.2. Let X be a smooth manifold. A map $f : X \to \Psi(\mathbb{R}^n)$ is *smooth* if the graph

$$\Gamma_f = \{(x, v) \in X \times \mathbb{R}^n | v \in f(x)\}$$

is a smooth topologically closed submanifold of $X \times \mathbb{R}^n$ such that the projection $p_f : \Gamma_f \to X$ is a submersion and the orientations of $f(x)$ vary continuously (i.e. assemble to an orientation on the kernel of $Dp_f : T\Gamma_f \to TX$).

In [2] and [1], a natural topology on $\Psi(\mathbb{R}^n)$ was defined, and the following property was proved.

LEMMA 1.3. *Any smooth map $X \to \Psi(\mathbb{R}^n)$ is continuous. Any continuous map $f : X \to \Psi(\mathbb{R}^n)$ can be perturbed to a smooth map; the perturbation can be assumed constant near any closed set on which f is already smooth.* □

"Perturbation" means a continuous extension of f to $X \times [0, 1)$ which is smooth on $X \times (0, 1)$. In these lectures we shall not prove in detail that a topology with this property exists, but we shall say some words about what the open sets are (and refer to [**1**] for more details). A neighborhood basis at $\emptyset \in \Psi(\mathbb{R}^n)$ is given by the sets

$$\mathcal{U}(K) = \{W \in \Psi(\mathbb{R}^n) | W \cap K = \emptyset\},$$

where K runs through compact subsets of $\mathbb{R}^n$. In particular, a sequence of points $W_i \in \Psi(\mathbb{R}^n)$ converges to $\emptyset$ if and only if the following condition holds: For each compact K, there exists an $N = N_K$ such that $W_i \cap K = \emptyset$ for $i > N$. A neighborhood basis at an arbitrary $W \in \Psi(\mathbb{R}^n)$ is defined similarly: It consists of manifolds which near a large compact subset of $\mathbb{R}^n$ looks like a small perturbation of W (where "small" is in a C^∞ sense.)

For example, the function $f : \mathbb{R} \to \Psi(\mathbb{R}^3)$ given by $f(t) = \{t^{-1}\} \times \mathbb{R}^2$ and $f(0) = \emptyset$ is continuous (in fact smooth). As another illustration of the compact-open flavor of the topology, let us prove that $\Psi(\mathbb{R}^n)$ is path connected for $n \geq 3$. In that case, given any $W \in \Psi(\mathbb{R}^n)$, we can pick a point $p \in \mathbb{R}^n - W$. For $t \in [0, 1]$ we let $W - tp$ denote the manifold W parallel translated by the vector $-tp$ and let $(1-t)^{-1}(W - tp)$ denote the result of scaling that manifold by the number $(1-t)^{-1}$, interpreted as $\emptyset$ when $t = 0$. Then $t \mapsto (1-t)^{-1}(W - tp)$ defines a continuous path which starts at W and ends at $\emptyset$, proving that there is a path from any element to the base point $\emptyset \in \Psi(\mathbb{R}^n)$.

The space $\Psi(\mathbb{R}^n)$ is related to surface bundles through the subspace

$$B_n = \{W \in \Psi(\mathbb{R}^n) | W \subseteq (0, 1)^n\}.$$

The exact relation to surface bundles is given by the following consequence of lemma 1.3. In these notes, "surface bundle" shall mean a smooth fiber bundle, equipped with an orientation of the fiberwise tangent bundle. We allow the fibers to have different diffeomorphism type over different path components of the base.

PROPOSITION 1.4. *Let X be a smooth k-dimensional manifold. For $n > 2k+5$, there is a bijection between $[X, B_n]$ and the set of isomorphism classes of surface bundles $E \to X$.*

PROOF. By lemma 1.3, any map $f : X \to B_n$ is homotopic to a smooth map, which is given by a smooth, topologically closed $\Gamma_f \subseteq X \times \mathbb{R}^n$, which is also contained in $X \times (0, 1)^n$. An exercise in point-set topology shows that this implies that the projection $\Gamma_f \to X$ is a surface bundle. If two maps $f_0, f_1 : X \to B_n$ are smoothly homotopic, there is an induced surface bundle over $X \times I$, proving that the two surface bundles are isomorphic. This gives the map in one direction.

Conversely, if $E \to X$ is a surface bundle and $n > 2k + 4$, we can (by a suitable version of Whitney's embedding theorem) pick an embedding $j : E \to X \times (0, 1)^n$ commuting with the projections to X, and define a smooth map $f : X \to B_n$ by the formula $j(E_x) = \{x\} \times f(x)$, where $E_x \subseteq E$ is the fiber over $x \in X$. If j' is another embedding of the same surface bundle, then j and j' are isotopic (smoothly homotopic through embeddings), so the resulting maps $X \to B_n$ become homotopic. This gives the map in the other direction, and it is easy to see that they are each other's inverse. □

The above definitions can be extended from finite n to $n = \infty$, by topologizing e.g. B_∞ as the direct limit of the B_n's. Using proposition 1.4, a cohomology class

$c \in H^*(B_\infty)$ then gives rise to a characteristic class of surface bundles. Explicitly, given a surface bundle $E \to X$, we may pick an embedding $j : E \to X \times (0,1)^n$ and let $f : X \to B_n \subseteq B_\infty$ be the resulting map, as in the above proof. The resulting cohomology class $f^*(c) \in H^*(X)$ depends only on the isomorphism class of the bundle $E \to X$ and is the characteristic class associated to c, evaluated on the bundle $E \to X$. In fact, we can say a little more.

COROLLARY 1.5. *For any smooth X, there is a natural bijection between $[X, B_\infty]$ and the set of isomorphism classes of surface bundles $E \to X$.* □

COROLLARY 1.6. *$H^*(B_\infty)$ is the ring of characteristic classes of surface bundles.*

SKETCH OF PROOF. If we extend the definition of surface bundle a little, to allow the identity map to classify a surface bundle over B_∞, this follows from Yoneda's lemma in the usual way. In our setup, where the base space of a surface bundle is required to be a (finite dimensional) manifold, a rigorous proof uses that B_∞ is weakly equivalent to a directed limit of spaces which are homotopy equivalent to manifolds (e.g. the system of finite subcomplexes of a CW approximation to B_∞). □

A slightly different point of view on corollary 1.5 is seen by noting a bijection

$$B_n = \coprod_W \mathrm{Emb}(W, (0,1)^n)/\mathrm{Diff}(W),$$

where the disjoint union is over oriented closed surfaces W, one of each diffeomorphism class. Giving the right hand side the quotient topology, this becomes a homeomorphism, and it is known that the quotient map $\mathrm{Emb}(W, (0,1)^n) \to \mathrm{Emb}(W, (0,1)^n)/\mathrm{Diff}(W)$ is a principal $\mathrm{Diff}(W)$-bundle for all $n \leq \infty$. It is easy to prove that $\mathrm{Emb}(W, (0,1)^\infty)$ is contractible, and hence a model for $E\mathrm{Diff}(W)$. It follows that we have models for $B\mathrm{Diff}(W)$, namely the path component of B_∞ consisting of manifolds diffeomorphic to W. Using these models we have

$$B_\infty = \coprod_W B\mathrm{Diff}(W). \tag{1}$$

We have explained how the subspace $B_n \subseteq \Psi(\mathbb{R}^n)$ is related to surface bundles and their characteristic classes, but we have not yet seen why the full space $\Psi(\mathbb{R}^n)$ is useful. Its relevance comes through the important construction in definition 1.7 below. We shall consider $\emptyset \in \Psi(\mathbb{R}^n)$ the basepoint, and as usual we let $\Omega^n\Psi(\mathbb{R}^n)$ be the n-fold loop space, i.e. the space of continuous based maps $S^n \to \Psi(\mathbb{R}^n)$. We shall consider S^n as the one-point compactification of $\mathbb{R}^n$.

DEFINITION 1.7. Let $\alpha : B_n \to \Omega^n\Psi(\mathbb{R}^n)$ be the map given by

$$\alpha(W)(v) = \begin{cases} W + v & \text{if } v \in \mathbb{R}^n \\ \emptyset & \text{if } v = \infty. \end{cases} \tag{2}$$

Since $W + v$ leaves any compact set as $|v| \to \infty$, $\alpha(W)$ is continuous at the basepoint of S^n, and in fact it can be seen that α is a continuous map. We can then let $n \to \infty$ (using the map $\Psi(\mathbb{R}^n) \to \Omega\Psi(\mathbb{R}^{n+1})$ given by $W \mapsto (t \mapsto W \times \{t\})$) and get a map

$$\alpha : B_\infty \to \Omega^\infty \mathbf{\Psi} = \operatorname*{colim}_{n\to\infty} \Omega^n\Psi(\mathbb{R}^n).$$

Neither the source nor the target of α are path connected. By (1), the path components of B_∞ are precisely the $B\mathrm{Diff}(W)$, and there is one path component for each diffeomorphism type of oriented 2-manifold W. The map α sends $B\mathrm{Diff}(W)$ to a path component of $\Omega^\infty \mathbf{\Psi}$; we shall (temporarily) write $\Omega^\infty_{[W]}\mathbf{\Psi}$ for that path component. We can now formulate Madsen–Weiss' theorem in a slightly unconventional form.

THEOREM 1.8 (Madsen-Weiss). *If W is a surface of genus g, the restricted map*

$$B\mathrm{Diff}(W) \to \Omega^\infty_{[W]}\mathbf{\Psi}$$

induces an isomorphism in integral cohomology through degree $2g/3$.

In the remaining part of this lecture, we will say more about the homotopy type of $\Psi(\mathbb{R}^n)$ and explain why theorem 1.8 is equivalent to the theorem stated in Wahl's lectures. To understand the homotopy type of $\Psi(\mathbb{R}^n)$, let us consider the Grassmannian $\mathrm{Gr}_2^+(\mathbb{R}^n)$ of oriented 2-planes in $\mathbb{R}^n$. There are two interesting vector bundles over that space, the *canonical bundle* γ_n and its orthogonal complement $\gamma_n^\perp$. A point in the total space of $\gamma_n^\perp$ is a pair (V, v) where $V \subseteq \mathbb{R}^n$ is an oriented 2-plane, and $v \in V^\perp$. We can think of V as an oriented 2-manifold and since it's topologically closed, it defines a point $V \in \Psi(\mathbb{R}^n)$. Translating it by the vector v, we get a map

$$\begin{aligned} \gamma_n^\perp &\to \Psi(\mathbb{R}^n) \\ (V, v) &\mapsto V + v. \end{aligned}$$

The manifold $V + v$ leaves all compact sets as $|v| \to \infty$, so the map extends to a continuous map

$$\begin{aligned} \mathrm{Th}(\gamma_n^\perp) &\xrightarrow{q} \Psi(\mathbb{R}^n) \\ (V, v) &\mapsto V + v \\ \infty &\mapsto \emptyset, \end{aligned}$$

where $\mathrm{Th}(\gamma_n^\perp)$ denotes the *Thom space* of $\gamma_n^\perp$, i.e. the one-point compactification of its total space.

We have the following result. Recall that a map $f : X \to Y$ is a weak equivalence if the induced map $\pi_k(X) \to \pi_k(Y)$ is an isomorphism for all k and all basepoints, and that homotopy equivalence implies weak equivalence which implies that $f_* : H_*(X) \to H_*(Y)$ is an isomorphism. For CW complexes, weak equivalence also implies homotopy equivalence.

THEOREM 1.9. *The map q is a weak equivalence.*

PROOF. This will be sketched in the exercises, using Lemma 1.3. □

The theorem above relates our definition of $\Omega^\infty\mathbf{\Psi}$ to the definition usually appearing in Madsen–Weiss' theorem, namely the space

$$\Omega^\infty MTSO(2) = \operatorname*{colim}_{n\to\infty} \Omega^n \mathrm{Th}(\gamma_n^\perp).$$

Theorem 1.9 above implies the following.

COROLLARY 1.10. *There is a weak equivalence*

$$\Omega^\infty MTSO(2) \to \Omega^\infty \mathbf{\Psi}.$$

PROOF. The theorem implies that $\Omega^n \mathrm{Th}(\gamma_n^\perp) \to \Omega^n \Psi(\mathbb{R}^n)$ induces an isomorphism in all homotopy groups. The direct systems on both sides consists of injective maps, and in this case homotopy groups commute with direct limit. □

The Madsen–Weiss theorem is usually stated in terms of the Thom spectrum $MTSO(2)$, instead of the weakly equivalent $\mathbf{\Psi}$. (The notation $\Omega^\infty MTSO(2)$ might seem a little heavy for just one space, but it has the advantage of generalizing easily to manifolds of other dimensions and with other structures, e.g. there's a space called $\Omega^\infty MT\mathrm{Spin}(3)$.)

1.2. Exercises for Lecture 1

(1) Recall that for a space X with base point $x \in X$ and $k \geq 0$, the homotopy group (set, if $k = 0$) $\pi_k(X, x)$ is defined as the set of basepoint preserving maps $S^k \to X$ modulo basepoint preserving homotopies. (Fine print: for $k = 0$ this does not depend on x and we also define $\pi_0(\emptyset) = \emptyset$.) Recall also that a map $f : X \to Y$ is called a weak equivalence if the induced map $\pi_k(X, x) \to \pi_k(Y, f(x))$ is a bijection for all $k \geq 0$ and all $x \in X$ Prove that f is a weak equivalence if and only if it satisfies the following condition: For any pair of maps $g_0 : \partial D^k \to X$ and $h_0 : D^k \to Y$ making the following diagram commute

$$\begin{array}{ccc} \partial D^k & \xrightarrow{g_0} & X \\ \big\downarrow & & \big\downarrow f \\ D^k & \xrightarrow[h_0]{} & Y, \end{array}$$

there exist homotopies $h_s : D^k \to Y$ and $g_s : \partial D^k \to X$, $s \in [0, 1]$ with $f \circ g_s = h_s|\partial D^k$, such that $g_1 : \partial D^k \to X$ extends to a map $G : D^k \to X$ with $f \circ G = h_1$. (In words: any such commutative diagram is homotopic, through commutative diagrams, to a diagram admitting a diagonal $D^k \to X$.)

(2) Prove that if X is a k-dimensional manifold, with $k < n - 2$, and $f : X \to \Psi(\mathbb{R}^n)$ is smooth, then there exists a point $p \in \mathbb{R}^n$, not contained in $f(x)$ for any $x \in X$. Use this to deduce that $\pi_k(\Psi(\mathbb{R}^n)) = 0$ for $k < n - 2$.

(3) Use theorem 1.9 to prove that

$$\pi_{n-2}(\Psi(\mathbb{R}^n)) = H_{n-2}(\Psi(\mathbb{R}^n)) = H_{n-2}(\mathrm{Th}(\gamma_n^\perp)).$$

Then use the Thom isomorphism to deduce $\pi_{n-2}(\Psi(\mathbb{R}^n)) = \mathbb{Z}$.

(4) The next exercises work through some examples with $n = 3$: First, prove that $\mathrm{Gr}_2^+(\mathbb{R}^3)$ is homeomorphic to S^2.

(5) Use theorem 1.9 to prove that $\Psi(\mathbb{R}^3)$ is weakly equivalent to $S^3 \vee S^1$, deduce that $\pi_1\Psi(\mathbb{R}^3) = \mathbb{Z}$ and find an explicit generator. Then deduce that $\pi_2(\Psi(\mathbb{R}^3)) = 0$ and that $\pi_3\Psi(\mathbb{R}^3)$ is a free abelian group of infinite rank. Describe an explicit map $S^1 \vee S^3 \to \Psi(\mathbb{R}^3)$.

(6) Reversing orientation gives map $\Psi(\mathbb{R}^n) \to \Psi(\mathbb{R}^n)$. Describe the induced maps on $\pi_1\Psi(\mathbb{R}^3)$ and $\pi_3\Psi(\mathbb{R}^3)$.

(7) Prove that if $T \subseteq (0, 1)^3$ is a torus, then $\alpha(T) \in \Omega^3\Psi(\mathbb{R}^3)$ is in the same path component as the basepoint. (Hint: Use that the Gauss map $T \to \mathrm{Gr}_2^+(\mathbb{R}^3)$ is null homotopic.)

(8) Let $f_1 : E_1 \to X$ and $f_2 : E_2 \to X$ be two surface bundles. Then $f_3 : E_3 = E_1 \amalg E_2 \to X$ is again a surface bundle. By proposition 1.4, they are represented by maps $g_i : X \to B_n$ for some n. Prove that the maps $\alpha \circ g_i \in [X, \Omega^n \Psi(\mathbb{R}^n)]$ satisfy $\alpha \circ g_3 = \alpha \circ g_1 + \alpha \circ g_2$, where $+$ denotes the group structure induced from concatenation of loops.

(9) Let X be a compact manifold and $f : X \to \Psi(\mathbb{R}^n)$ a smooth map. Prove that if $B(0, \varepsilon) \subseteq \mathbb{R}^n$ denotes the open ε ball, then $\{x \in X | f(x) \cap B(0, \varepsilon) = \emptyset\}$ is a closed subset of X. Let $U_\varepsilon \subseteq X$ denote its complement and prove that for small enough ε, there exists a smooth function $p : U_\varepsilon \to B(0, \varepsilon)$ such that $p(x) \in f(x)$ and that $|y| > |p(x)|$ for all $y \in f(x) - \{p(x)\}$. (I.e. $p(x)$ is the point in $f(x)$ which is closest to 0. Hint: use the tubular neighborhood theorem.) Use this to prove theorem 1.9.

LECTURE 2

Rational cohomology and outline of proof

The goals of this lecture are the following.

(1) Define $\kappa_i \in H^{2i}(\Omega^\infty \mathbf{\Psi}; \mathbb{Z})$.

(2) Explain why the induced ring homomorphism

$$\mathbb{Q}[\kappa_1, \kappa_2, \ldots] \to H^*(\Omega^\infty_\bullet \mathbf{\Psi}; \mathbb{Q})$$

is an isomorphism, where $\Omega^\infty_\bullet \mathbf{\Psi} \subseteq \Omega^\infty \mathbf{\Psi}$ denotes a path component.

(3) Outline the steps in the proof of Madsen–Weiss' theorem: $B\mathrm{Diff}(W) \to \Omega^\infty_\bullet \mathbf{\Psi}$ induces an isomorphism in H_k for $k < 2(g-1)/3$. (These steps will be carried out in lectures 3 and 4.)

2.1. Cohomology of $\Omega^\infty \mathbf{\Psi}$

The starting point in calculating the cohomology of $\Omega^\infty \mathbf{\Psi}$ is that we understand the cohomology of $\Psi(\mathbb{R}^n) \simeq \mathrm{Th}(\gamma_n^\perp)$. Firstly, the inclusion $\mathrm{Gr}_2^+(\mathbb{R}^n) \to \mathrm{Gr}_2^+(\mathbb{R}^\infty) \simeq \mathbb{C}P^\infty$ induces a map

$$\mathbb{Z}[e] = H^*(\mathrm{Gr}_2^+(\mathbb{R}^\infty)) \to H^*(\mathrm{Gr}_2^+(\mathbb{R}^n)), \tag{3}$$

which is an isomorphism in degrees $< (n-2)$. Secondly, the Thom isomorphism theorem gives an isomorphism

$$\begin{aligned} H^k(\mathrm{Gr}_2^+(\mathbb{R}^n)) &\to H^{k+(n-2)}(\mathrm{Th}(\gamma_n^\perp), *) \\ x &\mapsto x.u, \end{aligned} \tag{4}$$

where $u \in H^{n-2}(\mathrm{Th}(\gamma_n^\perp), *)$ denotes the Thom class, so at least we have a full understanding of $H^*(\mathrm{Th}(\gamma_n^\perp))$ in degrees $< 2n-4$.

To understand the effect of the functor Ω^n, we need the *suspension homomorphism*, a natural homomorphism

$$\sigma : H^{k+1}(X, *) \to H^k(\Omega X, *)$$

defined for any pointed space $(X, *)$. To define it, we use the evaluation map

$$\begin{aligned} \Sigma\Omega X &\xrightarrow{\mathrm{ev}} X \\ (t, \gamma) &\mapsto \gamma(t), \end{aligned}$$

where Σ denotes (unreduced) suspension. Then the suspension homomorphism is defined as the composition

$$H^{k+1}(X, *) \xrightarrow{\mathrm{ev}^*} H^{k+1}(\Sigma\Omega X, *) \cong H^k(\Omega X, *).$$

Letting $\sigma^n : H^{k+n}(X, *) \to H^k(\Omega^n X, *)$ denote the n-fold iteration, we can now give a definition of the κ-classes using (3) and (4).

DEFINITION 2.1. Let $\kappa_i \in H^{2i}(\Omega^n\Psi(\mathbb{R}^n))$ be the class defined as

$$\kappa_i = \sigma^n(e^{i+1}.u) \in H^{2i}(\Omega^n \mathrm{Th}(\gamma_n^\perp)) = H^{2i}(\Omega^n\Psi(\mathbb{R}^n)). \tag{5}$$

LEMMA 2.2. *There is a unique class $\kappa_i \in H^{2i}(\Omega^\infty\mathbf{\Psi})$ which restricts to the classes in definition 2.1 for all n.*

PROOF. This follows from the following two facts, which we leave to the reader as an exercise.

(i) The map $\Omega^n\mathrm{Th}(\gamma_n^\perp) \to \Omega^{n+1}\mathrm{Th}(\gamma_{n+1}^\perp)$ is $(n-4)$-connected. Hence

$$H^{2i}(\Omega^\infty\mathbf{\Psi}) \to H^{2i}(\Omega^n\Psi(\mathbb{R}^n))$$

is an isomorphism for $n > 2i+4$.

(ii) The classes $\kappa_i \in H^{2i}(\Omega^n\Psi(\mathbb{R}^n))$ in definition 2.1 are compatible with the map $\Omega^n\Psi(\mathbb{R}^n) \to \Omega^{n+1}\Psi(\mathbb{R}^{n+1})$.

□

We have defined classes κ_i in $\Omega^\infty\mathbf{\Psi}$, which can be pulled back to classes in B_∞. By corollary 1.6, these give rise to characteristic classes of surface bundles, which we shall also denote κ_i. The following lemma, whose proof we leave as an exercise, shows that our definition agrees with the "usual" definition of the κ classes (at least up to signs, which I shall be rather careless about in these lectures).

LEMMA 2.3. *Let $\pi : E \to X$ be a surface bundle with X a compact oriented n-manifold. Let $\pi_! : H^{k+2}(E) \to H^k(X)$ be the map Poincaré dual to $\pi_* : H_{n-k}(E, \partial E) \to H_{n-k}(X, \partial X)$, and let $T_\pi E$ be the fiberwise tangent bundle, i.e. the oriented 2-dimensional bundle* $\mathrm{Ker}(D\pi : TE \to TX)$. *Then $\kappa_i = \pi_!(e^{i+1}(T_\pi(E))$.* □

To calculate the entire ring $H^*(\Omega^\infty_\bullet\mathbf{\Psi};\mathbb{Q})$, it is helpful to again work more generally. For a *graded vector space* $V = \oplus_{n\geq 1}V_n$, we shall write $\mathbb{Q}[V]$ for the free graded-commutative $\mathbb{Q}$-algebra generated by V. If $(X,*)$ is a based space, and $\phi : V \to H^*(X,*;\mathbb{Q})$ is a homomorphism ($\mathbb{Q}$-linear and grading preserving), the cup product gives a unique extension to a $\mathbb{Q}$-algebra map

$$\mathbb{Q}[V] \to H^*(X;\mathbb{Q}). \tag{6}$$

We can also compose ϕ with the suspension to get a map

$$\sigma\circ\phi : V_{n+1} \to H^{n+1}(X,*) \to H^n(\Omega X, *),$$

and then the cup product in ΩX induces a $\mathbb{Q}$-algebra homomorphism

$$\mathbb{Q}[s^{-1}V] \to H^*(\Omega X;\mathbb{Q}), \tag{7}$$

where $(s^{-1}V)_n = V_{n+1}$ for $n \geq 1$ (and $s^{-1}(V)_0 = 0$). We will use the following general result about this situation. Again, we shall write $\Omega_\bullet X \subseteq \Omega X$ for the path component of the basepoint.

THEOREM 2.4. *Assume $\pi_1(X)$ is abelian and acts trivially on the rational cohomology of the universal cover. Let $V = \bigoplus_{n\geq 1} V_n$ be a graded vector space and $V \to H^*(X,*;\mathbb{Q})$ a homomorphism such that (6) is an isomorphism in degrees $\leq n$. Then (7) restricts to an isomorphism $\mathbb{Q}[s^{-1}V] \to H^*(\Omega_\bullet X;\mathbb{Q})$ in degrees $\leq (n-1)$.*

PROOF. This can be proved using the Serre spectral sequence for the path-loop fibration over the universal cover $\tilde{X}$. Alternatively, one can use the Eilenberg-Moore spectral sequence. □

LEMMA 2.5. *If $X = \Omega Y$, then $\pi_1(X)$ is abelian and acts trivially on the cohomology of the universal cover.*

PROOF. We leave it as an exercise to check that any deck transformation $\tilde{X} \to \tilde{X}$ is homotopic to the identity. □

We can now apply the general theory to calculate the ring $H^*(\Omega^n \mathrm{Th}(\gamma_n); \mathbb{Q})$ in the following way. Set $X = \mathrm{Th}(\gamma_n)$ and let V be the graded vector space with basis $e^{i+1}.u$ for all $i \geq -1$. Then the induced map

$$\mathbb{Q}[V] \to H^*(\mathrm{Th}(\gamma_n^\perp); \mathbb{Q})$$

is an isomorphism in degrees $< 2(n-4)$. (This is the range in which we previously calculated the *relative* cohomology $H^*(\mathrm{Th}(\gamma_n^\perp), *)$. The extra class in absolute H^0 corresponds to the unit $1 \in \mathbb{Q}[V]$ and there are no non-trivial products on the left hand side in the range considered.) If we apply theorem 2.4 n times, we see that we get a ring homomorphism

$$\begin{aligned} \mathbb{Q}[s^{-n}V] &\mapsto H^*(\Omega^n \mathrm{Th}(\gamma_n^\perp); \mathbb{Q}) \\ e^{i+1}.u &\mapsto \sigma^n(e^{i+1}.u) \end{aligned}$$

which is an isomorphism in degrees $< (n-4)$. By definition of the κ classes, the map can be rewritten as

$$\mathbb{Q}[\kappa_1, \kappa_2, \dots] \to H^*(\Omega^n \mathrm{Th}(\gamma_n^\perp); \mathbb{Q}).$$

2.2. Outline of proof

Finally, we will outline the steps in the proof of the Madsen–Weiss theorem. At the heart of the theorem is the map $\alpha : B_n \to \Omega^n \Psi(\mathbb{R}^n)$, obtained by moving compact manifolds around in all directions in $\mathbb{R}^n$. An essential step of the proof is to do this process in multiple steps, each of which moves manifolds around in only one coordinate direction in $\mathbb{R}^n$. To make this idea precise, we first need some definitions.

DEFINITION 2.6.

(i) Let $\psi(n,k) \subseteq \Psi(\mathbb{R}^n)$ be the subspace consisting of those $W \in \Psi(\mathbb{R}^n)$ which satisfy $W \subseteq \mathbb{R}^k \times (0,1)^{n-k}$.
(ii) Let $\alpha_k : \psi(n,k) \to \Omega\psi(n,k+1)$ be the map given by

$$\alpha_k(W)(t) = \begin{cases} W + te_{k+1} & \text{for } t \in \mathbb{R} \\ \emptyset & \text{for } t = \infty. \end{cases}$$

Each map α_k is continuous, and it is clear that the map (2) from lecture 1 can be decomposed as

$$B_n = \psi(n,0) \xrightarrow{\alpha_0} \psi(n,1) \xrightarrow{\Omega\alpha_1} \dots \xrightarrow{\Omega^{n-1}\alpha_{n-1}} \psi(n,n) = \Psi(\mathbb{R}^n).$$

We shall prove the Madsen–Weiss theorem in the following steps.

- For $1 \leq k \leq n-1$, the map $\alpha_k : \psi(n,k) \to \Omega\psi(n,k+1)$ is a weak equivalence.

- The map $\alpha_0 : B_n \to \Omega\psi(n,1)$ is compatible with letting $n \to \infty$. Restricting to a path component of the resulting map $B_\infty \to \Omega\psi(\infty, 1)$, we get a map $B\mathrm{Diff}(W) \to \Omega\psi(\infty, 1)$.
- For $W = \Sigma_g$ a surface of genus g, the resulting map
$$B\mathrm{Diff}(\Sigma_g) \to \Omega\psi(\infty, 1)$$
induces an isomorphism in homology through degree $2g/3$.

The last step actually has several sub-steps. An important one is to introduce surfaces with boundary, in order to take a limit as $g \to \infty$, cf. Nathalie's lectures.

2.3. Exercises for Lecture 2

(1) Let $c : B_\infty \to B_\infty$ be the map which reverses orientation. Prove that $c^*\kappa_i = (-1)^i \kappa_i$.
(2) Prove the assertion in lemma 2.5.
(3) Let X be a based space and let $\mu : \Omega X \times \Omega X \to \Omega X$ be the map which concatenates loops. For $x \in H^{k+1}(X, *)$ with $k \geq 1$, prove that the class $y = \sigma(x) \in H^k(\Omega X, *)$ has the property that $\mu^*(y) = y \otimes 1 + 1 \otimes y$.
(4) Deduce that the classes $\kappa_i \in H^{2i}(\Omega^n \Psi(\mathbb{R}^n))$ satisfy $\mu^*(\kappa_i) = \kappa_i \otimes 1 + 1 \otimes \kappa_i$. What does this imply about the characteristic classes $\kappa_i \in H^{2i}(B_\infty)$? (Hint: the question is about the behavior with respect to fiberwise disjoint union of surface bundles, cf. exercise 8 in lecture 1.)
(5) Use lemma 2.3 to prove that the value of $\kappa_0 \in H^0(\Omega^n \Psi(\mathbb{R}^n))$ at the point $\alpha(W) \in \Omega^n \Psi(\mathbb{R}^n)$ is the Euler characteristic $\chi(W)$.
(6) Prove the following property of the suspension homomorphism. Let $f : X \to \Omega Y$ have adjoint $g : \Sigma X \to Y$ (i.e. $g(t, x) = f(x)(t)$) and let $c \in H^{k+1}(Y)$. Then $f^*(\sigma c) \in H^k(X)$ and $g^*(c) \in H^{k+1}(\Sigma X)$ agree under the isomorphism $H^k(X) = H^{k+1}(\Sigma X)$.
(7) Let $u : \mathrm{Th}(\gamma_n^\perp) \to K(\mathbb{Z}/2, n-2)$ be a map representing the mod 2 Thom class. Prove that in mod 2 cohomology, $u^*(\mathrm{Sq}^2 \iota) = e.u$. Deduce that the image of the Hurewicz map $\pi_n(\mathrm{Th}(\gamma_n^\perp)) \to H_n(\mathrm{Th}(\gamma_n))$ vanishes in mod 2 homology. Then deduce that $\kappa_0 \in H^0(\Omega^n \Psi(\mathbb{R}^n))$ is divisible by 2.
(8) With notation as the previous exercise, prove that $u^*(\mathrm{Sq}^{2(i+1)}) = e^{i+1}.u$ and deduce that κ_i vanishes in mod 2 cohomology for all $i \geq 1$.
(9) Prove the assertions in lemma 2.2.
(10) Prove the assertion in lemma 2.3.

LECTURE 3

Topological monoids and the first part of the proof

The goal of this lecture is to prove that the map $\alpha_k : \phi(n,k) \to \Omega\phi(n,k+1)$ is a weak equivalence for $1 \leq k \leq n-1$. As a corollary, we get that the iterated map $\psi(n,1) \to \Omega^{n-1}\psi(n,n)$ is a weak equivalence, and hence

$$\Omega\psi(\infty,1) \simeq \Omega^\infty \mathbf{\Psi}.$$

This will reduce the Madsen–Weiss theorem to a property of the map $\alpha_0 : B_\infty \to \Omega\psi(\infty,1)$, which we shall study in the fourth lecture. For both lectures, a very important tool is *topological monoids* and their *classifying spaces*. We will discuss these now, omitting many proofs.

3.1. Topological monoids

A topological monoid is a space M with a multiplication $M \times M \to M$ which is associative but not necessarily commutative. We will not assume that M has a unit (although all our monoids will at least have homotopy units). Associated to such M is a space BM, called the classifying space of M. This is usually defined as the "geometric realization of the nerve of M"; we shall use the following more explicit (but equivalent, cf. exercise 9) definition, which fits well with our setup.

DEFINITION 3.1. Let BM be the set of pairs (A, f), where $A \subseteq \mathbb{R}$ is a finite subset, and $f : A \to M$ is a function. For the purpose of defining a topology on this set, we shall think of its points as "configurations of points in $\mathbb{R}$, labeled by M": Points in BM can be depicted as finitely many points on the real line, each labeled by an element of M.

Topologize this set by allowing the labels to move continuously in M and the points to move continuously in $\mathbb{R}$. Points are allowed to collide, in which case we multiply the labels (in the order they appear on the line), and to tend to ∞ or $-\infty$, in which case we forget the labels.

The point where $A = \emptyset$ gives the basepoint $\emptyset \in BM$.

To define the topology more rigorously, let $K \subseteq \mathbb{R}$ be a compact set, $V \subseteq M$ an open set, and $a < b \in \mathbb{R}$. Let $\mathcal{U}(K) \subseteq BM$ be the set of points satisfying $A \cap K = \emptyset$, and let $\mathcal{U}(a,b,V) \subseteq BM$ consist of those (A, f) such that if $A \cap (a,b) = (a_1 < \cdots < a_k)$, then $k \geq 1$ and $f(a_1)f(a_2)\dots f(a_k) \in V$. Declare the collection of sets $\mathcal{U}(K)$ and $\mathcal{U}(a,b,V)$ a subbasis for the topology.

There is a natural map $\beta : M \to \Omega BM$, given by

$$\beta(m)(t) = \begin{cases} (\{t\}, (t \mapsto m)) & \text{for } t \in \mathbb{R} \\ \emptyset & \text{for } t = \infty, \end{cases}$$

where we regard ΩBM as the space of pointed maps from the one-point compactification of $\mathbb{R}$. The following well known theorem shall be used without proof.

THEOREM 3.2. *$\beta : M \to \Omega BM$ is a weak equivalence if and only if M is* group-like *(i.e. that M has a homotopy unit and the monoid $\pi_0 M$ is a group).* □

This theorem suggests a very useful strategy for proving weak equivalences of the form $X \simeq \Omega Y$: If X admits a monoid structure, prove $BX \simeq Y$ instead; if not, find a monoid $M \simeq X$ and prove $BM \simeq Y$. It is often easier to follow this strategy than working directly with the loop space of Y. We shall apply this strategy to prove that $\alpha_k : \psi(n,k) \to \Omega\psi(n,k+1)$ is a weak equivalence. More precisely, we shall define a topological monoid M and construct a commutative diagram like the following

$$\begin{array}{ccc} M & \xrightarrow{\beta} & \Omega BM \\ \simeq \uparrow & & \downarrow \simeq \\ \psi(n,k) & \xrightarrow[\alpha_k]{} & \Omega\psi(n,k+1). \end{array} \tag{8}$$

The monoid M will of course depend on k and n, but we shall omit this from the notation. Let us define it.

DEFINITION 3.3. For $0 \le k < n$, let M denote the space

$$M = \{(t,W) \in (0,\infty) \times \psi(n,k+1) | W \subseteq \mathbb{R}^k \times (0,t) \times (0,1)^{n-k-1}\},$$

equipped with the multiplication

$$(t,W)(t',W') = (t+t', W \cup (W' + te_{k+1})).$$

In words, the product puts W and W' next to each other after making them disjoint by translating in the $(k+1)$st direction. The following lemma is obvious.

LEMMA 3.4. *The inclusion $\psi(n,k) \to M$ given by $W \mapsto (W,1)$ is a homotopy equivalence. (In fact a homeomorphism onto its image, which is a deformation retract of M.)* □

LEMMA 3.5. *The monoid M is grouplike for $k \ge 1$.*

PROOF. Let $m = (t,W) \in M$ with $W \subseteq \mathbb{R}^k \times (0,t) \times (0,1)^{n-k-1}$ with $k \ge 1$. We define $W' \subseteq \mathbb{R}^k \times (0,t) \times (0,1)^{n-k-1}$ by rotating W in the (e_k, e_{k+1}) plane around the point $(0, \frac{t}{2})$ and let $m' = (t, W')$. To see that mm' and $m'm$ are both in the path component of the empty set, we draw the cartoon in figure 1. The horizontal

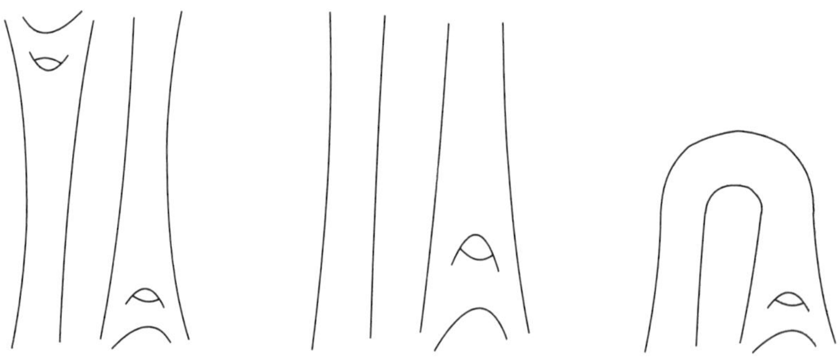

FIGURE 1. Path used in the proof of lemma 3.5

axis in the picture is e_{k+1} and the vertical is e_k. The first frame depicts mm', the disjoint union of W and W'. In the third frame we have instead "bent" the manifold W in the (e_k, e_{k+1})-plane. The bent manifold agrees with mm' whenever the kth coordinate is sufficiently negative. These two frames are connected by "stretching" the kth coordinate (moving the piece where they disagree away to $+\infty$ in the kth coordinate direction), resulting in the manifold depicted in the second frame. This describes a path from mm' to a manifold for which the kth coordinate is bounded above, but then we can push it to $-\infty$ in that direction and get a path to $\emptyset$. (If unconvinced by this description, you should look in [**1**] or better yet, make it rigorous yourself!). □

Next, we define the map $BM \to \psi(n, k+1)$, giving the right hand vertical map in the diagram (8). To this end, let us start with a point $(A, m) \in BM$ with $A = (a_1 < \cdots < a_p) \subseteq \mathbb{R}$ and labels $m_1 = (t_1, W_1), \ldots, m_p = (t_p, W_p)$. Let $b_i \geq a_i$ be the smallest possible numbers such that the intervals $(b_i, b_i + t_i)$ are disjoint (i.e. set $b_1 = a_1$ and inductively $b_{i+1} = \max(a_{i+1}, b_i + t_i)$). Define a subset $W \subseteq \mathbb{R}^n$ as the union

$$W = (W_1 + b_1 e_{k+1}) \cup \cdots \cup (W_p + b_p e_{k+1}).$$

Since $W_i \subseteq \mathbb{R}^k \times (0, t_i) \times (0,1)^{n-k-1}$ and the intervals $(b_i, b_i + t_i)$ are disjoint, W is the union of disjoint elements of $\psi(n, k+1)$, and hence $W \in \psi(n, k+1)$. It is easy to see that the resulting map $BM \to \psi(n, k+1)$ makes the diagram (8) commutative. Let us sketch a proof of its continuity.

LEMMA 3.6. *W depends continuously on $(A, m) \in BM$.*

SKETCH OF PROOF. There are three interesting events to check. The first is what happens when $a_i \in \mathbb{R}$ collides with a_{i+1}. It follows from the definition that W is independent of a_{i+1} as long as $a_{i+1} \leq a_i + t_i$, and that in this case, the value agrees with that of $(a_0 < \cdots < \widehat{a_{i+1}} < \ldots a_p)$ and $(m_1, \ldots, m_i m_{i+1}, \ldots, m_p)$.

The second is what happens when $a_1 \to -\infty$. In this case, W is eventually constant near any compact subset of $\mathbb{R}^n$, and converges to the value at $(a_2 < \cdots < a_p)$ and $(m_2, \ldots, m_p)$. The third interesting event is what happens when $a_p \to \infty$, but this is similar. □

We have defined all maps in the diagram (8), and it is easy to see that the diagram is commutative. We have proved that two of them are weak equivalences. The main result of this lecture is the following.

THEOREM 3.7. *The resulting map $BM \to \psi(n, k+1)$ is a weak equivalence for $k \geq 2$. For $k = 1$, it is a weak equivalence onto the path component of $\psi(n, 2)$ containing the empty manifold.*

Before embarking on the proof, let us point out the main consequence.

COROLLARY 3.8. *The map $\alpha_k : \psi(n, k) \to \Omega\psi(n, k+1)$ is a weak equivalence for $k \geq 1$. Consequently we get a weak equivalence $\Omega\psi(n, 1) \simeq \Omega^n\psi(n, n)$ and hence a weak equivalence*

$$\Omega\psi(\infty, 1) \simeq \Omega^\infty \mathbf{\Psi}.$$

PROOF. By theorem 3.7 and the commutative diagram (8), α_k is the composition of three weak equivalences. □

To motivate the proof of theorem 3.7, let us contemplate what we wish to achieve. To prove surjectivity in π_l, for example, we need to prove that any map $f : S^l \to \psi(n, k+1)$ is homotopic to a map which lifts to BM. In particular, for each $x \in S^l$, we need a path from $W = f(x) \in \psi(n, k+1)$ to a point in the image of $BM \to \psi(n, k+1)$. This is easy, provided W satisfies the following condition.

(9) There exists an $a \in \mathbb{R}$ such that $(\mathbb{R}^k \times \{a\} \times \mathbb{R}^{n-k-1}) \cap W = \emptyset$.

Namely, in that case we can pick a finite set $(a_1 < \cdots < a_{p+1})$ of real numbers with this property, let $t_i = a_{i+1} - a_i$, and let W_i be the part of W which is contained in $\mathbb{R}^k \times (a_i, a_{i+1}) \times \mathbb{R}^{n-k-1}$ (translated by $-a_i e_{k+1}$). Then the finite subset $A = (a_1 < \cdots < a_p)$, labeled by the elements $m_i = (t_i, W_i) \in M$, gives a point $(A, m) \in BM$, and there is an obvious path from $W \in \psi(n, k+1)$ to the image of (A, m) (the path translates $W \cap \mathbb{R}^k \times (a_{p+1}, \infty) \times \mathbb{R}^{n-k-1}$ by a bigger and bigger multiple of e_{k+1} and $W \cap \mathbb{R}^k \times (-\infty, a_1) \times \mathbb{R}^{n-k-1}$ by a bigger and bigger multiple of $-e_{k+1}$). As we shall explain in more detail below, this process can, with slightly more care, be performed for the manifolds $f(x)$ for all $x \in S^l$ at once, giving a continuous lift up to homotopy. Injectivity is similar: we wish to lift a map $f : [0,1] \times S^l \to \psi(n, k+1)$, with a prescribed lift over $\{0,1\} \times S^l$.

This discussion focuses attention on the condition (9), and it is convenient to have a name for it. For a map $f : X \to \psi(n, k+1)$, let us write $X_a \subseteq X$ for the set of elements $x \in X$ such that $W = f(x)$ satisfies (9). Let us say that a smooth map $f : X \to \psi(n, k+1)$ is *good* if $X = \cup_a \mathrm{int}(X_a)$. The following result is the main technical result underlying theorem 3.7.

LEMMA 3.9. *For $1 \leq k < n$ and X a compact manifold (possibly with boundary), any map $f : X \to \psi(n, k+1)$ with image in the basepoint component is homotopic to a good map. The homotopy can be taken constant near any closed set on which f is already good.*

Let us postpone the proof of this lemma until the end of the section. In order to formalize the above discussion of how this implies that $BM \to \psi(n, k+1)$ is a weak equivalence (to the basepoint component if $k = 1$), we define yet another space BP.

DEFINITION 3.10. Let BP be the space whose points are triples (A, C, W), where $A = (a_1 < a_2 < \cdots < a_p) \subseteq \mathbb{R}$, $C = (c_0 < \cdots < c_p) \subseteq \mathbb{R}$, and $W \in \psi(n, k+1)$ satisfies

$$W \cap (\mathbb{R}^k \times C \times \mathbb{R}^{n-k-1}) = \emptyset.$$

We write $a_0 = -\infty$ and $a_{p+1} = \infty$, and think of c_i as a label on the interval $(a_i, a_{i+1}) \subseteq \mathbb{R} - A$, and topologize BP so that the a_i's are allowed to collide and go to $\pm\infty$. (Both processes decrease the number of intervals in $\mathbb{R} - A$, and we forget the corresponding c_i.)

It is easy to factor the map $BM \to \psi(n, k+1)$ through BP, viz. we set $c_i = b_{i+1}$ and $c_p = b_p + t_p$.

LEMMA 3.11. *The forgetful map $BP \to \psi(n, k+1)$ is a weak equivalence.*

SKETCH OF PROOF. To prove that the induced map in π_l is surjective, we use Lemma 3.9 to represent an element of $\pi_l(n, k+1)$ by a good map $S^l \to \psi(n, k+1)$.

By compactness of S^l and by definition of goodness, we can find finitely many numbers $c \in \mathbb{R}$ such that the open sets

$$U_c = \operatorname{int}\{x \in S^l | f(x) \cap \mathbb{R}^k \times \{c\} \times \mathbb{R}^{n-k-1}\}$$

cover S^l. Then we use a partition of unity to define a map $S^l \to BP$ in the following way. Let $\lambda_c : S^l \to [0,1]$, $c \in \mathbb{R}$ be a (smooth, locally finite) partition of unity subordinate to the open sets U_c. For $x \in S^l$, we let $C = \{c \in \mathbb{R} | \lambda_c(x) > 0\}$ and $A = \{\rho(\sum_{c \leq t} \lambda_c(x)) | t \in \mathbb{R}\}$, where $\rho : (0,1) \to \mathbb{R}$ is an increasing homeomorphism. If we then define $S^l \to BP$ by mapping x to $(A, C, W) \in BP$, we have lifted our original good map $S^l \to \psi(n, k+1)$.

Injectivity is similar, using that if two good maps $S^l \to \psi(n, k+1)$ are homotopic, then there exists a homotopy $[0,1] \times S^l \to \psi(n, k+1)$ which is a good map. □

LEMMA 3.12. *The map $BM \to BP$ is a homotopy equivalence.*

PROOF. We first note that BM deformation retracts onto the subspace $B'M \subseteq BM$, defined by the inequalities $a_i + t_i \geq a_{i+1}$. The deformation increases t_i linearly until the inequality holds.

Secondly, we note that BP deformation retracts onto the subspace $B'P \subseteq BP$, defined by the requirements $a_1 = c_0$, $a_i \leq c_{i-1}$, and $W \subseteq \mathbb{R}^k \times (c_0, c_p) \times \mathbb{R}^{n-k-1}$. To see the deformation retraction, we write $t_i = c_i - c_{i-1}$, and let $c'_{i-1} \geq a_i$ be the smallest numbers such that $c'_i - c'_{i-1} \geq t_i$. The deformation retraction of BP onto $B'P$ deforms c_i to c'_i in a linear fashion. At the same time, it moves the part of W which lies in $\mathbb{R}^k \times (c_{i-1}, c_i) \times \mathbb{R}^{n-k-1}$ to $\mathbb{R}^k \times (c'_{i-1}, c'_{i-1} + t_i) \times \mathbb{R}^{n-k-1}$ in a linear fashion. The part of W which is in $\mathbb{R}^k \times (-\infty, c_0) \times \mathbb{R}^{n-k-1}$ is pushed away in the direction of $-e_{k+1}$, and the part of W which is in $\mathbb{R}^k \times (c_p, \infty) \times \mathbb{R}^{n-k-1}$ is pushed away in the direction of e_{k+1}.

Finally, we note that the map $BM \to BP$ restricts to a homeomorphism $B'M \to B'P$. The inverse is given by setting $t_i = c_i - c_{i-1}$ and letting W_i be the part of W which lies in $\mathbb{R}^k \times (c_{i-1}, c_i) \times \mathbb{R}^{n-k-1}$. □

PROOF OF LEMMA 3.9. We first give the proof in the case $f : X \to \psi(n, k+1)$ is a smooth map satisfying the following assumption: for all $x \in X$, the restriction of the projection $\mathbb{R}^n \to \mathbb{R}^{k+1}$ onto the first coordinates to the submanifold $f(x) \subseteq \mathbb{R}^n$ is *not* surjective. (This hypothesis is of course automatic in the case $k \geq 2$ since $f(x) \subseteq \mathbb{R}^n$ is a 2-dimensional manifold.) In that case, we pick for all x a point $q_x \in \mathbb{R}^{k+1}$ not in the image of the projection $f(x) \to \mathbb{R}^{k+1}$. We can also pick an $\varepsilon_x > 0$ so that the (closed) ε_x-disk around q_x is disjoint from the image of that projection. The same choices will work in a neighborhood U_x of x, so by compactness of X we can find $\varepsilon > 0$ and finitely many open sets $U_i \subseteq X$ with corresponding $q_i \in \mathbb{R}^{k+1}$ such that the ε-ball around q_i consists of points not in the image of $f(x) \to \mathbb{R}^{k+1}$ for all $x \in U_i$. Writing $q_i = (p_i, t_i) \in \mathbb{R}^k \times \mathbb{R}$, we can assume that the t_i are distinct and that the intervals $[t_i - \varepsilon, t_i + \varepsilon]$ are disjoint (after possibly shrinking ε). Then, we can find an isotopy of diffeomorphisms of $\mathbb{R}^k \times \mathbb{R}$ supported in $\cup_i \mathbb{R}^k \times (t_i - \varepsilon, t_i + \varepsilon)$ which starts at the identity and ends at a map that sends $(p_i, t_i) \mapsto (0, t_i)$. Using this isotopy to deform each $f(x) \subseteq \mathbb{R}^n$, we may assume that all $p_i = 0$, and hence that all $f(x) \subseteq \mathbb{R}^n$ is disjoint from some $B(0, \varepsilon) \times \{t_i\} \times \mathbb{R}^{n-k-1}$. Finally, we may pick an isotopy of embeddings $e_s : \mathbb{R}^k \to \mathbb{R}^k$ that starts at the identity and ends in a map with $e_1(\mathbb{R}^k)$ contained in the ε-ball. Then the isotopy of embeddings $\phi_s = e_s \times \mathrm{id} : \mathbb{R}^k \times \mathbb{R}^{n-k} \to \mathbb{R}^k \times \mathbb{R}^{n-k}$ gives a path $f_s(x) = \phi_s^{-1}(f(x)) \in \psi(n, k+1)$ from $f_0(x)$ to an element satisfying that $f_1(x) \subseteq \mathbb{R}^n$ is disjoint from $\mathbb{R}^k \times \{t_i\} \times \mathbb{R}^{n-k-1}$ for all $t \in U_i$. Thus, f_1 is good, and we have finished the proof in the case $k \geq 2$.

In the remaining case $k = 1$, we will prove that f can be homotoped to a map satisfying the extra non-surjectivity assumption in the first part of the proof. For clarity, let us first discuss the case where X is a point. (That case is of course trivial, but the point is to explain what the general proof does for each $x \in X$.) In that case we are given an element $W \in \psi(n, 2)$ in the path component of the basepoint. Thus $W \subseteq \mathbb{R}^n$ is a topologically closed 2-manifold contained in $\mathbb{R}^2 \times (0,1)^{n-2}$, and there is a smooth path $\mathbb{R} \to \psi(n, 2)$, constant near $(-\infty, 0]$ and $[1, \infty)$ given by a 3-manifold $E \subseteq \mathbb{R} \times \mathbb{R}^n$ with $E \cap \{0\} \times \mathbb{R}^n = \{0\} \times W$ and $E \cap \{1\} \times \mathbb{R}^n = \emptyset$. Projection onto the second and third coordinates gives a proper smooth map $W \to \mathbb{R}^2$, and we may pick a regular value (p, t). Then $(0, p, t) \in \mathbb{R}^3$ is a regular value for the projection $E \to \mathbb{R}^3$, and after an isotopy of self-diffeomorphisms of $\mathbb{R}^2$ we may assume $p = 0$. By properness, we may also pick an $\varepsilon > 0$ so that no critical value of $E \to \mathbb{R}^3$ is contained in the cube $(0,0,t) + [-\varepsilon, \varepsilon]^3$. We then pick a smooth function $\lambda : \mathbb{R} \to [0,1]$ which is 1 near 0 and has support in $(-\varepsilon, \varepsilon)$, and consider the maps $\phi_r : \mathbb{R}^n \to \mathbb{R} \times \mathbb{R}^n$, $r \in [0,1]$, given by

$$\begin{aligned} &\mathbb{R}^n \to \mathbb{R} \times \mathbb{R}^n \\ &x \mapsto (x_1 \sin(\tfrac{r\pi}{2}\lambda(x_2 - t)), x_1 \cos(\tfrac{r\pi}{2}\lambda(x_2 - t)), x_2, \dots, x_n) \end{aligned}$$

When $r = 0$, this is just the inclusion as $\{0\} \times \mathbb{R}^n$, but as $r \in [0,1]$, it rotates in the first two coordinates, dampened by the bump function, so that the rotation happens only near $x_2^{-1}(t)$ and that ϕ_r is independent of r unless $x_2 \in (t - \varepsilon, t + \varepsilon)$. We then consider the subsets $W(r) = \phi_r^{-1}(E) \subseteq \mathbb{R}^n$ for $r \in [0,1]$. We have $W(0) = W$, and all $W(r)$ are closed subsets of $\mathbb{R}^n$ contained in $\mathbb{R}^2 \times (0,1)^{n-2}$, but they are not necessarily smooth manifolds because ϕ_r need not be transverse to E. However, our assumptions imply that it is transverse near the subset

$$(\{0\} \times \mathbb{R} \cup \mathbb{R} \times \{t\}) \times \mathbb{R}^{n-2}$$

and hence $W(r)$ is smooth near that subset for all $r \in [0,1]$. After (carefully) pushing singularities to infinity in the x_1 direction, we get a family of manifolds $\tilde{W}(r) \in \psi(n, 2)$ starting at $\tilde{W}(0) = W$ and ending at $\tilde{W}(1)$ which satisfies $\tilde{W}(1) \cap \mathbb{R} \times \{0\} \times \mathbb{R}^{n-2} = \emptyset$ as desired. Precisely, "pushing singularities to infinity in the x_1 direction" means that we set $\tilde{W}(r) = (e \times \mathrm{id})^{-1}(W(r))$, where $e : \mathbb{R}^2 \to \mathbb{R}^2$ is an embedding which is isotopic to the identity and restricts to the identity near $\{0\} \times \mathbb{R} \cup \mathbb{R} \times \{t\}$.

The case of a general X is similar. First we use the hypothesis that f maps to the basepoint component of $\psi(n, 2)$ to pick, for each $x \in X$, a contractible open neighborhood $U_x \subseteq X$ and a null homotopy of $f|U_x$, given by a smooth map $h_x : U_x \times \mathbb{R} \to \psi(n, 2)$. For $y \in U_x$, we shall identify the smooth map $h_x(y, -) : \mathbb{R} \to \psi(n, 2)$ with its graph, which is a three-dimensional smooth submanifold $E_x(y) \subseteq \mathbb{R} \times \mathbb{R}^n$. The coordinates in $\mathbb{R} \times \mathbb{R}^n$ shall be written $(s, x_1, \dots, x_n)$, and we consider the restriction $(s, x_1, x_2) : E_x(y) \to \mathbb{R}^3$. We can pick a regular value of the form $(0, p_x, t_x)$ and an $\varepsilon_x > 0$ so that, after possibly shrinking U_x, no critical point of $(s, x_1, x_2) : E_x(y) \to \mathbb{R}^3$ is contained in the set

$$((0, p_x, t_x) + [-\varepsilon_x, \varepsilon_x]^3) \times \mathbb{R}^{n-2}.$$

We can then refine the U_x's to a cover of X by finitely many U_i with corresponding regular values $(0, p_i, t_i)$ and $\varepsilon_i = \varepsilon > 0$. As before, we can arrange that the t_i's are distinct, the intervals $[t_i - \varepsilon, t_i + \varepsilon]$ are disjoint, and that all $p_i = 0$. We then pick a smooth function $\lambda : \mathbb{R} \to [0, 1]$ which is 1 near 0 and has support in $(-\varepsilon, \varepsilon)$, and smooth functions $\rho_i : X \to [0, 1]$ with support in U_i and such that X is covered by the open sets $U_i' = \operatorname{int}(\rho_i^{-1}(1))$. Then consider the maps $\phi_{y,i,r} : \mathbb{R}^n \to \mathbb{R} \times \mathbb{R}^n$, given by

$$\begin{aligned} &\mathbb{R}^n \to \mathbb{R} \times \mathbb{R}^n \\ &\quad x \mapsto (x_1 \sin(\tfrac{r\pi}{2}\rho_i(y)\lambda(x_2 - t_i)), x_1 \cos(\tfrac{r\pi}{2}\rho_i(y)\lambda(x_2 - t_i)), x_2, \dots, x_n) \end{aligned}$$

and define subsets

$$W_i(y, r) = \phi_{y,i,r}^{-1}(E_i(y)) \subseteq \mathbb{R}^n.$$

Again, these are closed subsets contained in $\mathbb{R}^2 \times (0, 1)^{n-2}$ which agree with $f(y)$ outside $x_2^{-1}(t_i - \varepsilon, t_i + \varepsilon)$. We can therefore define a subset $W(y, r) \subseteq \mathbb{R}^n$ which agrees with $W_i(y, r)$ inside $x_2(t_i - \varepsilon, t_i + \varepsilon)$ and with $f(y)$ outside these sets. The subsets $W(y, r) \subseteq \mathbb{R}^n$ are closed and contained in $\mathbb{R}^2 \times (0, 1)^{n-2}$, but are not necessarily smooth manifolds, although they are smooth near $\{0\} \times \mathbb{R}^{n-1}$ and $\mathbb{R} \times \{t_i\} \times \mathbb{R}^{n-2}$ if $y \in U_i'$. After carefully pushing all singularities to infinity in the x_1 direction (by an analogue of what we did in the case X is a point), we obtain smooth manifolds $\tilde{W}(y, r) \in \psi(n, 2)$, giving a homotopy $X \times \mathbb{R} \to \psi(n, 2)$ from f to a map which satisfies the hypothesis in the first part of the proof. □

This finishes our proof of the weak equivalence $\psi(n, 1) \to \Omega^{n-1}\Psi(\mathbb{R}^n)$.

3.2. Exercises for Lecture 3

(1) Use theorem 3.2 to prove that if $f : M \to M'$ is a map of group-like topological monoids, and f is a weak equivalence (of the underlying topological spaces), then $Bf : BM \to BM'$ is a weak equivalence.

(2) Prove that if M is commutative, then BM is a commutative monoid (define a product on BM which takes union of finite subsets of $\mathbb{R}$, possibly multiplying labels). Explain why $B(BM)$ is the "space of configurations of points in $\mathbb{R}^2$, labeled by elements of M".

(3) Let $N = \{1, 2, 3, \dots\}$ have monoid structure given by addition and pick any homeomorphism $\lambda : S^1 = \mathbb{R} \cup \{\infty\} \to U(1)$ with $\lambda(\infty) = 1$. (For example $\lambda(t) = (t + i)(t - i)$.) Prove that there is a unique monoid map $BN \to U(1)$ which takes the point $(\{t\}, 1)$ to $\lambda(t) \in U(1)$. (The results of the following two exercises imply that this map is a homotopy equivalence.)

(4) With N as in the previous exercise, let $B'N \subseteq BN$ consist of labeled configurations where the sum of the labels of points in $(0, 1)$ is at most 1. Prove that $B'N \simeq S^1$. (Hint: "push to infinity".)

(5) With $B'N$ as in the previous exercise, prove that the inclusion $B'N \to BN$ is a homotopy equivalence. (Hint: There are several ways to construct a homotopy inverse, one is as follows: If the number $a_i \in \mathbb{R}$ have label n_i, we let $b_i \geq a_i$ be the smallest numbers such that the intervals $(b_i, b_i + n_i)$ are disjoint. If we give each point $b_i, b_i + 1, \dots, b_i + n_i - 1$ the label 1, we have a point in $B'N$.)

(6) Prove that the inclusion $N \to \mathbb{Z}$ induces a weak equivalence $BN \to B\mathbb{Z}$.

(7) Prove that diagram (8) is commutative.

(8) The "Moore loop space" of a based space X is the space of pairs (t, γ), where $t \geq 0$ and $\gamma : [0, t] \to X$ is a loop. This space is naturally a topological monoid (add the t's, concatenate the loops), and we denote it $\Omega' X$. Prove that the inclusion $\Omega X \to \Omega' X$ is a homotopy equivalence and that $\Omega' X$ is a grouplike topological monoid and hence (by theorem 3.2) that $\beta : \Omega' X \to \Omega B(\Omega' X)$ is a weak equivalence. Then construct a map $B\Omega' X \to X$ which is a weak equivalence onto the basepoint component of X.

(9) Let $\lambda : (0,1) \to \mathbb{R}$ be a homeomorphism and M a topological monoid. For $t = (t_0, \dots, t_p) \in \Delta^p$ we have points $a_n = \lambda(\sum_{i=0}^{n-1} t_i) \in [-\infty, \infty]$ for $1 \leq n \leq p$. If we are also given $(m_1, \dots, m_p) \in M^p$ and label a_n by m_n, we have defined a map $\phi_p : \Delta^p \times M^p \to BM$ (if some a_i's coincide, we multiply the labels; if some a_i are infinite, we forget their labels). Prove that ϕ_p is continuous. (Remark: BM is often defined as the "geometric realization of the nerve" of M. The maps defined in this exercise glue to a map from the thick realization of $N_\bullet M$ to the space BM defined in the text; the resulting map is a continuous bijection and a homotopy equivalence, with homotopy inverse constructed as follows. Using the functions $s_i = t_i / \max(t_i)$ as coordinates on Δ^p, the self-maps $s_i \mapsto \max(0, 2s_i - 1)$ induces a self-map of $\|N_\bullet M\|$, which precomposed with the inverse of the continuous bijection constructed in this exercise can be shown to be a (continuous!) homotopy inverse.)

LECTURE 4

Final step of the proof

It remains to study the map $\alpha_0 : \psi(n,0) \to \Omega\psi(n,1)$. Let us contemplate applying the same methods as we did for $k \geq 1$ and see where it goes wrong. The exact same proof as for $k > 0$ shows that $\psi(n,0)$ is homotopy equivalent to a monoid M. However, the monoid $\pi_0 M$ is the set of diffeomorphism classes of closed oriented 2-manifolds (at least for $n \geq 5$), where the monoid operation is disjoint union. Firstly, this monoid is not a group, so we won't have $M \simeq \Omega BM$. Secondly, the natural map $BM \to \psi(n,1)$ has no chance of being a weak equivalence, because BM is much too large: We will show in the exercises that $\pi_1\psi(n,1) = \mathbb{Z}$ while $\pi_1(BM)$ contains an abelian group of infinite rank.

In retrospect, it is also clear that $\psi(n,0)$ is not quite the right object: The Madsen–Weiss theorem concerns surfaces that are *connected* and have *high genus*, whereas $\psi(n,0)$ contain all surfaces. If we restrict to the subspace consisting of connected surfaces, we no longer have that $\psi(n,0)$ is homotopy equivalent to a monoid (the monoid operation is essentially disjoint union). The solution to these problems is to modify $\psi(n,0)$ in a way that we only have path connected surfaces, but still have a monoid operation. To achieve this, we will consider surfaces with boundary, and construct a monoid operation which glues surfaces along their boundary. More precisely, we make the following definition.

DEFINITION 4.1. Write $L_t = [0,t] \times [0,1] \subseteq \mathbb{R}^2$ and let M be the set of pairs (t, W) where $t > 0$ and $W \subseteq L_t \times (-1,1)^{n-2}$ is a compact, connected, oriented 2-dimensional submanifold which agrees with $L_t \times \{0\}$ near $(\partial L_t) \times \mathbb{R}^{n-2}$. Define a product operation on M as

$$(t, W)(t', W') = (t + t', W \cup (W' + te_1)).$$

In order to describe the topology, we note that M is in bijection with the set of (t, W), where $W \in \Psi(\mathbb{R}^n)$ agrees with $\mathbb{R}^2 \times \{0\}$ outside $L_t \times \mathbb{R}^{n-2}$ and has $W \subseteq \mathbb{R}^2 \times (-1,1)^{n-2}$. Then we topologize as a subspace of $\Psi(\mathbb{R}^n)$.

LEMMA 4.2. *M is a homotopy commutative topological monoid. If $n \geq 5$, $\pi_0 M = \mathbb{N}$.*

PROOF. Homotopy commutativity is proved using the same picture as one uses to prove that π_2 of a space is abelian.

The map $M \to \mathbb{N}$ which maps a connected surface to its genus gives a monoid map $\pi_0 M \to \mathbb{N}$ which is surjective for $n \geq 3$ (because a genus 1 surface can be embedded in $L_t \times \mathbb{R}$) and injective for $n \geq 5$ (because any two embeddings of a genus g surface are isotopic in that case). □

In fact, we can say a bit more when n is large. Indeed, Lemma 1.3 can again be used to interpret M as a classifying space for smooth surface bundles (at least in the limit where $n \to \infty$), but now it classifies bundles of connected surfaces with one parametrized boundary component. Thus, for $n = \infty$ we have

$$M = \coprod_{g \geq 0} B\mathrm{Diff}^{\partial}(\Sigma_{g,1}),$$

where $B\mathrm{Diff}^{\partial}(\Sigma_{g,1})$ classifies smooth surface bundles $E \to X$ whose fibers are connected genus g surfaces and $\partial E = X \times S^1$.

In this last lecture we shall prove the following result, which is the proper replacement of theorem 3.7 for $k = 0$.

THEOREM 4.3. *For $n = \infty$, there is a weak equivalence $BM \to \psi(n, 1)$.*

The $BM \to \psi(n, 1)$ can be seen to fit into a homotopy commutative diagram

$$\begin{array}{ccc} M & \xrightarrow{\beta} & \Omega BM \\ \downarrow & & \downarrow \\ \psi(n,0) & \xrightarrow[\alpha_0]{} & \Omega\psi(n,1) \end{array}$$

which is the analogue for theorem 4.3 of the diagram (8) for theorem 3.7. Using the theorem, we are *almost* ready to copy the steps involved in the case $k > 0$. The only caveat is that the monoid M still is not group-like, so $\beta : M \to \Omega BM$ is not a weak equivalence. A striking result, known as the "group completion" theorem (see [**4**]), describes the induced map $H_*(M) \to H_*(\Omega BM)$ as an algebraic localization. More precisely, let $m_0 \in M$ be a surface of genus 1, and let M_∞ be the mapping telescope of the direct system

$$M \xrightarrow{\cdot m_0} M \xrightarrow{\cdot m_0} \dots.$$

The mapping telescope is defined as the quotient space

$$(M \times \mathbb{N} \times [0, 1])/((m, n, 1) \sim (mm_0, n + 1, 0)).$$

and in our case, it can be rewritten as $\mathbb{Z} \times B\mathrm{Diff}_\infty$, where $B\mathrm{Diff}_\infty$ is the mapping telescope of the direct system

$$\dots \to B\mathrm{Diff}(\Sigma_{g,1}) \to B\mathrm{Diff}(\Sigma_{g+1,1}) \to \dots.$$

We regard $M \subset M_\infty$ as the subspace $M \times \{0\} \times \{0\}$. In this case, the group completion theorem implies that the canonical map $\beta : M \to \Omega BM$ extends to a map

$$\beta_\infty : M_\infty \to \Omega BM$$

which induces an isomorphism in integral homology. In the limit $n \to \infty$, we have

$$H_*(M) = \bigoplus H_*(B\mathrm{Diff}^{\partial}(\Sigma_{g,1}))$$

and

$$H_*(M_\infty) = H_*(M)[m_0^{-1}] = H_*(\mathbb{Z} \times B\mathrm{Diff}_\infty).$$

Combining with what we proved previously, we have a map

$$B\mathrm{Diff}_\infty \to \Omega_0^\infty \mathbf{\Psi},$$

inducing an isomorphism in integral homology. Together with homological stability, this proves the Madsen–Weiss theorem.

4.1. Proof of theorem 4.3

We now want to prove the weak equivalence $BM \simeq \psi(n,1)$. The natural map ends up in a modified space $\psi'(n,1)$, defined as follows.

DEFINITION 4.4. Let $\psi'(n,1)$ be the space of topologically closed submanifolds $W \subseteq \mathbb{R} \times [0,1] \times \mathbb{R}^{n-2}$ which are contained in $\mathbb{R} \times [0,1] \times (-1,1)^{n-2}$ and agree with $\mathbb{R} \times [0,1] \times \{0\}$ near $\mathbb{R} \times \partial[0,1] \times \mathbb{R}^{n-2}$.

LEMMA 4.5. *The map* $\psi(n,1) \to \psi'(n,1)$ *which maps* W *to* $W \cup \mathbb{R} \times [0,1] \times \{0\}$ *is a homotopy equivalence for* $n \geq 3$.

PROOF. There is a homotopy inverse constructed in the following way. First pick a compact 1-manifold $S \subseteq \mathbb{R}^2 - (0,1) \times (-1,1)$ with $\partial S = (\partial[0,1]) \times \{0\}$. For future use, we also assume that S is contained in $[-2,2] \times [-2,2]$, contains $[-1,1] \times \{-2\}$, and is collared near its boundary. For $W \in \psi'(n,1)$, the union $W \cup \mathbb{R} \times S \times \{0\} \subseteq \mathbb{R}^n$ will be a smooth manifold, and we can arrange (by choice of S) that it is contained in $\mathbb{R} \times (-3,3)^{n-1}$. Letting $\lambda : (-3,3) \to (0,1)$ be the increasing affine diffeomorphism, we then define a map $\psi'(n,1) \to \psi(n,1)$ as

$$W \mapsto (\mathrm{id} \times \lambda^{n-1})(W \cup \mathbb{R} \times S \times \{0\}).$$

The composition $\psi'(n,1) \to \psi(n,1) \to \psi'(n,1)$ sends W to a manifold depicted in the first frame of the cartoon in figure 1. The cartoon also shows how to find a

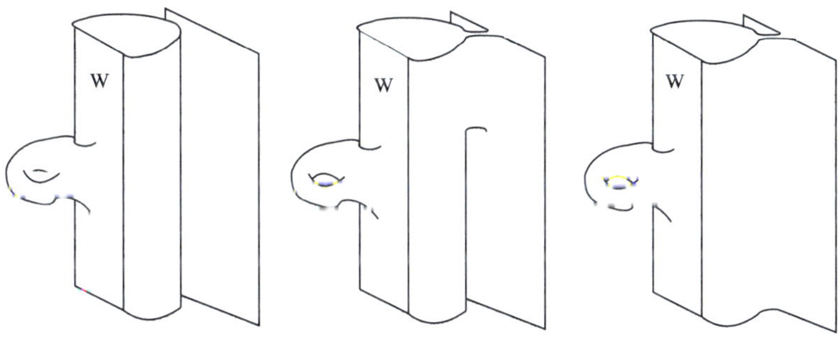

FIGURE 1. Path used in the proof of lemma 4.5

canonical path back to W: The image of W contains a piece which up to scaling of coordinates is the disjoint union of $\mathbb{R} \times S \times \{0\}$ and a plane (namely the complement of the original W). Then slide down a "saddle point" to get a manifold which can be canonically deformed to the original W. The other composition is similar but easier. □

With M the monoid of connected surfaces defined above, we define $BM \to \psi'(n,1)$ analogously to what we did for $k > 0$. In that case, an important notion was that of a "good map" $X \to \psi(n,k+1)$. It is key to the case $k=0$ to have the right notion of good map $X \to \psi'(n,1)$ in this case.

DEFINITION 4.6. Let X be a smooth manifold and $f : X \to \psi'(n,1)$ a smooth map. For $a \in \mathbb{R}$, let $X_a \subseteq X$ be the set of points x satisfying

$$f(x) \cap (\{a\} \times \mathbb{R}^{n-1}) = \{a\} \times [0,1] \times \{0\}.$$

Furthermore, let $X^{\mathrm{nc}} \subseteq X$ be the set of points x satisfying that no path component of $f(x) \subseteq \mathbb{R} \times [0,1] \times \mathbb{R}^{n-2}$ is compact. Let us say that f is good provided $X = X^{\mathrm{nc}} = \cup_a \mathrm{int}(X_a)$.

PROPOSITION 4.7. *Any smooth map $f : X \to \psi'(n,1)$ is smoothly homotopic to a good map if $n \geq 5$. The homotopy can be taken constant near a closed set on which f is already good.*

SKETCH OF PROOF. We first deform f to achieve that no path component of any $f(x) \subseteq \mathbb{R} \times (-1,1)^{n-1}$ is contained in $\{0\} \times (-1,1)^{n-1}$. Then we pick an isotopy of embeddings $e_s : \mathbb{R} \to \mathbb{R}$ which has $e_0 = \mathrm{id}$ and $e_1(\mathbb{R}) \subseteq (-\varepsilon, \varepsilon)$. Then we deform $f(x)$ through the path $f_s(x) = (e_s \times \mathrm{id})^{-1}(f(x))$. After this deformation, $f(x)$ has no compact path components.

Next, we want to change f in order to have that for each x there exists t such that $f(x) \cap x_1^{-1}(t)$ is diffeomorphic to an interval. We will explain how to do this for a single x at a time. If we pick a generic $t \in \mathbb{R}$, the 1-manifold $f(x) \cap x_1^{-1}(t)$ will be a disjoint union of an interval and a finite number of circles (since it is a compact 1-manifold with boundary $\partial[0,1]$). If there are no circle components, we are done, otherwise we pick a point in each circle component and join it with a tube to a point near the standard boundary, as in figure 2. The left picture shows the

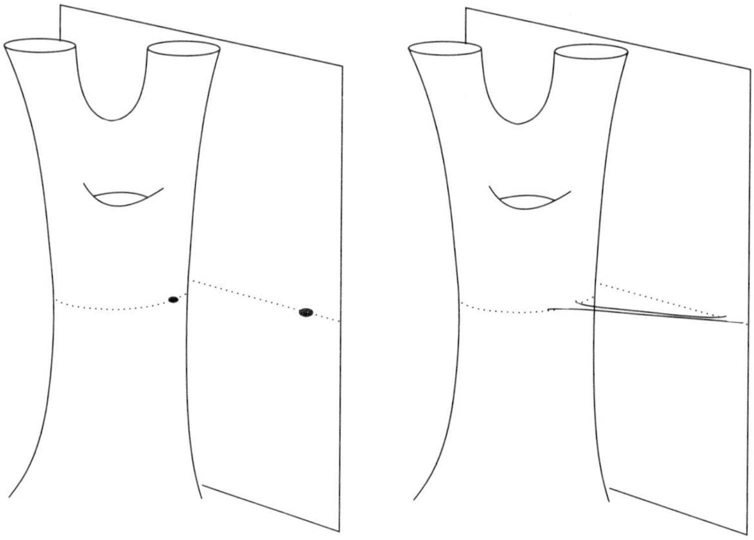

FIGURE 2. Construction in the proof of proposition 4.7

original manifold $W = f(x)$. The 1-manifold $W \cap x_1^{-1}(t)$ is in this case the disjoint union of an interval and a circle. We then pick a point in the circle component and one in the interval component (labeled by dots in the picture) and join them with a tube in the surrounding euclidean space (possible when $n \geq 4$). If we let W' denote the result of this procedure, then W' now satisfies that $W' \cap x_1^{-1}(t)$ is an interval (namely the connected sum of the circle and the interval in $W \cap x_1^{-1}(t)$). Finally we need a path from W to W', but this can be achieved by sliding the tube away to ∞ (this is possible since W has no compact components which could trap the tube).

Once we have achieved that $W' \cap x_1^{-1}(t)$ is diffeomorphic to an interval, we can move it to the standard interval $\{t\} \times [0,1] \times \{0\}$ using an isotopy of embeddings $W' \to \mathbb{R}^n$. (For $n \geq 5$ the space of embeddings is path connected.)

In general we need a *homotopy* from f to a good map f'. This is a little harder, but uses the same idea. Roughly speaking, we apply the above procedure "for each $f(x)$", but some care is required for the result to be a continuous homotopy. The details can be found in [**1**] or [**3**]. □

Using proposition 4.7 we finish the proof in the same way as for $k > 0$.

4.2. Exercises for Lecture 4

(1) Let $N = \{1, 2, \dots\}$ with addition as monoid structure. Describe the map $\beta : N \to \Omega BN$ up to homotopy, both explicitly (using the homotopy equivalence $BN \simeq S^1$ from the previous lecture's exercises) and using the "group completion theorem".

(2) Let M be the monoid which is homotopy equivalent to $\psi(n, 0)$ (i.e. pairs (t, W) with $W \subseteq (0, t) \times (0, 1)^{n-1}$). For each $g \geq 0$ construct a map $BM \to BN$ which "counts the number of genus g components" (i.e. the composition $M \to \Omega BM \to \Omega BN$ sends $[W] \in \pi_0 M$ to the element of $\pi_0 \Omega BN = \pi_1 BN = \mathbb{Z}$ which is the number of path components of W which have genus g. Use these maps to prove that $\pi_1 BM$ surjects to a free abelian group of countable rank.

(3) We proved earlier that $\kappa_0 \in H^0(\Omega^n \Psi(\mathbb{R}^n)) = H^0(\Omega \psi(n, 1))$ is divisible by 2. Prove that for large n, $\kappa_0/2$ gives rise to an isomorphism $\pi_1 \psi(n, 1) \to \mathbb{Z}$. (Hint: One method is to use the Serre spectral sequence to calculate H_{n-1} and H_n of the homotopy fiber of a map $u : \mathrm{Th}(\gamma_n^\perp) \to K(\mathbb{Z}, n-2)$ representing the Thom class.)

Bibliography

[1] S. Galatius, O. Randal-Williams: *Monoids of moduli spaces of manifolds*, Geom. Topol. **14** (2010), 1243–1302. MR2653727 (2011j:57047)

[2] S. Galatius: *Stable homology of automorphism groups of free groups*, Ann. of Math. **173** (2011), 705–768. MR2784914 (2012c:20149)

[3] A. Hatcher: *A short exposition of the Madsen–Weiss theorem*, preprint. arXiv:1103.5223v1 [math.GT]

[4] D. McDuff, G. Segal: *Homology fibrations and the "group completion" theorem*, Invent. Math. **31** (1975/76), no. 3, 279–284. MR0402733 (53:6547)

[5] I. Madsen, M. Weiss: *The stable moduli space of Riemann surfaces: Mumford's conjecture*, Ann. of Math. **165** (2007), 843–941. MR2335797 (2009b:14051)

The Torelli Group and Congruence Subgroups of the Mapping Class Group

Andrew Putman

IAS/Park City Mathematics Series
Volume 20, 2011

The Torelli Group and Congruence Subgroups of the Mapping Class Group

Andrew Putman

Introduction

Let $\Sigma_{g,n}$ be a compact oriented genus g surface with n boundary components. The *mapping class group* of $\Sigma_{g,n}$, denoted $\text{Mod}_{g,n}$, is the group of orientation-preserving diffeomorphisms of $\Sigma_{g,n}$ that restrict to the identity on $\partial\Sigma_{g,n}$, modulo isotopies that fix $\partial\Sigma_{g,n}$. The group $\text{Mod}_{g,n}$ plays a fundamental role in many areas of mathematics, ranging from low-dimensional topology to algebraic geometry. At least for large g, the cohomology of $\text{Mod}_{g,n}$ is well-understood due to the resolution of the Mumford conjecture by Madsen and Weiss [**24**].

However, the cohomology of finite-index subgroups of $\text{Mod}_{g,n}$ remains a mystery. In these notes, we will focus on one low-degree calculation. Consider $n \in \{0, 1\}$. For an integer p, the *level p congruence subgroup* of $\text{Mod}_{g,n}$, denoted $\text{Mod}_{g,n}(p)$, is the subgroup of $\text{Mod}_{g,n}$ consisting of mapping classes that act trivially on $\text{H}_1(\Sigma_{g,n}; \mathbb{Z}/p)$. Another description of $\text{Mod}_{g,n}(p)$ is as follows. The action of $\text{Mod}_{g,n}$ on $\text{H}_1(\Sigma_{g,n}; \mathbb{Z})$ preserves the algebraic intersection pairing. Since $n \leq 1$, this is a nondegenerate alternating form, so we obtain a representation $\text{Mod}_{g,n} \to \text{Sp}_{2g}(\mathbb{Z})$. Classically this representation was known to be surjective (see §1). Let $\text{Sp}_{2g}(\mathbb{Z}, p)$ be the subgroup of $\text{Sp}_{2g}(\mathbb{Z})$ consisting of matrices which equal the identity modulo p. Then $\text{Mod}_{g,n}(p)$ is the pullback of $\text{Sp}_{2g}(\mathbb{Z}, p)$ to $\text{Mod}_{g,n}$.

These notes will discuss the calculation of $\text{H}^2(\text{Mod}_{g,n}(p); \mathbb{Z})$. One motivation for this is the study of line bundles on the finite cover of the moduli space of curves associated to $\text{Mod}_{g,n}(p)$, which is known as the moduli space of curves with level p structures. The first Chern class of such a line bundle lies in $\text{H}^2(\text{Mod}_{g,n}(p); \mathbb{Z})$, and the determination of $\text{H}^2(\text{Mod}_{g,n}(p); \mathbb{Z})$ is the heart of the paper [**33**], which gives a complete classification of such line bundles. However, in these notes we will ignore this connection to algebraic geometry. Instead, we will use the computation of this cohomology group as an excuse to discuss a number of interesting topics related to the mapping class group.

The universal coefficients exact sequence for $\text{H}^2(\text{Mod}_{g,n}(p); \mathbb{Z})$ takes the form

$$0 \longrightarrow \text{Ext}(\text{H}_1(\text{Mod}_{g,n}(p); \mathbb{Z}), \mathbb{Z}) \longrightarrow \text{H}^2(\text{Mod}_{g,n}(p); \mathbb{Z}) \longrightarrow \text{Hom}(\text{H}_2(\text{Mod}_{g,n}(p); \mathbb{Z}), \mathbb{Z}) \longrightarrow 0.$$

Department of Mathematics, MS 136, Rice University, PO Box 1892, Houston, Texas 77251
E-mail address: andyp@rice.edu
Supported in part by NSF grant DMS-1005318

The third and fourth lecture will be devoted to calculating the kernel and cokernel of this exact sequence. They will be proceeded by two lectures on necessary background. Let us now give a more detailed description of the four lectures.

- Lecture 1 will be devoted to the *Torelli group.* Denoted $\mathcal{I}_{g,n}$, this is the subgroup of $\text{Mod}_{g,n}$ consisting of mapping classes that act trivially on $\text{H}_1(\Sigma_{g,n};\mathbb{Z})$. There are short exact sequences
$$1 \longrightarrow \mathcal{I}_{g,n} \longrightarrow \text{Mod}_{g,n} \longrightarrow \text{Sp}_{2g}(\mathbb{Z}) \longrightarrow 1$$
and
$$1 \longrightarrow \mathcal{I}_{g,n} \longrightarrow \text{Mod}_{g,n}(p) \longrightarrow \text{Sp}_{2g}(\mathbb{Z},p) \longrightarrow 1,$$
and the structure of $\text{Mod}_{g,n}(p)$ is a sort of mixture of the structure of $\mathcal{I}_{g,n}$ and $\text{Sp}_{2g}(\mathbb{Z},p)$.
- Lecture 2 will be devoted to the *Johnson homomorphism.* Set $\mathcal{H} = \text{H}_1(\Sigma_{g,n};\mathbb{Z})$. The Johnson homomorphism is a surjective homomorphism $\tau : \mathcal{I}_{g,1} \longrightarrow \wedge^3 \mathcal{H}$. A deep theorem of Johnson shows that the Johnson homomorphism gives the "rational part" of the abelianization of $\mathcal{I}_{g,1}$. More precisely, $\text{H}_1(\mathcal{I}_{g,1};\mathbb{Z}) \cong W \oplus \wedge^3 \mathcal{H}$, where W consists of torsion (in fact, 2-torsion). We will also construct a "mod p" version of the Johnson homomorphism which takes the form $\tau_p : \text{Mod}_{g,1}(p) \longrightarrow \mathcal{H}_p$, where $\mathcal{H}_p = \text{H}_1(\Sigma_{g,n};\mathbb{Z}/p)$.
- Lecture 3 is devoted to calculating $\text{H}_1(\text{Mod}_{g,1}(p);\mathbb{Z})$ for odd p. See the beginning of that lecture for why we restrict to odd p and do not consider the closed case. There are two basic pieces. The first comes from the mod p Johnson homomorphism and the second comes from the abelianization of $\text{Sp}_{2g}(\mathbb{Z},p)$.
- Lecture 4 is devoted to proving that $\text{H}_2(\text{Mod}_g(p);\mathbb{Q}) \cong \mathbb{Q}$. Of course, this implies that
$$\text{Hom}(\text{H}_2(\text{Mod}_{g,n}(p);\mathbb{Z}),\mathbb{Z}) \cong \mathbb{Z}.$$
The major work here is related to homological stability.

LECTURE 1

The Torelli group

The Torelli group was first considered by Nielsen and Magnus in the early 20th century. However, its study only really took off in the late '70's and early '80's thanks to work of several people, most especially Birman and Johnson. Johnson's work has proven particularly fundamental and influential, and his survey [**22**] cannot be recommended enough.

Throughout this lecture, we will fix some $n \in \{0, 1\}$.

The symplectic representation. Recall that $\mathcal{I}_{g,n}$ is the kernel of the representation $\text{Mod}_{g,n} \to \text{Sp}_{2g}(\mathbb{Z})$ arising from the action of $\text{Mod}_{g,n}$ on $\text{H}_1(\Sigma_{g,n}, \mathbb{Z})$. We will need the following fact about this action. If x is a simple closed curve on $\Sigma_{g,n}$, then let T_x denote the right Dehn twist about x. Also, let $i_a(\cdot, \cdot)$ denote the algebraic intersection pairing on $\text{H}_1(\Sigma_{g,n}; \mathbb{Z})$.

LEMMA 1.1. *Let x be a simple closed curve on x. Orient x in an arbitrary way, and let $[x] \in \text{H}_1(\Sigma_{g,1}; \mathbb{Z})$ denote the associated homology class. Then for $v \in \text{H}_1(\Sigma_{g,1}; \mathbb{Z})$, we have*

$$T_x(v) = v + i_a([x], v) \cdot [x].$$

REMARK. The Dehn twist T_x does not depend on an orientation on x. As a sanity check, you should verify that despite its appearance, the formula in Lemma 1.1 does not depend on the orientation of x.

EXERCISE 1.2. Prove Lemma 1.1.

Basic elements of Torelli. Lemma 1.1 allows us to construct some important elements of $\mathcal{I}_{g,1}$. First, if $[x] = 0$, then T_x acts trivially on $\text{H}_1(\Sigma_{g,1}; \mathbb{Z})$. This will happen exactly when x bounds an embedded subsurface of x (see Figure 1). We will call such elements of Torelli *separating twists*. Next, the formula Lemma 1.1 only depends on the homology class of the simple closed curve. Thus if y and y' are homologous, then T_y and $T_{y'}$ act the same on $\text{H}_1(\Sigma_{g,1}; \mathbb{Z})$, so $T_y T_{y'}^{-1} \in \mathcal{I}_{g,1}$. If y and y' are disjoint and homologous, then their union bounds an embedded subsurface (see Figure 1). If in addition to being disjoint neither y nor y' is separating, then we will call $T_y T_{y'}^{-1} \in \mathcal{I}_{g,1}$ a *bounding pair map*.

Generating sets for Torelli. We have the following theorem.

THEOREM 1.3. *For all $g \geq 1$, the group $\mathcal{I}_{g,1}$ is generated by bounding pair maps and separating twists. For $g \geq 3$, only bounding pair maps are needed.*

REMARKS.

(1) The fact that $\mathcal{I}_{g,1}$ is generated by bounding pair maps and separating twists was originally proven by Powell [**28**], using earlier work of Birman

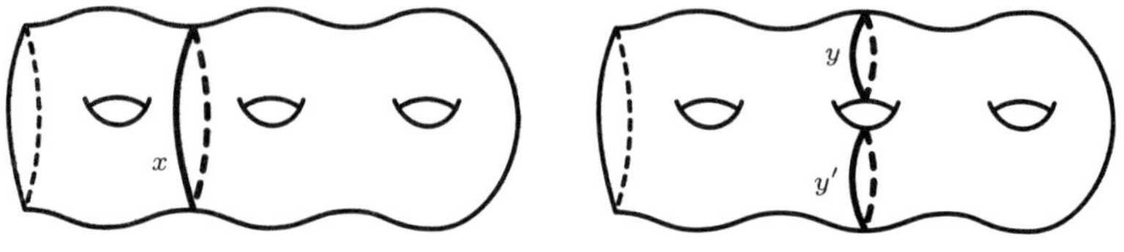

FIGURE 1. A separating twist T_x and a bounding pair map $T_y T_{y'}^{-1}$

[3]. This proof depended on some heroic calculations in the symplectic group whose details were omitted from the published papers. More recently, Putman [29] gave a modern proof using the curve complex. Even more recently, Hatcher and Margalit [14] have given an even shorter proof.

(2) The fact that for $g \geq 3$ only bounding pair maps are needed is due to Johnson [16]. He later proved a remarkable theorem which says that $\mathcal{I}_{g,1}$ is finitely generated for $g \geq 3$ (see [19]). The size of Johnson's generating set grows exponentially in g. Answering a conjecture of Johnson, Putman [35] has recently constructed a generating set for $\mathcal{I}_{g,1}$ that grows cubically in g. As we will discuss below, the abelianization of $\mathcal{I}_{g,1}$ has rank cubic in the genus, so one cannot do better.
(3) McCullough and Miller [25] proved that $\mathcal{I}_{2,n}$ is not finitely generated for $n \in \{0,1\}$. Later, in his thesis Mess [26] proved that $\mathcal{I}_2$ is an infinite rank free group.
(4) It is not known if $\mathcal{I}_{g,n}$ is finitely presentable for $g \geq 3$.

The Birman exact sequence. We will need to make several calculations in $\mathcal{I}_{g,n}$. For us, the key tool for making such calculations is the fact that $\mathcal{I}_{g,n}$ contains a large number of groups derived from surface groups. This follows from the *Birman exact sequence*, which takes the following form. Let $U\Sigma_g$ be the unit tangent bundle of Σ_g. For $g \geq 2$, we then have an exact sequence

$$1 \longrightarrow \pi_1(U\Sigma_g) \longrightarrow \mathrm{Mod}_{g,1} \longrightarrow \mathrm{Mod}_g \longrightarrow 1.$$

The terms here have the following meanings. Let β be the boundary component of $\Sigma_{g,1}$.

- The map $\mathrm{Mod}_{g,1} \to \mathrm{Mod}_g$ comes from gluing a disc to β and extending mapping classes over this disc by the identity.
- The subgroup $\pi_1(U\Sigma_g)$ of $\mathrm{Mod}_{g,1}$ is known as the "disc-pushing subgroup". The mapping class associated to $\gamma \in \pi_1(U\Sigma_g)$ "pushes" the boundary component β around the surface while allowing it to rotate.

Of course, the original version of the Birman exact sequence goes back to work of Birman [2]. The version here first appeared in [19]; see [10] for a textbook treatment.

The loop around the fiber. The fiber F_0 of $\mathrm{U}\,\Sigma_g$ over the basepoint satisfies $F_0 \cong S^1$. The orientation on Σ_g determines an orientation on F_0, so it makes sense to talk about "clockwise" and "counterclockwise" directions on F_0. The group $\pi_1(U\Sigma_g)$ contains a distinguished element δ_0 which goes once around F_0 in the clockwise direction. The element of the disc-pushing subgroup of $\mathrm{Mod}_{g,1}$ corresponding to δ_0 rotates the boundary component β by a full turn in the clockwise direction. Clearly this is simply T_β. Observe that $T_\beta \in \mathcal{I}_{g,1}$.

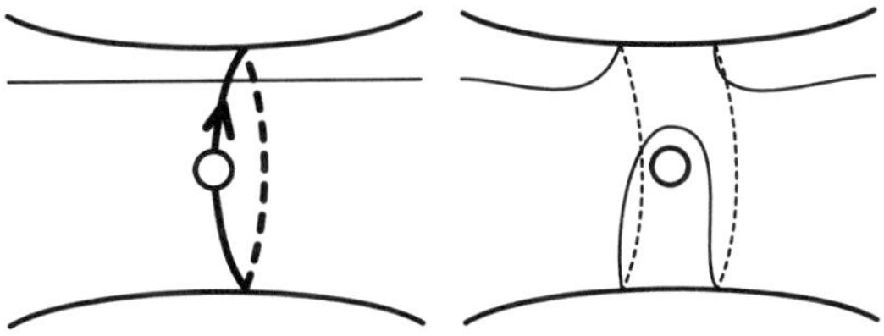

FIGURE 2. Pushing the boundary component around a simple closed curve induces a bounding pair map

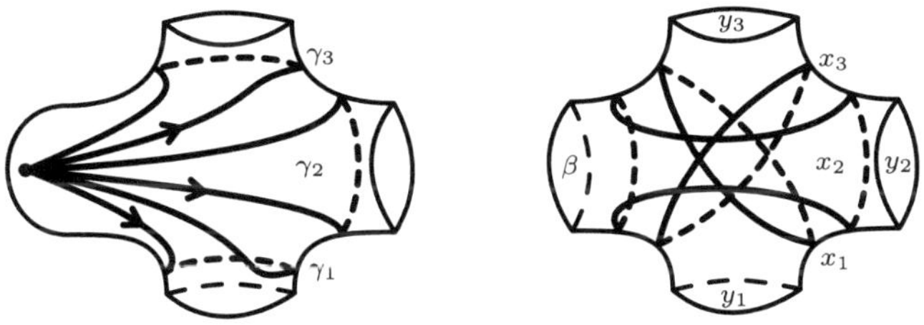

FIGURE 3. The lantern relation is $\tilde{\gamma}_1 \cdot \tilde{\gamma}_2 \cdot \tilde{\gamma}_3 = \delta_0^k$. To make the figure more attractive, the curves γ_i have what appears to be a singularity at the basepoint, but in reality one should imagine them rounded and smooth there. In terms of Dehn twists, the lantern relation is $(T_{x_3}T_{y_3}^{-1})(T_{x_2}T_{y_2}^{-1})(T_{x_1}T_{y_1}^{-1}) = T_\beta$

Calculating in the disc-pushing subgroup. If $\gamma \in \pi_1(\Sigma_g)$ can be realized by a simple closed curve, then γ can be realized by a smoothly embedded simple closed curve. The derivative of a smooth simple representative of γ is a path in the tangent bundle of γ which does not contain any zero vectors. For some fixed Riemannian metric on the surface, we can reparametrize γ so that its derivative is a loop $\tilde{\gamma}$ in the unit tangent bundle. If $\gamma \neq 1$, then the element $\tilde{\gamma} \in \pi_1(U\Sigma_g)$ is independent of the choice of a smoothly embedded representative of γ. Indeed, any two such realizations are smoothly homotopic (to test your understanding, you should verify that this fails if $\gamma = 1$).

Let $\tau_\gamma \in \mathrm{Mod}_{g,1}$ be the element of the disc-pushing subgroup corresponding to $\tilde{\gamma} \in \pi_1(U\Sigma_g)$. As is shown in Figure 2, the mapping class τ_γ is a bounding pair map, and hence lies in $\mathcal{I}_{g,1}$. Since the loop δ_0 around the fiber also corresponds to an element of $\mathcal{I}_{g,1}$, we deduce that the disc-pushing subgroup lies in $\mathcal{I}_{g,1}$. This implies that relations in $\pi_1(U\Sigma_g)$ yield relations in $\mathcal{I}_{g,1}$. Even more relations can be obtained by embedding $\mathcal{I}_{g,1}$ into $\mathcal{I}_{g',n}$ via a subsurface inclusion $\Sigma_{g,1} \hookrightarrow \Sigma_{g',n}$.

As an example, consider the relation $\gamma_1 \cdot \gamma_2 \cdot \gamma_3 = 1$ in $\pi_1(\Sigma_{g,1})$ depicted in Figure 3. We have $\tilde{\gamma}_1 \cdot \tilde{\gamma}_2 \cdot \tilde{\gamma}_3 = \delta_0^k$ for some $k \in \mathbb{Z}$.

EXERCISE 1.4. Prove that $k = 1$.

The associated relation $\tau_{\gamma_3}\tau_{\gamma_2}\tau_{\gamma_1} = \delta_0$ in $\mathcal{I}_{g,1}$ is the *lantern relation*

$$(T_{x_3}T_{y_3}^{-1})(T_{x_2}T_{y_2}^{-1})(T_{x_1}T_{y_1}^{-1}) = T_\beta; \tag{1}$$

here the curves x_i and y_i are as depicted in Figure 3.

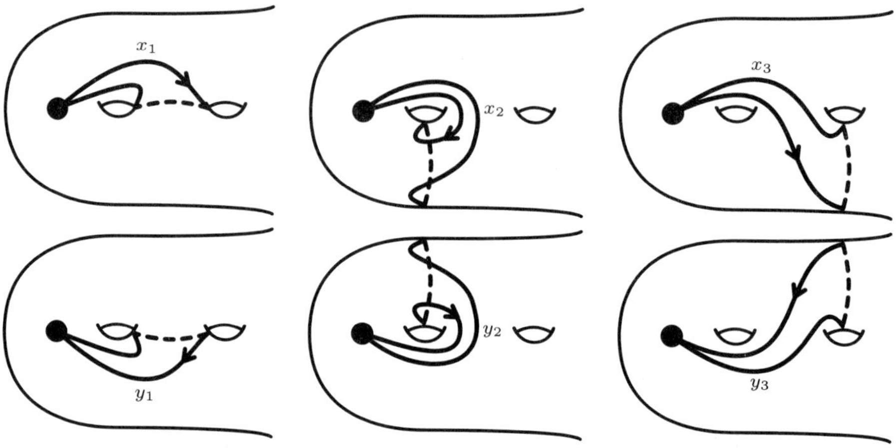

FIGURE 4. $x_1x_2x_3 = 1$ and $y_1y_2y_3 = 1$

REMARK. The order of the terms in (1) is the opposite of what one might expect because elements in the fundamental group are composed left to right but mapping classes are composed right to left.

Observe that if $g' \geq 3$, then this relation can be embedded in $\mathcal{I}_{g',n}$ to express a separating twist as a product of bounding pair maps (cf. Theorem 1.3).

Killing off separating twists. The lantern relation gives numerous ways of expressing T_β as a product of bounding pair maps. For $1 \leq i \leq 3$, let $x_i, y_i \in \pi_1(\Sigma_g)$ be the curves in Figure 4. Observe that $x_1x_2x_3 = 1$ and $y_1y_2y_3 = 1$, so we have two different lantern relations

$$T_\beta = \tau_{x_3}\tau_{x_2}\tau_{x_1} \quad \text{and} \quad T_\beta = \tau_{y_3}\tau_{y_2}\tau_{y_1}.$$

These curves have the property that x_i is homologous to y_i^{-1} for $1 \leq i \leq 3$. The group $\mathcal{I}_{g,1}$ acts on $\pi_1(\Sigma_g)$, and it is not hard to see that there exists some $f_i \in \mathcal{I}_{g,1}$ such that $f_i(x_i) = y_i^{-1}$. We thus have the following relation in $\mathcal{I}_{g,1}$.

$$\begin{aligned} T_\beta^2 &= (\tau_{x_3}\tau_{x_2}\tau_{x_1})(\tau_{y_3}\tau_{y_2}\tau_{y_1}) \\ &= (\tau_{x_3}\tau_{x_2}\tau_{x_1})(\tau^{-1}_{f_3(x_3)}\tau^{-1}_{f_2(x_2)}\tau^{-1}_{f_1(x_1)}) \\ &= (\tau_{x_3}\tau_{x_2}\tau_{x_1})(f_3\tau^{-1}_{x_3}f_3^{-1}f_2\tau^{-1}_{x_2}f_2^{-1}f_1\tau^{-1}_{x_1}f_1^{-1}) \end{aligned} \tag{2}$$

Upon abelianizing $\mathcal{I}_{g,1}$, the right hand side of (2) vanishes. Letting $[T_\beta] \in \mathrm{H}_1(\mathcal{I}_{g,1};\mathbb{Z})$ be the associated element of the abelianization, we obtain that $2[T_\beta] = 0$.

If x is a separating curve on a surface of genus at least 3, then we can embed the above relation into the surface to get that T_x^2 has to vanish upon abelianizing the Torelli group. We have proven the following.

LEMMA 1.5. *Fix $g \geq 3$ and $n \in \{0,1\}$. Let T_x be a separating twist in $\mathcal{I}_{g,n}$. Then the image $[T_x]$ of T_x in $\mathrm{H}_1(\mathcal{I}_{g,n};\mathbb{Z})$ satisfies $2[T_x] = 0$.*

Lemma 1.5 first appeared in [**21**]. The above is a version of Johnson's proof. For an alternate exposition of that proof which arranges the details a little differently, see [**32**, §7.2].

A preview. In Lecture 2, we will construct the important Johnson homomorphism. Letting $\mathcal{H} = \mathrm{H}_1(\Sigma_{g,n};\mathbb{Z})$, this is a surjective homomorphism

$$\tau : \mathcal{I}_{g,1} \longrightarrow \wedge^3 \mathcal{H}.$$

There is also a version for closed surfaces, but we will not discuss it. The key property of the Johnson homomorphism is that its kernel is exactly the subgroup generated by separating twists. Lemma 1.5 will then allow us to deduce the following theorem of Johnson [**21**].

THEOREM 1.6. *For* $g \geq 3$, *we have* $\mathrm{H}_1(\mathcal{I}_{g,1};\mathbb{Z}) \cong W \oplus \wedge^3 \mathcal{H}$, *where* W *consists of* 2*-torsion.*

REMARK. Johnson also calculated the 2-torsion W. The associated $\mathbb{Z}/2$-quotients of $\mathcal{I}_{g,1}$ come from the Rochlin invariants of homology 2-spheres. They were originally constructed by Birman and Craggs [**4**]. Later, in [**18**] Johnson packaged all of Birman and Craggs's homomorphisms together into a single homomorphism and determined exactly how many linearly independent quotients they had constructed.

As a prologue for the construction, we recommend performing the following exercise, which explains the appearance of $\wedge^3 \mathcal{H}$ in the Johnson homomorphism.

EXERCISE 1.7. Let $\mathbb{T}^n$ denote the n-torus $(S^1)^n$.

(1) Prove that the cohomology ring $\mathrm{H}^*(\mathbb{T}^n;\mathbb{Z})$ is isomorphic to the exterior algebra $\wedge^* \mathbb{Z}^n$.
(2) Let G be an abelian topological group. Define a product

$$\mathrm{H}_i(G;\mathbb{Z}) \otimes \mathrm{H}_j(G;\mathbb{Z}) \longrightarrow \mathrm{H}_{i+j}(G;\mathbb{Z})$$

via the composition

$$\mathrm{H}_i(G;\mathbb{Z}) \otimes \mathrm{H}_j(G;\mathbb{Z}) \xrightarrow{\phi} \mathrm{H}_{i+j}(G \times G;\mathbb{Z}) \xrightarrow{\psi} \mathrm{H}_{i+j}(G;\mathbb{Z}),$$

where ϕ is the map coming from the Künneth exact sequence and ψ is induced by the group product $G \times G \to G$. Prove that with this product structure, $\mathrm{H}_*(G;\mathbb{Z})$ is a graded-commutative algebra. We remark that this product is known as the *Pontryagin product.*
(3) The space $\mathbb{T}^n$ is an abelian topological group. Prove that the resulting graded-commutative ring $\mathrm{H}_*(\mathbb{T}^n;\mathbb{Z})$ is isomorphic to the exterior algebra $\wedge^* \mathbb{Z}^n$.

Appendix to Lecture 1 : the surjectivity of the symplectic representation

Recall that the action of $\mathrm{Mod}_{g,1}$ on $\mathrm{H}_1(\Sigma_{g,1};\mathbb{Z})$ preserves the algebraic intersection pairing and thus gives a representation $\pi : \mathrm{Mod}_{g,1} \to \mathrm{Sp}_{2g}(\mathbb{Z})$. In this appendix, we will give a sequence of exercises about the surjectivity of π.

A *symplectic basis* for $\mathrm{H}_1(\Sigma_{g,1};\mathbb{Z})$ is a basis $\{a_1, b_1, \ldots, a_g, b_g\}$ for $\mathrm{H}_1(\Sigma_{g,1};\mathbb{Z})$ such that

$$i_a(a_i, b_j) = \delta_{ij} \quad \text{and} \quad i_a(a_i, a_j) = i_a(b_i, b_j) = 0$$

for all $1 \leq i, j \leq g$. Let $\mathcal{S}$ be the set of symplectic bases for $\mathrm{H}_1(\Sigma_{g,1};\mathbb{Z})$. The following exercise should be straightforward.

EXERCISE 1.8. $\mathrm{Sp}_{2g}(\mathbb{Z})$ acts simply transitively on $\mathcal{S}$.

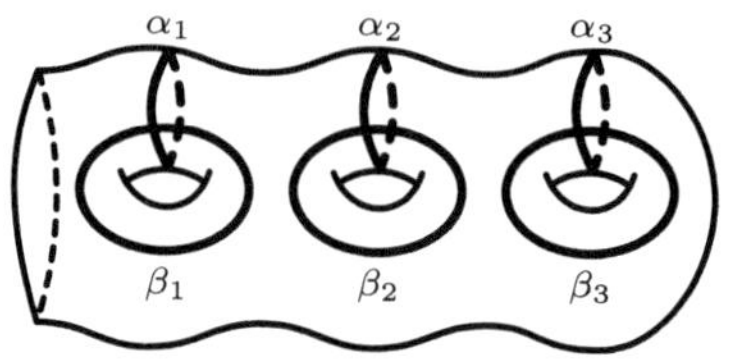

FIGURE 5. A geometric symplectic basis

If x and y are simple closed curves on $\Sigma_{g,1}$, then let $i_g(x,y)$ be their *geometric intersection number*; i.e. the minimal cardinality of $x' \cap y'$ as x' and y' range over all simple closed curves homotopic to x and y, respectively. A *geometric symplectic basis* (see Figure 5) is a collection $\{\alpha_1, \beta_1, \ldots, \alpha_g, \beta_g\}$ of simple closed curves on $\Sigma_{g,1}$ such that

$$i_g(\alpha_i, \beta_j) = \delta_{ij} \quad \text{and} \quad i_g(\alpha_i, \alpha_j) = i_g(\beta_i, \beta_j) = 0$$

for all $1 \leq i, j \leq g$. Let $\mathcal{G}$ be the set of geometric symplectic bases on $\Sigma_{g,1}$. We then have the following.

EXERCISE 1.9. $\text{Mod}_{g,1}$ acts transitively on $\mathcal{G}$. Hint : given two geometric symplectic bases, prove using the Euler characteristic that you get homeomorphic surfaces when you cut along them.

The following lemma is the heart of the fact that $\pi(\text{Mod}_{g,1}) = \text{Sp}_{2g}(\mathbb{Z})$.

LEMMA 1.10. *If $\{a_1, b_1, \ldots, a_g, b_g\}$ is a symplectic basis for $\text{H}_1(\Sigma_g; \mathbb{Z})$, then there exists a geometric symplectic basis $\{\alpha_1, \beta_1, \ldots, \alpha_g, \beta_g\}$ on Σ_g such that $[\alpha_i] = a_i$ and $[\beta_i] = b_i$ for $1 \leq i \leq g$.*

Proofs of Lemma 1.10 can be found in [**29**, Lemma A.3] and [**10**, 3rd proof of Lemma 6.4]; however, it is worthwhile to contemplate how one might prove it (though it is probably too hard for an exercise).

EXERCISE 1.11. Combine Lemma 1.10 with Exercises 1.8 and 1.9 to deduce that $\pi(\text{Mod}_{g,1}) = \text{Sp}_{2g}(\mathbb{Z})$.

LECTURE 2

The Johnson homomorphism

Let $\mathcal{H} = \mathrm{H}_1(\Sigma_g; \mathbb{Z})$. In this lecture, we will construct the *Johnson homomorphism*, which is a surjective homomorphism

$$\tau : \mathcal{I}_{g,1} \longrightarrow \wedge^3 \mathcal{H}.$$

This homomorphism can be constructed in a number of completely different ways. It was originally constructed in [**17**] by examining the action of $\mathcal{I}_{g,1}$ on the second nilpotent truncation of $\pi_1(\Sigma_{g,1})$. We explain this original construction in an appendix. In his survey [**22**], Johnson outlined several alternate constructions. We will use a definition in terms of mapping tori which was introduced in [**22**] and was first shown to be equivalent to the original definition by Hain [**11**]. Our exposition will follow the paper [**9**] of Church and Farb, which gives a more direct proof of this equivalence.

The construction. Consider $f \in \mathcal{I}_{g,1}$. Though it is an abuse of notation, we will regard f as a homeomorphism of $\Sigma_{g,1}$. Glue a disc to the boundary component of $\Sigma_{g,1}$ and extend f over this disc by the identity to obtain a homeomorphism F of Σ_g. Let $p_0 \in \Sigma_g$ be the center of the glued-in disc, so $F(p_0) = p_0$. Now let M_F be the *mapping torus* of F, i.e. the quotient $\Sigma_g \times I/\sim$, where $(x, 1) \sim (F(x), 0)$. Give M_F the basepoint $q_0 = (p_0, 0)$.

There is a distinguished element $t \in \pi_1(M_F, q_0)$ which traverses the embedded loop $p_0 \times I/\sim$ in M_F in the positive direction. Fix a standard generating set $S = \{s_1, \ldots, s_{2g}\}$ for $\pi_1(\Sigma_g)$ that satisfies the surface relation

$$[s_1, s_2] \cdots [s_{2g-1}, s_{2g}] = 1.$$

Since $F(p_0) = p_0$, the map F acts on $\pi_1(\Sigma_g, p_0)$. For $1 \leq i \leq 2g$, let w_i be an expression for $F_*(s_i)$ in terms of the generating set S. We then have a presentation

$$\pi_1(M_F, q_0) = \langle s_1, \ldots, s_{2g}, t \mid [s_1, s_2] \cdots [s_{2g-1}, s_{2g}] = 1, t s_i t^{-1} = w_i \text{ for } 1 \leq i \leq 2g \rangle.$$

For $\gamma \in \pi_1(\Sigma_g, p_0)$, let $[\gamma] \in \mathcal{H}$ be the associated element of the abelianization. Since $F \in \mathcal{I}_g$, we have $[s_i] = [w_i]$ for $1 \leq i \leq 2g$. This implies that we can define a homomorphism $\phi_* : \pi_1(M_F, q_0) \to \mathcal{H}$ such that $\phi_*(s_i) = [s_i]$ for $1 \leq i \leq 2g$ and such that $\phi_*(t) = 0$.

The space M_F is clearly a $K(\pi_1(M_F), 1)$. Let $\mathbb{T}^{2g}$ be the $2g$-torus. Fix an identification of $\pi_1(\mathbb{T}^{2g})$ with $\mathcal{H}$. Since $\mathbb{T}^{2g}$ is a $K(\mathcal{H}, 1)$, the standard properties of Eilenberg-MacLane spaces show that there is a canonical homotopy class of continuous maps $\phi : M_F \to \mathbb{T}^{2g}$ inducing the homomorphism ϕ_*. The space M_F is a closed 3-manifold, so it has a canonical class $[M_F] \in \mathrm{H}_3(M_F; \mathbb{Z})$. Define

$$\tau(f) = \phi_*([M_F]) \in \mathrm{H}_3(\mathbb{T}^{2g}; \mathbb{Z}) \cong \wedge^3 \mathcal{H}.$$

The last isomorphism here comes from Exercise 1.7

Summing up, we have constructed a map $\tau : \mathcal{I}_{g,1} \to \wedge^3 \mathcal{H}$. The following exercise is a good test of your understanding of the above construction.

EXERCISE 2.1. Prove that τ is independent of all the above choices except for the identification of $\pi_1(\mathbb{T}^{2g})$ with $\mathcal{H}$ (which is fixed). Next, prove that τ is a homomorphism.

Effect on generators. The following lemma calculates τ on the generators for $\mathcal{I}_{g,1}$. Recall that if S is a genus h surface with at most 1 boundary component, then a *symplectic basis* for $\mathrm{H}_1(S;\mathbb{Z}) \cong \mathbb{Z}^{2h}$ is a basis $\{a_1, b_1, \ldots, a_h, b_h\}$ for $\mathrm{H}_1(S;\mathbb{Z})$ such that

$$i_a(a_i, b_j) = \delta_{ij} \quad \text{and} \quad i_a(a_i, a_j) = i_a(b_i, b_j) = 0$$

for all $1 \leq i, j \leq h$. Here $i_a(\cdot, \cdot)$ is the algebraic intersection form.

LEMMA 2.2.

(1) *Let $T_x \in \mathcal{I}_{g,1}$ be a separating twist. Then $\tau(T_x) = 0$.*

(2) *Let $T_x T_{x'}^{-1}$ be a bounding pair map on $\mathcal{I}_{g,1}$. Let S be the component of $\Sigma_{g,1}$ cut along $x \cup x'$ that does not contain $\partial \Sigma_{g,1}$, so $S \cong \Sigma_{h,2}$ for some $h < g$. Let $S' \subset S$ be an embedded subsurface such that $S' \cong \Sigma_{h,1}$ and let $\{a_1, b_1, \ldots, a_h, b_h\}$ be a symplectic basis for $\mathrm{H}_1(S';\mathbb{Z}) \subset \mathrm{H}_1(\Sigma_{g,1};\mathbb{Z})$. Then*

$$\tau(T_x T_{x'}^{-1}) = \pm[x] \wedge (a_1 \wedge b_1 + \cdots + a_h \wedge b_g).$$

We will discuss the proof of this lemma at the end of this section. Right now, we suggest doing the following two exercises.

EXERCISE 2.3. Prove that the formula in Lemma 2.2 is independent of the choice of S' and its symplectic basis.

EXERCISE 2.4. Using Lemma 2.2, prove that τ is surjective.

Johnson's theorem. In [**20**], Johnson proved the following deep theorem, which is a sort of converse to part 1 of Lemma 2.2. For an alternate proof, see [**32**].

THEOREM 2.5. *The kernel of τ is generated by separating twists.*

As we indicated at the end of Lecture 1, this theorem together with Lemma 1.5 implies that $\mathrm{H}_1(\mathcal{I}_{g,1};\mathbb{Z}) \cong W \oplus \wedge^3 \mathcal{H}$ for $g \geq 3$, where W consists of 2-torsion.

The Johnson homomorphism mod p. Set $\mathcal{H}_p = \mathrm{H}_1(\Sigma_{g,1};\mathbb{Z}/p)$. We wish to construct a "mod p" Johnson homomorphism

$$\tau_p : \mathrm{Mod}_{g,n}(p) \to \mathcal{H}_p .$$

The construction goes exactly like the construction of the ordinary Johnson homomorphism. Consider $f \in \mathrm{Mod}_{g,1}(p)$. Regard f as a homeomorphism of $\Sigma_{g,1}$. Glue a disc to the boundary component of $\Sigma_{g,1}$ and extend f over this disc by the identity to obtain a homeomorphism F of Σ_g. Let $p_0 \in \Sigma_g$ be the center of the glued-in disc, so $F(p_0) = p_0$. Let M_F be the mapping torus of F and let $q_0 = (p_0, 0)$ be the basepoint for M_F.

There is a distinguished element $t \in \pi_1(M_F, q_0)$ which traverses the embedded loop $p_0 \times I / \sim$ in M_F in the positive direction. Fix a standard generating set $S = \{s_1, \ldots, s_{2g}\}$ for $\pi_1(\Sigma_g)$ that satisfies the surface relation

$$[s_1, s_2] \cdots [s_{2g-1}, s_{2g}] = 1.$$

Since $F(p_0) = p_0$, the map F acts on $\pi_1(\Sigma_g, p_0)$. For $1 \leq i \leq 2g$, let w_i be an expression for $F_*(s_i)$ in terms of the generating set S. We then have a presentation

$$\pi_1(M_F, q_0) = \langle s_1, \ldots, s_{2g}, t \mid [s_1, s_2] \cdots [s_{2g-1}, s_{2g}] = 1, ts_it^{-1} = w_i \text{ for } 1 \leq i \leq 2g \rangle.$$

For $\gamma \in \pi_1(\Sigma_g, p_0)$, let $[\gamma]_p \in \mathcal{H}_p$ be the associated element. Since $F \in \mathrm{Mod}_g(p)$, we have $[s_i]_p = [w_i]_p$ for $1 \leq i \leq 2g$. This implies that we can define a homomorphism $\phi_* : \pi_1(M_F, q_0) \to \mathcal{H}_p$ such that $\phi_*(s_i) = [s_i]_p$ for $1 \leq i \leq 2g$ and such that $\phi_*(t) = 0$.

The space M_F is clearly a $K(\pi_1(M_F), 1)$. Let Z be a $K(\mathcal{H}_p, 1)$. The standard properties of Eilenberg-MacLane spaces show that there is a canonical homotopy class of continuous maps $\phi : M_F \to Z$ inducing the homomorphism ϕ_*. Define

$$\tau_p'(f) = \phi_*([M_F]) \in \mathrm{H}_3(Z; \mathbb{Z}) \cong \mathrm{H}_3(\mathcal{H}_p; \mathbb{Z}),$$

where $M_F \in \mathrm{H}_3(M_F; \mathbb{Z})$ is the canonical class. It is not true that $\mathrm{H}_3(\mathcal{H}_p; \mathbb{Z}) \cong \wedge^3 \mathcal{H}_p$; however, $\mathrm{H}_3(\mathcal{H}_p; \mathbb{Z})$ does contain $\wedge^3 \mathcal{H}_p$ as a direct factor (see, for example, [**7**, Theorem V.6.4]). Let $\tau_p : \mathrm{Mod}_{g,1}(p) \to \wedge^3 \mathcal{H}_p$ be the composition of τ_p' with some (fixed for all time) projection of $\mathrm{H}_3(\mathcal{H}_p; \mathbb{Z})$ onto $\wedge^3 \mathcal{H}_p$. By the same argument used in Exercise 2.1, the map τ_p is a well-defined homomorphism.

From its construction, it is clear that the following diagram commutes.

$$\begin{array}{ccc} \mathcal{I}_{g,1} & \xrightarrow{\tau} & \wedge^3 \mathcal{H} \\ \downarrow & & \downarrow \\ \mathrm{Mod}_{g,n}(p) & \xrightarrow{\tau_p} & \wedge^3 \mathcal{H}_p \end{array}$$

Since τ is surjective for $g \geq 3$, we obtain the following.

THEOREM 2.6. *The map* $\tau_p : \mathrm{Mod}_{g,1}(p) \to \mathcal{H}_p$ *is surjective for* $g \geq 3$.

Images of mapping tori. To prove Lemma 2.2, one would need to be able to calculate the images of fundamental classes of mapping tori under maps to a torus. Instead of giving the details of the proof, we will prove an easier theorem which illustrates how one can do this. For the proof of Lemma 2.2, see [**9**].

The result we will prove is as follows. In its statement and proof, we will use the natural identification of the graded-commutative algebra structure on $\mathrm{H}_*(\mathbb{T}^n; \mathbb{Z})$ with $\wedge^* \mathbb{Z}^n$ which was identified in Exercise 1.7.

THEOREM 2.7. *Let* $\phi : \Sigma_g \to \mathbb{T}^n$ *be a continuous map. Choose a standard basis* $\{\alpha_1, \beta_1, \ldots, \alpha_g, \beta_g\}$ *for* $\pi_1(\Sigma_g)$, *so*

$$\pi_1(\Sigma_g) = \langle \alpha_1, \beta_1, \ldots, \alpha_g, \beta_g \mid [\alpha_1, \beta_1] \cdots [\alpha_g, \beta_g] = 1 \rangle.$$

Then

$$\phi_*([\Sigma_g]) = \pm \sum_{i=1}^{g} \phi_*(\alpha_i) \wedge \phi_*(\beta_i) \in \wedge^2 \mathbb{Z}^n \cong \mathrm{H}_2(\mathbb{T}^n; \mathbb{Z}).$$

PROOF. Assume first that $g = 1$, so $\Sigma_g \cong \mathbb{T}^2$. We will identify Σ_g with $\mathbb{T}^2$. Observe that $\pi_1(\mathbb{T}^2) \cong \mathbb{Z}^2$ and $\mathrm{H}_*(\mathbb{T}^2; \mathbb{Z}) \cong \wedge^*(\mathbb{Z}^2)$. Moreover, $\mathrm{H}_2(\mathbb{T}^2; \mathbb{Z}) \cong \wedge^2 \mathbb{Z}^2$ is generated by $\alpha_1 \wedge \beta_1$, so $[\mathbb{T}^2] = \pm \alpha_1 \wedge \beta_1$. By the naturality of the Pontryagin product we have

$$\phi_*([\mathbb{T}^2]) = \pm \phi_*(\alpha_1) \wedge \phi_*(\beta_1).$$

Now assume that $g > 1$. We will reduce to the case $g = 1$ by a method that will be familiar to algebraic geometers. Namely, we will "degenerate" our surface

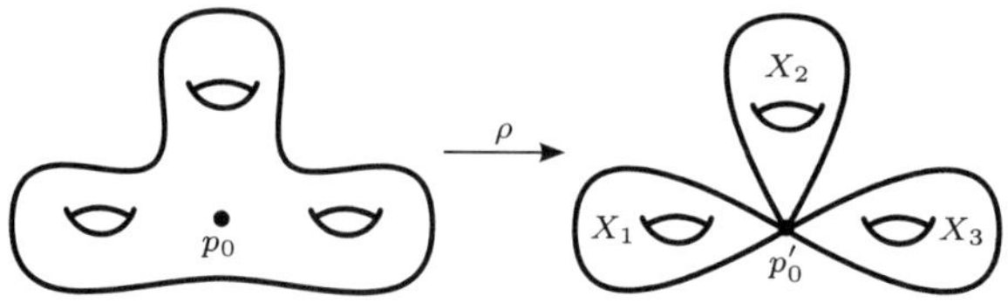

FIGURE 1. Degenerating our surface to a wedge of tori

to a nodal surface and make the computation there. Fix a basepoint p_0 for Σ_g. As shown in Figure 1, let (X, p_0') be the result of collapsing (Σ_g, p_0) to the wedge of g tori $X_1, \ldots, X_g$. Choose X such that under the collapse map $\rho : \Sigma_g \to X$, the curves α_i and β_i map to generators for $\pi_1(X_i, p_0') \subset \pi_1(X, p_0')$.

EXERCISE 2.8. Prove that the map $\phi_* : \pi_1(\Sigma_g, p_0) \to \pi_1(\mathbb{T}^n)$ factors as

$$\pi_1(\Sigma_g, p_0) \xrightarrow{\rho_*} \pi_1(X, p_0') \xrightarrow{\phi_*'} \pi_1(\mathbb{T}^n)$$

for some homomorphism $\phi_*' : \pi_1(X, p_0') \to \pi_1(\mathbb{T}^n)$. Hint : $\pi_1(\mathbb{T}^n)$ is abelian.

Since the spaces Σ_g and X and $\mathbb{T}^n$ are all Eilenberg-MacLane spaces, there exists some continuous map $\phi' : X \to \mathbb{T}^n$ inducing ϕ_*' such that ϕ is homotopic to $\phi' \circ \rho$.

Observe that $\rho_*([\Sigma_g]) = [X_1] + \cdots + [X_g]$. Moreover, $\rho_*(\alpha_i)$ and $\rho_*(\beta_i)$ are a standard basis for $\pi_1(X_i) \subset \pi_1(X)$. By the case $g = 1$, we have

$$(\phi'|_{X_i})_*([X_i]) = \pm\phi_*'(\rho_*(\alpha_i)) \wedge \phi_*'(\rho_*(\beta_i) = \pm\phi_*(\alpha_i) \wedge \phi_*(\beta_i).$$

Moreover, all the $\pm$ signs are identical. Adding everything up, we get the desired result. □

Appendix to Lecture 2 : the original construction of the Johnson homomorphism

The original construction of the Johnson homomorphism gave a homomorphism

$$\tau : \mathcal{I}_{g,1} \to \mathrm{Hom}(\mathcal{H}, \wedge^2 \mathcal{H}).$$

Johnson then calculated τ on generators and showed that its image was isomorphic to $\wedge^3 \mathcal{H}$. In this appendix, we will guide you through Johnson's original construction.

If G is a group, then let $\gamma_k(G)$ be the kth term in its *lower central series*, which is the inductively defined sequence of groups

$$\gamma_0(G) = G \quad \text{and} \quad \gamma_{k+1}(G) = [\gamma_k(G), G].$$

Set $\pi = \pi_1(\Sigma_{g,1}, *)$, where $* \in \partial\Sigma_{g,1}$. We will need the following fact. We will state it as an exercise, but you should only attempt it if you know some of the basics of group cohomology (otherwise, treat it as a black box).

EXERCISE 2.9. Prove that $\gamma_1(\pi)/\gamma_2(\pi) \cong \wedge^2 \mathcal{H}$. Hint : Apply the 5-term exact sequence in group homology to the short exact sequence

$$1 \longrightarrow \gamma_1(\pi) \longrightarrow \pi \longrightarrow \mathcal{H} \longrightarrow 1.$$

You will need the fact that $\mathrm{H}_2(\mathcal{H}; \mathbb{Z}) \cong \wedge^2 \mathcal{H}$.

Now consider the short exact sequence

$$1 \longrightarrow \gamma_1(\pi) \longrightarrow \pi \longrightarrow \mathcal{H} \longrightarrow 1.$$

To simplify this, we mod out by $\gamma_2(\pi) < \gamma_1(\pi)$. By the above exercise, we get a short exact sequence

$$1 \longrightarrow \wedge^2 \mathcal{H} \longrightarrow \Gamma \longrightarrow \mathcal{H} \longrightarrow 1, \tag{1}$$

where $\Gamma = \pi/\gamma_2(\pi)$.

EXERCISE 2.10. The subgroup $\mathcal{H} < \Gamma$ is central.

The mapping class group acts on π (this is where we use the fact that the basepoint is on the boundary component, so it is fixed by the mapping class group). This action preserves $\gamma_k(\pi)$ for all π (indeed, all automorphisms of π do this!). We thus get an action of $\mathrm{Mod}_{g,1}$ on Γ which preserves $\wedge^2 \mathcal{H} < \Gamma$.

Restrict this action to $\mathcal{I}_{g,1}$. The group $\mathcal{I}_{g,1}$ acts trivially on $\mathcal{H}$.

EXERCISE 2.11. The action of $\mathcal{I}_{g,1}$ on $\wedge^2 \mathcal{H} < \Gamma$ is trivial. This looks obvious, but you have to trace through the above definitions to see that the action of $\mathrm{Mod}_{g,1}$ on $\wedge^2 \mathcal{H} < \Gamma$ is what you think it is.

Fix $f \in \mathcal{I}_{g,1}$. For $x \in \Gamma$, observe that $f(x)$ and x project to the same element of $\mathcal{H}$, so $x(f(x))^{-1} \in \wedge^2 \mathcal{H}$. Define a set map $J'_f : \Gamma \to \wedge^2 \mathcal{H}$ by $J'_f(x) = x(f(x))^{-1}$.

EXERCISE 2.12. The map J'_f factors through a set map $J_f : \mathcal{H} \to \wedge^2 \mathcal{H}$.

EXERCISE 2.13. The set map J_f is a homomorphism.

We can thus define a set map

$$\tau : \mathcal{I}_{g,1} \to \mathrm{Hom}(\mathcal{H}, \wedge^2 \mathcal{H})$$

by $\tau(f) = J_f$.

EXERCISE 2.14. τ is a homomorphism.

LECTURE 3

The abelianization of $\mathrm{Mod_{g,n}(p)}$

In this lecture, we calculate $\mathrm{H}_1(\mathrm{Mod}_{g,n}(p);\mathbb{Z})$. To simplify our exposition, we will do the following.

- We will only consider the case of $\mathrm{Mod}_{g,1}(p)$. The case of $\mathrm{Mod}_g(p)$ can be dealt with in a similar way, but there are a few added complications.
- We will only consider the case where p is odd. This greatly simplifies both the statements of the results and their proofs.

Main theorem. Our main theorem is as follows.

THEOREM 3.1. *Fix $g, p \geq 3$ such that p is odd. Set $\mathcal{H}_p = \mathrm{H}_1(\Sigma_{g,1};\mathbb{Z}/p)$. There is then a short exact sequence*

$$0 \longrightarrow \wedge^3 \mathcal{H}_p \longrightarrow \mathrm{H}_1(\mathrm{Mod}_{g,1}(p);\mathbb{Z}) \longrightarrow \mathrm{H}_1(\mathrm{Sp}_{2g}(\mathbb{Z},p);\mathbb{Z}) \longrightarrow 0.$$

REMARKS.

(1) Theorem 3.1 was proven independently by Putman [**30**] and Sato [**36**]. At the same time, Perron [**27**] calculated $\mathrm{H}_1(\mathrm{Mod}_{g,1}(p);\mathbb{Z})$ up to a 2-torsion ambiguity.

(2) Sato [**36**] also calculated $\mathrm{H}_1(\mathrm{Mod}_{g,1}(2);\mathbb{Z})$. The answer is more complicated than that given in Theorem 3.1. This was later extended by Putman [**33**] to $\mathrm{H}_1(\mathrm{Mod}_{g,1}(p);\mathbb{Z})$ for p not divisible by 4. The case where p is divisible by 4 is still open – see the introduction of [**33**] for a discussion.

(3) It was originally proven by Hain [**11**] that $\mathrm{H}_1(\mathrm{Mod}_{g,1}(p);\mathbb{Z})$ is finite for $g \geq 3$.

(4) At the end of this lecture, we will calculate $\mathrm{H}_1(\mathrm{Sp}_{2g}(\mathbb{Z},p);\mathbb{Z})$.

Beginning of the proof. Of course, Theorem 3.1 is derived from the short exact sequence

$$1 \longrightarrow \mathcal{I}_{g,1} \longrightarrow \mathrm{Mod}_{g,1}(p) \longrightarrow \mathrm{Sp}_{2g}(\mathbb{Z},p) \longrightarrow 1.$$

We will need the following exercise.

EXERCISE 3.2. Consider a short exact sequence

$$1 \longrightarrow G_1 \longrightarrow G_2 \longrightarrow G_3 \longrightarrow 1$$

of groups. There is then a short exact sequence

$$0 \longrightarrow V \longrightarrow \mathrm{H}_1(G_2;\mathbb{Z}) \longrightarrow \mathrm{H}_1(G_3;\mathbb{Z}) \longrightarrow 0,$$

where V is the image of $\mathrm{H}_1(G_1;\mathbb{Z})$ in $\mathrm{H}_1(G_2;\mathbb{Z})$.

Applying Exercise 3.2, we see that it is enough to prove that the image of $\mathrm{H}_1(\mathcal{I}_{g,1};\mathbb{Z})$ in
$\mathrm{H}_1(\mathrm{Mod}_{g,1}(p);\mathbb{Z})$ is $\wedge^3 \mathcal{H}_p$. The key to this is the following lemma.

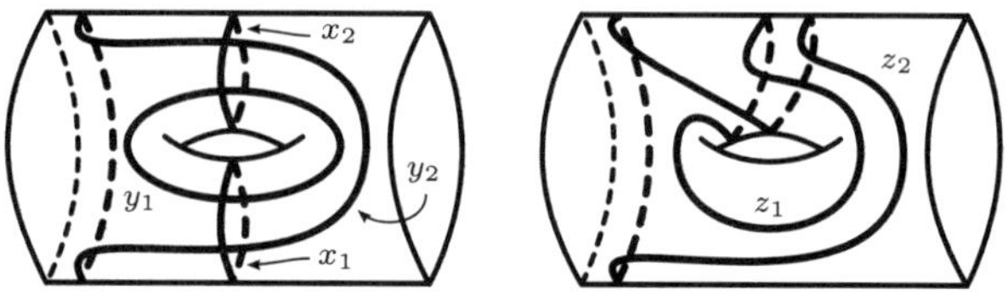

FIGURE 1. The crossed lantern relation $(T_{y_1}T_{y_2}^{-1})(T_{x_1}T_{x_2}^{-1}) = T_{z_1}T_{z_2}^{-1}$.

LEMMA 3.3. *For $g \geq 3$, let $v \in \mathrm{H}_1(\mathrm{Mod}_{g,1}(p);\mathbb{Z})$ be in the image of $\mathrm{H}_1(\mathcal{I}_{g,1};\mathbb{Z})$. Then $p \cdot v = 0$.*

Before proving Lemma 3.3, let use it to finish the proof of Theorem 3.1. Theorem 1.6 says that

$$\mathrm{H}_1(\mathcal{I}_{g,1};\mathbb{Z}) \cong W \oplus \wedge^3 \mathcal{H},$$

where W consists of 2-torsion. Lemma 3.3 and the fact that p is odd imply that the inclusion map $\mathrm{H}_1(\mathcal{I}_{g,1};\mathbb{Z}) \to \mathrm{H}_1(\mathrm{Mod}_{g,1}(p);\mathbb{Z})$ factors through

$$(W/p \cdot W) \oplus (\wedge^3 \mathcal{H})/(p \cdot \wedge^3 \mathcal{H}) = \wedge^3 \mathcal{H}_p.$$

Theorem 2.6 says that the image of $\mathrm{H}_1(\mathcal{I}_{g,1};\mathbb{Z})$ in $\mathrm{H}_1(\mathrm{Mod}_{g,1}(p);\mathbb{Z})$ contains $\wedge^3 \mathcal{H}_p$, so we conclude that the image equals $\mathcal{H}_p$, as desired.

The crossed lantern relation. To prove Lemma 3.3, we will need the following relation the mapping class group, which is known as the *crossed lantern relation.*

LEMMA 3.4. *For $i = 1, 2$, let x_i and y_i and z_i be the curves in Figure 1. Then $(T_{y_1}T_{y_2}^{-1})(T_{x_1}T_{x_2}^{-1}) = T_{z_1}T_{z_2}^{-1}$.*

PROOF. This rests on two key facts:

$$T_{z_i} = T_{x_2}(y_i) \quad \text{and} \quad T_{y_2}T_{y_1}^{-1}(x_2) = x_1.$$

Repeatedly using the fact that $fT_cf^{-1} = T_{f(c)}$ for any simple closed curve c and any mapping class f, we calculate

$$\begin{aligned} T_{x_1}T_{x_2}^{-1} &= \left((T_{y_2}T_{y_1}^{-1})T_{x_2}(T_{y_1}T_{y_2}^{-1})\right)T_{x_2}^{-1} \\ &= (T_{y_2}T_{y_1}^{-1})\left(T_{x_2}(T_{y_1}T_{y_2}^{-1})T_{x_2}^{-1}\right) \\ &= T_{y_2}T_{y_1}^{-1}T_{z_1}T_{z_2}^{-1}. \end{aligned}$$

Rearranging this gives the desired relation. □

EXERCISE 3.5. Give an alternate derivation of the crossed lantern relation by the same technique we used to construct the lantern relation in Lecture 1. Hint : The lantern relation comes from a relation $xyz = 1$ between simple closed curve in a surface group. Try to write down a different relation of the form $xy = z$ between simple closed curves. If you get stuck, see [**31**, §3.1.3].

The proof of Lemma 3.3. We are now in a position to prove Lemma 3.3. For $f \in \mathrm{Mod}_{g,1}(p)$, let $[f]$ be the corresponding element of $\mathrm{H}_1(\mathrm{Mod}_{g,1}(p);\mathbb{Z})$. Theorem 1.3 says that $\mathcal{I}_{g,1}$ is generated by bounding pair maps. Letting $T_{x_1}T_{x_2}^{-1}$ be a bounding pair map, it is thus enough to show that $p \cdot [T_{x_1}T_{x_2}^{-1}] = 0$. There exists an embedded subsurface S of $\Sigma_{g,1}$ containing $\{x_1, x_2\}$ such that $S \cong \Sigma_{1,2}$ and such that the curves $\{x_1, x_2\}$ are embedded in S as depicted in Figure 1.a. Let $\{y_1, y_2\}$ and $\{z_1, z_2\}$ be the curves in S depicted in Figures 1.a and 1.b, respectively. We thus have a crossed lantern relation

$$(1) \qquad (T_{y_1}T_{y_2}^{-1})(T_{x_1}T_{x_2}^{-1}) = T_{z_1}T_{z_2}^{-1}$$

Also, $z_i = T_{x_2}(y_i)$ for $i = 1, 2$.

The key observation is that for all $k \geq 0$, conjugating (1) by $T_{x_2}^k$ results in another crossed lantern relation

$$(T_{T_{x_2}^k(y_1)}T^{-1}_{T_{x_2}^k(y_2)})(T_{x_1}T_{x_2}^{-1}) = (T_{T_{x_2}^{k+1}(y_1)}T^{-1}_{T_{x_2}^{n+1}(y_2)}).$$

Since $T_{x_2}^p \in \mathrm{Mod}_{g,1}(p)$, we conclude that $[T_{y_1}T_{y_2}^{-1}]$ is equal to

$$\begin{aligned}
[T_{x_2}^p(T_{y_1}T_{y_2}^{-1})T_{x_2}^{-p}] &= [(T_{T_{x_2}^p(y_1)}T^{-1}_{T_{x_2}^p(y_2)})] \\
&= [T_{x_1}T_{x_2}^{-1}] + [(T_{T_{x_2}^{p-1}(y_1)}T^{-1}_{T_{x_2}^{p-1}(y_2)})] \\
&= 2[T_{x_1}T_{x_2}^{-1}] + [(T_{T_{x_2}^{p-2}(y_1)}T^{-1}_{T_{x_2}^{p-2}(y_2)})] \\
&\ \ \vdots \\
&= p \cdot [T_{x_1}T_{x_2}^{-1}] + [T_{y_1}T_{y_2}^{-1}],
\end{aligned}$$

so $p \cdot [T_{x_1}T_{x_2}^{-1}] = 0$, as desired.

The abelianization of $\mathrm{Sp}_{2g}(\mathbb{Z}, p)$. To complete our description of the abelianization of $\mathrm{Mod}_{g,1}(p)$, we need a description of $\mathrm{H}_1(\mathrm{Sp}_{2g}(\mathbb{Z}, p); \mathbb{Z})$. We first need some notation. Denote the $n \times n$ identity matrix by $\mathbb{I}_n$ and the $n \times n$ zero matrix by $\mathbb{O}_n$. Let Ω_g be the $2g \times 2g$ matrix $\begin{pmatrix} \mathbb{O}_g & \mathbb{I}_g \\ -\mathbb{I}_g & \mathbb{O}_g \end{pmatrix}$. By definition, for a commutative ring R the group $\mathrm{Sp}_{2g}(R)$ consists of all $2g \times 2g$ matrices M with entries in R such that $M^t\Omega_g M = \Omega_g$. Define $\mathfrak{sp}_{2g}(R)$ to be the additive group of $2g \times 2g$ matrices A with entries in R such that $A^t\Omega_g + \Omega_g A = 0$.

We then have the following theorem.

THEOREM 3.6. *Fix $g, p \geq 3$ such that p is odd. Then*

$$\mathrm{H}_1(\mathrm{Sp}_{2g}(\mathbb{Z}, p); \mathbb{Z}) \cong \mathfrak{sp}_{2g}(\mathbb{Z}/p) \quad \textit{and} \quad [\mathrm{Sp}_{2g}(\mathbb{Z}, p), \mathrm{Sp}_{2g}(\mathbb{Z}, p)] = \mathrm{Sp}_{2g}(\mathbb{Z}, p^2).$$

REMARKS.

(1) Theorem 3.6 was proved independently by Perron [**27**], Putman [**30**], and Sato [**36**]. All three papers were inspired by a basic result of Lee and Szczarba [**23**] which calculates the abelianizations of congruence subgroups of $\mathrm{SL}_n(\mathbb{Z})$. See Exercise 3.9 for a discussion of Lee and Szczarba's work.

(2) For p even, Sato [**36**] proved that there is a short exact sequence

$$0 \longrightarrow \mathcal{H}_2 \longrightarrow \mathrm{H}_1(\mathrm{Sp}_{2g}(\mathbb{Z}, p); \mathbb{Z}) \longrightarrow \mathfrak{sp}_{2g}(\mathbb{Z}/p) \longrightarrow 0.$$

Here $\mathcal{H}_2 = \mathrm{H}_1(\Sigma_{g,1}; \mathbb{Z}/2)$.

The map from $\mathrm{Sp}_{2g}(\mathbb{Z}, p)$ to $\mathfrak{sp}_{2g}(\mathbb{Z}/p)$. To prove Theorem 3.6, we first construct a surjective homomorphism

$$\psi : \mathrm{Sp}_{2g}(\mathbb{Z}, p) \longrightarrow \mathfrak{sp}_{2g}(\mathbb{Z}/p).$$

Consider $M \in \mathrm{Sp}_{2g}(\mathbb{Z},p)$. By definition, we can write $M = \mathbb{I}_{2g} + p \cdot A$ for some matrix A. Define $\psi(M) = A$ modulo p. We have $\psi(M) \in \mathfrak{sp}_{2g}(\mathbb{Z}/p)$; indeed, by definition we have $\Omega_g = M^t \Omega_g M$, so

$$\begin{aligned}\Omega_g &= (\mathbb{I}_{2g} + p \cdot A)^t \Omega_g (\mathbb{I}_{2g} + p \cdot A) \\ &= \Omega_g + p \cdot (A^t \Omega_g + \Omega_g A) + p^2 A^t \Omega_g A.\end{aligned}$$

This implies that modulo p, we have $A^t\Omega_g + \Omega_g A = 0$. Next, we prove that ψ is a homomorphism. Consider $M, N \in \mathrm{Sp}_{2g}(\mathbb{Z},p)$ with $M = \mathbb{I}_{2g} + pA$ and $N = \mathbb{I}_{2g} + pB$. We then have $MN = \mathbb{I}_{2g} + p(A+B) + p^2 AB$, so modulo p we have $\psi(MN) = A+B$.

It remains to show that ψ is surjective. If p is prime, then this is easy. Indeed, ψ is equivariant with respect to the conjugation actions of $\mathrm{Sp}_{2g}(\mathbb{Z})$ on $\mathrm{Sp}_{2g}(\mathbb{Z},p)$ and $\mathfrak{sp}_{2g}(\mathbb{Z}/p)$. The latter factors through $\mathrm{Sp}_{2g}(\mathbb{Z}/p)$, so the image of ψ is a $\mathrm{Sp}_{2g}(\mathbb{Z}/p)$-subrepresentation of $\mathfrak{sp}_{2g}(\mathbb{Z}/p)$. However, it is well-known that $\mathrm{Sp}_{2g}(\mathbb{Z}/p)$ is an irreducible $\mathrm{Sp}_{2g}(\mathbb{Z}/p)$-representation, so since ψ is not trivial it must be surjective (we remark that this uses the fact that p is odd – the abelian group $\mathfrak{sp}_{2g}(\mathbb{Z}/2)$ is *not* an irreducible representation of $\mathrm{Sp}_{2g}(\mathbb{Z}/2)$). We leave the general case as an exercise.

EXERCISE 3.7. Prove that ψ is surjective by constructing elements of $\mathrm{Sp}_{2g}(\mathbb{Z},p)$ that map to generators of $\mathfrak{sp}_{2g}(\mathbb{Z}/p)$.

Finishing the proof of Theorem 3.6. Observe now that $\ker(\psi) = \mathrm{Sp}_{2g}(\mathbb{Z},p^2)$. Since the target of ψ is abelian, this implies that

$$[\mathrm{Sp}_{2g}(\mathbb{Z},p), \mathrm{Sp}_{2g}(\mathbb{Z},p)] \subset \mathrm{Sp}_{2g}(\mathbb{Z},p^2).$$

Theorem 3.6 thus follows from the following lemma.

LEMMA 3.8. *If $g, p \geq 3$ and p is odd, then* $\mathrm{Sp}_{2g}(\mathbb{Z},p^2) \subset [\mathrm{Sp}_{2g}(\mathbb{Z},p), \mathrm{Sp}_{2g}(\mathbb{Z},p)]$.

The proof of Lemma 3.8 follows from a direct matrix calculation. Rather than giving the details, we will give an exercise which outlines a proof of a somewhat easier result for $\mathrm{SL}_n(\mathbb{Z})$. The proof of Lemma 3.8 follows the same basic pattern, though the details are more complicated.

EXERCISE 3.9. Let $\mathrm{SL}_n(\mathbb{Z},p)$ be the kernel of the natural map $\mathrm{SL}_n(\mathbb{Z}) \to \mathrm{SL}_n(\mathbb{Z}/p)$. In this exercise, you will prove a theorem of Lee and Szczarba [**23**] which says that for $n \geq 3$, we have $\mathrm{H}_1(\mathrm{SL}_n(\mathbb{Z},p);\mathbb{Z}) \cong \mathfrak{sl}_n(\mathbb{Z}/p)$. Here $\mathfrak{sl}_n(\mathbb{Z}/p)$ is the abelian group of $n \times n$ matrices over $\mathbb{Z}/p$ with trace 0. We remark that in this exercise, we do not need to assume that p is odd.

(1) Construct a surjective homomorphism $\mathrm{SL}_n(\mathbb{Z},p) \to \mathfrak{sl}_n(\mathbb{Z}/p)$ whose kernel is $\mathrm{SL}_n(\mathbb{Z}/p^2)$. Conclude that $[\mathrm{SL}_n(\mathbb{Z},p), \mathrm{SL}_n(\mathbb{Z},p)] \subset \mathrm{SL}_n(\mathbb{Z},p^2)$.
(2) Prove that $\mathrm{SL}_n(\mathbb{Z}/p^2) \subset [\mathrm{SL}_n(\mathbb{Z},p), \mathrm{SL}_n(\mathbb{Z},p)]$. You will need the following theorem of Bass, Milnor, and Serre [**1**]. For $1 \leq i,j \leq n$ such that $i \neq j$, let e_{ij} be the *elementary matrix* which is obtained from the $n \times n$ identity matrix by changing the entry at position (i,j) to 1. Bass, Milnor, and Serre proved that for $n \geq 3$ and $q \geq 2$, the group $\mathrm{SL}_n(\mathbb{Z},q)$ is normally generated (as a subgroup of $\mathrm{SL}_n(\mathbb{Z})$) by the set $\{e_{ij}^q \mid 1 \leq i,j \leq n,\ i \neq j\}$.

LECTURE 4

The second rational homology group of $\mathrm{Mod_g(p)}$

In this final lecture, we turn to the second homology group of $\mathrm{Mod}_g(p)$. To set the stage, let us first recall what happens for the first homology group. Powell [**28**] proved that $\mathrm{H}_1(\mathrm{Mod}_g;\mathbb{Z}) = 0$ for $g \geq 3$. See [**10**, §5.1] for an easier proof (due to Harer). Hain [**11**] later proved that $\mathrm{H}_1(\mathrm{Mod}_g(p);\mathbb{Q}) = 0$ for $g \geq 3$. In other words, over $\mathbb{Q}$ the first homology group of Mod_g does not change when you pass to the finite-index subgroup $\mathrm{Mod}_g(p)$. Our proof of Theorem 3.1 in Lecture 3 is essentially an elaboration of Hain's proof.

Harer [**12**] proved that $\mathrm{H}_2(\mathrm{Mod}_g;\mathbb{Z}) \cong \mathbb{Z}$ for $g \geq 4$. The following theorem says that a similar result holds for $\mathrm{Mod}_g(p)$ as long as we work over $\mathbb{Q}$.

THEOREM 4.1 (Putman, [**34**]). *For $g \geq 5$ and $p \geq 2$, we have* $\mathrm{H}_2(\mathrm{Mod}_g(p);\mathbb{Q}) \cong \mathbb{Q}$.

The first part of this lecture is devoted to motivating why one might expect such a theorem to hold. Next, we show that Theorem 4.1 is equivalent to a weak form of homological stability for $\mathrm{Mod}_g(p)$. Finally, we discuss how to prove this stability result.

The transfer map. What happens to the homology of a group when you pass to a finite-index subgroup? The following standard lemma says that over $\mathbb{Q}$, the homology can get larger but it cannot get smaller.

LEMMA 4.2. *If G is a finite-index subgroup of Γ, then the map $\mathrm{H}_k(G;\mathbb{Q}) \to \mathrm{H}_k(\Gamma;\mathbb{Q})$ is surjective for all $k \geq 1$.*

EXAMPLE. If Γ is a free abelian group and G is a finite-index subgroup of Γ, then there exists some $\ell \geq 1$ such that $\ell \cdot \Gamma < G < \Gamma$. Consequently, we have $G \cong \Gamma$ and it is easy to see that the map $\mathrm{H}_k(G;\mathbb{Q}) \to \mathrm{H}_k(\Gamma;\mathbb{Q})$ is an isomorphism for all $k \geq 1$.

EXAMPLE. If Γ is a free group of rank n and G is a proper finite-index subgroup of Γ, then G is a free group of rank strictly greater than n. Consequently, the rank of $\mathrm{H}_1(G;\mathbb{Q})$ is strictly greater than the rank of $\mathrm{H}_1(\Gamma;\mathbb{Q}) \cong \mathbb{Q}^n$.

REMARK. Lemma 4.2 is false over $\mathbb{Z}$. Indeed, if Γ is a finite group, then $G = 1$ is a finite-index subgroup of Γ. The homology groups of G are all trivial, but Γ can certainly have nontrivial homology groups. However, Lemma 4.2 does imply that the homology groups of Γ are all torsion.

PROOF OF LEMMA 4.2. Let $n = [\Gamma : G]$. The key to the proof is the *transfer map*. Letting R be a commutative ring and letting $i_* : \mathrm{H}_k(G;R) \to \mathrm{H}_k(\Gamma;R)$ be the map induced by the inclusion $i : G \hookrightarrow \Gamma$, the transfer map is a homomorphism $\tau_k : \mathrm{H}_k(\Gamma;R) \to \mathrm{H}_k(G;R)$ satisfying

$$i_*(\tau_k(x)) = nx \quad \text{for all } x \in \mathrm{H}_k(\Gamma;R). \tag{1}$$

The existence of τ_k immediately implies the lemma. Indeed, if $R = \mathbb{Q}$, then $\frac{1}{n}\tau_k$ is a right-inverse to i_*, so i_* is surjective.

The transfer map is *not* induced by a homomorphism $\Gamma \to G$; rather, it is constructed at the level of chains. The construction goes as follows. Let $B\Gamma$ be a $K(\Gamma, 1)$ and let $\rho : BG \to B\Gamma$ be the cover corresponding to $G < \Gamma$. Thus BG is a $K(G, 1)$ and ρ is a degree n cover. Define a map $\hat{\tau}_k : C_k(B\Gamma; R) \to C_k(BG; R)$ as follows. If $\sigma^k \in C_k(B\Gamma; R)$ is a singular k-simplex, then there are exactly n singular k-simplices $\tilde{\sigma}_1^k, \ldots, \tilde{\sigma}_n^k$ in BG satisfying $\rho_*(\tilde{\sigma}_i^k) = \sigma^k$. Define

$$\hat{\tau}_k(\sigma^k) = \tilde{\sigma}_1^k + \cdots + \tilde{\sigma}_n^k.$$

EXERCISE 4.3. The maps $\hat{\tau}_k$ commute with the boundary operators, so we have a commutative diagram

$$\begin{array}{ccccccc} \cdots & \xrightarrow{\partial} & C_k(\Gamma; R) & \xrightarrow{\partial} & C_{k-1}(\Gamma; R) & \xrightarrow{\partial} & \cdots \\ & & \downarrow{\scriptstyle \hat{\tau}_k} & & \downarrow{\scriptstyle \hat{\tau}_{k-1}} & & \\ \cdots & \xrightarrow{\partial} & C_k(G; R) & \xrightarrow{\partial} & C_{k-1}(G; R) & \xrightarrow{\partial} & \cdots \end{array}$$

By this exercise, the maps $\hat{\tau}_k$ induce maps $\tau_k : \mathrm{H}_k(\Gamma; R) \to \mathrm{H}_k(G; R)$ in homology. Clearly the maps τ_k satisfy (1), so we are done. □

Borel stability. Lemma 4.2 implies that $\mathrm{H}_2(\mathrm{Mod}_g(p); \mathbb{Q}) \cong \mathbb{Q}^m$ for some $m \geq 1$. Why should we expect that $m = 1$? There is a very fruitful analogy between the mapping class group and lattices in Lie groups. The Borel stability theorem says that for the classical sequences of arithmetic lattices, passing to finite-index subgroups does not change their rational homology in a stable range. For example, denote $\mathrm{SL}_n(\mathbb{Z})$ by Γ_n and define

$$\Gamma_n(p) = \ker(\mathrm{SL}_n(\mathbb{Z}) \to \mathrm{SL}_n(\mathbb{Z}/p))$$

for $p \geq 2$. Here the map $\mathrm{SL}_n(\mathbb{Z}) \to \mathrm{SL}_n(\mathbb{Z}/p)$ comes from reducing all the entries in the matrices modulo p. Borel's theorem then takes the following form.

THEOREM 4.4 (Borel, [**5**]). *For $k \geq 1$, there exists some $N_k \geq 1$ such that if $n \geq N_k$ and $p \geq 2$, then $\mathrm{H}_k(\Gamma_n(p); \mathbb{Q}) \cong \mathrm{H}_k(\Gamma_n; \mathbb{Q})$.*

One can view Theorem 4.1 as an analogue (for $k = 2$) of Theorem 4.4 for the mapping class group. As we mentioned, Hain [**11**] proved a similar theorem for $k = 1$. It would be very interesting to extend this to $k \geq 3$.

Reduction to stability. We now discuss the proof of Theorem 4.1. In [**13**], Harer proved that the homology groups of Mod_g satisfy homological stability. One special case of his theorem is as follows. Choose a subsurface S of Σ_g such that $S \cong \Sigma_{g-1,1}$ (see Figure 1). Extending mapping classes on S by the identity to Σ_g, we get a homomorphism $\mathrm{Mod}_{g-1,1} \to \mathrm{Mod}_g$. Harer proved that for each k, there exists some N_k such that if $g \geq N_k$, then the induced map $\mathrm{H}_k(\mathrm{Mod}_{g-1,1}; \mathbb{Z}) \to \mathrm{H}_k(\mathrm{Mod}_g; \mathbb{Z})$ is an isomorphism.

The following result asserts that something similar happens for $\mathrm{Mod}_g(p)$ for $k = 2$.

LEMMA 4.5. *For $g \geq 5$ and $p \geq 2$, let $i : \mathrm{Mod}_{g-1,1}(p) \to \mathrm{Mod}_g(p)$ be the restriction of the map $\mathrm{Mod}_{g-1,1} \to \mathrm{Mod}_g$ described above. Then the induced map $i_* : \mathrm{H}_2(\mathrm{Mod}_{g-1,1}(p); \mathbb{Q}) \to \mathrm{H}_2(\mathrm{Mod}_g(p); \mathbb{Q})$ is surjective.*

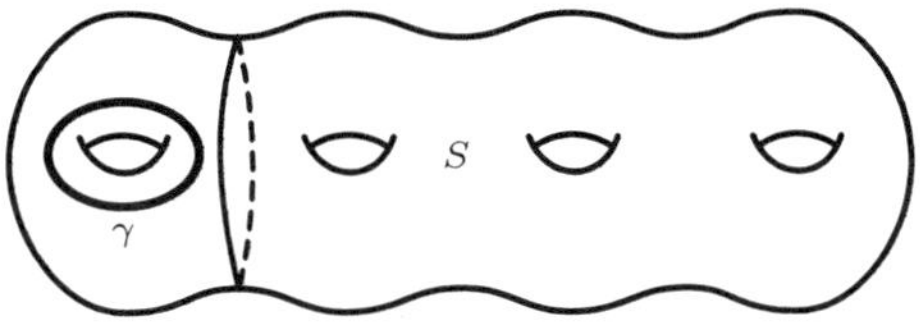

FIGURE 1. S is a subsurface of Σ_g satisfying $S \cong \Sigma_{g-1,1}$ and γ is a simple closed nonseparating curve contained in $\Sigma_g \setminus S$

Observe that Theorem 4.1 implies Lemma 4.5. Indeed, if Theorem 4.1 is true, then we have a commutative diagram of the form

$$\begin{array}{ccc} \mathrm{H}_2(\mathrm{Mod}_{g-1,1}(p);\mathbb{Q}) & \longrightarrow & \mathrm{H}_2(\mathrm{Mod}_g(p);\mathbb{Q}) \\ \downarrow & & \downarrow \cong \\ \mathrm{H}_2(\mathrm{Mod}_{g-1,1};\mathbb{Q}) & \longrightarrow & \mathrm{H}_2(\mathrm{Mod}_g;\mathbb{Q}) \end{array}$$

By Lemma 4.2, the map $\mathrm{H}_2(\mathrm{Mod}_{g-1,1}(p);\mathbb{Q}) \to \mathrm{H}_2(\mathrm{Mod}_{g-1,1};\mathbb{Q})$ is a surjection, and Harer's stability theorem say that the map $\mathrm{H}_2(\mathrm{Mod}_{g-1,1};\mathbb{Q}) \to \mathrm{H}_2(\mathrm{Mod}_g;\mathbb{Q})$ is an isomorphism as long as $g \geq 5$. We conclude that the map $\mathrm{H}_2(\mathrm{Mod}_{g-1,1}(p);\mathbb{Q}) \to \mathrm{H}_2(\mathrm{Mod}_g(p);\mathbb{Q})$ is a surjection, as desired.

Somewhat surprisingly, Lemma 4.5 also implies Theorem 4.1.

LEMMA 4.6. *Lemma* 4.5 *implies Theorem* 4.1.

Before proving Lemma 4.6, we need some more abstract nonsense. Let G be a finite-index normal subgroup of Γ. The conjugation action of Γ on G induces an action of Γ on $\mathrm{H}_k(G;\mathbb{Q})$. Recall that if M is a vector space upon which Γ acts, then the *coinvariants* of the action, denoted M_Γ, is the quotient M/I with I the subspace spanned by $\{x - g(x) \mid x \in M,\ g \in \Gamma\}$. In other words, M_Γ is the largest quotient of M upon which Γ acts trivially. Since Γ acts trivially on its own homology groups $\mathrm{H}_k(\Gamma;\mathbb{Q})$, the map $\mathrm{H}_k(G;\mathbb{Q}) \to \mathrm{H}_k(\Gamma;\mathbb{Q})$ factors through $(\mathrm{H}_k(G;\mathbb{Q}))_\Gamma$. We then have the following exercise.

EXERCISE 4.7. $\mathrm{H}_k(\Gamma;\mathbb{Q}) \cong (\mathrm{H}_k(G;\mathbb{Q}))_\Gamma$. Hint : carefully study the proof of Lemma 4.2. Alternatively, this can be proven using the Hochschild-Serre spectral sequence of the extension

$$1 \longrightarrow G \longrightarrow \Gamma \longrightarrow \Gamma/G \longrightarrow 1.$$

PROOF OF LEMMA 4.6. Assume that Lemma 4.5 holds. Since $\mathrm{H}_2(\mathrm{Mod}_g;\mathbb{Q}) \cong \mathbb{Q}$, to prove that $\mathrm{H}_2(\mathrm{Mod}_g(p);\mathbb{Q}) \cong \mathbb{Q}$, it is enough to prove that $\mathrm{H}_2(\mathrm{Mod}_g(p);\mathbb{Q}) \cong \mathrm{H}_2(\mathrm{Mod}_g;\mathbb{Q})$. By Exercise 4.6, this is equivalent to showing that the action of Mod_g on $\mathrm{H}_2(\mathrm{Mod}_g(p);\mathbb{Q})$ is trivial. This can be checked on a generating set. The group Mod_g is generated by the set of Dehn twists about nonseparating simple closed curves. Consider such a Dehn twist T_γ. As is shown in Figure 1, we can find a subsurface S of Σ_g such that $S \cong \Sigma_{g-1,1}$ and such that $\gamma \subset \Sigma_g \setminus S$. Let $i : \mathrm{Mod}_{g-1,1}(p) \to \mathrm{Mod}_g(p)$ be the map induced by the subsurface inclusion $S \hookrightarrow \Sigma_g$. Clearly T_γ commutes with $\mathrm{Im}(i)$, so T_γ acts trivially on $i_*(\mathrm{H}_2(\mathrm{Mod}_{g-1,1}(p);\mathbb{Q}))$. Lemma 4.5 implies that $i_*(\mathrm{H}_2(\mathrm{Mod}_{g-1,1}(p);\mathbb{Q})) = \mathrm{H}_2(\mathrm{Mod}_g(p);\mathbb{Q})$, so we conclude that T_γ acts trivially on $\mathrm{H}_2(\mathrm{Mod}_g(p);\mathbb{Q})$, as desired. $\square$

A weaker result suffices. In fact, we do not need the full strength of Lemma 4.5 to prove Theorem 4.1. If γ is the isotopy class of a nonseparating simple closed curve on Σ_g, then denote by $(\mathrm{Mod}_g(p))_\gamma$ the subgroup of $\mathrm{Mod}_g(p)$ which stabilizes γ. Examining the proof of Lemma 4.6, it is clear that the following lemma also implies Theorem 4.1.

LEMMA 4.8. *For $g \geq 5$ and $p \geq 2$, let γ be a nonseparating simple closed curve on Σ_g. Then the map $\mathrm{H}_2((\mathrm{Mod}_g(p))_\gamma; \mathbb{Q}) \to \mathrm{H}_2(\mathrm{Mod}_g(p); \mathbb{Q})$ is surjective.*

The homological stability machine. Homological stability theorems are known for many different sequences of groups, and there is now a standard procedure for proving them. This procedure goes back to Quillen. See [**15**] for an excellent discussion of how this machine applies to many different sequences of groups and [**37**] for a detailed exposition of how to apply it to the mapping class group. The input for this machine consists of a sequence of highly connected spaces upon which the groups in question act. For Mod_g, Harer proved homological stability by studying the action on the following space.

DEFINITION. The *nonseparating curve complex*, denoted $\mathcal{C}_g^{\mathrm{ns}}$, is the simplicial complex whose k-simplices are sets $\{\gamma_0, \ldots, \gamma_k\}$ of isotopy classes of simple closed curves on Σ_g that can be realized such that the γ_i are all disjoint and $\Sigma_g \setminus (\gamma_0 \cup \cdots \cup \gamma_k)$ is connected.

One's first impulse, of course, is to apply this machine to $\mathrm{Mod}_g(p)$ to try to prove Lemma 4.8. Unfortunately, it does not quite work. The problem is that if it did work, then it would prove that $\mathrm{H}_k(\mathrm{Mod}_g(p); \mathbb{Z})$ is stable, and this is false even for $k = 1$ (see Theorem 3.1). The machine does, however, give the following result.

LEMMA 4.9. *For $g \geq 5$ and $p \geq 2$, the map*

$$\bigoplus_{\gamma \in (\mathcal{C}_g^{ns})^{(0)}} \mathrm{H}_2((\mathrm{Mod}_g(p))_\gamma; \mathbb{Q}) \longrightarrow \mathrm{H}_2(\mathrm{Mod}_g(p); \mathbb{Q})$$

is surjective.

Rescuing the machine. To deduce Lemma 4.8 from Lemma 4.9, we need to prove that if $g \geq 5$ and $p \geq 2$, then for any two nonseparating simple closed curves γ and γ' on Σ_g, the images of $\mathrm{H}_2((\mathrm{Mod}_g(p))_\gamma; \mathbb{Q})$ and $\mathrm{H}_2((\mathrm{Mod}_g(p))_{\gamma'}; \mathbb{Q})$ in $\mathrm{H}_2(\mathrm{Mod}_g(p); \mathbb{Q})$ are the same. A standard result about surfaces called the "change of coordinates principle" (see [**10**]) says that there exists some $\phi \in \mathrm{Mod}_g$ such that $\phi(\gamma) = \gamma'$. This implies that

$$\phi(\mathrm{Mod}_g(p))_\gamma \phi^{-1} = (\mathrm{Mod}_g(p))_{\gamma'}.$$

Thus the action of Mod_g on $\mathrm{H}_2(\mathrm{Mod}_g(p); \mathbb{Q})$ takes the image of $\mathrm{H}_2((\mathrm{Mod}_g(p))_\gamma; \mathbb{Q})$ to the image of $\mathrm{H}_2((\mathrm{Mod}_g(p))_{\gamma'}; \mathbb{Q})$. We conclude that Lemma 4.9 would following from the following lemma.

LEMMA 4.10. *For $g \geq 5$ and $p \geq 2$, if γ is a nonseparating simple closed curve on Σ_g, then Mod_g acts trivially on the image of $\mathrm{H}_2((\mathrm{Mod}_g(p))_\gamma; \mathbb{Q})$ in $\mathrm{H}_2(\mathrm{Mod}_g(p); \mathbb{Q})$.*

The proof of Lemma 4.10 is quite complicated (see [**34**]). Rather than give it here, we will prove an analogous result about $\Gamma_n = \mathrm{SL}_n(\mathbb{Z})$. Recall that

$$\Gamma_n(p) = \ker(\mathrm{SL}_n(\mathbb{Z}) \longrightarrow \mathrm{SL}_n(\mathbb{Z}/p)).$$

There is a natural map $\Gamma_{n-1} \to \Gamma_n$ which takes a matrix $A \in \Gamma_{n-1}$ to the matrix $\left(\begin{smallmatrix} 1 & 0 \\ 0 & A \end{smallmatrix}\right) \in \Gamma_n$. In [**8**], Charney proves the following theorem.

THEOREM 4.11 (Charney, [**8**, Proposition 5.5]). *For $n \geq 2$ and $k, p \geq 1$, the group Γ_n acts trivially on the image of $\mathrm{H}_k(\Gamma_{n-1}(p); \mathbb{Q})$ in $\mathrm{H}_k(\Gamma_n(p); \mathbb{Q})$.*

Though she does not say it in her paper, it is not hard to combine this with the standard homological stability machinery to give a new proof of the Borel stability theorem for $\mathrm{SL}_n(\mathbb{Z})$. After the proof, we will comment on the relation between the proofs of Theorem 4.11 and Lemma 4.10.

PROOF OF THEOREM 4.11. For $1 \leq i, j \leq n$ such that $i \neq j$, let e_{ij} be the *elementary matrix* which is obtained from the $n \times n$ identity matrix by changing the entry at position (i, j) to 1. The group Γ_n is generated by the set $\{e_{ij} \mid 1 \leq i, j \leq n,\ i \neq j\}$. However, we do not need this entire set.

EXERCISE 4.12. For $2 \leq i, j \leq n$ such that $i \neq j$, prove that $[e_{i1}, e_{1j}] = e_{ij}$.

Consequently, we only need to prove that e_{1j} and e_{i1} act trivially for $2 \leq i, j \leq n$. We will give the details for e_{1j}. The other case is similar.

Consider the subgroups

$$G = \left\{ \left(\begin{array}{c|ccc} 1 & c_2 & \cdots & c_n \\ \hline 0 & & & \\ \vdots & & A & \\ 0 & & & \end{array} \right) \;\middle|\; A \in \Gamma_{n-1}(p),\ c_2, \ldots, c_n \in p\mathbb{Z} \right\}$$

and

$$\hat{G} = \left\{ \left(\begin{array}{c|ccc} 1 & c_2 & \cdots & c_n \\ \hline 0 & & & \\ \vdots & & A & \\ 0 & & & \end{array} \right) \;\middle|\; A \in \Gamma_{n-1}(p),\ c_2, \ldots, c_n \in \mathbb{Z} \right\}$$

of Γ_n. The conjugation action of e_{1j} on Γ_n takes G and $\hat{G}$ to themselves. We have $\Gamma_{n-1}(p) \subset G \subset \Gamma_n(p)$, so it is enough to show that e_{1j} acts trivially on $\mathrm{H}_k(G; \mathbb{Q})$. Since $e_{1j} \in \hat{G}$, the action of e_{1j} on $\mathrm{H}_k(\hat{G}; \mathbb{Q})$ is trivial. It is therefore enough to show the inclusion $G \hookrightarrow \hat{G}$ induces an isomorphism $\mathrm{H}_k(G; \mathbb{Q}) \cong \mathrm{H}_k(\hat{G}; \mathbb{Q})$.

Define

$$K = \left\{ \left(\begin{array}{c|ccc} 1 & c_2 & \cdots & c_n \\ \hline 0 & & & \\ \vdots & & 0 & \\ 0 & & & \end{array} \right) \;\middle|\; c_2, \ldots, c_n \in p\mathbb{Z} \right\} \subset G.$$

Observe that $K \cong \mathbb{Z}^{n-1}$. We have a short exact sequence

$$1 \longrightarrow K \longrightarrow G \longrightarrow \Gamma_n(p) \longrightarrow 1.$$

Similarly, setting

$$\hat{K} = \{\left(\begin{array}{c|ccc} 1 & c_2 & \cdots & c_n \\ \hline 0 & & & \\ \vdots & & \mathbf{0} & \\ 0 & & & \end{array}\right) \mid c_2, \ldots, c_n \in \mathbb{Z}\} \subset \hat{G}$$

we have a short exact sequence

$$1 \longrightarrow \hat{K} \longrightarrow \hat{G} \longrightarrow \Gamma_n(p) \longrightarrow 1.$$

These short exact sequences fit into a commutative diagram of the form

$$\begin{array}{ccccccccc} 1 & \longrightarrow & K & \longrightarrow & G & \longrightarrow & \Gamma_n(p) & \longrightarrow & 1 \\ & & \downarrow & & \downarrow & & \downarrow = & & \\ 1 & \longrightarrow & \hat{K} & \longrightarrow & \hat{G} & \longrightarrow & \Gamma_n(p) & \longrightarrow & 1. \end{array}$$

By Lemma 4.13 (a sort of 5-lemma for group homology), it is enough to show that the inclusion map $K \hookrightarrow \hat{K}$ induces an isomorphism $\mathrm{H}_k(K;\mathbb{Q}) \cong \mathrm{H}_k(\hat{K};\mathbb{Q})$. However, we have $\hat{K} \cong \mathbb{Z}^{n-1}$ and $K = p \cdot \hat{K}$, so this is immediate. We remark that this final step is the only place we use the fact that we are working over $\mathbb{Q}$. □

LEMMA 4.13 (5-lemma for group homology). *Fix a commutative ring R. Consider a commutative diagram*

$$\begin{array}{ccccccccc} 1 & \longrightarrow & A_1 & \longrightarrow & B_1 & \longrightarrow & C & \longrightarrow & 1 \\ & & \downarrow & & \downarrow & & \downarrow = & & \\ 1 & \longrightarrow & A_2 & \longrightarrow & B_2 & \longrightarrow & C & \longrightarrow & 1 \end{array}$$

of short exact sequences of groups. Assume that the map $A_1 \to A_2$ induces an isomorphism $\mathrm{H}_k(A_1;R) \cong \mathrm{H}_k(A_2;R)$ for all k. Then the map $B_1 \to B_2$ induces an isomorphism $\mathrm{H}_k(B_1;R) \cong \mathrm{H}_k(B_2;R)$ for all k.

PROOF. Associated each of our short exact sequences is a Hochschild-Serre spectral sequence in group homology. The assumptions in the lemma imply that the induced map between these spectral sequences is an isomorphism on the E^2-page. This implies that it converges an an isomorphism on the E^∞-pages, so we obtain isomorphisms $\mathrm{H}_k(B_1;R) \cong \mathrm{H}_k(B_2;R)$ for all k. □

Relation between Theorem 4.11 and Lemma 4.10. The proof of Lemma 4.10 in [**34**] is not nearly as short as the proof above of Theorem 4.11. However, they share some features. The key to the proof of Theorem 4.11 above is the short exact sequence

$$1 \longrightarrow K \longrightarrow G \longrightarrow \Gamma_n(p) \longrightarrow 1.$$

A similar role in the proof of Lemma 4.10 is played by various analogues for $\mathrm{Mod}_g(p)$ of the Birman exact sequence which was discussed in Lecture 1. However, the kernels of these Birman exact sequences are not free abelian, so passing to finite-index subgroups actually changes their isomorphism types. This necessitates making a number of rather intricate twisted group cohomology computations.

Bibliography

[1] H. Bass, J. Milnor and J.-P. Serre, Solution of the congruence subgroup problem for SL_n ($n \geq 3$) and Sp_{2n} ($n \geq 2$), Inst. Hautes Études Sci. Publ. Math. No. 33 (1967), 59–137. MR0244257 (39:5574)

[2] J. S. Birman, Mapping class groups and their relationship to braid groups, Comm. Pure Appl. Math. **22** (1969), 213–238. MR0243519 (39:4840)

[3] J. S. Birman, On Siegel's modular group, Math. Ann. **191** (1971), 59–68. MR0280606 (43:6325)

[4] J. S. Birman and R. Craggs, The μ-invariant of 3-manifolds and certain structural properties of the group of homeomorphisms of a closed, oriented 2-manifold, Trans. Amer. Math. Soc. **237** (1978), 283–309. MR0482765 (58:2818)

[5] A. Borel, Stable real cohomology of arithmetic groups, Ann. Sci. École Norm. Sup. (4) **7** (1974), 235–272 (1975). MR0387496 (52:8338)

[6] N. Broaddus, B. Farb and A. Putman, Irreducible Sp-representations and subgroup distortion in the mapping class group, Comment. Math. Helv. **86** (2011), 537-556. MR2803852 (2012e:57002)

[7] K. S. Brown, *Cohomology of groups*, corrected reprint of the 1982 original, Graduate Texts in Mathematics, 87, Springer, New York, 1994. MR1324339 (96a:20072)

[8] R. Charney, On the problem of homology stability for congruence subgroups, Comm. Algebra **12** (1984), no. 17-18, 2081–2123. MR747219 (86b:20053)

[9] T. Church and B. Farb, Parameterized Abel-Jacobi maps and abelian cycles in the Torelli group, J. Topol. 5 (2012), no. 1, 15–38. MR2897048

[10] B. Farb and D. Margalit, A primer on mapping class groups, Princeton Mathematical Series, 49. Princeton University Press, Princeton, NJ, 2012. MR2850125 (2012h:57032)

[11] R. Hain, Torelli groups and geometry of moduli spaces of curves, in *Current topics in complex algebraic geometry (Berkeley, CA, 1992/93)*, 97–143, Cambridge Univ. Press, Cambridge. MR1397061 (97d:14036)

[12] J. Harer, The second homology group of the mapping class group of an orientable surface, Invent. Math. **72** (1983), no. 2, 221–239. MR700769 (84g:57006)

[13] J. L. Harer, Stability of the homology of the mapping class groups of orientable surfaces, Ann. of Math. (2) **121** (1985), no. 2, 215–249. MR786348 (87f:57009)

[14] A. Hatcher and D. Margalit, Generating the Torelli group, Enseign. Math. 58 (2012), 165-188. MR2985015

[15] A. Hatcher and N. Wahl, Stabilization for mapping class groups of 3-manifolds, Duke Math. J. **155** (2010), no. 2, 205–269. MR2736166 (2012c:57001)

[16] D. Johnson, Homeomorphisms of a surface which act trivially on homology, Proc. Amer. Math. Soc. **75** (1979), no. 1, 119–125. MR529227 (80h:57008)

[17] D. Johnson, An abelian quotient of the mapping class group $\mathcal{I}_g$, Math. Ann. **249** (1980), no. 3, 225–242. MR579103 (82a:57008)

[18] D. Johnson, Quadratic forms and the Birman-Craggs homomorphisms, Trans. Amer. Math. Soc. **261** (1980), no. 1, 235–254. MR576873 (81k:57010)

[19] D. Johnson, The structure of the Torelli group. I. A finite set of generators for $\mathcal{I}$, Ann. of Math. (2) **118** (1983), no. 3, 423–442. MR727699 (85a:57005)

[20] D. Johnson, The structure of the Torelli group. II. A characterization of the group generated by twists on bounding curves, Topology **24** (1985), no. 2, 113–126. MR793178 (86i:57011)

[21] D. Johnson, The structure of the Torelli group. III. The abelianization of $\mathcal{T}$, Topology **24** (1985), no. 2, 127–144. MR793179 (87a:57016)

[22] D. Johnson, A survey of the Torelli group, in *Low-dimensional topology (San Francisco, Calif., 1981)*, 165–179, Contemp. Math., 20, Amer. Math. Soc., Providence, RI. MR718141 (85d:57009)

[23] R. Lee and R. H. Szczarba, On the homology and cohomology of congruence subgroups, Invent. Math. **33** (1976), no. 1, 15–53. MR0422498 (54:10485)

[24] I. Madsen and M. Weiss, The stable moduli space of Riemann surfaces: Mumford's conjecture, Ann. of Math. (2) **165** (2007), no. 3, 843–941. MR2335797 (2009b:14051)

[25] D. McCullough and A. Miller, The genus 2 Torelli group is not finitely generated, Topology Appl. **22** (1986), no. 1, 43–49. MR831180 (87h:57015)

[26] G. Mess, The Torelli groups for genus 2 and 3 surfaces, Topology **31** (1992), no. 4, 775–790. MR1191379 (93k:57003)

[27] B. Perron, Filtration de Johnson et groupe de Torelli modulo p, p premier, C. R. Math. Acad. Sci. Paris **346** (2008), no. 11-12, 667–670. MR2423275 (2009c:57033)

[28] J. Powell, Two theorems on the mapping class group of a surface, Proc. Amer. Math. Soc. **68** (1978), no. 3, 347–350. MR0494115 (58:13045)

[29] A. Putman, Cutting and pasting in the Torelli group, Geom. Topol. **11** (2007), 829–865. MR2302503 (2008c:57049)

[30] A. Putman The abelianization of the level L mapping class group, preprint 2008.

[31] A. Putman, An infinite presentation of the Torelli group, Geom. Funct. Anal. **19** (2009), no. 2, 591–643. MR2545251 (2010k:57069)

[32] A. Putman, The Johnson homomorphism and its kernel, preprint 2009.

[33] A. Putman, The Picard group of the moduli space of curves with level structures, Duke Math. J. 161 (2012), no. 4, 623–674. MR2891531

[34] A. Putman, The second rational homology group of the moduli space of curves with level structures, Adv. Math. 229 (2012), no. 2, 1205–1234. MR2855091 (2012i:57032)

[35] A. Putman, Small generating sets for the Torelli group, Geom. Topol. 16 (2012), no. 1, 111–125. MR2872579 (2012k:57003)

[36] M. Sato, The abelianization of the level d mapping class group, J. Topol. **3** (2010), no. 4, 847–882. MR2746340 (2012c:55016)

[37] N. Wahl, Homological stability for mapping class groups of surfaces, to appear in the Handbook of Moduli.

Tautological Algebras of Moduli Spaces of Curves

Carel Faber

IAS/Park City Mathematics Series
Volume 20, 2011

Tautological Algebras of Moduli Spaces of Curves

Carel Faber

Introduction

These are the lecture notes for my course at the 2011 Park City Mathematical Institute on moduli spaces of Riemann surfaces. The two lectures here correspond roughly to the first and second half of the course.

The subject of the first lecture is the tautological ring $R^*(M_g)$ of M_g. I recall Mumford's definition of the tautological classes and some of his results from [**48**]. Then I discuss my conjecture on $R^*(M_g)$ from [**10**] and the results obtained on it. Finally, I survey some recent developments indicating that the relations that suffice to prove the conjecture for $g \leq 23$ may not suffice for larger g.

The second lecture concerns mainly the tautological ring of $\overline{M}_{g,n}$. Some natural spaces in between $M_{g,n}$ and $\overline{M}_{g,n}$ are discussed as well. I close with some recent results regarding non-tautological cohomology classes and the cohomology of $\overline{M}_{g,n}$ in low genus.

Department of Mathematics, KTH Royal Institute of Technology, 100 44 Stockholm, Sweden
E-mail address: faber@kth.se
The author thanks Benson Farb, Dick Hain, and Eduard Looijenga for the organization of the Institute and Dan Petersen for being the TA for my course.

LECTURE 1

The tautological ring of M_g

Let $g \geq 2$ be an integer and let M_g be the moduli space of nonsingular curves of genus g. Denote by $A^*(M_g)$ the Chow ring of M_g with rational coefficients. The tautological ring $R^*(M_g)$ is defined as the $\mathbb{Q}$-subalgebra of $A^*(M_g)$ generated by the tautological classes (whose definition will be recalled in a moment). The image of $R^*(M_g)$ in the rational cohomology ring $H^*(M_g)$ will be denoted $RH^*(M_g)$. A reader who is unfamiliar with the Chow ring may substitute $RH^*(M_g)$ for $R^*(M_g)$ and will not lose much as a result. Throughout these notes, however, we will use the algebraic degree, which is half of the cohomological degree. (It should also be noted that it is not known whether the map from $R^*(M_g)$ to $RH^*(M_g)$ can have a nontrivial kernel.)

Mumford's definition [**48**, §4] of the tautological classes $\kappa_i \in R^i(M_g)$ in this setting is as follows. Let $C_g = M_{g,1}$ be the moduli space of 1-pointed curves of genus g. We view C_g as the universal curve. We have the map $\pi : C_g \to M_g$ that forgets the marked point. Over C_g, we have the cotangent line bundle $\mathbb{L}$ at the marked point, which equals the relative dualizing sheaf ω_π. We denote the first Chern class $c_1(\mathbb{L}) \in A^1(C_g)$ by K. Using the ring structure of $A^*(C_g)$, we obtain the powers $K^i \in A^i(C_g)$ for any $i \geq 0$. The map $\pi : C_g \to M_g$ is proper, so that we have push-forward maps $\pi_* : A^{i+1}(C_g) \to A^i(M_g)$. The tautological class κ_i is defined via

$$\kappa_i = \pi_*(K^{i+1}) \in A^i(M_g).$$

Clearly, $\kappa_0 = 2g - 2$; in the sequel, we will sometimes write κ_0 instead of $2g - 2$ and κ_{-1} instead of 0 to write relations in a genus independent form.

The tautological ring $R^*(M_g)$ is defined as the $\mathbb{Q}$-subalgebra of the rational Chow ring $A^*(M_g)$ generated by the κ_i $(i \geq 1)$.

A second set of very important classes is formed by the Chern classes of the Hodge bundle. The Hodge bundle $\mathbb{E} = \pi_*(\omega_\pi)$ is the vector bundle of rank g over M_g whose fiber over the isomorphism class $[C]$ is the space $H^0(C, \omega_C)$ of regular differentials on C. (In fact, $\mathbb{E}$ extends naturally to a vector bundle over $\overline{M}_g$, cf. [**48**, §4].) Put $\lambda_i = c_i(\mathbb{E}) \in A^i(M_g)$.

Using the Grothendieck-Riemann-Roch theorem, Mumford [**48**, §5] computed the Chern character of the Hodge bundle. The computation is done over the moduli space $\overline{M}_g$ of stable curves of genus g; restricted to M_g, the result becomes

$$ch(\mathbb{E}) = g + \sum_{i=1}^{\infty} \frac{B_{2i}\kappa_{2i-1}}{(2i)!}.$$

(The Bernoulli numbers B_{2i} are defined by $t/(e^t - 1) = \sum_{i=0}^{\infty} B_i t^i/(i!)$.) In particular, $ch_{2i}(\mathbb{E}) = 0$ for $i \geq 1$ (this is in fact true on $\overline{M}_g$). The vanishing (on $\overline{M}_g$) implies that $ch_j(\mathbb{E}) = 0$ for all $j \geq 2g$ (on $\overline{M}_g$, hence on M_g as well). Cf. Exc. 1.

One also finds the following formula for the lambda classes:

$$\sum_{i=0}^{\infty} \lambda_i t^i = \exp\Big(\sum_{i=1}^{\infty} \frac{B_{2i}\kappa_{2i-1}}{2i(2i-1)} t^{2i-1}\Big).$$

So the lambda classes are polynomials in the odd kappa classes.

Mumford also found further relations between the lambda and kappa classes using geometric properties of nonsingular curves. Both the nature of the relations and the fact that it is not at all clear how to extend these relations to $\overline{M}_g$ are features that they have in common with the relations that we will see later in this lecture.

The idea is to use the fact that on a smooth curve of genus $g \geq 2$ the canonical line bundle is generated by its global sections. This translates as the simple statement that the natural map $\pi^*\mathbb{E} \to \omega_\pi$ of locally free sheaves on C_g is surjective. The kernel Q_{g-1} of this map is therefore locally free of rank $g-1$, so its Chern classes $c_j(Q_{g-1})$ vanish in degrees $j \geq g$. In other words, the degree j components of the formal expression

$$\frac{\pi^*(1+\lambda_1+\lambda_2+\cdots+\lambda_g)}{1+K}$$

vanish for $j \geq g$. Expanding this as a power series, we obtain on C_g the relation

$$K^g - K^{g-1}\pi^*\lambda_1 + K^{g-2}\pi^*\lambda_2 + \cdots + (-1)^g \pi^*\lambda_g = 0,$$

as well as its multiples obtained by multiplying with powers of K. Pushing down to M_g, we find the relations

$$\kappa_{j-1} - \kappa_{j-2}\lambda_1 + \kappa_{j-3}\lambda_2 + \cdots + (-1)^g \kappa_{j-1-g}\lambda_g = 0,$$

for any $j \geq g$. Combining these relations with the earlier relations expressing the lambda classes as polynomials in the odd kappa classes, Mumford [**48**, §6] obtained the following result:

THEOREM 1.1. *$R^*(M_g)$ is generated by the first $g-2$ kappa classes $\kappa_1, \ldots, \kappa_{g-2}$.*

(The growth estimate for the Bernoulli numbers that Mumford uses in the final step of his argument can be replaced by a simple congruence property of these numbers; see Exc. 2.)

We now discuss another method for producing relations between the kappa classes. In fact, the method gives many relations between natural classes on moduli spaces of curves with marked points; pushing down these relations to M_g, one obtains relations between the kappa classes. It appears to be very difficult to carry out an exhaustive analysis of all the relations produced by the method. As we will see, however, a subset of the relations has been analysed to a substantial extent.

The idea is to study the loci of curves with divisors of a certain degree moving in a complete linear system of at least a given dimension, or more precisely, the loci of such curves together with the divisors, viewed as subvarieties of moduli spaces of pointed curves. In many cases, one can write formulas for the classes of such loci in terms of tautological classes. The loci typically have positive-dimensional fibers for the forgetful map to M_g and one cannot push forward the class of such a locus directly to find the class in M_g of the locus of curves with such divisors. But by cutting (i.e., intersecting) with various tautological divisors, one obtains loci that upon push-forward to M_g give the class in M_g with a certain multiplicity, which often can be computed. By cutting with different sets of divisors, one obtains different tautological expressions for the class in M_g; equating them, one finds relations between the kappa classes. In fact, since the goal was to obtain tautological relations, one can just as well push forward the class of a locus with positive-dimensional fibers and obtain relations right away.

To set this up, we consider the d-fold fiber product of C_g over M_g, which we denote by C_g^d. This is the moduli space of d-pointed curves $(C; p_1, \ldots, p_d)$, where C is nonsingular of genus g and the p_i are d ordered points of C, which may coincide. (We find it convenient to work in the situation where the points are ordered, even though we only study divisors, that is, unordered sums of points. It may very well be worthwhile to work out these relations on the symmetric fiber product, or perhaps on the universal Jacobian.) Write D for the effective divisor $p_1 + \cdots + p_d$ on C; by abuse of notation, I will sometimes write $(C; D)$ for a point of C_g^d.

We consider a natural map of vector bundles on C_g^d. The first bundle is the pull-back from M_g of the Hodge bundle. We will still denote this by $\mathbb{E}$ (suppressing the pull-back).

The second bundle is the vector bundle $\mathbb{F}_d$ of rank d. Its fiber at $(C; D)$ is the vector space $H^0(C, \omega_C/\omega_C(-D)) = H^0(C, \omega_C|_D)$. Formally, let $\pi : U \to C_g^d$ be the universal curve; U is isomorphic to C_g^{d+1} and the map π comes with d sections s_1, $\ldots$, s_d. Let $\Sigma \subset U$ be the subscheme that is the union (or sum) of the sections. Then $\mathbb{F}_d = \pi_*(\omega_\pi|_\Sigma)$. One can think of $\mathbb{F}_d$ as a universal d-pointed jet bundle.

Since $\mathbb{E} = \pi_*(\omega_\pi)$, we have a natural map

$$\phi_d : \mathbb{E} \to \mathbb{F}_d$$

of vector bundles over C_g^d. At a point $(C; D)$, the kernel of ϕ_d is the vector space $H^0(C, \omega_C(-D))$. By Riemann-Roch,

$$\dim H^0(C, \mathcal{O}(D)) - \dim H^0(C, \omega_C(-D)) = d - g + 1$$

So the rank of ϕ_d is at most $d - r$ if and only if $\dim H^0(C, \omega_C(-D)) \geq g - d + r$ if and only if $\dim H^0(C, \mathcal{O}(D)) \geq r + 1$.

We conclude that

$$\{\mathrm{rk}\, \phi_d \leq d - r\},$$

considered as a subscheme of C_g^d, equals the locus of d-pointed curves $(C; D)$ for which the complete linear system $|D|$ has dimension at least r. The *expected* codimension of this determinantal locus in C_g^d equals $(g-d+r)(d-d+r) = r(g-d+r)$, and the expected codimension of the locus in M_g of curves with a g_d^r (a linear system of dimension r and degree d) equals $r(g-d+r) - d + r = (r+1)(g-d+r) - g$ (the opposite of the Brill-Noether number ρ).

Porteous's formula (cf. [**2, 19**]) computes the class of the locus $\{\mathrm{rk}\, \phi_d \leq d - r\}$ if it is either empty or has the expected codimension. The formula is:

$$\text{class}\ (\{\mathrm{rk}\, \phi_d \leq d - r\}) = \Delta_{r, g-d+r}(c_t(\mathbb{F}_d)/c_t(\mathbb{E})).$$

Here, c_t stands for the total Chern class, viewed as a polynomial in t; the quotient $c_t(\mathbb{F}_d)/c_t(\mathbb{E})$ is viewed as a formal power series in t; and

$$\Delta_{p,q}\left(\sum_{i=0}^{\infty} c_i t^i\right) = \begin{vmatrix} c_p & c_{p+1} & \cdots & c_{p+q-1} \\ c_{p-1} & c_p & \cdots & c_{p+q-2} \\ \vdots & \vdots & \ddots & \vdots \\ c_{p-q+1} & c_{p-q+2} & \cdots & c_p \end{vmatrix}.$$

We have the following tautological classes on C_g^d: the kappa classes κ_i, pulled back from M_g (again, it will be more convenient to suppress the pull-backs in the notation), and the divisor classes K_i and $D_{i,j}$: the class K_i is the pull-back of K via the projection from C_g^d onto the ith factor, and $D_{i,j} = D_{j,i}$ (for $i \neq j$) is the class of the diagonal $p_i = p_j$. For $k > 1$, we also put $\Delta_k = D_{1,k} + D_{2,k} + \cdots + D_{k-1,k}$.

We know already that the Chern classes λ_i of $\mathbb{E}$ are tautological. In addition,

$$c_t(\mathbb{E})^{-1} = c_t(\mathbb{E}^\vee) = 1 - \lambda_1 t + \lambda_2 t^2 + \cdots + (-1)^g \lambda_g t^g$$

(see Exc. 1), so we can work with polynomials instead of power series in the formula for the class of the degeneracy locus. Fortunately, we also have a relatively simple formula for the total Chern class of $\mathbb{F}_d$ in terms of the tautological classes:

$$c(\mathbb{F}_d) = (1 + K_1)(1 + K_2 - \Delta_2)(1 + K_3 - \Delta_3) \cdots (1 + K_d - \Delta_d)$$

(see Exc. 3). This means that the classes of the degeneracy loci above are tautological — if the conditions in Porteous's formula are satisfied — and we can produce many relations between the kappa classes on M_g by means of the indicated method. Moreover, if a certain degeneracy locus is empty, we find a relation on C_g^d, and if a degeneracy locus has positive-dimensional fibers over C_g^k for some $k < d$, we find a relation on C_g^k by pushing forward directly.

So far, we have tacitly assumed that the push-forward of a tautological class by a morphism forgetting a point is tautological. Let us make this explicit:

(1) Every monomial in the classes K_i $(1 \leq i \leq d)$ and $D_{i,j}$ $(1 \leq i < j \leq d)$ on C_g^d can be rewritten as a monomial M pulled back from C_g^{d-1} times either a single diagonal $D_{i,d}$ or a power K_d^k of K_d by a repeated application of the following substitution rules:

$$\begin{cases} D_{i,d}D_{j,d} \to D_{i,j}D_{i,d} & (i < j < d); \\ D_{i,d}^2 \to -K_i D_{i,d} & (i < d); \\ K_d D_{i,d} \to K_i D_{i,d} & (i < d). \end{cases}$$

Note that $D_{i,d}D_{j,d} = D_{i,j}D_{i,d}$, $D_{i,d}^2 = -K_i D_{i,d}$, and $K_d D_{i,d} = K_i D_{i,d}$; the second equation follows from the self-intersection formula and the other ones are clear.

(2) For M a monomial pulled back from C_g^{d-1}:

$$\begin{cases} \pi_{d,*}(M \cdot D_{i,d}) = M; \\ \pi_{d,*}(M \cdot K_d^k) = M \cdot \kappa_{k-1}. \end{cases}$$

Here $\pi_d : C_g^d \to C_g^{d-1}$ forgets the dth point. Note that $\pi_{d,*}(M) = 0$, as it should be, since $\kappa_{-1} = 0$.

As a first example, consider the case of g_d^1's, one-dimensional linear systems. The expected codimension in C_g^d is $g - d + 1$ and the expected codimension in M_g of the locus of curves with a g_d^1 is $g - 2d + 2$. It is known that these are the actual codimensions, if $2 \leq d \leq (g+2)/2$. The class of the locus in C_g^d is given by Porteous's formula. The locus has 1-dimensional fibers over its image in M_g, so by pushing down directly, we obtain a relation between the kappa classes in codimension $g - 2d + 1$. Alternatively, we can compute the class of the image in M_g in two ways: (a) by cutting with $D_{1,2}$, with multiplicity $(2g - 2 + 2d)(d-2)!$; (b) by cutting with K_1, with multiplicity $(2g-2)(d-1)!$. (The multiplicities are easily explained: $2g - 2 + 2d$ is the degree of the ramification divisor, $2g - 2$ is the degree of the canonical divisor, and the factorials arise since the remaining points are ordered.) Equating the two expressions, one obtains a kappa relation in codimension $g-2d+2$. We find therefore one kappa relation in codimension c for each $c \leq g-2$.

Ionel [**34**] has proved that the first $\lfloor g/3 \rfloor$ kappa classes $\kappa_1, \ldots, \kappa_{\lfloor g/3 \rfloor}$ generate $R^*(M_g)$ using these relations. Earlier, Morita [**47**] had obtained the analogous result for $RH^*(M_g)$ using a completely different set of relations.

One could also consider the case $d=1$. The locus in C_g is empty, so we find a relation on C_g:

$$0 = \Delta_{1,g}(c_t(\mathbb{F}_1)/c_t(\mathbb{E})) = \Delta_{g,1}(c_t(\mathbb{E}^\vee)/c_t(\mathbb{F}_1^\vee)) = (-1)^g \Big(c_t(\mathbb{E})/c_t(\mathbb{F}_1) \Big)_g .$$

(The subscript g on the right means taking the coefficient of t^g.) This is exactly the same relation as Mumford's relation discussed above.

As a second example, consider the residual linear system of a g^1_2, namely, a g^{g-2}_{2g-4}, obtained from $g-2$ copies of the g^1_2 (the curve is necessarily hyperelliptic and the g^1_2 is unique). The expected codimension equals the actual codimension, since equality holds for a linear system if and only if it holds for the residual system. Porteous's formula gives the class of the locus of divisors moving in a g^{g-2}_{2g-4} as

$$\Delta_{g-2,2}(c_t(\mathbb{F}_{2g-4})/c_t(\mathbb{E})) = (c^2_{g-2} - c_{g-1}c_{g-3})(\mathbb{F}_{2g-4} - \mathbb{E}).$$

(The difference is taken in the Grothendieck group; this notation is sometimes more convenient.)

The locus has $(g-2)$-dimensional fibers over the hyperelliptic locus H_g in M_g (of codimension $g-2$). To obtain relations, we can cut with $c < g-2$ tautological divisors and push forward, to obtain a relation in codimension c; or we cut with $g-2$ divisors and obtain the class of H_g with some multiplicity. It is not hard to check that the number of essentially different ways in which one can cut with c non-overlapping diagonals equals the number $p(c)$ of partitions of c. Thus, this choice gives $p(c)$ kappa relations in codimension c for every $c < g-2$, and $p(g-2)-1$ relations in codimension $g-2$. As we will see later, this method may produce all kappa relations in codimensions up to $g-2$. As far as I know, it has not been studied in detail.

As a third example, consider the divisors moving in a g^g_{2g-1}. The locus is empty, and Porteous's formula gives the relation

$$c_g(\mathbb{F}_{2g-1} - \mathbb{E}) = 0$$

on C_g^{2g-1}. The bundle map $\mathbb{E} \to \mathbb{F}_{2g-1}$ is therefore injective with quotient bundle of rank $g-1$, so that $c_j(\mathbb{F}_{2g-1} - \mathbb{E}) = 0$ for all $j \geq g$. Analogously, $c_j(\mathbb{F}_d - \mathbb{E}) = 0$ for $d \geq 2g-1$ and $j \geq d-g+1$.

Let us concentrate on the relation $c_g(\mathbb{F}_{2g-1} - \mathbb{E}) = 0$ of codimension g on C_g^{2g-1}. If we push it forward to M_g, we land in negative codimension, so that the obtained relation is necessarily trivial. To have a chance of obtaining nontrivial relations on M_g, we should cut with $c+g-1$ tautological divisors for a relation in codimension c. As it turns out, this method produces many nontrivial relations. Moreover, this class of relations has several good properties, which makes it possible on the one hand to analyse them theoretically and on the other hand to compute them for quite large genus. In fact, the results of computing these relations for $g \leq 15$ were to me so convincing that I conjectured (in 1993) a precise description of $R^*(M_g)$. I will now recall this conjecture [**10**].

The first part of the conjecture is that $R^*(M_g)$ is a Gorenstein ring with socle in degree $g-2$. This means that $R^j(M_g) = 0$ for $j > g-2$, that $R^{g-2}(M_g)$ is isomorphic to $\mathbb{Q}$, and that the natural pairing

$$R^i(M_g) \times R^{g-2-i}(M_g) \to R^{g-2}(M_g)$$

is supposed to be perfect.

Secondly, I conjectured that $R^*(M_g)$ is generated by $\kappa_1, \ldots, \kappa_{\lfloor g/3 \rfloor}$ and that there are no relations in degrees $\leq g/3$.

Thirdly, I gave explicit formulas for the proportionalities in degree $g-2$. I will not recall these formulas here, but note that they determine $R^*(M_g)$ completely if the first part of the conjecture is true.

Thanks to the work of many people, large parts of the conjecture are now proved. As to the first part, Looijenga [**43**] proved the vanishing in degrees greater than $g-2$. He also proved that $R^{g-2}(M_g)$ is at most one-dimensional. I proved [**10, 14**] that κ_{g-2} is nonzero, so $R^{g-2}(M_g)$ is isomorphic to $\mathbb{Q}$. As to the second part, the generation statement was proved by Morita for $RH^*(M_g)$ and by Ionel for $R^*(M_g)$, as already discussed. From the work of Boldsen [**5**], it follows that there are no relations in degrees $\leq g/3$, so the second part of the conjecture is completely proved.[1] The third part is completely proved, and there are in fact three proofs: by Givental [**26**], following earlier work of Eguchi-Hori-Xiong [**9**] and Getzler-Pandharipande [**25**]; by Liu and Xu [**41**]; and by Buryak and Shadrin [**6**].

Finally, the complete conjecture has been proved for $g \leq 23$. For all these genera, the method of the third example above has produced sufficiently many relations (by cutting $c_g(\mathbb{F}_{2g-1} - \mathbb{E})$ with non-overlapping diagonals and pushing down to M_g).

The fact that $R^{g-2}(M_g)$ is one-dimensional, generated by κ_{g-2}, and that the third part of the conjecture is proved, means that it makes sense to talk about the Gorenstein quotient $G^*(M_g)$ of $R^*(M_g)$. This is one-dimensional in degree $g-2$ and is the quotient of $R^*(M_g)$ by the homogeneous ideal generated by all classes of pure degree that have zero pairing with all classes of the complementary degree.

The ring $G^*(M_g)$ can be studied independently from $R^*(M_g)$. This will of course not prove the Gorenstein conjecture. The point is rather to try to make the kappa relations in the Gorenstein quotient explicit.

Zagier and I have studied $G^*(M_g)$ in considerable detail. A first question is what the dimensions of the graded pieces are. Let $q(g,k)$ denote the dimension of $G^k(M_g)$. Then $q(g,k) = q(g, g-2-k)$ by definition, and $q(g,k) \leq p(k)$ (the number of partitions of k), with equality conjectured for $k \leq g/3$. We would like to know $p(k) - q(g,k)$ for $g/3 < k \leq \lfloor (g-2)/2 \rfloor$. These values are known for $g \leq 23$ by the truth of the conjecture and can be computed for further genera by linear algebra. One observes a remarkable property: $p(k) - q(g,k)$ turns out to be a function of $3k - g - 1$. Let us assume that this is indeed the case, so that

$$a(3k - g - 1) = p(k) - q(g,k)$$

for an unknown function a. One can compute several values of the function a:

ℓ	0	1	2	3	4	5	6	7	8	9
$a(\ell)$	1	1	2	3	5	6	10	13	18	24

[1] I am grateful to Randal-Williams for explaining to me how this follows when the degree equals $g/3$.

Based on these values, Zagier and I made the following guess: $a(m)$ equals the number of partitions of m in parts of the form

$$1, 2, 3, 4, 6, 7, 9, 10, 12, 13, 15, \ldots,$$

that is, the only part congruent to 2 modulo 3 that can be used is the part 2. Fairly recently, Liu and Xu [**42**] computed five more values of the function a (note that their $a(n)$ equals our $a(n-1)$). We see that the guess that Zagier and I made is compatible with these new values.

However, it may be more important to study the actual kappa relations in $G^*(M_g)$. The form of the (conjecturally) unique relation in codimension k and genus $3k-1$ was found, and later proved, long ago by Zagier and me (see [**10**], pp. 124–125). Later, we also found the form of the relation in codimension k and genus $3k-2$. These relations were also derived by Morita [**47**] and Ionel [**34**].

Continuing this work, Zagier and I wanted to find explicit expressions for the relations in 'the first half' of $G^*(M_g)$, i.e., in codimensions $k \leq (g-2)/2$, where the numbers of relations are directly governed by the function a (instead of by duality). Approximately ten years ago, we obtained the following result. Let

$$\mathbf{p} = \{p_1, p_3, p_4, p_6, p_7, p_9, p_{10}, \ldots\}$$

be a collection of variables indexed by the positive integers not congruent to 2 modulo 3. Let $\Psi(t, \mathbf{p})$ be the following formal power series:

$$\Psi(t,\mathbf{p}) = \sum_{i=0}^{\infty} t^i p_{3i} \sum_{j=0}^{\infty} \frac{(6j)!}{(3j)!(2j)!} t^j + \sum_{i=0}^{\infty} t^i p_{3i+1} \sum_{j=0}^{\infty} \frac{(6j)!}{(3j)!(2j)!} \frac{6j+1}{6j-1} t^j,$$

where $p_0 := 1$. Define rational numbers $C_r(\sigma)$, for σ any partition (of $|\sigma|$) with parts not congruent to 2 modulo 3, by the formula

$$\log(\Psi(t,\mathbf{p})) = \sum_{\sigma} \sum_{r=0}^{\infty} C_r(\sigma) t^r \mathbf{p}^\sigma ,$$

where $\mathbf{p}^\sigma$ denotes $p_1^{a_1} p_3^{a_3} p_4^{a_4} \ldots$ if σ is the partition $[1^{a_1} 3^{a_3} 4^{a_4} \ldots]$. Put

$$\gamma = \sum_{\sigma} \sum_{r=0}^{\infty} C_r(\sigma) \kappa_r t^r \mathbf{p}^\sigma ;$$

then the relation

$$[\exp(-\gamma)]_{t^r \mathbf{p}^\sigma} = 0$$

holds in the Gorenstein quotient when $g - 1 + |\sigma| < 3r$ and $g \equiv r + |\sigma| + 1 \pmod 2$. (Of course, $\kappa_0 = 2g-2$.)

Let me call these relations (in the Gorenstein quotient) the FZ-relations for brevity. Observe that we get the expected number of relations in every codimension less than or equal to $\lfloor (g-2)/2 \rfloor$, although we didn't prove that the obtained relations in such a codimension are independent. Our goal was precisely to understand the relations until the middle; just by looking at the numbers of relations, it is clear that the FZ-relations cannot give all relations in $G^*(M_g)$ for g large enough (more precisely, for odd $g \geq 25$ and even $g \geq 30$).

Nevertheless, one can study the actual rank of the FZ-relations, and Pandharipande and I looked at this a few years ago. The first result is that the FZ-relations give *all* relations for $g \leq 23$. For $g = 24$, however, one relation (in codimension 12) in the Gorenstein quotient is missing: the 41 FZ-relations in degree 12 admit an unexpected syzygy and span only a 40-dimensional space of relations. The quotient by the FZ-relations has rank $p(12) - 40 = 37$ in degree 12, whereas it has rank $p(10) - a(5) = 42 - 6 = 36$ in degree 10. Similarly, the quotient by the FZ-relations differs from the Gorenstein quotient by one relation in genera 25 and 26 (in codimension 12 resp. 13); for $g = 25$, this is the optimal outcome, but for $g = 26$, there is another unexpected syzygy.

The fact that the FZ-relations don't give all relations in the Gorenstein quotient is not surprising (as just mentioned). It is surprising that there are unforced syzygies between the FZ-relations in certain degrees, but on the other hand, in higher degrees there are very many syzygies between the FZ-relations. In any case, one wonders what can be said about the actual relations in $R^*(M_g)$.

Let us study the case of genus $g = 24$ and codimension $c = 12$. If we use the method of the third example above and restrict to cutting $c_g(\mathbb{F}_{2g-1} - \mathbb{E})$ with non-overlapping diagonals before pushing down to M_g, then the (a priori) distinct possibilities correspond to the partitions of $2g-1 = 47$ of length equal to $g-c = 12$. If the partition contains a part equal to 1, the relation is trivial (exercise), so we consider only partitions with all parts at least equal to 2. There are 1116 such partitions. So far, I have computed approximately 250 of the corresponding relations. The space spanned by these relations coincides exactly with the 40-dimensional space spanned by the 41 FZ-relations.

In the other codimensions in genus 24, the 'diagonal' method produces all relations. In other words, the spaces of diagonal and FZ-relations coincide for those codimensions. It is not known what happens in codimension 12, but it is hard to believe that the diagonal method will produce the missing Gorenstein relation. In genera 25 and 26, the situation is similar at the moment, although not as many computations have been done. Perhaps the diagonal method will always produce exactly the space of FZ-relations.

Recently, Pandharipande and Pixton [**53**] proved the strong result that the FZ-relations are actual relations in $R^*(M_g)$ (and not just in $G^*(M_g)$). The geometric origin of the relations used in the proof of this result is the moduli space of stable quotients, introduced and studied by Marian, Oprea, and Pandharipande [**45**]. The method of virtual localization on this space produces a wealth of relations. A subset of these relations could be worked out to some extent; and for a smaller subset, the relations could be made even more explicit. Finally, Pandharipande and Pixton obtained the striking result that this last subset of explicit relations is equivalent to the set of FZ-relations.

The current state of affairs allows several interpretations. Perhaps $R^*(M_g)$ is not Gorenstein for $g \geq 24$. One would then like to have a proof, at least in some cases, that a 'missing' Gorenstein relation is not an actual relation. On the other hand, perhaps $R^*(M_g)$ is Gorenstein after all. Then a new source of relations would seem to be necessary (and it should be possible to compute or analyse these relations for $g \geq 24$). Finally, the intriguing possibility remains that $R^*(M_g)$ and $RH^*(M_g)$ differ.

Exercises

1. Let V be a vector bundle of rank r. Review how one expresses the components $ch_j(V)$ of the Chern character in terms of the Chern classes $c_i(V)$. Use this to prove the equivalence of

$$ch_{2i}(V) = 0 \qquad (\forall i \geq 1)$$

and

$$c_t(V^\vee) = c_t(V)^{-1}.$$

Next, assume that these equivalent statements hold. Prove that

$$ch_i(V) = 0$$

for all $i \geq 2r$. Show also that $ch_{2r-1}(V)$ and $c_r(V)c_{r-1}(V)$ are multiples of each other. Finally, apply the results above to the Hodge bundle $\mathbb{E}$ of rank g on M_g and derive in particular the formula for the total Chern class given in the text.

2. The Bernoulli numbers B_{2i} are defined by $t/(e^t - 1) = \sum_{i=0}^{\infty} B_i t^i/(i!)$. They satisfy several well-known congruences; in particular, for $i > 0$,

$$B_{2i} + \sum_{\substack{p \text{ prime:} \\ (p-1)|2i}} \frac{1}{p}$$

is an integer. This implies immediately

$$v_2(B_{2i}) = -1$$

for $i > 0$. (Here $v_2 \colon \mathbb{Q}\backslash\{0\} \to \mathbb{Z}$ is the standard 2-adic valuation.) Prove Mumford's result that $\kappa_1, \ldots, \kappa_{g-2}$ generate $R^*(M_g)$ using the relations mentioned in the text. The only property of the Bernoulli numbers that you will need is their 2-adic valuation given above. (Distinguish between g even and odd. The case g odd is a little harder.)

3. Prove the formula for $c(\mathbb{F}_d)$ given in the text, for example as follows.

(a) Let Σ_k denote the subscheme of the universal curve $\pi : U \to C_g^d$ that is the (formal) sum of the images of the first k sections $s_1, \ldots, s_k$ of π. Prove that there is an exact sequence

$$0 \to \mathcal{O}_{s_d}(-s_1 - \cdots - s_{d-1}) \to \mathcal{O}_{\Sigma_d} \to \mathcal{O}_{\Sigma_{d-1}} \to 0.$$

(b) Tensor the exact sequence with ω_π and prove that the sheaf on the left is the push-forward via s_d of the line bundle with first Chern class $K_d - \Delta_d$.

(c) Obtain the result by applying π_* and using induction.

4. Show that the results of Boldsen in [**5**] imply that there are no relations between the kappa classes in (algebraic) degree $k < g/3$. With more work, show that this also holds in degree $g/3$.

5. Using the 'g_d^1-method' as discussed in the lectures (in any of the three incarnations in the notes, or in another one), derive a nontrivial relation between the kappa classes (in some genus).

6. Same problem as above, but now with the further condition that the degree k of the relation satisfies $g/3 < k < g - 2$ (so that $g \geq 4$).

7. Well-known classical formulas for plane curves (typically derived by studying the dual plane curve in the dual projective plane) say that a general (smooth) plane curve of degree d over the complex numbers (which has genus $g = \frac{1}{2}(d-1)(d-2)$) has $3d(d-2)$ inflection points (a.k.a. flexes — points where the tangent line has contact of order 3) and $\frac{1}{2}d(d-2)(d^2-9)$ bitangents (lines that are tangent in two distinct points).

Verify this for $d = 4$ by using the 'g_d^r-method' for $r = 2$ and $d = 4$ (hence $g = 3$). (Cut the class in C_3^4 with two appropriate diagonal divisors and push it down to M_3. The calculations simplify considerably by doing some of the push-downs in a clever way. For example, in the flex case, one can use a bundle on C_3^2 with total Chern class $(1+K_1)(1+2K_1)(1+3K_1)(1+K_2-3D_{1,2})$.)

If you feel courageous, compute the class of the locus of plane quintics in M_6 in two ways by the same method, and derive a relation in $R^3(M_6)$ as a result.

Tacitly, I have been assuming in this exercise that the conditions in Porteous's formula are fulfilled. If you are an algebraic geometer, verify this.

LECTURE 2

The tautological rings of $\overline{M}_{g,n}$ and of some natural partial compactifications of $M_{g,n}$

Let $M_{g,n}$ denote the moduli space of n-pointed nonsingular curves of genus g (where g and n are nonnegative integers with $2g - 2 + n > 0$). The n points are distinct and ordered. As is well-known [**8, 37, 38**], this moduli space admits a natural compactification: the Deligne-Mumford-Knudsen moduli space $\overline{M}_{g,n}$ of stable n-pointed curves of genus g. The curves are allowed to have ordinary double points as singularities, but are required to be connected. The n ordered points remain distinct and nonsingular. The n-pointed curves must have finite automorphism group; this is the stability condition.

The space $\overline{M}_{g,n}$ admits a natural stratification, by topological type: a stratum consists of all the n-pointed curves homeomorphic (as pointed curves with the complex topology) to a given one. (For a detailed discussion of the material contained in these introductory paragraphs, I refer to [**28**], Appendix.) The topological type is conveniently encoded in the dual graph, with n ordered legs and a genus function on the set V of vertices (which correspond to the normalizations of the irreducible components). The stratum is isomorphic to the quotient of the product $\prod_{v\in V} M_{g(v),n(v)}$ by the automorphism group of the dual graph. Here $n(v)$ equals the number of points on the smooth curve corresponding to v that map to a marked point or a node of the stable curve. (The stability condition is that $2g(v) - 2 + n(v) > 0$ for all $v \in V$.) The quotient of $\prod_{v\in V} \overline{M}_{g(v),n(v)}$ by the automorphism group of the graph is the normalization of the closure of the stratum (which is a union of strata).

How should the tautological ring $R^*(\overline{M}_{g,n})$ of $\overline{M}_{g,n}$ be defined? We certainly want the classes of the closures of the strata to be tautological. Considering the natural map from $\prod_{v\in V} \overline{M}_{g(v),n(v)}$ to $\overline{M}_{g,n}$ with image the closure of a stratum, we also want the push-forward of a product of tautological classes on the factors to be tautological. The most natural way of defining the κ-classes on $\overline{M}_{g,n}$ was given by Arbarello and Cornalba [**1**]:

$$\kappa_i = \pi_*(\psi_{n+1}^{i+1}),$$

where $\pi : \overline{M}_{g,n+1} \to \overline{M}_{g,n}$ is the map given by forgetting the $(n+1)$st point and stabilizing and ψ_j is the first Chern class of the cotangent line bundle at the jth point. (If the $(n+1)$st point lies on a 3-pointed curve of genus 0, the curve becomes unstable after forgetting the point. The component needs to be contracted; this is the process of stabilization.)

Continuing with this kind of considerations, one arrives eventually at a definition of $R^*(\overline{M}_{g,n})$, but it seems somewhat ad hoc. A perhaps more natural definition was given by Pandharipande and myself: we define the system

$$\{R^*(\overline{M}_{g,n})\}_{(g,n)}$$

as the system of minimal $\mathbb{Q}$-subalgebras of the rational Chow rings $A^*(\overline{M}_{g,n})$ closed under push-forward via all maps forgetting markings and all gluing maps (see [**16**], pp. 13–14). One shows that the psi and kappa classes are tautological, and therefore also the push-forward of a product of monomials in the psi and kappa classes on a product of compact moduli spaces as above. In fact, as shown in [**28**], Prop. 11, these classes generate $R^*(\overline{M}_{g,n})$ additively. The system of tautological rings is also closed under pull-back via the forgetting and gluing maps. It follows from Mumford's computation of the Chern character of the Hodge bundle [**48**, §5] that the lambda classes are tautological on $\overline{M}_g$.

Let U be an open subvariety of $\overline{M}_{g,n}$. We use the surjectivity of the restriction maps $A^k(\overline{M}_{g,n}) \to A^k(U)$ (cf. [**19**], §1.8) to define the tautological ring $R^*(U)$ as the image of $R^*(\overline{M}_{g,n})$.

For M_g, this agrees with the earlier definition. (An independent motivation is provided by the Madsen-Weiss theorem describing the stable cohomology. Recall that Harer [**30**] has proved that the cohomology groups $H^k(M_g)$ stabilize as $g \to \infty$. The resulting algebra is called the stable cohomology and may be denoted $H^*(M_\infty, \mathbb{Q})$. Madsen and Weiss [**44**] prove that $H^*(M_\infty, \mathbb{Q})$ equals the free polynomial algebra in the kappa classes.) Note that there is in general no guarantee that the kernel of the map $R^k(\overline{M}_{g,n}) \to R^k(U)$ is spanned by *tautological* classes supported on the complement of U.

The partial compactifications U of $M_{g,n}$ in whose tautological rings we are most interested are the moduli space $M^c_{g,n}$ of *curves of compact type*, the complement of the divisor Δ_{irr} of irreducible singular curves and their degenerations, and (for $g \geq 2$) the moduli space $M^{rt}_{g,n}$ of *curves with rational tails*, the inverse image of M_g for the map $\overline{M}_{g,n} \to \overline{M}_g$. The work of Graber and Vakil [**29**] shows that other partial compactifications are important as well: they prove that a class in $R^i(\overline{M}_{g,n})$ vanishes when restricted to the open set consisting of strata parameterizing curves with at most $i-g$ components of geometric genus zero.

It follows that $R^*(M_{g,n})$ vanishes in positive degrees $\geq g$, that $R^*(M^{rt}_{g,n})$ vanishes in degrees $\geq g-1+n$, and that $R^*(M^c_{g,n})$ vanishes in degrees $\geq 2g-2+n$. With more work, one shows that $R^{g-2+n}(M^{rt}_{g,n})$, $R^{2g-3+n}(M^c_{g,n})$ and $R^{3g-3+n}(\overline{M}_{g,n})$ are one-dimensional (cf. [**29**] and [**16**]). The classes λ_g and $\lambda_g\lambda_{g-1}$ play a crucial role in the nonvanishing statement: they vanish on the complement of $M^c_{g,n}$, respectively $M^{rt}_{g,n}$.

A fundamental result for the study of the tautological rings is Witten's conjecture [**58**]. It completely determines the intersection numbers (or integrals) of psi classes on all the moduli spaces $\overline{M}_{g,n}$. It was first proved by Kontsevich [**39**] and there are by now several proofs (e.g., by Mirzakhani [**46**] and Okounkov-Pandharipande [**49**]; Mirzakhani's proof is the subject of Wolpert's lectures and notes [**59**]).

The psi integrals on $\overline{M}_{g,n}$ determine the kappa integrals on $\overline{M}_g$, and vice versa. The recipes for both transitions are relatively simple (cf. [**1, 35**]). In fact, the psi integrals determine the full intersection pairing on $R^*(\overline{M}_{g,n})$ (cf. [**28**], Props. 10, 11). The situation is entirely analogous for the psi integrals against λ_g or $\lambda_g\lambda_{g-1}$; they determine the intersection pairing on $R^*(M^c_{g,n})$, respectively $R^*(M^{rt}_{g,n})$.

The psi integrals against λ_g satisfy amazingly simple proportionalities:

$$\int_{\overline{M}_{g,n}} \psi_1^{a_1} \cdots \psi_n^{a_n} \lambda_g = \binom{2g-3+n}{a_1, \ldots, a_n} \int_{\overline{M}_{g,1}} \psi_1^{2g-2} \lambda_g ,$$

where the integral on the right side equals 1 by definition for $g = 0$. The proportionalities were derived by Getzler and Pandharipande [**25**] from the Virasoro conjecture of Eguchi, Hori, and Xiong [**9**]. They were first proved in [**15**]; see also [**40, 27**]. The not quite as simple proportionalities satisfied by the psi integrals against $\lambda_g \lambda_{g-1}$ are part of my conjecture describing $R^*(M_g)$. They were proved directly by Liu and Xu [**41**] and later by Buryak and Shadrin [**6**]. A more roundabout proof for both sets of proportionalities is obtained from Givental's proof of the Virasoro conjecture for projective spaces [**26**]. The kappa integrals against $\lambda_g \lambda_{g-1}$ are not nearly as simple; yet they govern the Gorenstein quotient $G^*(M_g)$.

In analogy with the conjecture for $R^*(M_g)$, it has been speculated that the rings $R^*(\overline{M}_{g,n})$, $R^*(M^c_{g,n})$ and $R^*(M^{rt}_{g,n})$ all are Gorenstein. It has long been known that $R^*(\overline{M}_{0,n})$ is Gorenstein [**36**]. Recently, Tavakol [**55, 56**] proved that $R^*(M^c_{1,n})$ and $R^*(M^{rt}_{2,n})$ are Gorenstein.[1] Besides this, the Gorenstein property is known in only a handful of cases. One of the main difficulties (at least in the case of $\overline{M}_{g,n}$) is the large number of tautological classes. Another one is that the undetermined boundary terms of a tautological relation on an open set are rarely known to be tautological; Tavakol's codimension 3 relation on $R^*(M^{rt}_{2,6})$ provides a current example. However, see [**53**] for an encouraging exception.

There is considerable room for further exploration of these speculations. Yang [**60**] has computed the dimensions of the graded pieces of the Gorenstein quotients in quite a few cases. To prove that one of the rings is Gorenstein, one needs to produce the required number of tautological relations. See [**12**] for a discussion of the conjectures and the relation to a conjecture of Hain and Looijenga. Pandharipande [**51**] has obtained strong results on the kappa subrings of $R^*(M^c_{g,n})$ (see also [**52**]). Finally, Cavalieri and Yang [**7**] show that the Gorenstein property fails for the tautological rings of certain partial compactifications in between $M^c_{g,n}$ (respectively $M^{rt}_{g,n}$) and $\overline{M}_{g,n}$, while the top graded pieces are one-dimensional.

Many tautological relations, i.e., between decorated strata classes, have immediate implications for Gromov-Witten theory. Well-known examples are the relations on $\overline{M}_{0,4}$, Getzler's relation on $\overline{M}_{1,4}$, and the Belorousski-Pandharipande relation on $\overline{M}_{2,3}$ (cf. [**21, 50, 3**]). Tommasi [**57**] has found a codimension 3 relation between tautological cohomology classes on $M_{3,2}$. It is not known whether this relation holds in $R^3(\overline{M}_{3,2})$. Pandharipande and the author found a codimension 3 relation in $R^*(\overline{M}_4)$, predicted on M^c_4 by the Gorenstein conjecture; Yang [**60**] computed its boundary terms. It is a priori possible that nontrivial relations hold between the push-forwards along a gluing map of decorated strata classes that are not push-forwards of relations. Such relations would be of considerable interest.

The length of a partition is its number of parts. There exists a standard correspondence between the irreducible representations of the symmetric group Σ_n and the partitions of n (see, e.g., [**20**]; the partition $[n]$ corresponds to the trivial representation and the partition $[1^n]$ to the alternating one). In [**17**], Pandharipande

[1]Note added in proof: Petersen [**54**] has proved that $R^*(\overline{M}_{1,n})$ is Gorenstein for all $n \geq 1$. Even more recently, Yin [**61**] has proved that $R^*(M_{g,1})$ is Gorenstein for $g \leq 19$. He conjectures that $R^*(M_g)$ and $R^*(M_{g,1})$ are not Gorenstein in general.

and the author prove certain length bounds on the partitions of n corresponding to the irreducible representations of Σ_n occurring in $R^*(\overline{M}_{g,n})$. Perhaps the most interesting result is the following:

$$\ell(R^k(\overline{M}_{g,n})) \leq \min\left(k+1, 3g-2+n-k, \left\lfloor \tfrac{2g-1+n}{2} \right\rfloor\right),$$

where the left side denotes the maximum of the lengths. Using this bound, we have with near certainty established (loc. cit., §3) the existence of lots of non-tautological cohomology classes of Tate type on $\overline{M}_{2,21}$; this relies on a conjectural description (proved to a large extent and supported by extensive point counting data) of the entire cohomology of $\overline{M}_{2,n}$ found by van der Geer and the author (cf. [**13**]). In fact, in §5 of [**17**], explicit non-tautological algebraic cohomology classes are found on $\overline{M}_{2,21}$, the existence of which implies the non-tautologicality of the explicit classes on $\overline{M}_{2,22}$ found earlier by Graber and Pandharipande [**28**].

With all the interest for the tautological algebras, it is easy to forget that there are in general many more cohomology classes than just the tautological ones. Only in finitely many cases beyond genus zero will the entire cohomology be tautological, it would seem. One should expect odd cohomology, or cohomology not of (p,p)-type, or (on a non-compact moduli space) non-pure cohomology of (p,p)-type; and, probably slightly less often, algebraic classes which are not tautological.

A successful method for obtaining a lot of information about the entire cohomology of moduli spaces of curves and abelian varieties, used in joint work with Bergström and van der Geer, has been to count their numbers of points over finite fields. For a survey, see §§2 and 3 of [**17**]. For genus 1, one finds cohomology related to elliptic cusp forms; Getzler [**22, 23**] has shown how the classical Eichler-Shimura theory can be used to determine the cohomology of $\overline{M}_{1,n}$. The generalization by Faltings and Chai [**18**] of Eichler-Shimura theory gives strong results on the cohomology of irreducible symplectic local systems on the moduli space A_g of principally polarized abelian varieties of dimension g. The same holds for the theory of automorphic forms and representations, but it is not easy to obtain explicit formulas. Explicit conjectural formulas are now available for $g = 2$ (see [**13**]) and $g = 3$ (see [**4**]). The former is proved in the case of a local system corresponding to a regular weight. The formulas are in terms of Galois representations (or Hodge structures) associated to Siegel cusp forms of degree g and products or Tate twists of Galois representations associated to Siegel cusp forms of lower degree. This leads to formulas for the pull-backs of the local systems to M_2, and from there (as pointed out by Getzler) to the Σ_n-equivariant cohomology of $M_{2,n}$, and (by the work of Getzler and Kapranov [**24**]) to the Σ_n-equivariant cohomology of $\overline{M}_{2,n}$. The symmetric and exterior squares of the Galois representations associated to elliptic cusp forms occur now as well. For genus 3, the formula for A_3 gives at best half of the answer for M_3, since the map of stacks $M_3 \to A_3$ is $2:1$ onto its image, ramified along the hyperelliptic locus. We have proved that new Galois representations occur, not expressible in terms of the representations associated to Siegel cusp forms. These Galois representations are so far completely mysterious, but appear to be associated to Teichmüller modular forms (vector-valued in general). Scalar-valued Teichmüller modular forms have been studied in detail by Ichikawa (e.g., [**31, 32, 33**]).

Exercises

1. Prove that the undecorated strata classes additively generate $R^*(\overline{M}_{g,n})$ for $g \leq 1$. Conclude that $R^*(\overline{M}_{0,n})$ is multiplicatively generated by divisor classes. Does this hold for $g = 1$?

2. Study $R^*(\overline{M}_{g,n})$ in some cases (with $2g-2+n$ small) where it is multiplicatively generated by divisor classes, for example with the program **MgnF**, described in [**11**] and available from `http://math.stanford.edu/~vakil/programs/index.html` .

3. Find a case in Yang's tables [**60**] where the Gorenstein property is not yet known and try to prove (or disprove) that it holds.

Bibliography

[1] E. Arbarello and M. Cornalba, *Combinatorial and algebro-geometric cohomology classes on the moduli spaces of curves*, J. Algebraic Geom. **5** (1996), 705–749. MR1486986 (99c:14033)

[2] E. Arbarello, M. Cornalba, P.A. Griffiths, and J. Harris, *Geometry of algebraic curves.* Vol. I. Grundlehren der Mathematischen Wissenschaften, 267. Springer-Verlag, New York, 1985. MR770932 (86h:14019)

[3] P. Belorousski and R. Pandharipande, *A descendent relation in genus 2.* Ann. Scuola Norm. Sup. Pisa Cl. Sci. (4) **29** (2000), no. 1, 171–191. MR1765541 (2001g:14044)

[4] J. Bergström, C. Faber, and G. van der Geer, *Siegel modular forms of degree three and the cohomology of local systems.* Preprint, arXiv:1108.3731.

[5] S. Boldsen, *Improved homological stability for the mapping class group with integral or twisted coefficients.* Math. Z. **270** (2012, no. 1–2, 297–329. MR2875835 (2012k:57026)

[6] A. Buryak and S. Shadrin, *A new proof of Faber's intersection number conjecture.* Adv. Math. **228** (2011), no. 1, 22–42. MR2822226 (2012j:14040)

[7] R. Cavalieri and S. Yang, *Tautological pairings on moduli spaces of curves.* Proc. Amer. Math. Soc. **139** (2011), no. 1, 51–62. MR2729070 (2012a:14065)

[8] P. Deligne and D. Mumford, *The irreducibility of the space of curves of given genus.* Inst. Hautes Études Sci. Publ. Math. No. 36 1969 75–109. MR0262240 (41:6850)

[9] T. Eguchi, K. Hori, and C.-S. Xiong, *Quantum cohomology and Virasoro algebra.* Phys. Lett. B **402** (1997), no. 1-2, 71–80. MR1454328 (98j:14035)

[10] C. Faber, *A conjectural description of the tautological ring of the moduli space of curves*, in *Moduli of Curves and Abelian Varieties (The Dutch Intercity Seminar on Moduli)* (C. Faber and E. Looijenga, eds.), 109–129, Aspects of Mathematics E 33, Vieweg, Wiesbaden 1999. MR1722541 (2000j:14044)

[11] C. Faber, *Algorithms for computing intersection numbers on moduli spaces of curves, with an application to the class of the locus of Jacobians.* New trends in algebraic geometry (Warwick, 1996), 93–109, London Math. Soc. Lecture Note Ser., **264**, Cambridge Univ. Press, Cambridge, 1999. MR1714822 (2000m:14032)

[12] C. Faber, *A remark on a conjecture of Hain and Looijenga.* Preprint, arXiv:0812.3631. To appear in Ann. Inst. Fourier (Grenoble).

[13] C. Faber and G. van der Geer, *Sur la cohomologie des systèmes locaux sur les espaces des modules des courbes de genre 2 et des surfaces abéliennes*, I, II. C.R. Acad. Sci. Paris, Sér. I, **338** (2004), 381–384, 467–470. MR2057161 (2005a:14033a)

[14] C. Faber and R. Pandharipande (with an appendix by D. Zagier), *Logarithmic series and Hodge integrals in the tautological ring*, Michigan Math. J. **48** (2000), 215–252. MR1786488 (2002e:14041)

[15] C. Faber and R. Pandharipande, *Hodge integrals, partition matrices, and the λ_g conjecture*, Ann. of Math. **157** (2003), 97–124. MR1954265 (2004b:14095)

[16] C. Faber and R. Pandharipande, *Relative maps and tautological classes*, J. Eur. Math. Soc. **7** (2005), 13–49. MR2120989 (2005m:14046)

[17] C. Faber and R. Pandharipande, *Tautological and non-tautological cohomology of the moduli space of curves.* Preprint, arXiv:1101.5489. To appear in the *Handbook of Moduli.*

[18] G. Faltings and C.-L. Chai, *Degeneration of abelian varieties.* With an appendix by David Mumford. Ergebnisse der Mathematik und ihrer Grenzgebiete (3), 22. Springer-Verlag, Berlin, 1990. MR1083353 (92d:14036)

[19] W. Fulton, *Intersection theory.* Ergebnisse der Mathematik und ihrer Grenzgebiete (3), 2. Springer-Verlag, Berlin, 1984. MR732620 (85k:14004)

[20] W. Fulton and J. Harris, *Representation theory. A first course.* Graduate Texts in Mathematics, 129. Readings in Mathematics. Springer-Verlag, New York, 1991. xvi+551 pp. MR1153249 (93a:20069)

[21] E. Getzler, *Intersection theory on* $\overline{M}_{1,4}$ *and elliptic Gromov-Witten invariants.* J. Amer. Math. Soc. **10** (1997), no. 4, 973–998. MR1451505 (98f:14018)

[22] E. Getzler, *Resolving mixed Hodge modules on configuration spaces.* Duke Math. J. **96** (1999), no. 1, 175–203. MR1663927 (2000j:14020)

[23] E. Getzler, *The semi-classical approximation for modular operads.* Comm. Math. Phys. **194** (1998), no. 2, 481–492. MR1627677 (99d:14017)

[24] E. Getzler and M. Kapranov, *Modular operads.* Compositio Math. **110** (1998), no. 1, 65–126. MR1601666 (99f:18009)

[25] E. Getzler and R. Pandharipande, *Virasoro constraints and the Chern classes of the Hodge bundle.* Nuclear Phys. B **530** (1998), no. 3, 701–714. MR1653492 (2000b:14073)

[26] A. Givental, *Gromov-Witten invariants and quantization of quadratic Hamiltonians.* Mosc. Math. J. **1** (2001), no. 4, 551–568, 645. MR1901075 (2003j:53138)

[27] I. Goulden, D. Jackson, and R. Vakil, *A short proof of the* λ_g*-conjecture without Gromov-Witten theory: Hurwitz theory and the moduli of curves.* J. Reine Angew. Math. **637** (2009), 175–191. MR2599085 (2011f:14087)

[28] T. Graber and R. Pandharipande, *Constructions of nontautological classes on moduli spaces of curves.* Michigan Math. J. **51** (2003), no. 1, 93–109. MR1960923 (2004e:14043)

[29] T. Graber and R. Vakil, *Relative virtual localization and vanishing of tautological classes on moduli spaces of curves.* Duke Math. J. **130** (2005), no. 1, 1–37. MR2176546 (2006j:14035)

[30] J.L. Harer, *Stability of the homology of the mapping class groups of orientable surfaces.* Ann. of Math. (2) **121** (1985), no. 2, 215–249. MR786348 (87f:57009)

[31] T. Ichikawa, *Teichmüller modular forms of degree* 3. Amer. J. Math. **117** (1995), no. 4, 1057–1061. MR1342841 (96e:11067)

[32] T. Ichikawa, *Theta constants and Teichmüller modular forms.* J. Number Theory **61** (1996), no. 2, 409–419. MR1423061 (98e:11060)

[33] T. Ichikawa, *Generalized Tate curve and integral Teichmüller modular forms.* Amer. J. Math. **122** (2000), no. 6, 1139–1174. MR1797659 (2001m:11101)

[34] E. Ionel, *Relations in the tautological ring of* M_g, Duke Math J. **129** (2005), 157–186. MR2155060 (2006c:14040)

[35] R. Kaufmann, Yu. I. Manin, D. Zagier, *Higher Weil-Petersson volumes of moduli spaces of stable n-pointed curves.* Comm. Math. Phys. **181** (1996), no. 3, 763–787. MR1414310 (98i:14029)

[36] S. Keel, *Intersection theory of moduli space of stable n-pointed curves of genus zero.* Trans. Amer. Math. Soc. **330** (1992), no. 2, 545–574. MR1034665 (92f:14003)

[37] F.F. Knudsen, *The projectivity of the moduli space of stable curves. II. The stacks* $M_{g,n}$. Math. Scand. 52 (1983), no. 2, 161–199. MR702953 (85d:14038a)

[38] F.F. Knudsen, *The projectivity of the moduli space of stable curves. III. The line bundles on* $M_{g,n}$*, and a proof of the projectivity of* $\overline{M}_{g,n}$ *in characteristic* 0. Math. Scand. 52 (1983), no. 2, 200–212. MR702954 (85d:14038b)

[39] M. Kontsevich, *Intersection theory on the moduli space of curves and the matrix Airy function.* Comm. Math. Phys. **147** (1992), no. 1, 1–23. MR1171758 (93e:32027)

[40] C.-C. M. Liu, K. Liu, and J. Zhou, *Mariño-Vafa formula and Hodge integral identities.* J. Algebraic Geom. **15** (2006), no. 2, 379–398. MR2199062 (2007e:14090)

[41] K. Liu and H. Xu, *A proof of the Faber intersection number conjecture.* J. Differential Geom. **83** (2009), no. 2, 313–335. MR2577471 (2011d:14051)

[42] K. Liu and H. Xu, *Descendent integrals and tautological rings of moduli spaces of curves.* Geometry and analysis no. 2, 137–172, Adv. Lect. Math. (ALM), 18, Int. Press, Somerville, MA, 2011. MR2882445 (2012j:14043)

[43] E. Looijenga, *On the tautological ring of* M_g. Invent. Math. **121** (1995), 411–419. MR1346214 (96g:14021)

[44] I. Madsen and M. Weiss, *The stable moduli space of Riemann surfaces: Mumford's conjecture.* Ann. of Math. (2) 165 (2007), no. 3, 843–941. MR2335797 (2009b:14051)

[45] A. Marian, D. Oprea, and R. Pandharipande, *The moduli space of stable quotients.* Geom. Topol. **15** (2011), no. 3, 1651–1706. MR2851074 (2012h:14138)

[46] M. Mirzakhani, *Weil-Petersson volumes and intersection theory on the moduli space of curves.* J. Amer. Math. Soc. **20** (2007), no. 1, 1–23. MR2257394 (2007g:14029)

[47] S. Morita, *Generators for the tautological algebra of the moduli space of curves.* Topology **42** (2003), 787–819. MR1958529 (2004g:14029)

[48] D. Mumford, *Towards an enumerative geometry of the moduli space of curves.* In *Arithmetic and Geometry* (M. Artin and J. Tate, eds.), Part II, Birkhäuser, 1983, 271–328. MR717614 (85j:14046)

[49] A. Okounkov and R. Pandharipande, *Gromov-Witten theory, Hurwitz numbers, and matrix models.* In *Algebraic geometry—Seattle 2005*, Part 1, Proc. Sympos. Pure Math. **80**, Amer. Math. Soc., Providence, RI, 2009, 325–414. MR2483941 (2009k:14111)

[50] R. Pandharipande, *A geometric construction of Getzler's elliptic relation.* Math. Ann. **313** (1999), no. 4, 715–729. MR1686935 (2000d:14059)

[51] R. Pandharipande, *The kappa ring of the moduli of curves of compact type: I, II.* Acta Math. **208** (2012), no. 2, 335–388. MR2931383

[52] R. Pandharipande, *The kappa classes on the moduli spaces of curves.* Preprint, arXiv:1108.5984.

[53] R. Pandharipande and A. Pixton, *Relations in the tautological ring.* Preprint, arXiv:1101.2236.

[54] D. Petersen, *The structure of the tautological ring in genus one.* Preprint, arXiv:1205.1586.

[55] M. Tavakol, *The tautological ring of $M^{ct}_{1,n}$.* Preprint, arXiv:1007.3091.

[56] M. Tavakol, *The tautological ring of the moduli space $M^{rt}_{2,n}$.* Preprint, arXiv:1101.5242.

[57] O. Tommasi, *Rational cohomology of $M_{3,2}$.* Compos. Math. **143** (2007), no. 4, 986–1002. MR2339836 (2008d:14043)

[58] E. Witten, *Two-dimensional gravity and intersection theory on moduli space.* Surveys in differential geometry (Cambridge, MA, 1990), 243–310, Lehigh Univ., Bethlehem, PA, 1991. MR1144529 (93e:32028)

[59] S.A. Wolpert, *Lectures and notes: Mirzakhani's volume recursion and approach for the Witten-Kontsevich theorem on moduli tautological intersection numbers.* IAS/Park City Math. Ser., Vol. 20, Amer. Math. Soc., Providence, RI., 2013.

[60] S. Yang, *Calculating intersection numbers on moduli spaces of pointed curves.* Preprint, arXiv:0808.1974.

[61] Q. Yin, *On the tautological rings of $M_{g,1}$ and its universal Jacobian.* Preprint, arXiv:1206.3783.

Mirzakhani's Volume Recursion and Approach for the Witten-Kontsevich Theorem on Moduli Tautological Intersection Numbers

Scott A. Wolpert

IAS/Park City Mathematics Series
Volume 20, 2011

Mirzakhani's Volume Recursion and Approach for the Witten-Kontsevich Theorem on Moduli Tautological Intersection Numbers

Scott A. Wolpert

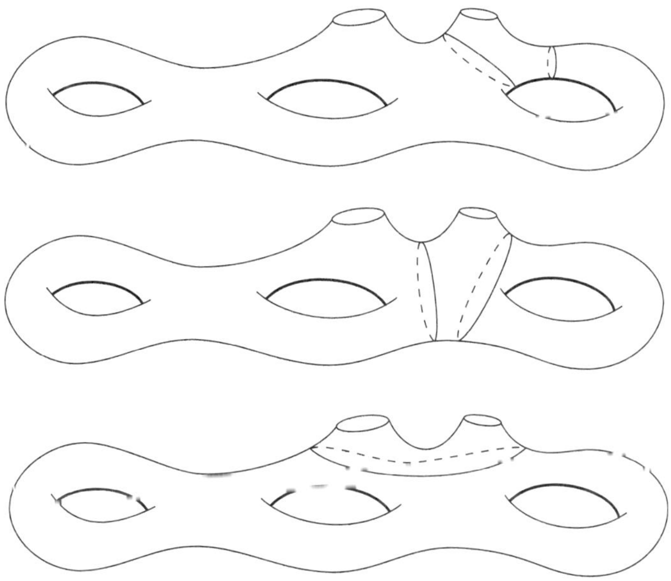

FIGURE 1. The boundary pants configurations for the length identity.

The following materials were presented in a short course at the 2011 Park City Mathematics Institute, Graduate Summer School on Moduli Spaces of Riemann Surfaces. Brad Safnuk consulted on the lectures, as well as assisted in the preparation for and running of the course. It is my pleasure to thank Brad for his assistance.

Department of Mathematics, University of Maryland
E-mail address: swolpert@umd.edu
Partially supported by National Science Foundation grant DMS - 100585

Prelude

1. Introduction

The papers.
Simple geodesics and Weil-Petersson volumes of moduli spaces of bordered Riemann surfaces. Invent. Math., 167(1):179-222, 2007.
Weil-Petersson volumes and intersection theory on the moduli space of curves. J. Amer. Math. Soc., 20(1):1-23, 2007.

The goals of the papers.
1. Derive an explicit recursion for WP moduli space volume polynomials.
2. Apply symplectic reduction to show that the polynomial coefficients are intersection numbers.
3. Show that the volume recursion satisfies the Virasoro relations - Witten's conjecture.

The trio of Maryam Mirzakhani papers [**Mir07a, Mir07b, Mir08**] are distinguished for involving a large number of highly developed considerations. The first work requires a detailed description of Teichmüller space, the action of the mapping class group, formulas for Weil-Petersson (WP) symplectic geometry, classification of simple geodesic arcs on a pair of pants, the length infinite sum identity and exact calculations of integrals. The second work involves a description of the moduli space tautological classes κ_1 and ψ, as characteristic classes for S^1 principal bundles in hyperbolic geometry, the moment map and exact symplectic reduction, as well as combinatorial calculations. The third work uses the PL structure of Thurston's space of measured geodesic laminations $\mathcal{MGL}$, the train-track symplectic form and Masur's result that the mapping class group acts ergodically on $\mathcal{MGL}$. The third work establishes a *prime simple geodesic theorem* and is not part of the considerations for Witten-Kontsevich. A fine feature of the works is that they showcase important aspects of the geometry, topology and deformation theory of Riemann/hyperbolic surfaces. Mirzakhani's recursion for volume is applied in the three works and in [**Mir10**]. A discussion placing Mirzakhani's work into the context of approaches to Witten-Kontsevich theory is given in the first section of the Mulase-Safnuk paper [**MS08**]. A brief exposition of Kontsevich's original solution [**Kon92**] of Witten's conjecture, including the basic geometry of tautological classes on moduli space and the ribbon graph expansion of matrix integrals is given in the Bourbaki Seminar of Looijenga [**Loo93**]. A brief exposition of Mirzakhani's volume recursion, solution of Witten-Kontsevich and applications of WP volume limits are given in [**Do11**]. An overall exposition of Mirzakhani's prime simple geodesic theorem [**Mir08**] is given in [**Wlp10**, Chaps. 9, 10]. Mirzakhani's work is just one part of a subject with high activity and many active researchers; Google Scholar shows

200 citations to the three Mirzakhani works, and much more generally over 1100 citations to Witten's original papers [**Wit91, Wit92**] on two-dimensional gravity and gauge theories.

First paper brief. A symplectic fibered product decomposition for covers of the moduli space of bordered Riemann/hyperbolic surfaces is combined with the $d\ell\wedge d\tau$ formula for the WP symplectic form and a universal identity for sums of geodesic-lengths to derive an explicit recursion for computing volume. The volume of the moduli space of genus g, n boundaries surfaces is shown to be a polynomial with positive coefficients in surface boundary lengths of total degree $6g-6+2n$.
Second paper brief. Twist Hamiltonian flows are combined with the boundary length moment map to apply symplectic reduction for the family of moduli spaces of bordered hyperbolic surfaces. The consequence is that the coefficients of the volume polynomials are moduli space characteristic numbers. A geometric construction shows that the characteristic numbers are tautological intersection numbers. The volume recursion is shown to satisfy Virasoro algebra constraints.

The following theorems are the immediate take away results of the papers.

THEOREM. The WP volume polynomials. *The volume polynomials are determined recursively from the volume polynomials of smaller total degree,* [**Mir07a**, Formulas (5.1) to (5.4) & Theorem 8.1]. *The volume $V_{g,n}(L_1,\dots,L_n)$ of the moduli space of genus g, n boundaries, hyperbolic surfaces with boundary lengths $L=(L_1,\dots,L_n)$ is a polynomial*

$$V_{g,n}(L) = \sum_{\substack{\alpha \\ |\alpha|\le 3g-3+n}} C_\alpha\, L^{2\alpha},$$

for multi-index $\alpha=(\alpha_1,\dots,\alpha_n)$ and where $C_\alpha>0$ lies in $\pi^{6g-6+2n-2|\alpha|}\mathbb{Q}$, [**Mir07a**, Theorems 1.1 & 6.1]. *The coefficients are intersection numbers given as*

$$C_\alpha = \frac{2^{\delta_{1g}\delta_{1n}}}{2^{|\alpha|}\alpha!(3g-3+n-|\alpha|)!}\int_{\overline{\mathcal{M}}_{g,n}} \psi_1^{\alpha_1}\cdots\psi_n^{\alpha_n}\omega^{3g-3+n-|\alpha|},$$

where ψ_j is the Chern class for the cotangent line along the jth puncture, ω is the WP symplectic form, $\alpha! = \prod_{j=1}^n \alpha_j!$, and δ_{**} is the Kronecker indicator delta,[1] [**Mir07b**, Theorem 4.4].

THEOREM ([**Mir07a**, Theorems 6.3 & 6.4]). Recursive relations for the volume polynomial leading coefficients. *For a multi index α, define*

$$(\alpha_1,\dots,\alpha_n)_g = C_\alpha \times 2^{-\delta_{1g}\delta_{1n}} \times \prod_{i=1}^n \alpha_i! \times 2^{|\alpha|},$$

then for $n>0$ and $\sum_i \alpha_i = 3g-3+n$,

$$\textit{the dilaton equation} \qquad (1,\alpha_1,\dots,\alpha_n)_g = (2g-2+n)(\alpha_1,\dots,\alpha_n)_g$$

and for $n>0$ and $\sum_i \alpha_i = 3g-2+n$,

$$\textit{the string equation} \qquad (0,\alpha_1,\dots,\alpha_n)_g = \sum_{\alpha_i\neq 0}(\alpha_1,\dots,\alpha_i-1,\dots,\alpha_n)_g.$$

[1]The factor $2^{\delta_{1g}\delta_{1n}}$ is discussed in Lecture 3 and is the result of the general elliptic curve involution.

For the intersection number generating function

$$\mathbf{F}(\lambda, t_0, t_1, \dots) = \sum_{g=0}^{\infty} \lambda^{2g-2} \sum_{\{d_j\}} \langle \prod_{j=1}^{\infty} \tau_{d_j} \rangle_g \prod_{r \geq 0} t_r^{n_r}/n_r!,$$

with $n_r = \#\{j \mid d_j = r\}$, *and*

$$\langle \tau_{d_1} \cdots \tau_{d_n} \rangle_g = \int_{\overline{\mathcal{M}}_{g,n}} \prod_{j=1}^{n} \psi_j^{d_j},$$

for $\sum_{j=1}^{n} d_j = 3g - 3 + n$ *and the product* $\langle \tau_* \rangle$ *otherwise zero, the exponential* $e^{\mathbf{F}}$ *satisfies Virasoro algebra constraints,* [**Mir07b**, Theorem 6.1].

2. The organizational outline and reading guide.

The following outline of the considerations combines [**Mir07a, Mir07b**] with the exposition of [**Wlp10**, Chapter 9].

- Teichmüller spaces, moduli spaces, mapping class groups and the symplectic geometry.
  - The Teichmüller space $\mathcal{T}_{g,n}$ and moduli space $\mathcal{M}_{g,n}$; the Teichmüller space $\mathcal{T}_g(L_1, \dots, L_n)$ and moduli space $\mathcal{M}_g(L_1, \dots, L_n)$ of prescribed length geodesic bordered hyperbolic surfaces; the augmented Teichmüller space and Deligne-Mumford type compactifications; Dehn twists and the mapping class group MCG action.
  - The WP symplectic geometry, [**Wlp10**].
    - The symplectic form $\omega = 2\omega_{\text{WP Kähler}}$ and normalizations.
    - Geodesic-length functions ℓ_α, Fenchel-Nielsen infinitesimal twist deformations t_α and the duality formula $\omega(\ , t_\alpha) = d\ell_\alpha$.
    - Fenchel-Nielsen (FN) twist-length coordinates (τ_j, ℓ_j) for Teichmüller space and the formula $\omega = \sum_j d\ell_j \wedge d\tau_j$.
  - The intermediate moduli space $\mathcal{M}_g^\gamma(L)$ of pairs (R, γ) - a surface and a weighted multicurve $\gamma = \sum_j c_j \gamma_j$.
    - The covering tower

$$\mathcal{T}_g(L) \longrightarrow \mathcal{M}_g^\gamma(L) \longrightarrow \mathcal{M}_g(L).$$

    - The stabilizer subgroup $\text{Stab}(\gamma) \subset$ MCG for a weighted multicurve. The MCG deck cosets for the covering tower.
    - Symplectic structures for $\mathcal{M}_g(L)$.
  - The $\mathcal{T}_g(L)$ and $\mathcal{M}_g^\gamma(L)$ level sets of the total length $\ell = \sum_j c_j \ell_{\gamma_j}$

    LEMMA ([**Mir07a**, Lemma 7.2]. Preparation for volume recursion and symplectic reduction). *A finite cover of $\mathcal{M}_g^\gamma(L)$ is a fibered product of symplectic planes and lower dimensional moduli spaces.*

- The McShane-Mirzakhani length identity.
  - The set $\mathcal{B}$ of homotopy classes rel boundary of simple arcs with endpoints on the boundary. Classification of geodesic arcs normal to a boundary: simple geodesics normal to a boundary at each endpoint $\Longleftrightarrow$ disjoint pairs of boundary intervals $\Longleftrightarrow$ wire frames for pants.
  - Birman-Series: the union of simple geodesics has measure zero.

   - The rational exponential function H; the hyperbolic trigonometric functions $\mathcal{D}$ and $\mathcal{R}$; relations.
   - The length identity.

   THEOREM ([**Mir07a**, Theorems 1.3 & 4.2] and [**TWZ06**, Thrm. 1.8]). *For a hyperbolic surface R with boundaries β_j with lengths L_j,*

$$L_1 = \sum_{\alpha_1,\alpha_2} \mathcal{D}(L_1, \ell_{\alpha_1}(R), \ell_{\alpha_2}(R)) + \sum_{j=2}^{n} \sum_{\alpha} \mathcal{R}(L_1, L_j, \ell_\alpha(R)),$$

   where the first sum is over all unordered pairs of simple closed geodesics with $\beta_1, \alpha_1, \alpha_2$ bounding an embedded pair of pants, and the double sum is over simple closed geodesics with β_1, β_j, α bounding an embedded pair of pants.

   - [**Mir07a**, Section 8] - Recognizing and understanding the identity as a smooth analog to a MCG fundamental domain. Reducing to the action of smaller MCG groups.
- The Mirzakhani volume recursion.
   - A covolume formula - writing a moduli integral as a length level set integral - the role of the intermediate moduli space $\mathcal{M}_g^\gamma(L)$ - [**Mir07a**, Theorem 7.1] and [**Wlp10**, Theorem 9.5].
   - The application for the McShane-Mirzakhani identity.
      * The integrals $\mathcal{A}_{g,n}^{con}$, $\mathcal{A}_{g,n}^{dcon}$ and $\mathcal{B}_{g,n}$ of lower-dimensional moduli volumes.
      * The corresponding connected and disconnected boundary pants configurations.
      * Combining the length identity and covolume formula for the main result; see [**Wlp10**, pg. 91 bottom, pg. 92].

      THEOREM. [**Mir07a**, Section 5 and Theorem 8.1]. *For $(g,n) \neq (1,1), (0,3)$, the volume $V_{g,n}(L)$ satisfies*

$$\frac{\partial}{\partial L_1} L_1 V_{g,n}(L) = \mathcal{A}_{g,n}^{con}(L) + \mathcal{A}_{g,n}^{dcon}(L) + \mathcal{B}_{g,n}(L).$$

      * The integrals $\mathcal{A}_*^*(L)$ and $\mathcal{B}_*(L)$ are polynomials in boundary lengths with coefficients sums of special values of Riemann zeta; the coefficients are positive rational multiples of powers of π.
- Symplectic reduction and the Duistermaat-Heckman theorem.
   - The WP kappa equation $\omega = 2\pi^2\kappa_1$ on Deligne-Mumford [**Wlp90**].
   - The geometry and topology of the Teichmüller space

$$\widehat{\mathcal{T}}_g(L) = \widehat{\mathcal{T}}_g(L_1, \ldots, L_n)$$

   of hyperbolic surfaces with geodesic boundaries with points.
   - A symplectic structure for $\widehat{\mathcal{T}}_g(L)$ by summing on almost tight pants. The boundary length moment map $R \in \widehat{\mathcal{T}}_g(L) \mapsto L^2/2 \in \mathbb{R}_+^n$. Twisting boundary points as Hamiltonian flows.

– Symplectic reduction, [**FO06**]. $\mathcal{T}_g(L)$ as the reduced space $\widehat{\mathcal{T}}_g(L)/(S^1)^n$. $\widehat{\mathcal{T}}_g(L)$ as a principal $(S^1)^n$ bundle over $\mathcal{T}_{g,n}(L)$ and the small L equivalence $\mathcal{T}_g(L) \approx \mathcal{T}_{g,n}(0)$,

$$\begin{array}{ccc} (S^1)^n & \longrightarrow & \widehat{\mathcal{T}}_g(L) \\ & & \downarrow \\ & & \mathcal{T}_g(L) \approx \mathcal{T}_{g,n}(0)\,. \end{array}$$

– Points on circles, cotangent $\mathbb{C}$-lines at punctures, circle bundles and homotopic structure groups. The tautological cotangent line class ψ.
– MCG equivariant maps and quotients. Deligne-Mumford type compactifications and finite covers. Contributions of the general elliptic curve involution.
– The Duistermaat-Heckman normal form, [**Mir07b**, Theorem 3.2], [**MS08**, Section 2.5] and [**Wlp10**, pg. 95],

$$2\omega_{\overline{\widehat{\mathcal{M}}_g(L)/(S^1)^n}} \equiv 2\omega_{\overline{\mathcal{M}}_{g,n}(0)} + \sum_j \frac{L_j^2}{2}\psi_j\,.$$

– The consequence of symplectic reduction, [**Mir07b**, Theorem 4.4].

THEOREM. *The volume polynomial $V_{g,n}(L)$ coefficients are $\overline{\mathcal{M}}_{g,n}$ intersection numbers given as*

$$C_\alpha = \frac{2^{\delta_{1g}\delta_{1n}}}{2^{|\alpha|}\alpha!(3g-3+n-|\alpha|)!}\int_{\overline{\mathcal{M}}_{g,n}} \psi_1^{\alpha_1}\cdots\psi_n^{\alpha_n}\omega^{3g-3+n-|\alpha|},$$

where ψ_j is the Chern class for the cotangent line along the jth puncture, ω is the WP symplectic form and δ_{**} is the Kronecker indicator.

- The pattern of intersection numbers.
  – The general intersection number symbol

$$\langle \kappa_1^{d_0}\tau_{d_1}\cdots\tau_{d_n}\rangle_g = \int_{\overline{\mathcal{M}}_{g,n}} \prod_{j=1}^{n}\psi_j^{d_j}\kappa_1^{d_0}$$

and volume polynomial expansions.
  – Examples: $V_{1,1}(L) = \frac{\pi^2}{12} + \frac{L^2}{48}$, $V_{0,4}(L_1,L_2,L_3,L_4) = (4\pi^2 + L_1^2 + L_2^2 + L_3^2 + L_4^2)/2$ and $V_{2,1}(L) = \frac{1}{2211840}(L^2+4\pi^2)(L^2+12\pi^2)(5L^4 + 384\pi^2L^2 + 6960\pi^4)$.
  – The partition function $\mathbf{F} = \sum_g \langle e^{\sum_j t_j\tau_j}\rangle_g$ and Virasoro constraint differential operators $\mathbf{L}_k$ [**Mir07b**]. The partition function $\mathbf{G} = \sum_g \langle e^{s\kappa_1 + \sum_j t_j\tau_j}\rangle_g$ and Virasoro constraint differential operators $\mathbf{V}_k$ [**MS08**, Theorem 1.1].
  – The Do *remove a boundary* relation, [**Do08, DN09**],

$$\frac{\partial V_{g,n+1}}{\partial L_{n+1}}(L, 2\pi i) = 2\pi i(2g-2+n)V_{g,n}(L).$$

– The Manin-Zograf volumes generating function, [**MZ00**]. The punctures asymptotic - there exist positive constants c, a_g, such that for genus fixed and n large,

$$V_{g,n} = c^n n! \, n^{(5g-7)/2}(a_g + O(1/n)).$$

– The Schumacher-Trapani genera asymptotic, [**Gru01, ST01**] - for n fixed, there exist positive constants c_1, c_2 such that for g large,

$$c_1^g(2g)! < V_{g,n} < c_2^g(2g)!.$$

– The Zograf conjecture, [**MZ11, Zog08**] - there exist constants c_n such that for n fixed and g large,

$$V_{g,n} = (4\pi^2)^{2g+n-3}(2g+n-3)!\frac{1}{\sqrt{g\pi}}\Big(1 + \frac{c_n}{g} + O(1/g^2)\Big).$$

Expected values of geometric invariants [**Mir10**].

3. The lectures.

The following is an exposition of Mirzakhani's proof of the Witten-Kontsevich theorem, including the immediate background material on Teichmüller theory, moduli space theory and on symplectic reduction. The lectures are presented as from a graduate text - the exposition follows the development of concepts, and does not consider the historical development of the material. The goals are general treatment of the material and overall understanding for the reader. In places, the approaches of several authors are combined for a simpler treatment of the material. Only immediate references to the literature are included. The reader should consult the literature for the historical development, for complete references and for consequences and applications of the material.

Lecture 1. The background and overview.

Lecture 2. The McShane-Mirzakhani identity

Lecture 3. The covolume formula and recursion.

Lecture 4. Symplectic reduction, principal S^1 bundles and the normal form.

Lecture 5. The pattern of intersection numbers and Witten-Kontsevich.

LECTURE 1

The background and overview

General background.

By the Uniformization Theorem, for a surface of negative Euler characteristic, a conformal structure is equivalent to a complete hyperbolic structure. We consider Riemann surfaces R of finite topological type with hyperbolic metrics, possibly with punctures and geodesic boundaries. Fix a topological reference surface F, and consider a marking, an orientation preserving homeomorphism $f : F \to R$ up to homotopy. If boundaries are present, homotopy is rel boundary setwise. Write $\mathcal{T}$ for the Teichmüller space of R - the space of equivalence classes of pairs $\{(f, R)\}$, where pairs (f, R) and (h, S) are equivalent if there is a conformal map (a hyperbolic isometry) $k : R \to S$ with $k \circ f$ homotopic to h.

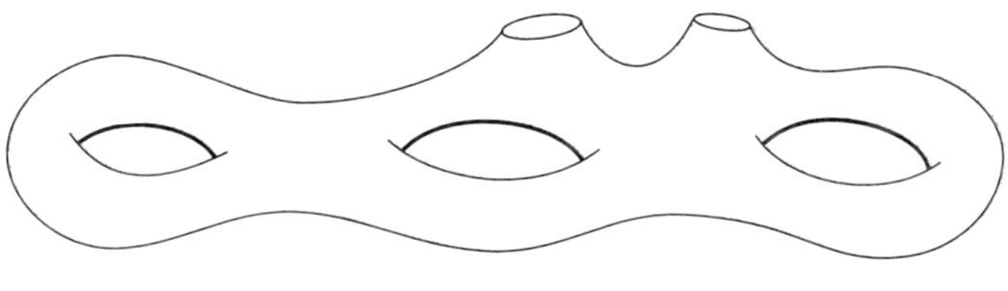

FIGURE 2. A genus 3 surface with 2 geodesic boundaries.

We consider the following Teichmüller spaces $\mathcal{T}$: $\mathcal{T}_g$ - for closed genus g surfaces; $\mathcal{T}_g(L)$ - for genus g surfaces with labeled geodesic boundaries of prescribed lengths $L = (L_1, \ldots, L_n)$; $\mathcal{T}_{g,n}$ - for genus g surfaces with n labeled punctures. In the case of $\mathcal{T}_g(L)$, homotopies of surfaces are rel boundaries setwise. $\mathcal{T}_g$ and $\mathcal{T}_{g,n}$ are complex manifolds, while $\mathcal{T}_g(L)$ is a real analytic manifold.

A non trivial, non puncture peripheral, free homotopy class α on F has a unique geodesic representative for $f(\alpha)$ on the surface R - the geodesic length $\ell_\alpha(R)$ provides a natural function on Teichmüller space. Collections of geodesic-length functions provide local coordinates and global immersions to Euclidean space for $\mathcal{T}$. The differential of geodesic-length for a simple curve is nowhere vanishing. At each point of $\mathcal{T}$, real multiples of the differentials of geodesic-lengths of simple curves are dense in the cotangent bundle.

A surface can be cut open on a simple closed geodesic - the boundaries are isometric circles. Since a neighborhood of a simple geodesic has an S^1 symmetry, the boundaries can be reassembled with a relative rotation to form a new hyperbolic structure. The deformation is the Fenchel-Nielsen (FN) twist. The infinitesimal deformation for unit speed hyperbolic displacement of initial adjacent points, is the Fenchel-Nielsen infinitesimal twist vector field t_α on $\mathcal{T}$. (A positive twist corresponds to displacing to the right when crossing the geodesic.)

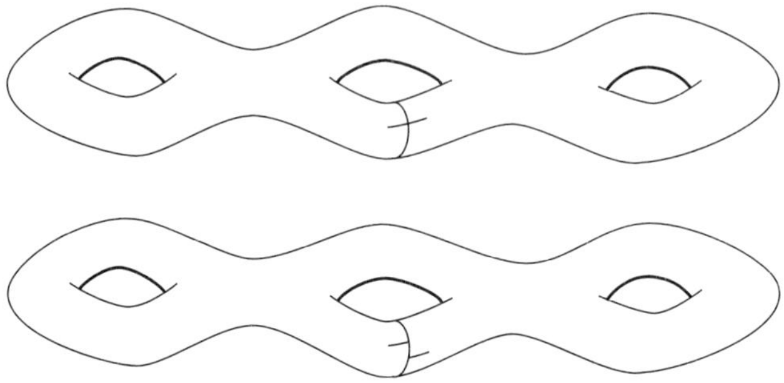

FIGURE 3. A positive Fenchel-Nielsen twist deformation.

Geodesic boundaries of hyperbolic surfaces of common length can be assembled to form new surfaces. Given boundary reference points, the relative rotation is measured in terms of arc length. A hyperbolic pair of pants is a genus zero surface with three geodesic boundaries. For pants, boundary reference points are provided by considering the unique orthogonal connecting geodesics between boundaries. At a gluing, the common boundary length ℓ and relative rotation, the FN twist parameter τ, are unrestricted parameters (τ is defined by continuation from an initial configuration). The length ℓ varies in $\mathbb{R}_{>0}$ and twist τ varies in $\mathbb{R}$. Each finite topological type hyperbolic surface can be assembled from pairs of pants.

THEOREM 1. Fenchel-Nielsen coordinates. *Fixing the topological type of a pants decomposition and an initial configuration, the FN parameters* $\prod_{j=1}^{3g-3+n}(\ell_j, \tau_j)$ *define a real analytic equivalence of* $\mathcal{T}$ *to* $\prod_{j=1}^{3g-3+n}\mathbb{R}_{>0}\times\mathbb{R}$.

The Weil-Petersson (WP) metric is Kähler. The symplectic geometry begins with the symplectic form $\omega = 2\omega_{\text{WP Kähler}}$ and the basic twist-length duality

$$\omega(\ ,t_\alpha) = d\ell_\alpha.$$

It follows from the Lie derivative equation $L_X\omega(\ ,\ ) = d\omega(X,\ ,\ ) + d(\omega(X,\ ))$ that the form ω is invariant under all twist flows. It follows that geodesic-length functions are Hamiltonian potentials for FN infinitesimal twists. Symmetry reasoning shows that ℓ and τ provide action-angle coordinates for ω.

THEOREM 2. (W), [**Wlp10**]. The $d\ell\wedge d\tau$ formula. *The WP symplectic form is*

$$\omega = \sum_{j=1}^{3g-3+n} d\ell_j\wedge d\tau_j.$$

The symplectic structure is independent of pants decomposition.

Frontier spaces are adjoined to $\mathcal{T}$ corresponding to allowing $\ell_j = 0$ with the FN angle $\theta_j = 2\pi\tau_j/\ell_j$ then undefined (in polar coordinates, angle is undefined at the origin). The vanishing length describes pairs of pants with corresponding boundaries represented by punctures - the equation $\ell_\alpha = 0$ describes hyperbolic structures with α represented by pairs of punctures. For a subset of indices $J\subset\{1,\dots,3g-3+n\}$, the J-null stratum is $\mathcal{S}(J) = \{R \text{ degenerate} \mid \ell_j(R) = 0 \text{ iff } j\in J\}$. Each null stratum is a product of lower dimensional Teichmüller spaces. A basis of neighborhoods in $\mathcal{T}\cup\mathcal{S}(J)$ is defined in terms of the parameters $(\ell_k,\theta_k,\ell_j)_{k\notin J,\, j\in J}$.

The augmented Teichmüller space is the stratified space

$$\overline{\mathcal{T}} = \mathcal{T} \cup_{pants\ decompositions\ \mathcal{P}} \cup_{J\subset\mathcal{P}} \mathcal{S}(J).$$

The space $\overline{\mathcal{T}}$ is also described as the Chabauty topology closure of the discrete faithful type-preserving representations of $\pi_1(F)$ into $PSL(2;\mathbb{R})$, modulo $PSL(2;\mathbb{R})$ conjugation. The augmentation construction is valid for $\mathcal{T}_g, \mathcal{T}_g(L)$ and $\mathcal{T}_{g,n}$. $\overline{\mathcal{T}}$ is a Baily-Borel type partial compactification. $\overline{\mathcal{T}}$ is never locally compact. The $d\ell\wedge d\tau$ formula provides for the extension of the symplectic structure to the augmented Teichmüller space $\overline{\mathcal{T}}$. Each stratum is symplectic.

The mapping class group (MCG) $Homeo^+(F)/Homeo_0(F)$ acts on markings by precomposition and thus acts on $\mathcal{T}$. For $Homeo^+(F)$ we consider type-preserving (boundary point, boundary curve), boundary label preserving, orientation preserving homeomorphisms. $Homeo_0(F)$ is the normal subgroup of elements homotopic to the identity rel boundary setwise. A Dehn twist is a homeomorphism that is the identity on the complement of an annular neighborhood of a simple closed curve, non trivial in homotopy, and rotates one boundary of the annular neighborhood relative to the other. Dehn twist classes generate MCG. MCG acts properly discontinuously on $\mathcal{T}$ and by biholomorphisms for $\mathcal{T}_g$ and $\mathcal{T}_{g,n}$. Except for a finite number of topological types the action is effective. Finite MCG subgroups act with fixed points. The symplectic form ω is MCG invariant and the quotients $\mathcal{T}_g/$ MCG and $\mathcal{T}_{g,n}/$ MCG are symplectic orbifolds. MCG acts on the stratified space $\overline{\mathcal{T}}$. Bers observed that there are constants $b_{g,n}$, depending on topological type, such that a genus g, n punctured hyperbolic surface has a pants decomposition with lengths at most $b_{g,n}$. It follows that the domain $\{\ell_j \leq b_{g,n},\ 0 < \tau_j \leq \ell_j\}$ in FN coordinates is a rough fundamental set - each MCG orbit intersects the domain a bounded positive number of times. The FN parameters provide $\overline{\mathcal{T}}/$ MCG with the structure of a compact real analytic orbifold; $\overline{\mathcal{T}}_g/$ MCG and $\overline{\mathcal{T}}_{g,n}/$ MCG are topologically the Deligne-Mumford stable curve compactifications of $\mathcal{M}_g$ and $\mathcal{M}_{g,n}$. The Bers fundamental set observation combines with the $d\ell\wedge d\tau$ formula to provide that the WP volume of $\mathcal{M}$ is finite.

The Bers fiber space $\mathcal{C}$ (specifically $\mathcal{C}_g$ and $\mathcal{C}_{g,n}$) is the complex disc holomorphic bundle over $\mathcal{T}$ with fiber over $\{(f,R)\}$ the universal cover $\widetilde{R}$. A point on a fiber can be considered as a puncture and determines a curve from basepoint for the fundamental group - so $\mathcal{C}_g \approx \mathcal{T}_{g,1}$ and $\mathcal{C}_{g,n} \approx \mathcal{T}_{g,n+1}$. An extension MCG$_{\mathcal{C}}$ of MCG(F) by the fundamental group $\pi_1(F)$ acts properly discontinuously and holomorphically on $\mathcal{C}$. The group MCG$_{\mathcal{C}}$ is isomorphic to MCG$_{g,n+1}$. For the epimorphism from MCG$_{\mathcal{C}}$ to MCG(F), the first group acts equivariantly on the fibration of $\mathcal{C}$ over $\mathcal{T}$. The resulting map $\pi : \mathcal{C}/\,\mathrm{MCG}_{\mathcal{C}} \to \mathcal{T}/\,\mathrm{MCG}(F)$ describes an orbifold bundle, the universal curve, with orbifold fibers - the fibers are Riemann surfaces modulo their automorphism group. Manifold finite local covers of the orbifolds $\mathcal{C}/$ MCG$_{\mathcal{C}}$ and $\mathcal{T}/$ MCG(F) are described by starting with a surface with locally maximal symmetries and introducing a local trivialization of $\mathcal{C}$ by canonical (extremal, harmonic) maps of surfaces.

The augmentation construction applies to the Bers fiber space to give $\overline{\mathcal{C}}$. MCG$_{\mathcal{C}}$ acts on the stratified space $\overline{\mathcal{C}}$. The augmentation quotient $\overline{\mathcal{C}}/$ MCG$_{\mathcal{C}}$ is an orbifold and almost an orbifold bundle over $\overline{\mathcal{M}}$ - at a node (a pair of punctures) of a Riemann surface, the fiber becomes vertical - the local model of the fibration is the germ at the origin of the projection $\{(z,w)\} \to \{t = zw\}$, the family of complex

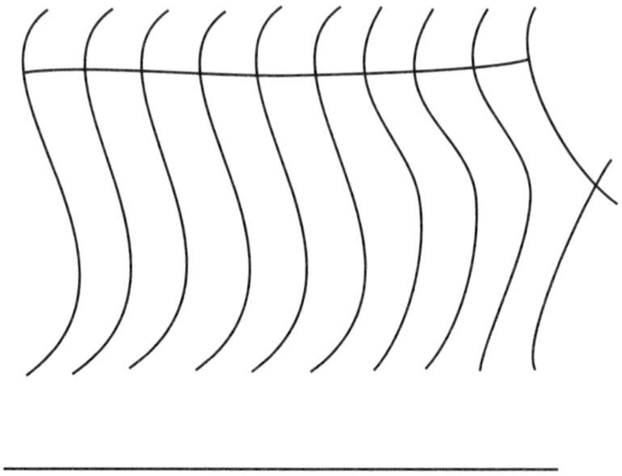

FIGURE 4. A puncture section of the universal curve $\overline{\mathcal{C}}/\operatorname{MCG}_{\mathcal{C}}$ over $\overline{\mathcal{T}}/\operatorname{MCG}(F)$.

hyperbolas. The turning of the fibers of the almost orbifold bundle $\overline{\mathcal{C}}/\operatorname{MCG}_{\mathcal{C}} \to \overline{\mathcal{M}}$ is measured by line bundles on $\overline{\mathcal{M}}$. The family of tangent $\mathbb{C}$-lines to the fibers $(Ker\, d\pi)$ is the tangent bundle along a smooth Riemann surface and the relative dualizing sheaf along a noded Riemann surface. The hyperbolic metrics of the individual fibers provide a line bundle metric for $(Ker\, d\pi)$ on $\overline{\mathcal{C}}$, that although not smooth is sufficiently regular for calculation of the Chern form $\mathbf{c}_1$. The kappa forms/cohomology classes $\kappa_k = \int_{\pi^{-1}(\{R\})} \mathbf{c}_1^{k+1}$ given by integration over fibers are basic to moduli geometry. The geometry and algebra of the kappa classes is studied in the Carel Faber lectures. Explicit calculation of the Chern form and integration provides the following.

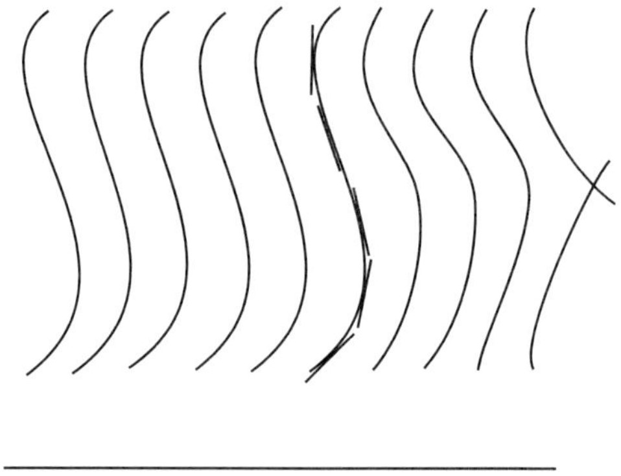

FIGURE 5. Tangents along a fiber of the universal curve $\overline{\mathcal{C}}/\operatorname{MCG}_{\mathcal{C}}$.

THEOREM 3. (W), [**Wlp90**]. *For the hyperbolic metrics on fibers,* $2\pi^2\kappa_1 = \omega$ *pointwise on* $\mathcal{M}$ *and in cohomology on* $\overline{\mathcal{M}}$.

A conformal structure has a unique extension to fill in a puncture. A labeled puncture defines a section s of $\overline{\mathcal{C}}_{g,n}/\operatorname{MCG}_{\mathcal{C}} \to \overline{\mathcal{M}}_{g,n}$. A section satisfies $\pi \circ s = id$, differentiating gives $d\pi \circ ds = d\, id$. At a node, $d\pi$ vanishes in the node opening direction - for $\pi(z,w) = t$ then $d\pi = wdz + zdw$ vanishes at the origin. Sections of $\overline{\mathcal{C}}/\operatorname{MCG}_{\mathcal{C}}$ over $\overline{\mathcal{M}}$ are consequently disjoint from nodes. Along a puncture section $s : \overline{\mathcal{M}} \to \overline{\mathcal{C}}/\operatorname{MCG}_{\mathcal{C}}$, we consider the family of tangent lines $(Ker\, d\pi)|_s$ or the dual family of cotangent lines $(Ker\, d\pi)^*|_s$. In the Carel Faber lectures, the Chern class of the family is denoted as K. — The Chern class of the pullback to $\overline{\mathcal{M}}$ by a puncture section s of the dual family $(Ker\, d\pi)^*$ is the moduli geometry canonical psi class ψ.

In these lectures, to emphasize concepts and the underlying geometry, we will at times informally interchange a line bundle and its Chern class, informally refer to the moduli space as a manifold, the universal curve as a fiber bundle, and at times refer to the open moduli space when actually the augmentation quotient is required. Our goal is to discuss the central matters. In spite of the informal approach, an experienced reader will find that the treatment is complete.

Basic references for the above material are [**Bus92**] and [**Wlp10**].

Volume results overview.

Mirzakhani shows that the WP orbifold volume $V_{g,n}(L) = V(\overline{\mathcal{M}}_g(L))$ is a polynomial in L, with coefficients given by the intersection numbers of powers of κ_1 and powers of ψ. She further shows that her recursion for determining the volume polynomials satisfies the defining relations for the Witten-Kontsevich conjecture. The following theorems are the immediate results of the two papers.

THEOREM 4. The WP volume polynomials. *The volume polynomials are determined recursively from the volume polynomials of smaller total degree,* [**Mir07a**, Formulas (5.1) to (5.4) & Theorem 8.1]. *The volume $V_{g,n}(L_1,\dots,L_n)$ of the moduli space of genus g, n boundaries, hyperbolic surfaces with boundary lengths $L = (L_1,\dots,L_n)$ is a polynomial*

$$V_{g,n}(L) = \sum_{\substack{\alpha \\ |\alpha|\le 3g-3+n}} C_\alpha\, L^{2\alpha},$$

for multi-index $\alpha = (\alpha_1,\dots,\alpha_n)$ and where $C_\alpha > 0$ lies in $\pi^{6g-6+2n-2|\alpha|}\mathbb{Q}$, [**Mir07a**, Theorems 1.1 & 6.1]. *The coefficients are intersection numbers given as*

$$C_\alpha = \frac{2^{\delta_{1g}\delta_{1n}}}{2^{|\alpha|}\alpha!(3g-3+n-|\alpha|)!}\int_{\overline{\mathcal{M}}_{g,n}} \psi_1^{\alpha_1}\cdots\psi_n^{\alpha_n}\omega^{3g-3+n-|\alpha|},$$

where ψ_j is the Chern class for the cotangent line along the jth puncture, ω is the WP symplectic form, $\alpha! = \prod_{j=1}^n \alpha_j!$, and δ_{**} is the Kronecker indicator delta, [**Mir07b**, Theorem 4 4].

THEOREM 5 ([**Mir07a**, Theorems 6.3 & 6.4]). Recursive relations for the volume polynomial leading coefficients. *For a multi index α, define*

$$(\alpha_1,\dots,\alpha_n)_g = C_\alpha \times 2^{-\delta_{1g}\delta_{1n}} \times \prod_{i=1}^n \alpha_i! \times 2^{|\alpha|},$$

then for $n > 0$ and $\sum_i \alpha_i = 3g-3+n$,

the dilaton equation $\qquad (1,\alpha_1,\dots,\alpha_n)_g = (2g-2+n)(\alpha_1,\dots,\alpha_n)_g$

and for $n > 0$ and $\sum_i \alpha_i = 3g - 2 + n$,

$$\textit{the string equation} \qquad (0, \alpha_1, \ldots, \alpha_n)_g = \sum_{\alpha_i \neq 0} (\alpha_1, \ldots, \alpha_i - 1, \ldots, \alpha_n)_g.$$

For the intersection number generating function

$$\mathbf{F}(\lambda, t_0, t_1, \ldots) = \sum_{g=0}^{\infty} \lambda^{2g-2} \sum_{\{d_j\}} \langle \prod_{j=1}^{\infty} \tau_{d_j} \rangle_g \prod_{r \geq 0} t_r^{n_r}/n_r!,$$

with $n_r = \#\{j \mid d_j = r\}$, and

$$\langle \tau_{d_1} \cdots \tau_{d_n} \rangle_g = \int_{\overline{\mathcal{M}}_{g,n}} \prod_{j=1}^{n} \psi_j^{d_j},$$

for $\sum_{j=1}^{n} d_j = 3g - 3 + n$ and the product $\langle \tau_* \rangle$ otherwise zero, the exponential $e^{\mathbf{F}}$ satisfies Virasoro algebra constraints, [**Mir07b**, Theorem 6.1].

A fine structure for volumes is suggested by the Zograf conjecture

$$V_{g,n} = (4\pi^2)^{2g+n-3}(2g + n - 3)! \frac{1}{\sqrt{g\pi}} \left(1 + \frac{c_n}{g} + O(1/g^2)\right),$$

for fixed n and g tending to infinity.

As an application of the method for recursion of volumes and intersection numbers, Do derives a *remove a boundary* relation for the volume polynomials [**Do08**],

$$\frac{\partial V_{g,n+1}}{\partial L_{n+1}}(L, 2\pi i) = 2\pi i(2g - 2 + n)V_{g,n}(L).$$

The relation gives the compact case volume V_g.

Statement of the volume recursion [**Mir07a**, Sec. 5]**.** The WP orbifold volume $V_g(L_1, \ldots, L_n)$ of the moduli space $\mathcal{T}_g(L_1, \ldots, L_n)/\operatorname{MCG}$ is a symmetric function of boundary lengths as follows.

- For $L_1, L_2, L_3 \geq 0$, formally set

$$V_{0,3}(L_1, L_2, L_3) = 1$$

  and

$$V_{1,1}(L_1) = \frac{\pi^2}{12} + \frac{L_1^2}{48}.$$

- For $L = (L_1, \ldots, L_n)$, let $\widehat{L} = (L_2, \ldots, L_n)$ and for $(g, n) \neq (1, 1)$ or $(0, 3)$, the volume satisfies

$$\frac{\partial}{\partial L_1} L_1 V_g(L) = \mathcal{A}_g^{con}(L) + \mathcal{A}_g^{dcon}(L) + \mathcal{B}_g(L)$$

  where

$$\mathcal{A}_g^*(L) = \frac{1}{2} \int_0^{\infty} \int_0^{\infty} \widehat{\mathcal{A}}_g^*(x, y, L)\, xy\, dxdy$$

  and

$$\mathcal{B}_g(L) = \int_0^{\infty} \widehat{\mathcal{B}}_g(x, L)\, x\, dx.$$

The quantities $\widehat{\mathcal{A}}_g^{con}$, $\widehat{\mathcal{A}}_g^{dcon}$ are defined in terms of the function

$$H(x,y) = \frac{1}{1+e^{\frac{x+y}{2}}} + \frac{1}{1+e^{\frac{x-y}{2}}}$$

and moduli volumes for subsurfaces

$$\widehat{\mathcal{A}}_g^{con}(x,y,L) = H(x+y,L_1)V_{g-1}(x,y,\widehat{L})$$

and surface decomposition sum

$$\widehat{\mathcal{A}}_g^{dcon}(x,y,L) = \sum_{\substack{g_1+g_2=g \\ I_1 \amalg I_2 = \{2,\ldots,n\}}} H(x+y,L_1)V_{g_1}(x,L_{I_1})V_{g_2}(y,L_{I_2}),$$

where in the second sum only decompositions for pairs of hyperbolic structures are considered and the unordered sets I_1, I_2 provide a partition. The third quantity $\widehat{\mathcal{B}}_g$ is defined by the sum

$$\frac{1}{2}\sum_{j=2}^{n}\left(H(x,L_1+L_j)+H(x,L_1-L_j)\right)V_g(x,L_2,\ldots,\widehat{L_j},\ldots,L_n),$$

where L_j is omitted from the argument list of V_g.

The basic point: the volume $V_g(L_1,\ldots,L_n)$ is an appropriate integral of volumes for surfaces formed with *one* fewer pair of pants.

LECTURE 2

The McShane-Mirzakhani identity

In 1991, Greg McShane discovered a universal identity for a sum of lengths of simple geodesics for a once punctured torus, [**McS98**]. A generalization of the identity serves as the analog of a partition of unity for the action of the mapping class group. The identity enables reduction of the action to the actions of smaller mapping class groups. Consideration of the identity begins with a surface with geodesic boundaries and a study of arcs from the boundary to itself.

Introduce $\mathcal{B}$, the set of non trivial free homotopy classes of simple curves from the boundary to the boundary, homotopy rel the boundary. We illustrate the approach by considering simple curves with endpoints on a common boundary β; the analysis is similar for simple curves connecting distinct boundaries. Each homotopy class contains a unique shortest geodesic, orthogonal to β at end points - refer to these geodesics as *ortho boundary geodesics.* If the surface is doubled across its boundary, then the ortho boundary geodesics double to simple closed geodesics.

The set $\mathcal{B}$ is in bijection to the set of topological pants embedded in the surface with β as one boundary - refer to these pants as β-cuff pants. First note that the endpoints of an ortho boundary geodesic $\gamma, [\gamma] \in \mathcal{B}$ are distinct. A small neighborhood/thickening of $\gamma \cup \beta$ is the corresponding topological pair of pants. Geometrically, the curve γ separates β into proper sub arcs; the union of each sub arc with γ is a simple curve, that defines a free homotopy class containing a unique geodesic. The corresponding geometric pair of pants $\mathcal{P}$ has boundaries β and the two determined geodesics. We will see below that a geometric pair of pants contains a unique ortho boundary geodesic. The unions of ortho boundary geodesics and β are the spines, the wire frames, for the embedded geometric β-cuff pants.

We now describe how the behavior of geodesics emanating orthogonally from β defines a Cantor subset of β. The Cantor set will have measure zero and the length identity is simply the sum of lengths of the complementary intervals. The following description follows the analysis by Tan-Wong-Zhang, [**TWZ06**].

Consider the maximal continuations of geodesics emanating orthogonally from β - refer to these geodesics as *ortho emanating geodesics.* In addition to the ortho boundary geodesics, there are three types of ortho emanating geodesics: non simple, simple infinite length and simple crossing the boundary obliquely at a second endpoint. We will see that the types are detected by considering initial segments in a pair of pants.

Consider the geometric pants $\mathcal{P}$, obtained from an ortho boundary geodesic γ (see Figure 6). The boundaries are β and the two defined geodesics α and λ (in the special case $(g, n) = (1, 1)$ then $\alpha = \lambda$). A *spiral* is an infinite simple geodesic ray that accumulates to a simple closed geodesic. Two ortho emanating geodesics are spirals with accumulation set α and two are spirals with accumulation set λ. The two spirals accumulating to a boundary wind in opposite directions around the boundary. The $\mathcal{P}$ *main gaps* are the two disjoint subarcs of β that each contain in their interior an endpoint of γ and have spiral initial points as endpoints. The main gaps will be the components of a Cantor set complement corresponding to the pants $\mathcal{P}$. From the geometry of pants, geodesics ortho emanating from the main gaps are either the spiral endpoints, γ, non simple with self intersection in $\mathcal{P}$ or simple crossing β obliquely at a second endpoint. The complement in β of the main gaps are a pair of open intervals. From the geometry of pants, for a given open interval all ortho emanating geodesics exit the pants by crossing one of the boundaries α or λ. For a given open interval, the initial segments in $\mathcal{P}$ are simple and these geodesics are classified by their subsequent behavior elsewhere on the surface, by their behavior on some other pair of pants. The ortho boundary geodesic connecting β to α is contained in one of the open intervals, and the ortho boundary geodesic from β to λ is contained in the other. A pair of pants has an equatorial reflection, stabilizing each boundary setwise. The equatorial reflection acts naturally on the set of geodesics emanating from β and on the decomposition of β, interchanging or stabilizing the items described above and below.

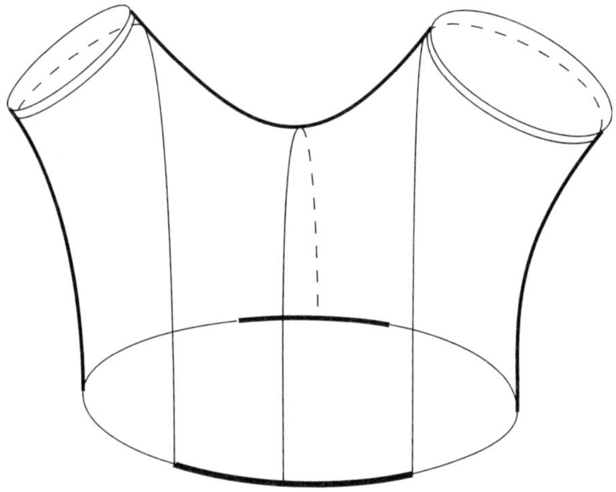

FIGURE 6. A pair of pants with equators, main gaps, two spirals and an orthoboundary geodesic γ.

In the above, associated to a main gap are the ortho emanating geodesics that self intersect in the pants, and the simple geodesics that obliquely cross β a second time - refer to the second type geodesics as *boundary oblique*. The next observation is that the associations can be reversed, the associations define bijections between main gaps and geodesics with particular behaviors on the surface.

A boundary oblique geodesic and β form a crooked wire frame that determines a pair of pants, similar to an ortho boundary geodesic determining a pair of pants. Boundary oblique geodesics come in continuous families with each family limiting to an ortho boundary geodesic. A family and its limit determine the same pair of pants. The initial points (the β orthogonal points) of family elements lie in a common main gap interval - this observation reverses the association of segments of main gaps to boundary oblique geodesics.

Next we describe reversing the association of segments of main gaps to non simple ortho emanating geodesics. The first self intersection of such a geodesic is contained in a unique embedded pair of pants. To see this, consider the lasso subarc beginning at β and ending where the geodesic passes through its first self intersection point a second time. The boundary of a small neighborhood/thickening of the lasso is the union of a simple closed curve and an element of $\mathcal{B}$. A geometric argument shows that the self intersection point is contained in the pants determined by the element of $\mathcal{B}$ and the lasso initial point lies in the main gap for the pants. This observation reverses the association of segments of main gaps to non simple ortho emanating geodesics.

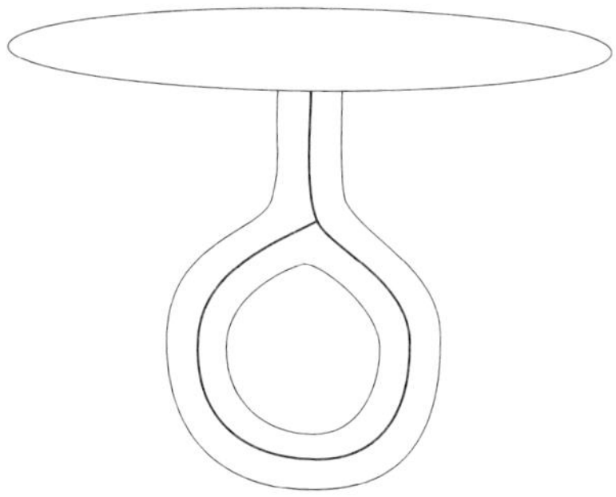

FIGURE 7. A small neighborhood of a lasso.

We recall that non simple with interior intersection is an open condition on the space of geodesics and an open condition on the space of ortho emanating geodesics. By considering the double of the surface, simple with all boundary intersections orthogonal is a closed condition on the space of geodesics. The set *simple with orthogonal single boundary intersection* is a Cantor set. The classification of ortho emanating geodesics is complete.

THEOREM 6. (Tan-Wong-Zhang [**TWZ06**], and Mirzakhani [**Mir07a**], all following McShane [**McS98**].) *There is a Cantor set partition of boundary points by the behavior of ortho emanating geodesics:*

$$\begin{aligned}\beta = &\{\textit{simple with orthogonal single boundary intersection}\}\\ &\cup \{\textit{ortho boundary geodesics}\} \cup \{\textit{simple boundary oblique geodesics}\}\\ &\cup \{\textit{non simple}\}.\end{aligned}$$

An ϵ-neighborhood of the simple complete geodesics, orthogonal to the boundary at intersections, is a countable union of thin corridors. In the universal cover the corridors are described by reduced bi infinite words in the fundamental group. By analyzing the asymptotics of the number of corridors for small ϵ, Birman-Series show that the set is very thin.

THEOREM 7. (Birman-Series, [**BS85**].) Simple geodesics have measure zero. *The set $\mathcal{S}$ of simple complete geodesics, orthogonal to the boundary at intersections, has Hausdorff dimension* 1. *The intersection of $\mathcal{S}$ and the boundary has Hausdorff dimension and measure* 0.

The basic summand for the length identity is a rational exponential function. Define the function H on $\mathbb{R}^2$ by

$$H(x,y) = \frac{1}{1+e^{\frac{x+y}{2}}} + \frac{1}{1+e^{\frac{x-y}{2}}} \tag{1}$$

and the corresponding functions $\mathcal{D}, \mathcal{R}$ on $\mathbb{R}^3$ by

$$\mathcal{D}(x,y,z) = 2\log\left(\frac{e^{\frac{x}{2}} + e^{\frac{y+z}{2}}}{e^{\frac{-x}{2}} + e^{\frac{y+z}{2}}}\right) \quad \text{and} \tag{2}$$

$$\mathcal{R}(x,y,z) = x - \log\left(\frac{\cosh\frac{y}{2} + \cosh\frac{x+z}{2}}{\cosh\frac{y}{2} + \cosh\frac{x-z}{2}}\right).$$

The functions $\mathcal{D}$ and $\mathcal{R}$ are related to H as follows,

$$\frac{\partial}{\partial x}\mathcal{D}(x,y,z) = H(y+z,x), \quad \mathcal{D}(0,0,0) = 0 \quad \text{and} \tag{3}$$

$$2\frac{\partial}{\partial x}\mathcal{R}(x,y,z) = H(z,x+y) + H(z,x-y), \quad \mathcal{R}(0,0,0) = 0.$$

THEOREM 8 ([**Mir07a**, Theorems 1.3 & 4.2] and [**TWZ06**, Thrm. 1.8]. The Mirzakhani-McShane identity). *For a hyperbolic surface R with boundaries β_j with lengths L_j,*

$$L_1 = \sum_{\alpha_1,\alpha_2} \mathcal{D}(L_1, \ell_{\alpha_1}(R), \ell_{\alpha_2}(R)) + \sum_{j=2}^{n}\sum_{\alpha} \mathcal{R}(L_1, L_j, \ell_\alpha(R)),$$

where the first sum is over all unordered pairs of simple closed geodesics with $\beta_1, \alpha_1, \alpha_2$ bounding an embedded pair of pants, and the double sum is over simple closed geodesics with β_1, β_j, α bounding an embedded pair of pants.

PROOF. By the above Theorems, a boundary length ℓ_β equals the sum over embedded β-cuff pants of main gap lengths and the counterpart lengths for double boundary cuff pants. To find the main gap lengths, begin with a formula for the lengths of the complementary intervals. In a pair of pants, the ortho boundary geodesic δ_α from β to α bisects a complementary interval (see Figure 6). Let β_α be one of the resulting half intervals. The segment β_α has the geodesic δ_β emanating at one end and a spiral σ to α emanating at the other end. In the universal cover, consider contiguous lifts $\widetilde{\delta_\alpha}$, $\widetilde{\beta_\alpha}$ and $\widetilde{\sigma}$. The three lifts and a half infinite ray lift $\widetilde{\alpha}$ of α, combine to form a quadrilateral $\widetilde{\alpha}, \widetilde{\delta_\alpha}, \widetilde{\beta_\alpha}, \widetilde{\sigma}$ with angles $\pi/2, \pi/2, \pi/2$ and 0 between $\widetilde{\sigma}$ and $\widetilde{\alpha}$. By hyperbolic trigonometry of quadrilaterals [**Bus92**], it follows that

$$\tanh \ell_{\beta_\alpha} = \operatorname{sech} \delta_\alpha = \frac{\sinh(\ell_\beta/2)\sinh(\ell_\alpha/2)}{\cosh(\ell_\lambda/2) + \cosh(\ell_\beta/2)\cosh(\ell_\alpha/2)}.$$

The complementary interval length is $2\ell_{\beta_\alpha} = \ell_\beta - \mathcal{R}(\ell_\beta, \ell_\lambda, \ell_\alpha)$. The formula for main gap lengths now follows from the general relation $\mathcal{R}(x,y,z) + \mathcal{R}(x,z,y) = x + \mathcal{D}(x,y,z)$. For double boundary cuff pants, the main gap lengths are added to the complementary interval length. The result is $\mathcal{R}(\ell_\beta, \ell_\alpha, \ell_\lambda)$. $\square$

LECTURE 3

The covolume formula and recursion

The main step is application of the length identity to reduce the action of the mapping class group to an action of smaller mapping class groups, and consequently express the volume as an integral over a length level set. The result is an integral of products of lower dimensional volume functions - the recursion.

The approach is illustrated by computing the genus one, one boundary, volume. The length identity is

$$L = \sum_{\alpha\ simple} \mathcal{D}(L, \ell_\alpha, \ell_\alpha).$$

Introduce $\mathrm{Stab}(\alpha) \subset \mathrm{MCG}$, the stabilizer for MCG acting on free homotopy classes. A torus is an elliptic curve with universal cover $\mathbb{C}$ with involution $z \to -z$ stabilizing the deck transformation lattice. The involution acts on tori and tori with one puncture or boundary. The involution reverses orientation for the free homotopy class of each simple closed geodesic and the stabilizer $\mathrm{Stab}(\alpha)$ is the semi direct product of the Dehn twists by the involution $\mathbb{Z}/2\mathbb{Z}$ subgroup. The involution acts trivially on Teichmüller space. (The torus is one of the exceptional cases where the MCG action on $\mathcal{T}$ is not effective. We will also discuss the torus case below, where a multiplicity is involved.) A Dehn twist acts on the Teichmüller space in FN coordinates by $(\ell, \tau) \to (\ell, \tau + \ell)$. The sector $\{0 \leq \tau < \ell\}$ is a fundamental domain for the $\mathrm{Stab}(\alpha)$ action. A mapping class $h \in \mathrm{MCG}$ acts on a geodesic-length function by $\ell_\alpha \circ h^{-1} = \ell_{h(\alpha)}$.

Write the length identity as

$$L = \sum_{\alpha} \mathcal{D}(L, \ell_\alpha, \ell_\alpha) = \sum_{h \in \mathrm{MCG}\,/\,\mathrm{Stab}(\alpha)} \mathcal{D}(L, \ell_{h(\alpha)}, \ell_{h(\alpha)}),$$

use the MCG action on geodesic-length functions, to find

$$LV(L) = \int_{\mathcal{T}(L)/\,\mathrm{MCG}} \sum_{\mathrm{MCG}\,/\,\mathrm{Stab}(\alpha)} \mathcal{D}(L, \ell_\alpha \circ h^{-1}, \ell_\alpha \circ h^{-1})\,\omega,$$

change variables on $\mathcal{T}$ by $p = h(q)$ to find

$$\sum_{h \in \mathrm{MCG}\,/\,\mathrm{Stab}(\alpha)} \int_{h(\mathcal{T}(L)/\,\mathrm{MCG})} \mathcal{D}(L, \ell_\alpha, \ell_\alpha)\, d\tau d\ell = \int_{\mathcal{T}(L)/\,\mathrm{Stab}(\alpha)} \mathcal{D}(L, \ell_\alpha, \ell_\alpha)\, d\tau d\ell,$$

and use the $\mathrm{Stab}(\alpha)$ fundamental domain, to obtain the integral

$$\int_0^\infty \int_0^\ell \mathcal{D}(L, \ell, \ell)\, d\tau d\ell.$$

The integral in τ gives a factor of ℓ.

The derivatives $\partial\mathcal{D}(x,y,z)/\partial x$ and $\partial\mathcal{R}(x,y,z)/\partial x$ are simpler than the original functions $\mathcal{D}$ and $\mathcal{R}$ - apply this observation and differentiate in L to obtain a formula for the derivative of $LV(L)$,

$$\frac{\partial}{\partial L}LV(L) = \int_0^\infty \frac{1}{1+e^{\ell+\frac{L}{2}}} + \frac{1}{1+e^{\ell-\frac{L}{2}}}\,\ell d\ell = \frac{\pi^2}{6} + \frac{L^2}{8}.$$

The formula $V(L) = \frac{\pi^2}{6} + \frac{L^2}{24}$ results.

We prepare for the general case. In algebraic geometry intersection theory, the elliptic involution gives rise to multiplying elliptic intersection counts by a factor of $1/2$. The factor corresponds to the generic fiber of the universal elliptic curve being the quotient of the elliptic curve by its involution. Along this line, the general volume recursion is simplified if $V_{1,1}(L)$ is formally defined to be $1/2$ of the given value $V(L)$. In mapping class group theory, the elliptic involution appears as the *half Dehn twist* for simple closed curves bounding a torus. In particular, consider the fundamental group $\pi_1(R)$ of a surface, with the standard presentation $a_1b_1a_1^{-1}b_1^{-1}\cdots a_gb_ga_g^{-1}b_g^{-1}c_1\cdots c_n = 1$, with c_j a loop about the jth boundary. The half Dehn twist about the curve $a_1b_1a_1^{-1}b_1^{-1}$ is the automorphism of $\pi_1(R)$ given by: $a_1 \to b_1a_1^{-1}b_1^{-1}, b_1 \to b_1^{-1}; a_j \to a_1^{-1}a_ja_1, b_j \to a_1^{-1}b_ja_1$, for $j = 2\ldots g$, and $c_j \to a_1^{-1}c_ja_1$, for $j = 1\ldots n$. The square of a half Dehn twist is a Dehn twist and a half Dehn twist acts on the associated FN parameters by $(\ell,\tau) \to (\ell, \tau + \ell/2)$.

We now set up for the covolume formula. Let R be a hyperbolic surface with geodesic boundaries $\beta_1,\ldots,\beta_n$. Consider a weighted multicurve

$$\gamma = \sum_{j=1}^m a_j\gamma_j,$$

where a_j are real weights and γ_j are distinct, disjoint, simple closed geodesics. Define $\operatorname{Stab}(\gamma) \subset \mathrm{MCG}$ to be the mapping classes stabilizing the collection of unlabeled, weighted geodesics - elements of $\operatorname{Stab}(\gamma)$ may permute components of the multicurve with equal weights. Write $\operatorname{Stab}(\gamma_j)$ for the stabilizer of an individual geodesic and $\operatorname{Stab}_0(\gamma_j)$ for the subgroup of elements preserving orientation.

Write $R(\gamma)$ for the surface cut open along the γ - each γ_j gives rise to two new boundaries - $R(\gamma)$ may be disconnected. Write $\mathcal{T}(R(\gamma);\mathbf{x})$, $\mathbf{x} = (x_1,\ldots,x_m)$ for the (product) Teichmüller space of the cut open surface with the pair of boundaries for γ_j having length x_j. Denote by $\mathrm{MCG}(R(\gamma))$ the product of mapping class groups of the components of $R(\gamma)$ and by $\mathcal{T}(R(\gamma);\mathbf{x})/\mathrm{MCG}(R(\gamma))$ the corresponding product of moduli spaces. For the product of symplectic forms on $\mathcal{T}(R(\gamma);\mathbf{x})$ corresponding to the components of $R(\gamma)$, the volume $V(R(\gamma);\mathbf{x})$ is the product of volumes of the component moduli spaces, where again the pair of boundaries for γ_j have common length x_j. Considerations also involve the finite symmetry group

$$\operatorname{Sym}(\gamma) = \operatorname{Stab}(\gamma)/\cap_j \operatorname{Stab}_0(\gamma_j)$$

of mapping classes that possibly permute and reverse orientation of the γ elements.

Summing the translations of a function over a group gives a group action invariant function. Begin with a function f, suitably small at infinity, and introduce the MCG sum

$$f_\gamma(R) = \sum_{\mathrm{MCG}\,/\,\operatorname{Stab}(\gamma)} f\Big(\sum_{j=1}^m a_j\ell_{h(\gamma_j)}(R)\Big). \tag{4}$$

The next theorem expresses the moduli space integral

$$\int_{\mathcal{M}(R)} f_\gamma \, dV$$

as a weighted integral of lower dimensional moduli space volumes.

THEOREM 9. [**Mir07a**, Thrm. 7.1] The covolume formula. *For a weighted* $\gamma = \sum_{j=1}^m a_j\gamma_j$ *and the* MCG *sum of a function* f, *small at infinity, then*

$$\int_{\mathcal{T}(R)/\,\mathrm{MCG}} f_\gamma \, dV = (|Sym(\gamma)|)^{-1} \int_{\mathbb{R}^m_{>0}} f(|\mathbf{x}|)V(R(\gamma);\mathbf{x})\,\mathbf{x}\cdot d\mathbf{x}$$

where $|\mathbf{x}| = \sum_j a_j x_j$ *and* $\mathbf{x}\cdot d\mathbf{x} = x_1\cdots x_m dx_1\cdots dx_m$.

PROOF. Corresponding to the components R' of the cut open surface $R(\gamma)$, consider the short exact sequence for mapping class groups,

$$1 \longrightarrow \prod_j \mathrm{Dehn}(\gamma_j) \longrightarrow \bigcap_j \mathrm{Stab}_0(\gamma_j) \longrightarrow \prod_{R(\gamma)\ components} \mathrm{MCG}(R') \longrightarrow 1,$$

and the associated fibration of Teichmüller spaces from Fenchel-Nielsen coordinates,

$$\begin{array}{ccc} \prod_{R(\gamma)\ components} \mathcal{T}(R') & \hookrightarrow & \mathcal{T}(R) \\ & & \downarrow \\ & & \prod_{\gamma_j} \mathbb{R}_{>0}\times\mathbb{R}\,. \end{array}$$

(The short exact sequence places half Dehn twists in the mapping class groups of the tori with single boundaries.) The $d\ell\wedge d\tau$ formula provides that the fibration is a fibration of symplectic manifolds.

To establish the formula, first write for coset sums

$$\sum_{\mathrm{MCG}\,/\cap_j \mathrm{Stab}_0(\gamma_j)} f = \sum_{\mathrm{MCG}\,/\,\mathrm{Stab}(\gamma)} \ \sum_{\mathrm{Stab}(\gamma)/\cap_j \mathrm{Stab}_0(\gamma_j)} f = |\,\mathrm{Sym}(\gamma)|\, f_\gamma,$$

using that f_γ is $\mathrm{Sym}(\gamma)$ invariant for the second equality. Substitute the resulting formula for f_γ into the integral, and unfold the sum (express the $\mathrm{MCG}\,/\cap_j \mathrm{Stab}_0(\gamma_j)$ translation sum as a sum of translates of a MCG fundamental domain) to obtain the equality

$$\int_{\mathcal{T}(R)/\,\mathrm{MCG}} f_\gamma \, dV = (|\,\mathrm{Sym}(\gamma)|)^{-1} \int_{\mathcal{T}(R)/\cap_j \mathrm{Stab}_0(\gamma_j)} f\, dV.$$

Substitute the fibration

$$\begin{array}{ccc} \prod_{R(\gamma)\ components} \mathcal{T}(R')/\,\mathrm{MCG}(R') & \hookrightarrow & \mathcal{T}(R)/\bigcap_j \mathrm{Stab}_0(\gamma_j) \\ & & \downarrow \\ & & \prod_{\gamma_j} (\mathbb{R}_{>0}\times\mathbb{R})/\mathrm{Dehn}_*(\gamma_j)\,, \end{array}$$

where $\mathrm{Dehn}_*(\gamma_j)$ is generated by a half twist if the curve bounds a torus with a single boundary and otherwise is generated by a simple twist. Substitute the factorization of the volume element

$$dV = \prod_{R(\gamma)\ components} dV(R') \times \prod_{\gamma_j} d\ell_j \wedge d\tau_j.$$

The function f depends only on the values $\mathbf{x}$. For the values $\mathbf{x}$ fixed, perform the integration over $\prod \mathcal{T}(R')/\operatorname{MCG}(R')$ to obtain the product volume $V(R(\gamma);\mathbf{x})$. Finally $\operatorname{Dehn}_*(\gamma_j)$ acts only on the variable τ_j with fundamental domain $0<\tau_j<\ell_j/2$ if γ_j bounds a torus with a single boundary or otherwise with fundamental domain $0<\tau_j<\ell_j$. For a torus with a single boundary, the action is accounted for by using the volume value that is $1/2$ the original $V(L)$. The right hand side of the formula is now established. □

We are now ready to apply the covolume formula to the length identity. The application follows the genus one example. Again - it is essential to use the MCG geodesic-length function action $\ell_\alpha \circ h^{-1} = \ell_{h(\alpha)}$, to consider a sum over topological configurations as a sum of MCG translates of a function. Consider the action on configurations. The mapping class group naturally acts on $\mathcal{B}$, the set of non trivial free homotopy classes of simple curves with endpoints on the boundary, rel the boundary. Recall the correspondences: elements of $\mathcal{B} \Longleftrightarrow$ wire frames $\Longleftrightarrow$ boundary pants $\mathcal{P}$. Viewing Figure 1, the MCG orbits on $\mathcal{B}$ are of three types, describing location of the boundary pants $\mathcal{P}$.

- Orbits for simple curves from β_1 to β_1.
  - A single orbit for $R-\mathcal{P}$ connected, with $\operatorname{Sym}(\beta_1,\alpha_1,\alpha_2)=\mathbb{Z}/2\mathbb{Z}$.
  - A collection of orbits for $R-\mathcal{P}$ disconnected. The orbits are classified by the joint partitions of genus $g=g_1+g_2$ and of labeled boundaries $\{\beta_2,\dots,\beta_n\}$, with each resulting component with negative Euler characteristic. In general $\operatorname{Sym}(\beta_1,\alpha_1,\alpha_2)=1$, except in the special case of one boundary and $g_1=g_2$.
- A collection of orbits, one for each choice of a second boundary. In particular, an orbit for simple curves from β_1 to β_j, $j\neq 1$. A resulting surface $R-\mathcal{P}$ is connected, with $\operatorname{Sym}(\beta_1,\beta_j,\alpha)=1$.

Consider Theorem 8, and integrate each side of the identity

$$L_1 = \sum_{\alpha_1,\alpha_2} \mathcal{D}(L_1,\ell_{\alpha_1}(R),\ell_{\alpha_2}(R)) + \sum_{j=2}^{n}\sum_{\alpha} \mathcal{R}(L_1,L_j,\ell_\alpha(R)),$$

over the moduli space of R relative to the volume dV. Form the L_1 partial derivative of each side to simplify the quantities $\mathcal{D}$ and $\mathcal{R}$. Apply formulas (3) for the right hand side. Express the right hand side as individual sums for given orbit types. Apply Theorem 9 for each orbit type to find integrals in terms of lower dimensional moduli volumes as follows.

- For the sum over simple curves from β_1 to β_1 with $R-\mathcal{P}$ connected, the summand function is $\mathcal{D}$ with

$$\frac{\partial \mathcal{D}}{\partial L_1} = H(\ell_{\alpha_1}+\ell_{\alpha_2},L_1),$$

  a function of length of a multicurve, and the resulting integral is

$$\int_0^\infty\int_0^\infty H(x+y,L_1)V_{g-1}(x,y,\widehat{L})\,xy\,dxdy,$$

  for $\widehat{L}=(L_2,\dots,L_n)$.

- For the sum over simple curves from β_1 to β_1 with $R - \mathcal{P}$ disconnected, the summand function is $\mathcal{D}$ with
$$\frac{\partial \mathcal{D}}{\partial L_1} = H(\ell_{\alpha_1} + \ell_{\alpha_2}, L_1),$$
a function of length of a multicurve, and the resulting integral is
$$\int_0^\infty \int_0^\infty \sum_{\substack{g_1+g_2=g \\ I_1 \amalg I_2 = \{2,\ldots,n\}}} H(x+y, L_1) V_{g_1}(x, L_{I_1}) V_{g_2}(y, L_{I_2})\, xy\, dxdy.$$
- For the sum over simple curves from β_1 to β_j, $j \neq 1$, the summand function is $\mathcal{R}$ with
$$\frac{\partial \mathcal{R}}{\partial L_1} = \frac{1}{2} H(\ell_\alpha, L_1 + L_j) + \frac{1}{2} H(\ell_\alpha, L_1 - L_j),$$
a sum of functions of weighted length of a multicurve, and the resulting integral is
$$\int_0^\infty \frac{1}{2} \sum_{j=2}^{n} \big(H(x, L_1 + L_j) + H(x, L_1 - L_j)\big) V_g(x, L_2, \ldots, \widehat{L_j}, \ldots, L_n)\, x\, dx,$$
where L_j is omitted from the argument list of V_g.

Compare to the end of Lecture 1 - the volume recursion is established. The volume *function* $V_{g,n}(L)$ is recursively determined.

What type of function is $V_{g,n}(L)$? The recursion involves two elementary integrals, see formula (1) above for the definition of H,
$$\int_0^\infty x^{2j+1} H(x,t)\, dx \qquad \text{and} \qquad \int_0^\infty \int_0^\infty x^{2j+1} y^{2k+1} H(x+y, t)\, dxdy.$$
By direct calculation, see [**Mir07a**, formula (6.2) and Lemma 6.2], each integral is a polynomial in t^2 with each coefficient a product of factorials and the Riemann zeta function at a non negative even integer, each coefficient is a positive rational multiple of an appropriate power of π. The first polynomial has degree $j+1$ in t^2, while the second has degree $i+j+2$. The first part of Theorem 4 is now established. The second part is the subject of the next lecture.

Symplectic reduction, principal S^1 bundles and the normal form

The goal of the lecture is to establish the following theorem. The formula combines with Theorem 3, the WP kappa equation $\omega = 2\pi^2\kappa_1$, to provide that the coefficients of the volume polynomial $V_g(L)$ are tautological intersection numbers. The result completes the proof of Theorem 4.

THEOREM 10. *For* $d = \frac{1}{2}\dim_{\mathbb{R}} \mathcal{T}_g(L) - \dim_{\mathbb{C}} \mathcal{T}_{g,n}$, *then*

$$V_g(L) = \frac{1}{d!}\int_{\mathcal{T}_g(L)/\,\mathrm{MCG}} \omega^d_{\mathcal{T}_g(L)} = \frac{1}{d!}\int_{\mathcal{T}_{g,n}/\,\mathrm{MCG}} \Big(\omega + \sum_{j=1}^{n} \frac{L_j^2}{2}\psi_j\Big)^d.$$

The proof is essentially by establishing a cohomology equivalence between symplectic spaces - combining symplectic reduction, the Duistermaat-Heckman theorem and explicit geometry to obtain the formula.

The considerations of the lecture are presented for the appropriate Teichmüller spaces $\mathcal{T}$ and the open moduli spaces $\mathcal{T}/\,\mathrm{MCG}$. The constructions are compatible with the augmentation construction. The results are valid for the appropriate Teichmüller spaces and compactified moduli spaces. The compactified moduli spaces are orbifolds. The following results are for cohomology statements over $\mathbb{Q}$; cohomology arguments over $\mathbb{Q}$ for manifolds are in general also valid for orbifolds. Alternatively, the orbifold matter can be bypassed by applying the general result that compactified moduli spaces have manifold finite covers, [**BP00, Loo94**].

The Teichmüller spaces. We consider the trio.

- $\mathcal{T}_g(L)$ - the space of marked genus g hyperbolic surfaces, with geodesic boundaries $\beta_1, \dots, \beta_n$ of prescribed lengths $L_1, \dots, L_n$. A hyperbolic surface can be doubled across its geodesic boundary to obtain a compact hyperbolic surface of higher genus. Accordingly, $\mathcal{T}_g(L)$ can be considered as a locus in $\mathcal{T}_{2g+n-1}$. The symplectic form ω of the image Teichmüller space restricts to the locus and defines a symplectic form on the locus. A pants decomposition for a surface with boundary, can be doubled to give a pants decomposition for a doubled surface. The doubled pants decomposition is characterized by containing the geodesics $\beta_1, \dots, \beta_n$ and being symmetric. Fenchel-Nielsen coordinates and the $d\ell \wedge d\tau$ formula are applied for doubled decompositions to obtain a description of the locus $\mathcal{T}_g(L) \subset \mathcal{T}_{2g+n-1}$, and to define a symplectic from $\omega_{\mathcal{T}_g(L)}$. The symplectic form is given as $\sum_j d\ell_j \wedge d\tau_j$ (without boundary parameters) for any pants decomposition of a surface with boundaries. MCG invariance is immediate.
- $\widehat{\mathcal{T}}_{g,n}$ - the space of marked genus g hyperbolic surfaces, with pointed geodesic boundaries $\beta_1, \dots, \beta_n$ - boundary lengths are allowed to vary *and*

a variable point is given on each boundary. The $\mathbb{R}$ dimension of $\widehat{\mathcal{T}}_{g,n}$ is $2n$ greater than the $\mathbb{R}$ dimension of $\mathcal{T}_g(L)$. Symplectic reduction requires a symplectic form on $\widehat{\mathcal{T}}_{g,n}$, that is equivalent to $\omega_{\mathcal{T}_g(L)}$ on L level sets and is invariant under rotating points on boundaries. A form is given by describing $\widehat{\mathcal{T}}_{g,n}$ as a higher dimensional Teichmüller space.

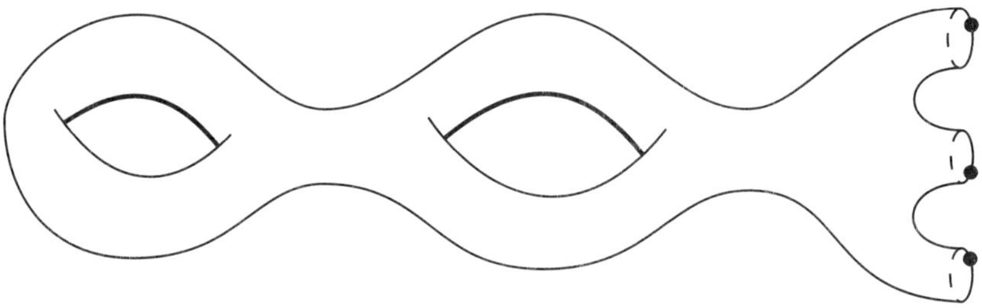

FIGURE 8. A genus 2 surface with three pointed boundaries.

To this purpose, introduce almost tight pants - pairs of pants with two labeled boundaries being punctures (length zero) and a third boundary of prescribed length. An almost tight pants will be glued to each surface boundary β_j. The pants equatorial reflection defines symmetric points on the pants boundary; the puncture labeling uniquely determines an equatorial boundary point. A standard model for a surface with pointed geodesic boundaries is given by gluing on almost tight pants - at each boundary glue on an almost tight pants with matching boundary length and the equatorial point aligned with the point on the boundary. The construction does not involve choices, so is natural with respect to marking homeomorphisms and the MCG action.

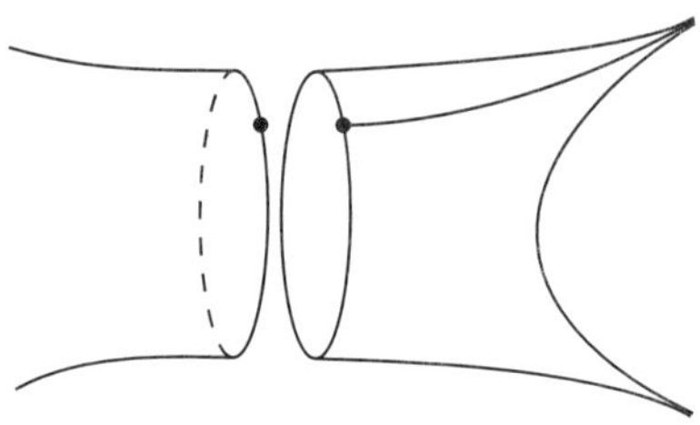

FIGURE 9. Aligning boundary and equatorial points to glue on almost tight pants.

For a surface R with labeled, pointed boundaries, write $\widehat{R}$ for the standard model surface with glued on almost tight pants. The punctures of $\widehat{R}$ are labeled in pairs. By the general hyperbolic collar result, small length geodesics are necessarily disjoint [**Bus92, Wlp10**]. For the lengths $L_1, \ldots, L_n$ suitably small, the labeled geodesics $\beta_1, \ldots, \beta_n$ are uniquely determined on the surface $\widehat{R}$ by having small length and bounding labeled punctures. The pointed boundary, marked surface R *is equivalent* to the marked surface $\widehat{R}$ modulo Dehn twists about the β_j (Dehn twists, since the boundary points are given on a circle). In particular, for c suitably small, the equivalence is between the open subset $\{L < c\}$ of $\widehat{\mathcal{T}}_{g,n}$ and the open subset $\{L < c\}$ in $\mathcal{T}_{g,2n}/\prod_j \mathrm{Dehn}(\beta_j)$.

Definition and equivalence of S^1 principal bundles are next. Considerations begin with the short exact sequence from Theorem 9,

$$1 \to \mathrm{Dehn}(\beta) = \prod_j \mathrm{Dehn}(\beta_j) \to \mathrm{Stab}(\beta) = \bigcap_j \mathrm{Stab}(\beta_j) \to \mathrm{MCG}(R) \to 1,$$

(now $\mathrm{Stab}(\beta_j) = \mathrm{Stab}_0(\beta_j)$, since an orientation preserving pants homeomorphism preserves boundary orientation). The geodesics $\beta_1, \dots, \beta_n$ define subsets of the Riemann surface bundles (the universal curves) over $\widehat{\mathcal{T}}_{g,n}$ and over $\mathcal{T}_{g,2n}/\mathrm{Dehn}(\beta)$. The subsets define oriented circle bundles, provided automorphisms of the Riemann surfaces act at most as rotations on the individual geodesics. The small lengths $L_1, \dots, L_n$ and labeled boundaries provide the condition. The geodesics define circle bundles over Teichmüller bases. We see below that rotation along geodesics defines an S^1 principal structure. Next, from the above short exact sequence and the definition of marking - the equivalence between geodesics $\beta \subset R$ and $\beta \subset \widehat{R}$, and the projections of circle bundles to bases, commute with the actions of $\mathrm{MCG}(R) \approx \mathrm{Stab}(\beta)/\mathrm{Dehn}(\beta)$. The geodesics define equivalent orbifold S^1 principal bundles over the quotients $\{L < c\} \subset \widehat{\mathcal{T}}_{g,n}/\mathrm{MCG}(R) \times (S^1)^n$ (the $\mathrm{MCG}(R)$ and $(S^1)^n$ actions on $\widehat{\mathcal{T}}_{g,n}$ commute) and $\{L < c\} \subset \mathcal{T}_{g,2n}/\mathrm{Stab}(\beta)$.

The symplectic form of $\mathcal{T}_{g,2n}$ defines a symplectic form $\omega_{\widehat{\mathcal{T}}_{g,n}}$ on the open subset $\{L < c\}$. Fenchel-Nielsen coordinates and the $d\ell \wedge d\tau$ formula are applied. The form $\omega_{\widehat{\mathcal{T}}_{g,n}}$ is given as $\sum_k d\ell_k \wedge d\tau_k$ for any pants decomposition of $\widehat{R}$ containing the multicurve β. Importantly, the form $\omega_{\widehat{\mathcal{T}}_{g,n}}$ is given for surfaces R with pointed boundaries β_j, by an extended interpretation of the $d\ell \wedge d\tau$ formula, with a term for each boundary, now with the interpretation that $\tau(\beta_j)$ parameterizes the location of the specified point. See Figure 3, the parameter $\tau(\beta_j)$ increasing corresponds to the point moving on the boundary with the surface interior on the right. $\mathrm{MCG}(R)$ invariance of the symplectic form is immediate. Restriction of the form to L level sets and invariance under rotating boundary points are discussed below.

- $\mathcal{T}_{g,n}$ - the space of marked genus g hyperbolic surfaces with n punctures. $\mathcal{T}_{g,n}$ has the MCG invariant symplectic form ω. $\mathcal{T}_{g,n}$ coincides with the Teichmüller space $\mathcal{T}_g(0)$, where surface boundary lengths are zero.

We will relate the three symplectic manifolds.

Symplectic reduction for $\widehat{\mathcal{T}}_{g,n}$. We consider the Hamiltonian geometry of FN twists, geodesic-lengths and especially the moment map

$$\widehat{\mathcal{T}}_{g,n} \xrightarrow{\mu} \widehat{L} = (L_1^2/2, \dots, L_n^2/2) \in \mathbb{R}^n_{\geq 0}.$$

Write t_j for the unit speed infinitesimal rotation of the point on the boundary β_j; t_j is a vector field on $\widehat{\mathcal{T}}_{g,n}$. In terms of the standard model surfaces, t_j is the FN infinitesimal twist vector field for β_j, and t_j is the infinitesimal rotation of the jth almost tight pants. By twist-length duality, we have $\omega_{\widehat{\mathcal{T}}_{g,n}}(-t_j, \ ) = dL_j$ and the scaled $-L_j t_j$ is unit infinitesimal rotation (unit time flow is a full rotation). The function $\frac{1}{2}L_j^2$ is the corresponding Hamiltonian potential, since $\omega_{\widehat{\mathcal{T}}_{g,n}}(-L_jt_j,\ ) = d(\frac{1}{2}L_j^2)$ (the momentum $\frac{1}{2}L^2$ determines the twist sign/orientation). The vector fields $-L_jt_j$ are the infinitesimal generators for the $(S^1)^n$ action on $\widehat{\mathcal{T}}_{g,n}$, given by rotating the boundary points. The symplectic form $\omega_{\widehat{\mathcal{T}}_{g,n}}$ is twist invariant and we are ready for symplectic reduction, ready to consider the quotient $\widehat{\mathcal{T}}_{g,n}/(S^1)^n$.

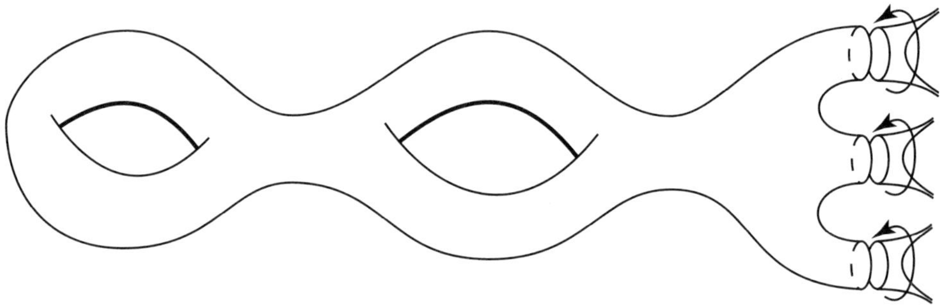

FIGURE 10. Positive rotations for an $(S^1)^3$ action.

A level set of the moment map $\mu : \widehat{\mathcal{T}}_{g,n} \longrightarrow \mathbb{R}^n$ is a locus of prescribed β length hyperbolic surfaces. The group $(S^1)^n$ acts on level sets by rotating almost tight pants. The quotient of a level set by the group is naturally $\mathcal{T}_g(L)$ - the level set prescribes the boundary lengths and the group action removes the location information for the points.

PROPOSITION 11. Symplectic reduction. *For small L, the reduced level set $\mu^{-1}(\widehat{L})/(S^1)^n$ is diffeomorphic to $\mathcal{T}_g(L)$ with*

$$\omega_{\widehat{\mathcal{T}}_{g,n}}\Big|_{\mu^{-1}(\widehat{L})}/(S^1)^n \approx \omega_{\mathcal{T}_g(L)}.$$

PROOF. The form $\omega_{\widehat{\mathcal{T}}_{g,n}}$ is given by the $d\ell \wedge d\tau$ formula for any pants decomposition of a standard model surface containing the multicurve β. The differentials dL_j vanish on μ level sets and the formula reduces to the sum for a pants decomposition of a surface with boundary, a sum without boundary parameters - the $\omega_{\mathcal{T}_g(L)}$ formula. □

S^1 **principal bundles.** We review basics about characteristic classes.

DEFINITION 12. *Let $\pi : P \longrightarrow M$ be a smooth circle bundle over a smooth compact manifold M. The bundle is S^1 principal provided,*

(1) *S^1 acts freely on P,*
(2) *$\pi(p_1) = \pi(p_2)$ if and only if there exists $s \in S^1$, such that $p_1 \cdot s = p_2$.*

A connection for an S^1 principal bundle is a smooth distribution $\mathcal{H} \subset \mathbf{T}P$ of tangent subspaces such that,

(1) *$\mathbf{T}_pP = \mathcal{H}_p \oplus \ker \pi_*\big|_p$, for each $p \in P$,*
(2) *$s^*\mathcal{H}_p = \mathcal{H}_{p\cdot s}$.*

A connection is uniquely given as $\mathcal{H} = \ker A$, for a 1-form A on P, provided A is S^1 invariant and $A(\dot{s}) = 1$. An S^1 invariant inner product $\langle\ ,\ \rangle$ provides an example of an invariant 1-form by $A(v) = \langle v, \dot{s}\rangle/\langle \dot{s}, \dot{s}\rangle$. The curvature 2-form on P for a connection is $\Phi(v,w) = dA(\operatorname{hor} v, \operatorname{hor} w)$, for hor the *horizontal* projection of $\mathbf{T}P$ to $\mathcal{H}$.

THEOREM 13 ([**MS74**]). *There exists a unique closed 2-form Ω on M, such that $\Phi = \pi^*\Omega$. The cohomology class of Ω is independent of the choice of S^1 principal connection for P and the first Chern class is $c_1(P) = [\Omega] \in H^2(M,\mathbb{Z})$.*

As above, the variable point on the boundary β_j of the surface R defines an S^1 principal bundle $\widehat{\beta}_j$ over $\mathcal{T}_g(L)$; S^1 acts by moving the point with the surface interior on the left. A choice of connection for the bundle gives a first Chern class $c_1(\widehat{\beta}_j)$.

Applying the Duistermaat-Heckman theorem. We must extend the definition of $\widehat{\mathcal{T}}_{g,n}$ to include $L = 0$; geodesic boundaries of R can be replaced with punctures. Hyperbolic structures converge for boundary lengths tending to zero; in particular collar regions converge to cusp regions. The extension of $\widehat{\mathcal{T}}_{g,n}$ is given by parameterizing boundary points by points on a collar/cusp region boundary.

We recall basics about collars and cusps. For a geodesic α of length ℓ_α, the standard collar in the upper half plane $\mathbb{H}$ is $\mathcal{C}(\ell_\alpha) = \{d(z, i\mathbb{R}_+) \leq w(\alpha)\}$, for the half width $w(\alpha)$ given by $\sinh w(\alpha) \sinh \ell_\alpha/2 = 1$. The quotient cylinder $\{d(z, i\mathbb{R}_+) \leq w(\alpha)\}/\langle z \mapsto e^{\ell_\alpha} z\rangle$ embeds into R to give a collar neighborhood of the geodesic. For a cusp, the standard cusp in $\mathbb{H}$ is $\mathcal{C}_\infty = \{\Im z \geq 1/2\}$. The quotient cylinder $\{\Im z \geq 1/2\}/\langle z \mapsto z+1\rangle$ embeds into R to give a cusp region. The boundary of a collar, for ℓ_α bounded, and boundary of a cusp region have length approximately 2. Collars and cusp regions are foliated by geodesics normal to the boundary. For geodesic-lengths tending to zero, half collar neighborhoods Gromov-Hausdorff converge to a cusp region (convergence is uniform on bounded distance neighborhoods of the boundary); boundaries and geodesics normal to the boundary converge

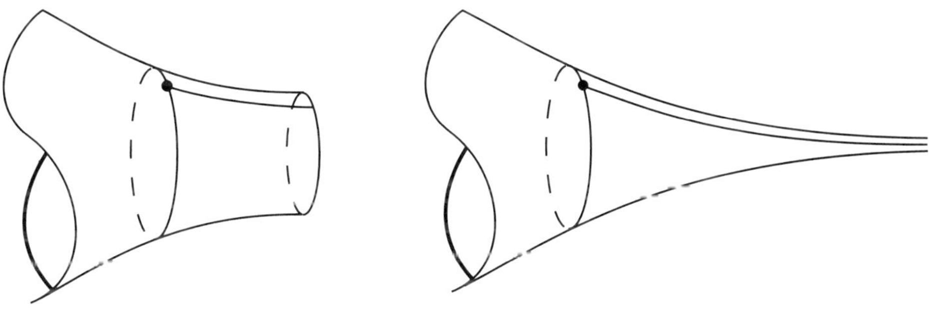

FIGURE 11. Projecting along geodesics to a collar and a cusp region boundary.

The geodesics normal to the boundary of a collar provide a projection from the core geodesic to each collar boundary. The projection is used to note that prescribing a point on a geodesic boundary of R is equivalent to prescribing a point on the boundary of the half collar neighborhood of the geodesic. Since collars and their boundaries converge to a cusp region and its boundary, for core geodesic length tending to zero, we have a description for the extension of the definition of $\widehat{\mathcal{T}}_{g,n}$ to include surfaces with collections of lengths L_j zero.

The standard cusp region is uniformized by the variable $w = e^{2\pi i z}$. A point on the cusp region boundary $\Im z = 1/2$ corresponds to a point on $|w| = e^{-\pi}$ and given the factor $e^{-\pi}$, a point on the circle corresponds to a tangent vector at the origin. The variable w is unique modulo multiplication by a unimodular number; the identification of the circle with tangent vectors at the origin is canonical. For an S^1 infinitesimal generator $-L_j t_j$, displacement is to the left when crossing the geodesic, (compare to the Figure 3 positive twist, right displacement) the reference point moves with the surface interior on its left, the tangent vector at the origin rotates clockwise, and a dual cotangent vector rotates counter clockwise (the positive direction for a $\mathbb{C}$-line). Combining equivalences, the S^1 principal bundle of a point on a cusp region boundary is equivalent to a non zero vector in the cotangent line for the puncture. -Combining considerations, the 0 level set of μ in $\widehat{\mathcal{T}}_{g,n}$ is equivalent to an $(S^1)^n$ bundle over $\mathcal{T}_{g,n}$, the collection of normalized cotangent vectors at punctures of surfaces.

Proof of Theorem 10. The $\mathrm{MCG}(R)$ and $(S^1)^n$ actions on $\widehat{\mathcal{T}}_{g,n}$ commute; consider the quotient $\widehat{\mathcal{T}}_{g,n}/\mathrm{MCG}(R)\times(S^1)^n$. By the Duistermaat-Heckman theorem, [**CdS01**, Chapter 30, Theorem 30.8], for small values of $\widehat{L}$, including 0, the reduced level sets $\mu^{-1}(\widehat{L})/(S^1)^n$ are mutually diffeomorphic. Furthermore by Duistermaat-Heckman, the L level set reduced symplectic form $\omega_{\widehat{\mathcal{T}}_{g,n}}\big|_{\mu^{-1}(\widehat{L})}/(S^1)^n$ is cohomologous to the sum of the 0 level set reduced form $\omega_{\widehat{\mathcal{T}}_{g,n}}\big|_{\mu^{-1}(0)}/(S^1)^n$ and the contributions $(L_j^2/2)\,c_1(\widehat{\beta}_j)$, for $c_1(\widehat{\beta}_j)$ the first Chern class for the S^1 principal bundle of a point varying on the jth cusp region boundary. Combining with Proposition 11, gives the desired cohomology equivalence,

$$\omega_{\mathcal{T}_g(L)} \equiv \omega_{\mathcal{T}_g(0)} + \sum_{j=1}^{n}\frac{L_j^2}{2}c_1(\widehat{\beta}_j).$$

By the description of collars and cotangent lines at punctures, the circle bundle $\widehat{\beta}_j$ is topologically equivalent to the cotangent lines at punctures (see Lecture 1) with equality of first Chern classes. The proof is finished.

LECTURE 5

The pattern of intersection numbers and Witten-Kontsevich

We begin with the discussion of Harris-Morrison [**HM98**, pgs. 71-75]. For a finite sequence of non negative integers $\{\alpha_j\}$, define the top ψ-intersection number by

$$\langle \tau_{\alpha_1} \tau_{\alpha_2} \cdots \tau_{\alpha_n} \rangle_g = \int_{\overline{\mathcal{M}}_{g,n}} \psi_1^{\alpha_1} \psi_2^{\alpha_2} \cdots \psi_n^{\alpha_n}.$$

For a non trivial pairing, the genus g, number of punctures n, and exponents α_j are related by $3g-3+n = \sum_{j=1}^{n} \alpha_j$, otherwise the pairing is defined as zero. More generally using exponents to denote powers (repetitions) of the variables τ, define

$$\langle \tau_0^{d_0} \tau_1^{d_1} \cdots \tau_m^{d_m} \rangle_g = \int_{\overline{\mathcal{M}}_{g,d}} \prod_{j=0}^{m} \prod_{k=1}^{d_j} \psi_{(j,k)}^j.$$

($\tau_j^{d_j}$ denotes that for d_j punctures, the associated ψ is raised to the jth power; the subscripts (j,k) are distinct puncture labels.) For the second pairing, the formal count of punctures is $d = \sum_{j=0}^{m} d_j$, and the formal degree of the product is the count of ψ factors $\sum_{j=0}^{m} jd_j$. For a non trivial pairing, the genus, number of punctures and degree are related by $3g-3+d = \sum_{j=0}^{m} jd_j$, otherwise the pairing is zero. The psi classes are known to be positive - integrals of products over subvarieties are positive; the non trivial pairings are positive, matching Mirzakhani's positivity of volume polynomial coefficients, see Theorem 4.

Witten considered a partition function (probability of states), for two-dimensional gravity. For an infinite vector $\mathbf{t} = (t_0, t_1, \dots, t_n, \dots)$, and γ the formal sum $\gamma = \sum_{j=0}^{\infty} t_j \tau_j$, Witten introduced a genus g generating function for τ products,

$$F_g(\mathbf{t}) = \sum_{n=0}^{\infty} \frac{\langle \gamma^n \rangle_g}{n!},$$

in which the numerator is defined by monomial expansion, resulting in the formal power series

$$F_g(\mathbf{t}) = \sum_{\{d_j\}} \langle \prod_{j=0}^{\infty} \tau_j^{d_j} \rangle_g \prod_{j=0}^{\infty} \frac{t_j^{d_j}}{d_j!},$$

where the sum is over all sequences of non negative integers $\{d_j\}$ with only finitely many non zero terms. By Theorem 4, the intersection numbers $\langle \tau_0^{d_0} \tau_1^{d_1} \cdots \tau_m^{d_m} \rangle_g$ are the coefficients of the leading terms of the volume polynomials $V_{g,n}(L)$.

The quantum gravity partition function is

$$\mathbf{F}(\lambda, \mathbf{t}) = \sum_{g=0}^{\infty} \lambda^{2g-2} F_g(\mathbf{t}).$$

Based on a realization of the function in terms of matrix integrals, Witten conjectured that the partition function should satisfy two forms of the Korteweg-deVries (KdV) equations. Kontsevich gave a proof of the conjecture using a cell decomposition of the moduli spaces $\mathcal{M}_{g,n}$, [**Kon92**]. Cells are enumerated by ribbon graphs/fat graphs. Kontsevich encoded the intersection numbers in an enumeration of trivalent ribbon graphs. He then used Feynman diagram techniques and a matrix Airy integral to establish Witten's conjectures.

Two basic relations for the intersection numbers are: for $n > 0$ and $\sum_i \alpha_i = 3g - 2 + n > 0$, the

$$\text{string equation} \qquad \langle \tau_0 \tau_{\alpha_1} \cdots \tau_{\alpha_n} \rangle_g = \sum_{\alpha_i \neq 0} \langle \tau_{\alpha_1} \cdots \tau_{\alpha_i - 1} \cdots \tau_{\alpha_n} \rangle_g,$$

and for $n \geq 0$ and $\sum_i \alpha_i = 3g - 3 + n \geq 0$, the

$$\text{dilaton equation} \qquad \langle \tau_1 \tau_{\alpha_1} \cdots \tau_{\alpha_n} \rangle_g = (2g - 2 + n) \langle \tau_{\alpha_1} \cdots \tau_{\alpha_n} \rangle_g.$$

The first equation is for adding a new puncture without an associated factor of ψ in the product, while the second equation is for adding a new puncture with a single associated factor of ψ.

Similar to setting $V_{0,3}(L) = 1$, the intersection symbol for the thrice punctured sphere is normalized to $\langle \tau_0^3 \rangle_0 = 1$. The general genus 0 formula is

$$\langle \tau_{\alpha_1} \cdots \tau_{\alpha_n} \rangle_0 = \left(\frac{(n-3)!}{\alpha_1! \cdots \alpha_n!} \right),$$

with the right hand side a multinomial coefficient for $n - 3$. The genus 0 string equation is simply Pascal's multinomial neighbor relation.

For genus 1, Theorem 10 and the WP kappa equation, Theorem 3, give $V(L) = \int_{\overline{\mathcal{M}}_{1,1}} 2\pi^2 \kappa_1 + \frac{L^2}{2}\psi$. The formula combines with the Lecture 3 calculation $V_{1,1}(L) = \frac{\pi^2}{12} + \frac{L^2}{48}$ (now including the elliptic involution $\frac{1}{2}$ factor) to provide the evaluations,

$$\frac{1}{2} \int_{\overline{\mathcal{M}}_{1,1}} \kappa_1 = \langle \tau_1 \rangle_1 = \frac{1}{24}.$$

General genus 1 evaluations are found from the single evaluation by applying the string and dilaton equations. A consequence of the Witten conjecture is that all $\langle \tau \rangle$ intersections can be calculated from the initial values $\langle \tau_0^3 \rangle_0 = 1$ and $\langle \tau_1 \rangle_1 = \frac{1}{24}$, using the Virasoro equations $L_n(e^{\mathbf{F}}) = 0$ for the partition function described below.

In the volume recursion, leading coefficients are obtained from leading coefficients - the recursion specializes to leading coefficients, see [**Mir07b**, Lemma 5.3]. We now sketch the application of the specialized recursion to relations for the partition function and a solution of Witten's conjecture.

Relations come from the Virasoro Lie algebra. The Witt subalgebra is generated by the differential operators $\mathcal{L}_n = -z^{n+1}\partial/\partial z$, $n \geq -1$, with commutators $[\mathcal{L}_n, \mathcal{L}_m] = (n-m)\mathcal{L}_{n+m}$. The string and dilaton equations can be written as linear homogeneous differential equations for the exponential $e^{\mathbf{F}}$ of the partition function.

The differential operator for the string equation is

$$L_{-1} = -\frac{\partial}{\partial t_0} + \frac{\lambda^{-2}}{2} t_0^2 + \sum_{j=0}^{\infty} t_{j+1} \frac{\partial}{\partial t_j},$$

and the differential operator for the dilaton equation is

$$L_0 = -\frac{3}{2}\frac{\partial}{\partial t_1} + \sum_{j=0}^{\infty} \frac{2j+1}{2} t_j \frac{\partial}{\partial t_j} + \frac{1}{16}.$$

With simple conditions, there is a unique way to extend operator definitions to obtain a representation of the $\{\mathcal{L}_n\}$ subalgebra. The general operator is

$$\begin{aligned} L_n = & -\frac{(2n+3)!!}{2^{n+1}} \frac{\partial}{\partial t_{n+1}} + \sum_{j=0}^{\infty} \frac{(2j+2n+1)!!}{(2j-1)!!\,2^{n+1}} t_j \frac{\partial}{\partial t_{j+n}} \\ & + \frac{\lambda^2}{2} \sum_{j=0}^{n-1} \frac{(2j+1)!!(2n-2j-1)!!}{2^{n+1}} \frac{\partial^2}{\partial t_j \partial t_{n-j-1}}, \end{aligned}$$

with commutator $[L_n, L_m] = (n-m)L_{n+m}$.

THEOREM 14 ([**Mir07b**, Theorem 6.1]). The Witten-Kontsevich conjecture: Virasoro constraints. *For* $n \geq -1$, *then* $L_n(e^{\mathbf{F}}) = 0$.

PROOF. For an exponents multi index $\mathbf{k} = (k_1, \ldots, k_n)$, the volume recursion formula becomes the coefficient relation

$$(2k_1+1)V_{g,n}(L)[\mathbf{k}] = \mathcal{A}_{g,n}^{con}(L)[\mathbf{k}] + \mathcal{A}_{g,n}^{dcon}(L)[\mathbf{k}] + \mathcal{B}_{g,n}(L)[\mathbf{k}].$$

The leading coefficient relation takes the following explicit form (following the labeling of boundaries, the punctures are labeled $1, \ldots, n$)

$$\begin{aligned} &(2k_1+1)!!\langle \tau_{k_1} \cdots \tau_{k_n} \rangle \\ &\quad = \frac{1}{2} \sum_{i+j=k_1-2} (2i+1)!!(2j+1)!! \sum_{I \subset \{2,\ldots,n\}} \langle \tau_i \tau_{\mathbf{k}_I} \rangle \langle \tau_j \tau_{\mathbf{k}_{I^c}} \rangle \\ &\quad\quad + \frac{1}{2} \sum_{i+j=k_1-2} (2i+1)!!(2j+1)!! \langle \tau_i \tau_j \tau_{k_2} \cdots \tau_{k_n} \rangle \\ &\quad\quad + \sum_{j=2}^{n} \frac{(2k_1+2k_j-1)!!}{(2k_j-1)!!} \langle \tau_{k_2} \cdots \tau_{k_1+k_j-1} \cdots \tau_{k_n} \rangle. \end{aligned}$$

Rearranging the explicit relation provides that $L_{k_1-1}(e^{\mathbf{F}}) = 0$. $\square$

Mulase and Safnuk consider a generating function for the intersections of combinations of the κ_1 and ψ classes [**MS08**]

$$\mathbf{G}(s, t_0, t_1, \ldots) = \sum_g \langle e^{s\kappa_1 + \sum t_j \tau_j} \rangle_g = \sum_g \sum_{m,\{d_j\}} \langle \kappa_1^m \tau_0^{d_0} \tau_1^{d_1} \cdots \rangle_g \frac{s^m}{m!} \prod_{j=0}^{\infty} \frac{t_j^{d_j}}{d_j!},$$

where again products, other than $3g-3+n$-products, are defined as zero. Mulase and Safnuk use the volume recursion and rearrangement of terms to prove the following.

THEOREM 15 ([**MS08**, Thrm. 1.1], Virasoro constraints). *For each $k \geq -1$, define*

$$\mathcal{V}_k = -\frac{1}{2}\sum_{i=0}^{\infty}(2(i+k)+3)!!\frac{(-2s)^i}{(2i+1)!}\frac{\partial}{\partial t_{i+k+1}} + \frac{1}{2}\sum_{j=0}^{\infty}\frac{(2(j+k)+1)!!}{(2j-1)!!}t_j\frac{\partial}{\partial t_{j+k}}$$
$$+\frac{1}{4}\sum_{\substack{d_1+d_2=k-1\\ d_1,d_2\geq 0}}(2d_1+1)!!(2d_2+1)!!\frac{\partial^2}{\partial t_{d_1}\partial t_{d_2}} + \frac{\delta_{k,-1}t_0^2}{4} + \frac{\delta_{k,0}}{48},$$

for the double factorial and Kronecker delta function $\delta_{*,*}$. Then

- *the operators $\mathcal{V}_k$ satisfy the Virasoro commutator relations $[\mathcal{V}_n, \mathcal{V}_m] = (n-m)\mathcal{V}_{n+m}$;*
- *the generating function* **G** *satisfies $\mathcal{V}_k(e^{\mathbf{G}}) = 0$ for $k \geq -1$.*

The initial conditions and second system of equations uniquely determine the generating function.

In a direct display that the intersection numbers for κ_1 and ψ classes are equivalent intersection numbers for ψ classes, Mulase and Safnuk show that

$$\mathbf{G}(s, t_0, t_1, t_2, t_3 \dots) = \mathbf{F}(t_0, t_1, t_2 + \gamma_2, t_3 + \gamma_3, \dots),$$

where $\gamma_j = -(-s)^{j-1}/(2j+1)j!$ [**MS08**, Thrm. 1.2]. An explicit proof of the relation also comes from a formula of Faber, expressing kappa classes in terms of psi classes on moduli spaces for a greater number of punctures.

In his thesis [**Do08**], Norman Do presents a

$$\text{generalized string equation} \qquad V_{g,n+1}(L, 2\pi i) = \sum_{k=1}^{n}\int L_k V_{g,n}(L)\, dL_k,$$

and

$$\text{generalized dilaton equation} \qquad \frac{\partial V_{g,n+1}}{\partial L_{n+1}}(L, 2\pi i) = 2\pi i(2g-2+n)V_{g,n}(L),$$

where on the left hand side, the value $2\pi i$ is substituted for the $(n+1)^{st}$ boundary length and $L = (L_1, \dots, L_n)$. By Theorem 4, the second equation, for appropriate non negative multi indices $\alpha = (\alpha_1, \alpha_2, \dots, \alpha_n)$ and integers m, is equivalent to the relations

$$\int_{\overline{\mathcal{M}}_{g,n+1}} \psi_1^{\alpha_1}\psi_2^{\alpha_2}\cdots\psi_n^{\alpha_n}\psi_{n+1}(\kappa_1 - \psi_{n+1})^m = (2g-2+n)\int_{\overline{\mathcal{M}}_{g,n}} \psi_1^{\alpha_1}\psi_2^{\alpha_2}\cdots\psi_n^{\alpha_n}\kappa_1^m.$$

A proof of the generalized equations is based on the pullback relations for psi and kappa classes, and general considerations for images of classes. In particular for $\pi : \overline{\mathcal{M}}_{g,n+1} \longrightarrow \overline{\mathcal{M}}_{g,n}$, the morphism of forgetting the last puncture, then the classes $\widetilde{\kappa}_m$, $\widetilde{\psi}_k$ on $\overline{\mathcal{M}}_{g,n+1}$ and κ_m, ψ_k on $\overline{\mathcal{M}}_{g,n}$, satisfy $\widetilde{\kappa}_m = \pi^*\kappa_m + \psi_{n+1}^m$ and $\widetilde{\psi}_k = \pi^*\psi_k, +D_k$ (for D_k the divisor of the kth puncture on the universal curve over $\overline{\mathcal{M}}_{g,n}$), [**HM98**].

A second proof of the equations is based on exact formulas for the operations in the volume recursion. For the generalized dilaton equation, consider the following four operators acting on the ring $\mathbb{C}[x^2, y^2, L_1^2, \ldots, L_n^2, \ldots]$,

$$2\frac{\partial}{\partial L_1}L_1[\cdot], \qquad \frac{\partial}{\partial L_{n+1}}[\cdot]\Big|_{L_{n+1}=2\pi i}, \qquad \int_0^\infty\int_0^\infty xyH(x+y, L_1)[\cdot]dxdy,$$
$$\text{and} \quad \int_0^\infty x\big(H(x, L_1+L_k) + H(x, L_1-L_k)\big)[\cdot]dx.$$

Formulas for the operators are developed. For the proof, the generalized dilaton equation is written using the second operator, the volume recursion is applied for the left hand side, operator formulas are applied, and terms are gathered to give the right hand side. The generalized dilaton equation gives WP volumes for the compact case, including the following examples

$$V_{2,0} = \frac{43\pi^6}{2160}, \qquad V_{3,0} = \frac{176557\pi^{12}}{1209600},$$
$$V_{4,0} = \frac{1959225867017\pi^{18}}{493807104000} \quad \text{and} \quad V_{5,0} = \frac{84374265930915479\pi^{24}}{355541114880000}.$$

We close the lecture by noting that there is extensive research on Witten's conjecture for the moduli space. The Kontsevich and Mirzakhani approaches are analytic in nature. Okounkov and Pandharipande [**OP09**] transformed the question to counting Hurwitz numbers, topological types of branched covers of the sphere, and used a combinatorial approach to count factorizations of permutations into transpositions. Their combinatorial approach gives Kontsevich's formula. There is an even greater body of research on intersection numbers and relations on the moduli space [**HM98**].

Questions for the problem sessions

Lecture 1 questions.

(1) Let $z_1, z_2, z_3, z_4 \in \mathbb{CP}^1$ (where we think of z_i as either complex numbers or ∞). Define the cross ratio by

$$(z_1, z_2, z_3, z_4) = \frac{(z_1 - z_3)(z_2 - z_4)}{(z_2 - z_3)(z_1 - z_4)}.$$

(a) Show that the cross ratio is invariant under Möbius transformations of $\mathbb{CP}^1$, hence defines a map $\mathcal{M}_{0,4} \to \mathbb{CP}^1$.
(b) What is the image of the cross ratio map?
(c) Show that the map is holomorphic.
(d) Determine what happens in any of the limits $z_i \to z_j$, and hence argue that $\overline{\mathcal{M}}_{0,4} = \mathbb{CP}^1$.

(2) Let R be a hyperbolic surface with disjoint simple closed geodesics γ_1 and γ_2. Let $\tau_{\gamma_i,\theta_i}(R)$ be the new hyperbolic surface obtained from R by twisting on the geodesic γ_i an amount θ_i. Show that $\tau_{\gamma_1,\theta_1} \circ \tau_{\gamma_2,\theta_2} = \tau_{\gamma_2,\theta_2} \circ \tau_{\gamma_1,\theta_1}$.

(3) How many pairs of pants are needed to decompose a surface of genus g with n boundaries?

(4) In a brief paragraph, using only a small amount of notation, describe the results of Theorem 4. Repeat the exercise with Theorem 5.

(5) Use Bers' constant to produce a rough upper bound for the WP volume of moduli space. (Bers proved that there is a universal constant $C_{g,n}$ for each distinct topological type (g, n) of surface, such that if R is a hyperbolic surface of type (g, n), there exists a pair of pants decomposition for R such that the pants boundary lengths in the decomposition are no greater than $C_{g,n}$.)

(6) Show that $\overline{\mathcal{T}}_{g,n}$ is not locally compact.

(7) Use the simple description below Theorem 3, to prove that a section of the universal curve cannot pass through a node.

(8) Show and describe the situation for which an isometry of a punctured Riemann surface can setwise fix the geodesics of a pants decomposition.

(9) (extra credit) Consider the following elementary family of punctured Riemann surfaces R_t. Begin with a fixed surface R. For a coordinate chart z, mapping an open set U to a neighborhood of the origin, the surface R_t has punctures at 0 and at t, for t small. Show that the family R_t can be described as a $zw = t$ plumbing construction. (Introduce a new coordinate for a neighborhood of 0 that keeps the punctures apart; then relate the new and old coordinates.)

Lecture 2 questions.

(1) Draw a picture of how a family of ortho emanating non simple geodesics can limit to a boundary oblique geodesic.

(2) Draw a schematic picture of the Cantor set partition of a boundary circle, according to the types of ortho emanating geodesics. Include six main gap intervals in your picture.

(3) Show that the equatorial reflection of a pair of pants stabilizes the ortho boundary geodesic for the boundary β and interchanges the spirals for the boundary β.

(4) For a sphere with 5 labeled punctures, show that for any pants decomposition there is exactly one pair of pants containing one puncture. Use this fact, to show that up to homeomorphism, there are 15 distinct pants decompositions of a sphere with 5 labeled punctures.

(5) Check the last two sentences of the proof of Theorem 8 - check that the lengths of the main gaps plus the length of a complementary interval add to $\mathcal{R}$.

Lecture 3 questions.

(1) Suppose that $\pi : \widetilde{M} \to M$ is a smooth covering space map between smooth n-manifolds $\widetilde{M}$ and M. Let dV be a volume form on M, and $f : \widetilde{M} \to \mathbb{R}$ a function such that $f\,\pi^* dV$ is integrable on $\widetilde{M}$. Define the function $f_\pi : M \to \mathbb{R}$ by

$$f_\pi(x) = \sum_{y \in \pi^{-1}(x)} f(y).$$

Show that

$$\int_M f_\pi \, dV = \int_{\widetilde{M}} f \pi^* dV.$$

(2) Repeat the above, with $\pi : \widetilde{M} \to M$ an orbifold covering map between smooth orbifolds.

(3) Let G be a group acting on a manifold M, and let N be a normal subgroup of G. Assume that $f : M \to \mathbb{R}$ is N-invariant and that dV is a G-invariant volume form on M with $f\,dV$ integrable. Show that

$$\int_{X/G} \sum_{[h] \in G/N} f \circ h \, dV = \int_{X/N} f \, dV,$$

and show that everything is well defined. (G and N may be infinite.)

(4) For all $g \leq 3$, $n \leq 4$, list the cases of the McShane-Mirzakhani length identity, and identify which cases have an automorphism.

(5) Calculate

$$\int_0^\infty \frac{x^{2k+1}\,dx}{1+e^x}.$$

Hint: Use a geometric series to expand the denominator.

(6) Give a brief written description of the following integral using only a small amount of notation,

$$\int_0^\infty \int_0^\infty \sum_{\substack{g_1+g_2=g \\ I_1 \amalg I_2 = \{2,\dots,n\}}} H(x+y, L_1) V_{g_1}(x, L_{I_1}) V_{g_2}(y, L_{I_2})\, xy \, dx dy.$$

(7) Use Mirzakhani's recursion formula to calculate $V_{0,4}(L)$.
(8) Write a Maple or Mathematica program implementing Mirzakhani's recursion formula.

Lecture 4 questions.

(1) (a) Show that the standard area form on the unit sphere is a symplectic form with a Hamiltonian circle action
(b) Calculate the Duistermaat-Heckman measure on the image of the moment map for the sphere
(2) Consider $\mathbb{C}^{n+1}$ with the standard symplectic form

$$\omega = \sum dx_i \wedge dy_i.$$

Consider the S^1 action on $\mathbb{C}^{n+1}$ given by multiplication by complex numbers of the unit circle. Show that the action is Hamiltonian, with moment map

$$\mu(z) = \frac{1}{2} - \frac{1}{2}\sum |z_i|^2.$$

Show that the symplectic quotient at level 0 is $\mathbb{CP}^n$ with the Fubini-Study symplectic form.
(3) Show the equivalence between describing a connection for a circle bundle as a splitting of the tangent bundle of the total space versus an invariant 1-form on the total space of the circle bundle.
(4) Show that for any two connection 1-forms of a circle bundle, their associated curvature 2-forms are in the same cohomology class.
(5) Consider a Riemann surface R with a point p. The complex structure gives an orientation to R, as well as an orientation of $T_p(R)$ (which can be naturally identified with the holomorphic tangent space). Consider a small disc $D \subset R$ centered at p. The boundary of $R \setminus D$ has an induced orientation, and there is also a map $\partial(R \setminus D) \to T_p(R)$, with image the tangent unit circle at p. Is this map orientation preserving or reversing?
(6) In the third paragraph of Applying the Duistermaat-Heckman theorem, the definition of $\hat{\mathcal{T}}_{g,n}$ is extended to include $L = 0$. Describe the $L = 0$ level set in $\hat{\mathcal{T}}_{g,n}$ as a bundle over an appropriate base.

Lecture 5 questions.

(1) The following are not complete rigorous statements, but interested readers (especially in algebraic geometry) can find solace in *Moduli of curves* by Harris and Morrison, [**HM98**].
(a) There is a natural map (the forgetful morphism)

$$\pi : \mathcal{M}_{g,n+1} \to \mathcal{M}_{g,n},$$

obtained by forgetting the $n + 1^{st}$ marked point. Convince yourself that it can be extended to a map of compactified moduli spaces. (Limiting noded Riemann surface genus 0 components with only two marked points are collapsed.)
(b) There are also n *section* maps

$$\sigma_i : \overline{\mathcal{M}}_{g,n} \to \overline{\mathcal{M}}_{g,n+1},$$

defined for the Riemann surface/stable curve, representing the point of the first moduli space, by attaching a triply punctured sphere to the i-th marked point to obtain a point of the second moduli space. Show that $D_i = \sigma_i(\overline{\mathcal{M}}_{g,n})$ is a divisor.

(c) For the tautological line bundles $\psi_{i,n} \to \overline{\mathcal{M}}_{g,n}$ (the line bundle subscript i refers to a marked point and the subscript n refers to the moduli space). Show that

$$\psi_{i,n+1} = \pi^* \psi_{i,n} \otimes D_i.$$

Hint: Consider a local, non-zero holomorphic section of $\psi_{i,n}$. Show that this section pulls back to a section of $\psi_{i,n+1}$ with a simple zero along D_i.

(d) Use the relation from part (c) to prove the *string equation*:

$$\int_{\overline{\mathcal{M}}_{g,n+1}} \psi_1^{k_1} \cdots \psi_n^{k_n} = \sum_{j=1}^{n} \int_{\overline{\mathcal{M}}_{g,n}} \psi_1^{k_1} \cdots \psi_j^{k_j-1} \cdots \psi_n^{k_n},$$

(ψ now refers to the first Chern class of the line bundle).

(e) The string equation can be reframed as a differential equation satisfied by the intersection number generating function $F(t_0, t_1, \ldots)$. What is the equation?

(f) Find a similar differential equation coming from the *dilaton equation*:

$$\int_{\overline{\mathcal{M}}_{g,n+1}} \psi_1^{k_1} \ldots \psi_n^{k_n} \psi_{n+1} = (2g-2+n) \int_{\overline{\mathcal{M}}_{g,n}} \psi_1^{k_1} \cdots \psi_n^{k_n}.$$

Readers with sufficient algebraic geometry background should prove the dilaton equation, following similar reasoning as for the string equation.

Note: The following questions have been adapted from Harris-Morrison, [**HM98**].

(2) Use the string equation and the initial condition $\langle \tau_0^3 \rangle_0 = 1$ to prove that

$$\langle \prod_{i=1}^{n} \tau_{k_i} \rangle_0 = \frac{(n-3)!}{\prod_{i=1}^{n} k_i!}.$$

(3) Show that the string equation, the dilaton equation and the initial condition $\langle \tau_1 \rangle_1 = \frac{1}{24}$ determine all the genus 1 products.

(4) Define differential operators

$$L_n = -\frac{(2n+3)!!}{2^{n+1}} \frac{\partial}{\partial t_{n+1}} + \sum_{i=0}^{\infty} \frac{(2i+2n+1)!!}{(2i-1)!!2^{n+1}} t_i \frac{\partial}{\partial t_{i+n}} + \frac{\lambda^2}{2} \sum_{i+j=n-1} \frac{(2i+1)!!(2j+1)!!}{2^{n+1}} \frac{\partial^2}{\partial t_i \partial t_j},$$

for all $n \geq -1$, where $(2n-1)!! = 1 \cdot 3 \cdot 5 \cdots (2n-1)$, $(-1)!! = 1$.

(a) Show that $[L_n, L_m] = (n-m)L_{n+m}$.

(b) Assuming that $L_n e^F = 0$ (this is one version of the Witten-Kontsevich Theorem), translate this differential equation into a relationship amongst the tautological intersection numbers (rather like the string and dilaton equations).

(c) Interpret the equations from part (b) as recursive formulas for calculating intersection numbers, and compare to the recursion structure in Mirzakhani's volume formula.

Bibliography

[BP00] M. Boggi and M. Pikaart. Galois covers of moduli of curves. *Compositio Math.*, 120(2):171–191, 2000. MR1739177 (2002a:14025)

[BS85] J. Birman and C. Series. Simple curves have Hausdorff dimension one. *Topology*, 24(1):217–225, 1985.

[Bus92] Peter Buser. *Geometry and spectra of compact Riemann surfaces*, volume 106 of *Progress in Mathematics*. Birkhäuser Boston Inc., Boston, MA, 1992. MR1183224 (93g:58149)

[CdS01] Ana Cannas da Silva. *Lectures on symplectic geometry*, volume 1764 of *Lecture Notes in Mathematics*. Springer-Verlag, Berlin, 2001. MR1853077 (2002i:53105)

[DN09] Norman Do and Paul Norbury. Weil-Petersson volumes and cone surfaces. *Geom. Dedicata*, 141:93–107, 2009. MR2520065 (2010c:32020)

[Do08] Norman Do. Intersection theory on moduli spaces of curves via hyperbolic geometry. *PhD Thesis, The University of Melbourne*, 2008.

[Do11] Norman Do. Moduli spaces of hyperbolic surfaces and their Weil-Petersson volumes. Arxiv:math/1103.4674, 2011.

[FO06] José Figueroa-O'Farrill. PG minicourse: BRST cohomology, Lecture 2: Symplectic reduction. Edinburgh Mathematical Physics Group, `http://empg.maths.ed.ac.uk/Activities/BRST/Lect2.pdf`, 2006

[Gru01] Samuel Grushevsky. An explicit upper bound for Weil-Petersson volumes of the moduli spaces of punctured Riemann surfaces. *Math. Ann.*, 321(1):1–13, 2001. MR1857368 (2002h:14040)

[Har77] W. J. Harvey, editor. *Discrete groups and automorphic functions*. Academic Press [Harcourt Brace Jovanovich Publishers], London, 1977. MR0466535 (57:6413)

[HM98] Joe Harris and Ian Morrison. *Moduli of curves*, volume 187 of *Graduate Texts in Mathematics*. Springer-Verlag, New York, 1998. MR1631825 (99g:14031)

[IT92] Y. Imayoshi and M. Taniguchi. *An introduction to Teichmüller spaces*. Springer-Verlag, Tokyo, 1992. Translated and revised from the Japanese by the authors. MR1215481 (94b:32031)

[Kon92] Maxim Kontsevich. Intersection theory on the moduli space of curves and the matrix Airy function. *Comm. Math. Phys.*, 147(1):1–23, 1992. MR1171758 (93e:32027)

[Loo93] Eduard Looijenga. Intersection theory on Deligne-Mumford compactifications (after Witten and Kontsevich). *Astérisque*, (216):Exp. No. 768, 4, 187–212, 1993. Séminaire Bourbaki, Vol. 1992/93. MR1246398 (95b:32033)

[Loo94] Eduard Looijenga. Smooth Deligne-Mumford compactifications by means of Prym level structures. *J. Algebraic Geom.*, 3(2):283–293, 1994. MR1257324 (94m:14029)

[McS98] Greg McShane. Simple geodesics and a series constant over Teichmuller space. *Invent. Math.*, 132(3):607–632, 1998. MR1625712 (99i:32028)

[Mir07a] Maryam Mirzakhani. Simple geodesics and Weil-Petersson volumes of moduli spaces of bordered Riemann surfaces. *Invent. Math.*, 167(1):179–222, 2007. MR2264808 (2007k:32016)

[Mir07b] Maryam Mirzakhani. Weil-Petersson volumes and intersection theory on the moduli space of curves. *J. Amer. Math. Soc.*, 20(1):1–23 (electronic), 2007. MR2257394 (2007g:14029)

[Mir08] Maryam Mirzakhani. Growth of the number of simple closed geodesics on hyperbolic surfaces. *Ann. of Math. (2)*, 168(1):97–125, 2008. MR2415399 (2009c:32027)

[Mir10] Maryam Mirzakhani. Growth of Weil-Petersson volumes and random hyperbolic surfaces of large genus. Arxiv:math/1012.2167, 2010.

[MS74] John W. Milnor and James D. Stasheff. *Characteristic classes.* Princeton University Press, Princeton, N. J., 1974. Annals of Mathematics Studies, No. 76. MR0440554 (55:13428)

[MS08] Motohico Mulase and Brad Safnuk. Mirzakhani's recursion relations, Virasoro constraints and the KdV hierarchy. *Indian J. Math.*, 50(1):189–218, 2008. MR2379144 (2009b:14053)

[MZ00] Yuri I. Manin and Peter Zograf. Invertible cohomological field theories and Weil-Petersson volumes. *Ann. Inst. Fourier (Grenoble)*, 50(2):519–535, 2000. MR1775360 (2001g:14046)

[MZ11] Maryam Mirzakhani and Peter Zograf. Towards large genus asymtotics of intersection numbers on moduli spaces of curves. Arxiv:math/1112.1151, 2011.

[OP09] A. Okounkov and R. Pandharipande. Gromov-Witten theory, Hurwitz numbers, and matrix models. In *Algebraic geometry—Seattle 2005. Part 1*, volume 80 of *Proc. Sympos. Pure Math.*, pages 325–414. Amer. Math. Soc., Providence, RI, 2009. MR2483941 (2009k:14111)

[ST01] Georg Schumacher and Stefano Trapani. Estimates of Weil-Petersson volumes via effective divisors. *Comm. Math. Phys.*, 222(1):1–7, 2001. MR1853862 (2002h:14047)

[TWZ06] Ser Peow Tan, Yan Loi Wong, and Ying Zhang. Generalizations of McShane's identity to hyperbolic cone-surfaces. *J. Differential Geom.*, 72(1):73–112, 2006. MR2215456 (2007a:53087)

[Wit91] Edward Witten. Two-dimensional gravity and intersection theory on moduli space. In *Surveys in differential geometry (Cambridge, MA, 1990)*, pages 243–310. Lehigh Univ., Bethlehem, PA, 1991. MR1144529 (93e:32028)

[Wit92] Edward Witten. Two-dimensional gauge theories revisited. *J. Geom. Phys.*, 9(4):303–368, 1992. MR1185834 (93m:58017)

[Wlp82] Scott A. Wolpert. The Fenchel-Nielsen deformation. *Ann. of Math. (2)*, 115(3):501–528, 1982. MR657237 (83g:32024)

[Wlp83] Scott A. Wolpert. On the symplectic geometry of deformations of a hyperbolic surface. *Ann. of Math. (2)*, 117(2):207–234, 1983. MR690844 (85e:32028)

[Wlp90] Scott A. Wolpert. The hyperbolic metric and the geometry of the universal curve. *J. Differential Geom.*, 31(2):417–472, 1990. MR1037410 (91a:32030)

[Wlp10] Scott A. Wolpert. *Families of Riemann surfaces and Weil-Petersson Geometry*, volume 113 of *CBMS Regional Conference Series in Mathematics.* Published for the Conference Board of the Mathematical Sciences, Washington, DC, 2010. MR2641916 (2011c:32020)

[Zog08] Peter Zograf. On the large genus asymptotics of Weil-Petersson volumes. Arxiv:math/0812.0544, 2008.

Teichmüller Curves, Mainly from the Viewpoint of Algebraic Geometry

Martin Möller

IAS/Park City Mathematics Series
Volume 20, 2011

Teichmüller Curves, Mainly from the Viewpoint of Algebraic Geometry

Martin Möller

1. Introduction

Teichmüller curves are algebraic curves $C \to \mathcal{M}_g$ in the moduli space of curves that are totally geodesic for the Teichmüller metric. They are generated by very special flat surfaces, i.e. compact Riemann surfaces with a flat metric (up to finitely many cone-type singularities), called Veech surfaces. Studying the geometry and dynamics of Veech surfaces and connection to billiards was the original motivation to introduce Teichmüller curves. Here we consider Teichmüller curves as intrinsically interesting curves and relate their geometry to the geometry of $\mathcal{M}_g$.

If we take a more distant look and consider just any curve C in $\mathcal{M}_g$ there are various quantities that one can associate algebraically with such a curve. Cohomologically, one can decompose the variation of Hodge structures of the family over C into its irreducible summands. Intersection theory with various natural divisors (and line bundles) on $\mathcal{M}_g$ gives a collection of numbers that one can attach to C. The most prominent of them in algebraic geometry is the slope. Finally, dynamics attaches to a Teichmüller curve some characteristic numbers, namely Lyapunov exponents.

The purpose of these lecture notes is to relate the quantities 'decomposition of the variation of Hodge structures', 'slope' and 'Lyapunov exponents' on Teichmüller curves. Of course, Teichmüller curves are not just arbitrary curves in $\mathcal{M}_g$ and their origin from flat geometry allows special techniques, thus giving e.g. restrictions on their slope. Towards the end of these lecture notes we broaden the picture and highlight that all these quantities in fact make sense for any curve in $\mathcal{M}_g$ and most do even for any curve in the moduli space of abelian varieties. They hence deserve to be studied also in this broadened context!

We outline three guiding questions of the field: Can one classify Teichmüller curves, in particular those that are primitive, i.e. that do not arise via covering constructions? If one performs covering constructions of a surface generating a Teichmüller curve, how do the above quantities 'slope' and 'Lyapunov exponents' change? What can one say about the values for slopes and Lyapunov exponents

---

Institut für Mathematik, Goethe-Universität Frankfurt, Robert-Mayer-Str. 6-8, 60325 Frankfurt am Main, Germany

E-mail address: moeller@math.uni-frankfurt.de

Thanks to Ralf Lehnert, André Kappes and Christian Weiss for comments on preliminary versions of the manuscript.

appearing for the Teichmüller curves with fixed type (e.g. of singularities)? More detailed versions of these questions appear in the open problem sections.

The main results presented in this text are the following. First, the variation of Hodge structures over a Teichmüller curve decomposes into r rank two pieces, where $r \leq g$ is a field extension degree. One of this pieces has 'maximal degree' and the presence of a Hodge decomposition with such a maximal piece characterizes Teichmüller curves. (Theorem 5.5 and Theorem 5.6). This decomposition implies real multiplication on an r-dimensional abelian subvariety of the family of Jacobians and has proven to be a major tool towards classification of Teichmüller curves. Second, to a Teichmüller curve and to a generic flat surface one can associate the Lyapunov exponents (Theorem 6.1). The sum of Lyapunov exponents for Teichmüller curves can be calculated as slopes by a Riemann-Roch argument (Proposition 6.4). Finally, we show how the geometry of the moduli space allows to calculate the sum of Lyapunov exponents for all Teichmüller curves in low genus (Theorem 6.8).

As guideline for the reader, we indicate that Section 2, Section 3, and Section 4 contain background material. The core of what has been indicated above starts with the definition of Teichmüller curves in Section 5. To get started there we suggest to have read Section 2 and Section 3.1 and to jump back to the other introductory sections on demand.

Plan of the lecture series: The first lecture will introduce Teichmüller curves state the VHS decomposition. Proofs of this decomposition along with some background on Hodge theory are part of the second lecture. The third lecture introduces Lyapunov exponents showing how their sum can be related to quantities that are well-known in algebraic geometry. The forth lecture introduces the slope of a family of curves and how properties of the slope lead to non-varying results for sums of Lyapunov exponents. Moreover, it connects to some open research problems, in particular characterizing other curves in moduli space, such as Shimura curves, by Lyapunov exponents.

2. Flat surfaces and $\mathrm{SL}_2(\mathbb{R})$-action

2.1. Flat surfaces and translation structures

Let $\mathcal{M}_g$ be the moduli space of genus g compact algebraic curves. Sometimes we also need the moduli space $\mathcal{M}_{g,n}$ parametrizing curves with n ordered marked points and we let $\mathcal{M}_{g,[n]}$ be the moduli space of curves with n unordered marked points. Let $\Omega\mathcal{M}_g$ denote the vector bundle of holomorphic one-forms over $\mathcal{M}_g$ minus the zero section. Points in $\Omega\mathcal{M}_g$, called *flat surfaces*, are usually written as a pair (X, ω) for a holomorphic one-form ω on the Riemann surface X.

We use the terminology (algebraic) curve and (Riemann or flat) surface interchangeably. When talking of objects of complex dimension two we emphasize this by speaking of 'complex surfaces'. Fibered surfaces introduced in Section 3.1 are by definition of complex dimension two.

A *translation structure* on a Riemann surfaces X^0 is an atlas of complex charts $\{(U_\alpha, g_\alpha)\}_{\alpha \in I}$, all whose transition functions are locally translations. The group $\mathrm{SL}_2(\mathbb{R})$ acts on the set of translation structures by postcomposing the chart maps g_α with the linear map (thereby identifying $\mathbb{C}$ with $\mathbb{R}^2$). Since $\mathrm{SL}_2(\mathbb{R})$ normalizes the subgroup of translations within the affine group of $\mathbb{R}^2$, this action is well-defined.

For a flat surface (X,ω) we let $Z(\omega)$ be the set of zeros of ω. From the viewpoint of translation structures it is natural to call $Z(\omega)$ the set of *singularities* of (X,ω). Note that the algebraic curve X is non-singular at $Z(\omega)$.

The proof of the next proposition is straightforward, given that a translation structure knows about the winding number of a loop around a singularity.

PROPOSITION 2.1. *If (X,ω) is a flat surface then $X \setminus Z(\omega)$ has a translation structure. Conversely, suppose X^0 is a Riemann surface obtained by removing from a compact topological surface X a finite number of points. If X^0 has a translation structure, such that X is the completion of X^0 with respect to the flat metric, then there is a flat surface (X,ω), such that $X^0 = X \setminus Z(\omega)$ and such that the translation structure associated with (X,ω) is just the given translation structure.*

COROLLARY 2.2. *There is an action of $\mathrm{SL}_2(\mathbb{R})$ on the moduli space of flat surfaces $\Omega\mathcal{M}_g$. This action preserves the stratification by number and type of zeros.*

PROOF. To define the $\mathrm{SL}_2(\mathbb{R})$-action we remove $Z(\omega)$, use the action of $\mathrm{SL}_2(\mathbb{R})$ on translation surfaces and use the converse statement in the previous proposition to glue the missing points back in. It is immediate to check that for any given $A \in \mathrm{SL}_2(\mathbb{R})$ the hypothesis on the metric completion still holds. □

Various one-parameter subgroups of $\mathrm{SL}_2(\mathbb{R})$ thus define flows on $\Omega\mathcal{M}_g$. The diagonal subgroup $g_t = \mathrm{diag}(e^t, e^{-t})$ is in fact the Teichmüller *geodesic flow* and the action of $h_t = \begin{pmatrix} 1 & t \\ 0 & 1 \end{pmatrix}$ defines a flow that is called *horocyclic flow*.

PROPOSITION 2.3. *The $\mathrm{SL}_2(\mathbb{R})$-action preserves the subvariety of hyperelliptic flat surfaces*

PROOF. The hyperelliptic involution acts as (-1) on all one-forms, hence on ω. In the flat coordinates of X given by $\Re(\omega)$ and $\Im(\omega)$, the hyperelliptic involution acts by the matrix $-\mathrm{Id}$. Since $-\mathrm{Id}$ is in the center of $\mathrm{SL}(2,\mathbb{R})$, we conclude that if (X,ω) admits a hyperelliptic involution, so does $A \cdot (X,\omega)$ for any $A \in \mathrm{SL}(2,\mathbb{R})$. □

Some flat geometry. The holomorphic one-form ω on X defines a flat metric and consequently for any given direction θ we can define the straight line flow $\phi_t^\theta(x)$ in the direction θ starting at $x \in X$, where t denotes the flow distance in the metric $|\omega|$. Consider the geodesics on a flat surface (X,ω) in a fixed direction θ. If such a geodesic γ ends forward and backward in finite time in a singularity (i.e. in $Z(\omega)$), then γ is called a *saddle connection*. If in a given direction θ all geodesics are either periodic or saddle connections, then θ is called a periodic direction. A maximal union of homotopic closed geodesics on (X,ω) is called a *cylinder*. The *width* is the length of a core geodesic, the *height* is the length of a straight segment perpendicular to a core geodesic crossing the cylinders once. We define the *modulus* of a cylinder to be the ratio height over width. If the moduli of all cylinders in a periodic direction are commensurable, the direction is called *parabolic*. (The reason for this terminology will become apparent once we define affine groups.)

Every flat surface has a saddle connection, more precisely, the set of directions of saddle connection vectors is dense in S^1. Here and in the sequel we consider the torus with one artificial singularity. Moreover, the set of saddle connection vectors (with multiplicities) is discrete in $\mathbb{R}^2$. In particular, there exists a finite number of shortest saddle connections on every flat surface. Proofs of these elementary, but fundamental properties can be found in [**Vor96**] and in [**MT02**].

Half-translation structures and quadratic differentials. A *half-translation structure* on a surface is an atlas whose transition functions are compositions of $\pm\mathrm{Id}$ and translations. Similarly to translation structures, one can set up a correspondence between half-translation structures and quadratic differentials. Except for hyperelliptic loci (defined below) we will disregard here the question whether the results on translation surfaces extend to half-translation surfaces or not. Sometimes this is easy, often there are problems. We sometimes give references to the corresponding result for quadratic differentials.

2.2. Affine groups and the trace field

One of the basic invariants of a flat surface is the *affine group* $\mathrm{SL}(X,\omega)$ (also called *Veech group*) defined as follows. Let $\mathrm{Aff}^+(X,\omega)$ be the group of orientation-preserving homeomorphisms of X, that are affine diffeomorphisms on $X\setminus Z(\omega)$ with respect to the charts defined by integrating ω. (We will abusively call elements of $\mathrm{Aff}^+(X,\omega)$ *affine diffeomorphisms*, although they are not differentiable at the zeros of ω.) The matrix part of the affine map is independent of the charts and provides a map

$$D : \mathrm{Aff}^+(X,\omega) \to \mathrm{SL}_2(\mathbb{R}).$$

The image of D is called the affine group $\mathrm{SL}(X,\omega)$.

PROPOSITION 2.4. *The image of D is a discrete group that is never cocompact.*

PROOF. Discreteness follows from the discreteness of the set of saddle connection vectors. If $\mathrm{SL}(X,\omega)$ was cocompact, suppose the horizontal direction has a saddle connection. Then there exists an unbounded sequence of times t_n and elements $\varphi_n \in \mathrm{SL}(X,\omega)$ such that $\varphi(g_{t_n}(X,\omega))$ converges. This contradicts the lower bound for norms of saddle connection vectors. □

Let $K = \mathbb{Q}(\mathrm{tr}(\varphi), \varphi \in \mathrm{SL}(X,\omega))$ denote the *trace field* of the affine group. The field extension $K/\mathbb{Q}$ has a priori no reason to be Galois and we let $L/\mathbb{Q}$ be the Galois closure of $K/\mathbb{Q}$.

An important restriction on the trace field is the following bound.

PROPOSITION 2.5. *The degree of the trace field of the affine group $\mathrm{SL}(X,\omega)$ is bounded by the genus $g(X)$.*

PROOF. Write $K = \mathbb{Q}(t)$ for some $t = \sum a_i \mathrm{tr}(D\varphi_i)$ with $a_i \in \mathbb{Q}$ and $\varphi_i \in \mathrm{SL}(X,\omega)$ using the theorem of the primitive element and the fact that $\mathrm{tr}(A)\mathrm{tr}(B) = \mathrm{tr}(AB) + \mathrm{tr}(AB^{-1})$ for A, $B \in \mathrm{SL}_2(\mathbb{R})$. Now consider

$$T = \sum a_i(\varphi_i^* + (\varphi_i^{-1})^*) \quad \in \mathrm{End}(H^1(X,\mathbb{Q})).$$

On the subspace $L = \langle \Re(\omega), \Im(\omega)\rangle$ we have $T|_L = t\cdot\mathrm{Id}$. Hence t is an eigenvalue of T of multiplicity at least two. The square of the characteristic polynomial of t thus divides the polynomial $\det(xI_g - T)$, which is of degree $2g$, since $\dim H^1(X,\mathbb{Q}) = 2g$. □

The Veech group of a general flat surface in a given stratum is of order two or trivial, depending on whether the stratum is hyperelliptic or not ([**Möl09**]). We will eventually be most interested in surfaces where the opposite extreme holds, i.e. where the affine group is as large as possible.

We briefly recall Thurston's classification of surface homeomorphisms. A homeomorphism φ of X is called *elliptic*, if it is isotopic to a diffeomorphism of finite order. It is easy to see that an affine diffeomorphism φ of (X, ω) is elliptic, if it is of finite order. In particular $D(\varphi)$ is of finite order. Conversely, if $D(\varphi)$ is of finite order, then φ is of finite order, since $\mathrm{Ker}(D)$ consists of holomorphic diffeomorphisms of X and consequently $\mathrm{Ker}(D)$ is finite by Hurwitz' theorem.

A diffeomorphism φ is called *reducible*, if it is isotopic to a diffeomorphism fixing a (real) simple closed curve on X. If φ is neither reducible nor elliptic, then φ is called *pseudo-Anosov*. By [**HM79**] there exists a pair (X, q) such that φ is an affine diffeomorphism of (X, q). As stated above, we will restrict to the case that $q = \omega^2$. Moreover, (X, ω) can be chosen such that φ stretches the horizontal lines by some factor $\lambda > 1$, called *dilatation coefficient*, and contracts the vertical lines by the same factor λ. Thus, $|\mathrm{tr}D(\varphi)| > 2$ for an affine pseudo-Anosov diffeomorphism.

Consequently, an affine diffeomorphism φ with $|\mathrm{tr}D(\varphi)| = 2$, i.e. such that $D(\varphi)$ is parabolic, is a reducible affine diffeomorphism. We briefly recall the structure of such a *parabolic diffeomorphism*. Say the horizontal direction is the eigendirection of $D(\varphi)$. Then some power of φ fixes all the finitely many horizontal saddle connections. The complement of these saddle connections has to consist of metric cylinders and φ acts as (power of a) Dehn twist along the core curves of the cylinders. Since φ has to be affine globally, the moduli have to be commensurable.

Conversely, composing Dehn twists in cylinders, we obtain the following proposition.

PROPOSITION 2.6. *If the horizontal direction of (X, ω) decomposes into cylinders of moduli m_i that are commensurable and $m = \mathrm{lcm}(m_1^{-1}, \dots, m_n^{-1})$, then there is an affine diffeomorphism φ with $D(\varphi) = \begin{pmatrix} 1 & m \\ 0 & 1 \end{pmatrix}$.*

This idea is elaborated on in the following construction.

Thurston-Veech construction. The following construction first appears in the famous 1976 preprint of Thurston ([**Thu88**]), see also [**Vee89**], [**HL06a**] and [**McM06a**] for recent versions and presentations.

A *multicurve* A on a surface Σ_g of genus g is a union of disjoint essential simple closed curves, no two of which bound an annulus. A pair (A, B) of multicurves *fills* (or *binds*) the surface if for each curve in A and each curve in B the geometric intersection number is minimal in their homotopy classes and if the complement $\Sigma_g \setminus (A \cup B)$ is a simply connected polygonal region with at least 4 sides. Such a set of curves is shown in Figure 1.

We index the components of A and B such that $A = \cup_{i=1}^{a} \gamma_i$ and $B = \cup_{i=a+1}^{a+b} \gamma_i$ and let C be the (unsigned) intersection matrix of A and B, i.e. for $i \neq j$ we have $C_{ij} = |\gamma_i \cap \gamma_j|$ and $C_{jj} = 0$ for all j.

As additional input datum for the construction we fix a set of multiplicities $d_i \in \mathbb{N}$ for $i = 1, \dots, a+b$. Since (A, B) fills Σ_g, the intersection graph is connected and the matrix $(d_i C_{ij})$ is a Perron-Frobenius matrix. Hence there is a unique

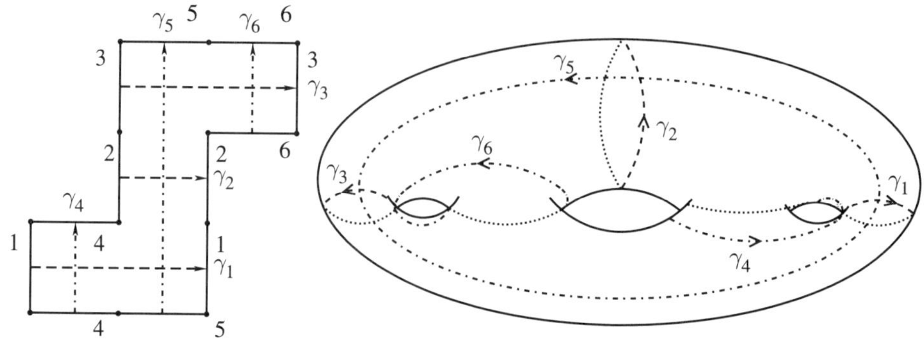

FIGURE 1. Binding curves and the surface resulting from the Thurston-Veech construction

positive eigenvector (h_i) up to scale such that

$$\mu h_i = \sum_{j=1}^{j=a+b} d_i C_{ij} h_j \tag{1}$$

for some positive eigenvalue μ.

We now glue a surface X from rectangles $R_p = [0, h_i] \times [0, h_j] \subset \mathbb{C}$ for each intersection point $p \in \gamma_i \cap \gamma_j$. Namely, glue R_p to R_q along the vertical (resp. horizontal) sides whenever p and q are joined by an edge in A (resp. B) of the graph $A \cup B$. The differentials dz^2 on each rectangle glue to a global quadratic differential q on X. The resulting surface is also shown in Figure 1, where sides with the same label have to be identified by parallel translations.

Let τ_i be the Dehn twist around γ_i and define

$$\begin{aligned} \tau_A &= \textstyle\prod_{i=1}^{a} \tau_i^{d_i} \\ \tau_B &= \textstyle\prod_{i=a+1}^{a+b} \tau_i^{d_i}. \end{aligned}$$

THEOREM 2.7 ([**Thu88**],[**Vee89**]). *The flat surface (X, q) constructed above contains affine diffeomorphisms τ_A and τ_B with derivatives*

$$D\tau_A = \begin{pmatrix} 1 & \mu \\ 0 & 1 \end{pmatrix} \quad \text{and} \quad D\tau_B = \begin{pmatrix} 1 & 0 \\ -\mu & 1 \end{pmatrix}$$

In particular the elements $\tau_A^n \tau_B$ are pseudo-Anosov diffeomorphisms for n large enough.

PROOF. By construction the modulus m_i of the cylinder with core curve γ_i is d_i/μ. Hence the powers of the Dehn twists occurring in the definition of τ_A and τ_B have linear part as claimed. They fix the boundary of the horizontal resp. vertical cylinders and together define affine diffeomorphisms.

In order to check the last claim, one has to recall that an affine diffeomorphism is pseudo-Anosov if and only if the absolute value of its trace is greater than two. □

Since we are dealing exclusively with flat surfaces in the sequel, we remark that the quadratic differential has a square root, i.e. $q = \omega^2$ if and only if for a suitable orientation of the γ_i their geometric and algebraic intersection numbers coincide.

Totally real fields. A field is called totally real if all its complex embeddings factor through $\mathbb{R}$. Totally real trace fields play an important role for the classification of Teichmüller curves. The following result was first established for Teichmüller curves in [**Möl06b**]. Later Hubert and Lanneau gave the following short proof, showing that trace fields are totally real in a much more general context.

THEOREM 2.8 ([**HL06a**]). *Let (X, ω) be a flat surface. If $\mathrm{SL}(X, \omega)$ contains a parabolic and a hyperbolic element, then the trace field of $\mathrm{SL}(X, \omega)$ is totally real.*

PROOF. In the first step we show that such a surface (X, ω) arises via the Thurston-Veech construction. Take the curves in A to be the core curves of the cylinders of the parabolic element P. Let H be the hyperbolic element. Then HPH^{-1} is again parabolic and its fixed direction is different from the fixed direction of P. Take B to be the core curves of the cylinders of HPH^{-1}. (We may conjugate $\mathrm{SL}(X, \omega)$ within $\mathrm{SL}_2(\mathbb{R})$ so that these fixed directions become horizontal and vertical.) Then (A, B) fills the surface. Take the d_i for $i = 1, \ldots, |A|$ resp. for $i = |A|+1, \ldots, |A \cup B|$ to be the least common multiples of the ratios of the moduli of the cylinders in the horizontal resp. vertical direction. Then (X, ω) is just the surface obtained by the Thurston-Veech construction using the data (A, B, d_i).

In the second step we show that any surface that arise via the Thurston-Veech construction has totally real trace field. We continue to use the notation of that section of Theorem 2.7. Let D be the diagonal matrix with entries d_i. The square of the largest eigenvalue of the matrix C is the largest eigenvalue of the matrix C^2. Hence we have to show that all the eigenvalues of $(DC)^2$ are real.

Suppose first for simplicity $d_i = 1$ for all i. Since for some matrix C_0 we have

$$DC = C = \begin{pmatrix} 0 & C_0 \\ C_0^T & 0 \end{pmatrix}, \quad \text{hence} \quad (DC)^2 = C^2 = \begin{pmatrix} C_0 C_0^T & 0 \\ 0 & C_0^T C_0 \end{pmatrix}.$$

Since C^2 is symmetric, all its eigenvalues are real. Thus $\mathbb{Q}(\mu^2)$ is totally real.

If the d_i are no longer identically one, $(DC)^2$ is still similar to a symmetric matrix: Split D into two pieces D' and D'' of size a resp. b and let $D'_{\surd}$ resp. $D''_{\surd}$ denote the diagonal matrix with entries $\sqrt{d_i}$. Then

$$(DC)^2 = \begin{pmatrix} D'C_0D''C_0^T & 0 \\ 0 & D''C_0^TD'C_0 \end{pmatrix}.$$

The upper block decomposes as

$$D'C_0D''C_0^T = D'_{\surd}(D'_{\surd}C_0D''_{\surd})(D'_{\surd}C_0D''_{\surd})^T(D'_{\surd})^{-1}$$

and for the lower block the same trick works. The above conclusion about the eigenvalues thus still holds. □

Example: The affine group of a regular n-gon is a triangle group generated by a rotation around the center of the n-gon and a parabolic element ([**Vee89**]). There are examples of surfaces whose Veech group contains pseudo-Anosovs but no parabolic elements, in particular this pseudo-Anosov does not arise from the Thurston-Veech construction ([**HL06a**]). A survey of examples of affine groups that can appear is [**Möl09**].

2.3. Strata of $\Omega\mathcal{M}_g$ and hyperelliptic loci

The space $\Omega\mathcal{M}_g$ parameterizing flat surfaces is stratified according to the zeros of one-forms. For $m_i \geq 1$ and $\sum_{i=1}^k m_i = 2g-2$, let $\Omega\mathcal{M}_g(m_1,\ldots,m_k)$ denote the stratum parameterizing one-forms that have k distinct zeros of order $m_1,\ldots,m_k$.

For $d_i \geq -1$ and $\sum_{i=1}^s d_i = 4g-4$, let $\mathcal{Q}(d_1,\ldots,d_s)$ denote the *moduli space of quadratic differentials* that have s distinct zeros or poles of order $d_1,\ldots,d_s$. For $d_i \geq 0$ this is a subset of the vector bundle of holomorphic quadratic differentials over $\mathcal{M}_g$. The condition $d_i \geq -1$ ensures that the quadratic differentials in $\mathcal{Q}(d_1,\ldots,d_s)$ have at most simple poles and thus finite volume. Namely, $\mathcal{Q}(d_1,\ldots,d_s)$ parametrizes pairs (X,q) of a Riemann surface X and a meromorphic section q of $\omega_X^{\otimes 2}$ with the prescribed type of zeros and poles.

If the quadratic differential is not a global square of a one-form, there is a natural double covering $\pi : Y \to X$ such that $\pi^*q = \omega^2$. This covering is ramified precisely at the zeros of odd order of q and at its poles. It gives a map

$$\phi : \mathcal{Q}(d_1,\ldots,d_s) \to \Omega\mathcal{M}_g(m_1,\ldots,m_k),$$

where the signature $(m_1,\ldots,m_k)$ is determined by the ramification type (see [**KZ03**] for more details).

If the domain and the range of the map ϕ have the same dimension for some signature, we call the image a *component of hyperelliptic flat surfaces* of the corresponding stratum. This can only happen, if the domain of ϕ parametrizes genus zero curves, thus justifying the terminology. More generally, if the domain of ϕ parametrizes genus zero curves, we call the image a *locus of hyperelliptic flat surfaces* in the corresponding stratum. These loci are often called hyperelliptic loci, e.g. in [**KZ03**] and [**EKZb**]. We prefer to reserve the expression *hyperelliptic locus* for the subset of $\mathcal{M}_g$ (or its closure in $\overline{\mathcal{M}}_g$) parameterizing hyperelliptic curves and thus specify with 'flat surfaces' if we speak of subsets of $\Omega\mathcal{M}_g$.

2.4. Spin structures and connected components of strata

A *spin structure* (or *theta characteristic*) on a smooth curve X is a line bundle $\mathcal{L}$ whose square is the canonical bundle, i.e. $\mathcal{L}^{\otimes 2} \sim K_X$. The *parity of a spin structure* is given by $\dim H^0(X,\mathcal{L}) \mod 2$. This parity is well-known to be a deformation invariant. The *moduli space of spin curves* $\mathcal{S}_g$ parametrizes pairs (X,η), where η is a theta characteristic of X. It has two components $\mathcal{S}_g^-$ and $\mathcal{S}_g^+$ distinguished by the parity of the spin structure. The spin structures on stable curves are defined such that the morphisms $\pi : \mathcal{S}_g^- \to \overline{\mathcal{M}}_g$ and $\pi : \mathcal{S}_g^+ \to \overline{\mathcal{M}}_g$ are finite of degree $2^{g-1}(2^g-1)$ and $2^{g-1}(2^g+1)$, respectively. If $(X,\omega) \in \Omega\mathcal{M}_g(2\ell_1,\ldots,2\ell_k)$ with zeros of ω being $P_1,\ldots,P_k$, then the line bundle $\mathcal{L} = \mathcal{O}_X(\sum_{i=1}^k \ell_i P_i)$ naturally defines a spin structure.

We can now recall the classification of connected components of strata in $\Omega\mathcal{M}_g$.

THEOREM 2.9 ([**KZ03**]). *The strata of $\Omega\mathcal{M}_g$ have up to three connected components, distinguished by the parity of the spin structure and by being hyperelliptic or not. For $g \geq 4$, the strata $\Omega\mathcal{M}_g(2g-2)$ and $\Omega\mathcal{M}_g(2\ell,2\ell)$ with an integer $\ell = (g-1)/2$ have three components, the component of hyperelliptic flat surfaces and two components with odd or even parity of the spin structure but not consisting exclusively of hyperelliptic curves.*

The stratum $\Omega\mathcal{M}_3(4)$ has two components, $\Omega\mathcal{M}_3(4)^{\text{hyp}}$ and $\Omega\mathcal{M}_3(4)^{\text{odd}}$. The stratum $\Omega\mathcal{M}_3(2,2)$ also has two components, $\Omega\mathcal{M}_3(2,2)^{\text{hyp}}$ and $\Omega\mathcal{M}_3(2,2)^{\text{odd}}$.

Each stratum $\Omega\mathcal{M}_g(2\ell_1, \ldots, 2\ell_r)$ for $r \geq 3$ or $r = 2$ and $\ell_1 \neq (g-1)/2$ has two components determined by even and odd spin structures.

Each stratum $\Omega\mathcal{M}_g(2\ell - 1, 2\ell - 1)$ for $\ell \geq 2$ has two components, the component of hyperelliptic flat surfaces $\Omega\mathcal{M}_g(2\ell - 1, 2\ell - 1)^{\mathrm{hyp}}$ and the other component $\Omega\mathcal{M}_g(2\ell - 1, 2\ell - 1)^{\mathrm{non-hyp}}$.

In all the other cases, the stratum is connected.

Consider the partition $(2, \ldots, 2)$. For $(X, \omega) \in \Omega\mathcal{M}_g(2, \ldots, 2)^{\mathrm{odd}}$ with $\mathrm{div}(\omega) = 2\sum_{i=1}^{g-1} p_i$, the line bundle $\eta = \mathcal{O}_X(\sum_{i=1}^{g-1} p_i)$ is an odd theta characteristic. Therefore, we have a natural morphism

$$f : \Omega\mathcal{M}_g(2, \ldots, 2)^{\mathrm{odd}}/\mathbb{C}^* \to \mathcal{S}_g^- .$$

Note that f contracts the locus where $h^0(\eta) > 1$. Note also that for $g = 3$ Clifford's theorem implies that always $h^0(\eta) = 1$, i.e. f is an isomorphism in this case.

2.5. Stable differentials and Deligne-Mumford compactification

The space $\Omega\mathcal{M}_g$ is not compact for two reasons. First, it is a vector bundle (minus the zero section), so we should rather use the bundle of projective spaces $\mathbb{P}\Omega\mathcal{M}_g = \Omega\mathcal{M}_g/\mathbb{C}^*$. (This projective space bundle is also useful when dealing with Teichmüller curves, since they will be naturally (complex) curves in $\mathbb{P}\Omega\mathcal{M}_g$, whereas in $\Omega\mathcal{M}_g$ they are objects of real dimension four.)

Second, the moduli space of curves itself is not compact. Denote by $\overline{\mathcal{M}}_g$ the Deligne-Mumford compactification of $\mathcal{M}_g$. Points in the boundary of $\overline{\mathcal{M}}_g$ are *stable curves*, i.e. projective algebraic curves with at most nodes as singularities and such that each irreducible component isomorphic to a projective line has at least three nodes.

The bundle of holomorphic one-forms extends over $\overline{\mathcal{M}}_g$, parameterizing *stable one-forms* or equivalently sections of the dualizing sheaf. We denote the total space of this extension by $\Omega\overline{\mathcal{M}}_g$. The stable one-forms are holomorphic except for simple poles at the nodes with the conditions that residues at the two branches of the node add up to zero. An example of a stable one-form is the form dz/z on the projective line $\mathbb{P}^1$ with puncture at zero, at ∞ and maybe some other points. If we view the projective line $\mathbb{P}^1$ as an infinitely high cylinder with waist curve of length one (i.e. obtained by identifying $\Re(z) = 0$ with $\Re(z) = 1$ in the complex plane) and with the points at $i\infty$ and $-i\infty$ glued together, this one-form becomes dz. This metric picture of one infinite cylinder (or two half-cylinders if we moreover cut along the real axis) should be kept in mind when understanding the boundary points of Teichmüller curves.

For a stable curve X, denote the dualizing sheaf by ω_X. We will stick to the notation that points in $\Omega\overline{\mathcal{M}}_g$ are given by a pair (X, ω) with $\omega \in H^0(X, \omega_X)$, although this notations may result in confusion since dropping the subscript X drastically changes the meaning.

3. Curves and divisors in $\mathcal{M}_g$

The aim of this section is a short introduction to the Picard group of $\mathcal{M}_g$. This will be used to attach to any curve or divisor in $\mathcal{M}_g$ the quantity slope.

3.1. Curves and fibered surfaces

The aim of this section is to show how to associate with a map $C \to \mathcal{M}_g$ from a smooth algebraic curve C to the moduli space a fibered surface and to discuss various models of that fibered surface.

Let $f : \mathcal{X} \to C$ be a smooth family of curves of genus g over the smooth curve C, i.e. a smooth morphism with connected fibers, which are smooth curves of genus g. By definition of the moduli space, such a family yields a moduli map $m : C \to \mathcal{M}_g$, but the converse does not quite hold. So we will pass to finite unramified covers. More precisely, let $\mathcal{M}_g^{[n]}$ be a finite cover, that is isomorphic to the quotient of Teichmüller space by some subgroup $\Gamma_g^{[n]}$ of finite index in the mapping class group without torsion elements. Technically the notation $\mathcal{M}_g^{[n]}$ refers to level-n-structure and the claim on torsion elements holds for any $n \geq 3$, but we will not need the precise definition.

The moduli space $\mathcal{M}_g^{[n]}$ carries a universal family $f_{\text{univ}} : \mathcal{X}_{\text{univ}} \to \mathcal{M}_g^{[n]}$ that we can pullback via any map $B \to \mathcal{M}_g^{[n]}$ to obtain a family of curves $f : \mathcal{X} \to B$.

If we start with $C \to \mathcal{M}_g$, we can take B to be the preimage of C in $\mathcal{M}_g^{[n]}$, or rather if the map is not an embedding we obtain B as the fiber product $C \times_{\mathcal{M}_g} \mathcal{M}_g^{[n]}$. In general this is not an unramified cover of C, but if we provide C with an orbifold structure such that $C \to \mathcal{M}_g$ factors through the moduli stack, then $B \to C$ is unramified. This will happen in all the cases we need in the sequel.

Given a (not necessarily compact) curve C (or B) as above, we denote by $\overline{C}$ (resp. $\overline{B}$) its closure, i.e. the corresponding smooth projective curve. Since $\overline{\mathcal{M}}_g$ is projective, there is a map $\overline{m} : \overline{C} \to \overline{\mathcal{M}}_g$ extending the map $m : C \to \mathcal{M}_g$. We denote by $\Delta \subset \overline{C}$ (or Δ_C and Δ_B if we need to distinguish) the preimage of the boundary of $\overline{\mathcal{M}}_g$.

The stable reduction theorem states that after a further covering of $\overline{B}$, unramified outside Δ the pullback of the map $f : \mathcal{X} \to B$ can be completed to a family of stable curves. Since B is already such kind of covering of C, we stick to the notation of B for the base curve and denote by $f : \overline{\mathcal{X}} \to \overline{B}$ this family of stable curves. Moreover we can suppose that the monodromy around the cusps of B (see Section 4) is unipotent after a further finite base change unramified outside Δ. Again, we assume this in the sequel but stick to the letter B for the base curve.

The total space $\overline{\mathcal{X}}$ is, in general, not a smooth complex surface. Indeed it has singularities at some of the singular points of the singular fibers of $\overline{f}$. If the node in the fiber is given by the equation $x \cdot y = 0$ and t is a local parameter on the base B, then the local equation of $\overline{\mathcal{X}}$ is $x \cdot y = t^n$ for some n, and this is smooth if and only if $t = 1$. In general this is a singularity of type A_n ([**BHPVdV04**]).

One can resolve these singularities and obtain a smooth surface $\widetilde{\mathcal{X}}$ together with a birational map $\widetilde{\mathcal{X}} \to \overline{\mathcal{X}}$. The price we pay for that is that the induced fibration $\tilde{f} : \widetilde{\mathcal{X}} \to \overline{B}$ has no longer stable but only semistable fibers. The fiber of the map $\widetilde{\mathcal{X}} \to \overline{\mathcal{X}}$ over a singular point of type A_n is a chain of $n-1$ rational curves. See [**HM98**] Proposition 3.47 and Proposition 3.48 for an algorithm how to compute the stable and semi-stable models and the references in loc. cit. for a general proof.

With a view towards Teichmüller curves, the advantage of the stable model $\overline{\mathcal{X}}$ is its direct relation to the geometry of flat surfaces (see Section 5.4), whereas calculations of intersection numbers work without correction terms on $\widetilde{\mathcal{X}}$ only.

3.2. Picard groups of moduli spaces

We write Pic$(\cdot)$ for the rational Picard group $\mathrm{Pic}_{\mathrm{fun}}(\cdot)_{\mathbb{Q}}$ of a moduli stack (see [**HM98**] for more details). Since the quantities we are interested in, the sum of Lyapunov exponents and slopes, are invariant under coverings unramified in the interior of $\overline{\mathcal{M}}_g$, this is the group we want to calculate intersections with, not the Picard group of the coarse moduli space.

We fix some standard notation for elements in the Picard group. Let λ denote the first Chern class of the Hodge bundle. Let δ_i, $i = 1, \ldots, \lfloor g/2 \rfloor$ be the boundary divisor of $\overline{\mathcal{M}}_g$ whose generic element is a smooth curve of genus i joined at a node to a smooth curve of genus $g - i$. The generic element of the boundary divisor δ_0 is an irreducible nodal curve of geometric genus $g - 1$. In the literature sometimes δ_0 is denoted by δ_{irr}. We write δ for the total boundary class.

For moduli spaces with marked points we denote by ω_{rel} the relative dualizing sheaf of $\overline{\mathcal{M}}_{g,1} \to \overline{\mathcal{M}}_g$ and $\omega_{i,\mathrm{rel}}$ its pullback to $\overline{\mathcal{M}}_{g,n}$ via the map forgetting all but the i-th marked point. For a set $S \subset \{1, \ldots, n\}$ we let $\delta_{i;S}$ denote the boundary divisor whose generic element is a smooth curve of genus i joined at a node to a smooth curve of genus $g - i$ and the sections in S lying on the first component.

THEOREM 3.1 ([**AC87**]). *The rational Picard group of $\overline{\mathcal{M}}_g$ is generated by λ and the boundary classes δ_i, $i = 0, \ldots, \lfloor g/2 \rfloor$.*

More generally, the rational Picard group of $\overline{\mathcal{M}}_{g,n}$ is generated by λ, $\omega_{i,\mathrm{rel}}$, $i = 1, \ldots, n$, by δ_0 and by $\delta_{i;S}$, $i = 0, \ldots, \lfloor g/2 \rfloor$, where $|S| > 1$ if $i = 0$ and $1 \in S$ if $i = g/2$.

Alternatively, we define $\psi_i \in \mathrm{Pic}(\overline{\mathcal{M}}_{g,n})$ to be the class with value $-\pi_*(\sigma_i^2)$ on the universal family $\pi : \mathcal{X} \to C$ with section σ_i corresponding to the i-th marked point. We have the relation

$$\omega_{i,\mathrm{rel}} = \psi_i - \sum_{i \in S} \delta_{0;S}.$$

Consequently, a generating set of $\mathrm{Pic}(\overline{\mathcal{M}}_{g,n})$ can also be formed by the ψ_i, λ and boundary classes.

For a divisor class $D = a\lambda - \sum_{i=0}^{\lfloor g/2 \rfloor} b_i\delta_i$ in $\mathrm{Pic}(\overline{\mathcal{M}}_g)$, define its *slope* to be

$$s(D) = \frac{a}{b_0}. \tag{2}$$

3.3. Special divisors on moduli spaces

In the application for Teichmüller curves we do not care about the coefficients of δ_i for $i \geq 1$ in the divisor classes in $\mathrm{Pic}(\overline{\mathcal{M}}_g)$, since Teichmüller curves do not intersect these components (see Corollary 5.11). As shorthand, we use δ_{other} to denote some linear combination of δ_i for $i \geq 1$. Similarly, in $\overline{\mathcal{M}}_{g,n}$ we use δ_{other} to denote some linear combination of all boundary divisors but δ_0. For the same reason we do not distinguish between $\omega_{i,\mathrm{rel}}$ and ψ_i for a divisor class, since they only differ by boundary classes in δ_{other}.

The hyperelliptic locus in $\overline{\mathcal{M}}_3$. Denote by $H \subset \overline{\mathcal{M}}_g$ the closure of locus of genus g hyperelliptic curves. We call H the hyperelliptic locus in $\overline{\mathcal{M}}_g$. Note that H is a divisor if and only if $g = 3$. A stable curve X lies in the boundary of H if there is an *admissible cover* of degree two $\tilde{X} \to \mathbb{P}^1$, for some nodal curve $\tilde{X}$ whose stabilization is X. We refer to [**HM98**] for the definition of admissible covers.

The class of the hyperelliptic locus $H \subset \overline{\mathcal{M}}_3$ calculated e.g. in [**HM98**, (3.165)] is given as follows:

$$H = 9\lambda - \delta_0 - 3\delta_1, \tag{3}$$

hence it has slope $s(H) = 9$.
The Brill-Noether divisors. For a divisor D of degree d on a curve X, denote by

$$|D| = \{\mathrm{div}(s) + D \mid s \in H^0(X, \mathcal{L}(D)) \setminus \{0\}\}$$

the set of all effective divisors linearly equivalent to D. $|D|$ naturally has the structure of a projective space. A g_d^r on X is a projective linear subspace of $|D|$ of dimension r. A g_d^1 is called a *pencil.*

The Brill-Noether locus BN_d^r in $\overline{\mathcal{M}}_g$ parametrizes curves X that possesses a g_d^r. If the Brill-Noether number

$$\rho(g, r, d) = g - (r+1)(g - d + r) = -1,$$

then BN_d^r is indeed a divisor.

There are pointed versions of this divisor. Let $\underline{w} = (w_1, \ldots, w_n)$ be a tuple of integers. Let $BN_{d,\underline{w}}^r$ be the locus in $\overline{\mathcal{M}}_{g,n}$ of pointed curves $(X, p_1, \ldots, p_n)$ with a line bundle $\mathcal{L}$ of degree d such that $\mathcal{L}$ admits a g_d^r and $h^0(\mathcal{L}(-\sum w_i p_i)) \geq r$. This Brill-Noether locus is a divisor, if the generalized Brill-Noether number

$$\rho(g, r, d, \underline{w}) = g - (r+1)(g - d + r) - r(|\underline{w}| - 1) = -1.$$

The hyperelliptic divisor and e.g. the Weierstrass divisor in $\mathcal{M}_{g,1}$ can also be interpreted as Brill-Noether divisors.

The class of these pointed divisors has been calculated in many special cases, in particular in [**Log03**] and later in [**Far09b**]. We give two examples. The class of the classical Brill-Noether divisor was calculated in [**HM82**], in particular

$$BN_3^1 = 8\lambda - \delta_0 - \delta_{\mathrm{other}} \quad \text{for} \quad g = 5. \tag{4}$$

If all $w_i = 1$ and $n = r+1$ the Brill-Noether divisor specializes to the divisor Lin calculated in [**Far09b**, Sec. 4.2]. In particular [**Far09b**, Thm. 4.6] gives

$$\mathrm{Lin}_3^1 = BN_{3,(1,1)}^1 = -\omega_{1,\mathrm{rel}} - \omega_{2,\mathrm{rel}} + 8\lambda - \delta_0 - \delta_{\mathrm{other}} \quad \text{for} \quad g = 4. \tag{5}$$

We will illustrate the method of *test curves* for calculating the class of a divisor (see e.g. [**HM98**] for more examples). For instance, using certain test curves including a Teichmüller curve we can determine (partially, but sufficiently for our purposes) the class of Lin_3^1. We have to use some terminology that is introduced below. The reader is invited skip over it at a first reading and later check that we do not use circular reasoning.

PROPOSITION 3.2. *The class of* Lin_3^1 *equals*

$$\mathrm{Lin}_3^1 = k(-\omega_{1,\mathrm{rel}} - \omega_{2,\mathrm{rel}} + 8\lambda - \delta_0 - \delta_{\mathrm{other}})$$

for some constant k.

Before proceeding to the proof, we recall some facts from algebraic geometry. If X is a non-hyperelliptic curve of genus 4, then its canonical image in $\mathbb{P}^3$ is contained in a unique irreducible quadric. Up to isomorphism there are two types of quadrics in $\mathbb{P}^3$: smooth quadrics, e.g. $xw = yz$ and singular quadrics, e.g. $xy = z^2$. In particular, in the smooth case, the quadric Q is isomorphic to $\mathbb{P}^1 \times \mathbb{P}^1$, and its Picard group is therefore isomorphic to $\mathbb{Z} \oplus \mathbb{Z}$. It is generated by the class

$(1,0) = \{\text{pt}\} \times \mathbb{P}^1$ and the class $(0,1) = \mathbb{P}^1 \times \{\text{pt}\}$. The intersection product of two divisors $D_1 = (a,b)$ and $D_2 = (a',b')$ is therefore

$$D_1 \cdot D_2 = ab' + a'b.$$

Note also that the canonical divisor K_Q has class $(-2,-2)$.

PROOF. Suppose that

$$\text{Lin}_3^1 = a_1\omega_{1,\text{rel}} + a_2\omega_{2,\text{rel}} + b\lambda - c\delta_0 - \delta_{\text{other}}$$

for some unknown coefficients a_1, a_2, b, c. By symmetry we have $a_1 = a_2$.

To construct the first test curve, we start with a general pencil B in the complete linear system $|(3,3)|$ on a smooth quadric $Q \subset \mathbb{P}^3$. Note that $|(3,3)|$ is a projective space of dimension 15. Via the isomorphism $Q \cong \mathbb{P}^1 \times \mathbb{P}^1$, its elements can be viewed as polynomials $f(x_0, x_1, y_0, y_1)$ which are bihomogenuous of degree 3 in the x and y-coordinates. Therefore a basis of the underlying vector space is given by $x_0^i x_1^{3-i} y_0^j y_1^{3-j}$ for $i,j = 0,\dots,3$. The elements of B are curves X_t on Q cut out by an equation of the form $f_t = \sum_{i,j} a_t x_0^i x_1^{3-i} y_0^j y_1^{3-j} = 0$ where $t \in \mathbb{P}^1$.

First note that without loss of generality, the generic member of B is non-singular and irreducible. For the singular curves in $|(3,3)|$ form a subset of projective dimension 3 which is determined by the vanishing of at least one partial derivative $\partial f/\partial x_i$, $\partial f/\partial y_i$, $i = 1,2$. Likewise, irreducibility is an open condition, and there is at least one irreducible curve in $|(3,3)|$. For a smooth projective curve X on a surface, we can compute its genus using the adjunction formula

$$2g(X) - 2 = X_t \cdot (X + K_Q) = (3,3) \cdot ((3,3) + (-2,-2)) = -6,$$

whence $g(X) = 4$.

Recall that a base point of the linear system B is a point $x \in Q$ that is contained in all divisors in B. We wish to determine the number of base points. Take two curves X_0 and X_t in B. They intersect in $(3,3) \cdot (3,3) = 18$ points, counted with multiplicity. Take any of the intersection points x_t on X_0. Now let t vary and trace the image of $x_t \in X_0 \cap X_t$. If x_t moved then we would obtain a non-constant map $\mathbb{P}^1 \to X_0$. But if we choose a smooth irreducible member of B as X_0 then $g(X_0) = 4$, which yields a contradiction.

There are 18 base points in the pencil. Choose two of them as the marked points p, q. Since B is general, p, q are not contained in any ruling of Q, namely there is no section of a linear series g_3^1 that contains both p and q. It implies that B and Lin_3^1 are disjoint in $\overline{\mathcal{M}}_{4,2}$, i.e.

$$B \cdot \text{Lin}_3^1 = 0.$$

Blowing up the 18 base points, we obtain a surface $S \subset \mathbb{P}^1 \times Q$, which is a one-parameter family of genus 4 curves over $B \cong \mathbb{P}^1$. Let $\chi_{\text{top}}(\cdot)$ denote the topological Euler characteristic. We have

$$\chi_{\text{top}}(S) = \chi_{\text{top}}(\mathbb{P}^1) \cdot \chi_{\text{top}}(X) + D \cdot \delta_0,$$

where X has genus equal to 4. We know that

$$\chi_{\text{top}}(\mathbb{P}^1) = 2, \quad \chi_{\text{top}}(X) = -6,$$

$$\chi_{\text{top}}(S) = \chi_{\text{top}}(Q) + 18 = \chi_{\text{top}}(\mathbb{P}^1) \cdot \chi_{\text{top}}(\mathbb{P}^1) + 18 = 22.$$

All together it implies there are 34 irreducible nodal curves in the family B, namely,

$$B \cdot \delta_0 = 34.$$

Let $\omega_{S/B}$ denote the relative dualizing sheaf of S over B. Since S has class $(1;3,3)$ in $\mathbb{P}^1 \times Q$, one checks that

$$\begin{aligned}\omega_{S/B} &= \omega_S - f^*\omega_B \\ &= (\omega_{\mathbb{P}^1\times Q} + S)|_S - f^*\omega_B \\ &= (-2;-2,-2) + (1;3,3) - (-2;0,0) \\ &= (1;1,1)\end{aligned}$$

on S, where $f : S \to B$ is the projection. Then $f_*(c_1^2(\omega_{S/B}))$ on B is equal to the top intersection

$$(1;1,1)\cdot(1;1,1)\cdot(1;3,3) = 14$$

on $\mathbb{P}^1 \times Q$. Using the Noether formula $12\lambda = f_*(c_1^2(\omega_{S/B})) + \sum_{i=0}^{\lfloor g/2 \rfloor} \delta_i$, we get

$$B\cdot\lambda = \frac{1}{12}(34+14) = 4.$$

Moreover, let Γ_p and Γ_q be the exceptional curves corresponding to the blow-up of the two marked points. For $i = 1,2$ we have $B \cdot \omega_{i,\mathrm{rel}} = -1$, since this is the self-intersection of Γ_p (resp. Γ_q). Note that B does not intersect any boundary divisors except for δ_0. Putting the above intersection numbers together, we obtain a relation

$$-a_1 + 2b + 17c = 0.$$

As the second test curve, we take a Teichmüller curve C generated by a flat surface in the stratum $\Omega\mathcal{M}_4(3,3)^{\text{non-hyp}}$, e.g. the square-tiled surface given by the permutations $(\pi_r = (123456789\,10), \pi_u = (19568))$. Using the algorithm in [**EKZb**] along with equation (13) we find for this particular curve that the sum of Lyapunov exponents equals $L(C) = 2$, i.e. the slope $s(C) = 33/4$. Using Proposition 5.12 with $\kappa = 5/8$, we obtain another relation

$$-a_1 + 4b + 33c = 0.$$

The two relations imply that

$$b = -8c, \ a_1 = c$$

and this concludes the proof. □

3.4. Slopes of divisors and of curves in $\mathcal{M}_g$

We summarize in this section some results on slopes of divisors and of curves. With the exception of one consequence of the Noether formula that we prove below, they are not strictly needed in the sequel. They are meant to compare the interest in calculating (the sum of) Lyapunov exponents below with a topic that is classic in algebraic geometry.

For $g \geq 4$ the canonical bundle of $\mathcal{M}_g$ has class

$$K_{\mathcal{M}_g} = 13\lambda - 2\delta_0 - 3\delta_1 - 2\delta_2 - \cdots - 2\delta_{\lfloor g/2 \rfloor}.$$

If there is a effective divisor D in $\overline{\mathcal{M}}_g$ with slope $s(D) < 13/2 = s(K_{\mathcal{M}_g})$, then one can show that $\overline{\mathcal{M}}_g$ is of general type. This observation on the birational geometry of $\overline{\mathcal{M}}_g$ initiated the quest for divisors of low slope and lead Harris-Morrison to conjecture

$$s(D) \geq 6 + \frac{12}{g+1}$$

for all effective divisors in $\mathcal{M}_g$. They could show that this bound holds in small genera, but the conjecture is known to be false. Still there is no divisor with $s(D) < 6$ known and the best known lower bounds for the slope are in the order of $1/g$.

We now turn our attention to curves and fibered surfaces. Let $s(C)$ be the *slope* of a curve $C \to M_g$ defined by

$$s(C) = \frac{\overline{C} \cdot \delta}{\overline{C} \cdot \lambda},$$

where $\delta = \sum_{i=0}^{\lfloor g/2 \rfloor} \delta_i$ is the total boundary. (This is not exactly dual to the standard definition of the slopes of divisors in (2), but because of Corollary 5.11 the difference does not matter for the applications we have in mind.)

There are numerous results on the slopes of a curve in the moduli space of curves. We will later be interested in slopes of Teichmüller curves in connection with dynamical properties. Just to put this into the right perspective we cite some results on slopes of curves from a geometric perspective.

PROPOSITION 3.3. *Let $f : \mathcal{X} \to C$ be a family of curves, giving rise to a curve $C \to M_g$. Then the slope satisfies the inequality*

$$s(C) \leq 12.$$

PROOF. On the smooth minimal model $\tilde{f} : \widetilde{\mathcal{X}} \to \overline{C}$ we have the Noether equality

$$12\chi(\mathcal{O}_{\widetilde{\mathcal{X}}}) - c_2(\omega_{\widetilde{\mathcal{X}}}) = c_1(\omega_{\widetilde{\mathcal{X}}})^2. \tag{6}$$

For a fibered surface with fiber genus g and base genus b we have by Riemann-Roch and the Leray spectral sequence (see e.g. [**Xia85**])

$$\begin{aligned} \chi(\mathcal{O}_{\widetilde{\mathcal{X}}}) &= \deg \tilde{f}_*\omega_{\widetilde{\mathcal{X}}/\overline{C}} + (g-1)(b-1), \\ c_2(\omega_{\widetilde{\mathcal{X}}}) &= \textstyle\sum_{F \text{ sing.}} \Delta\chi_{\text{top}}(F) + 4(g-1)(b-1), \\ c_1(\omega_{\widetilde{\mathcal{X}}})^2 &= \omega^2_{\widetilde{\mathcal{X}}/\overline{C}} + 8(g-1)(b-1), \end{aligned}$$

where $\Delta\chi_{\text{top}}(F)$ denotes the differences of the topological Euler characteristics of the given singular fiber F and a smooth fiber. We will use the equality in the form

$$12 \deg \tilde{f}_*\omega_{\widetilde{\mathcal{X}}/\overline{C}} - \sum_{F \text{ sing.}} \Delta\chi_{\text{top}}(F) = \omega^2_{\widetilde{\mathcal{X}}/\overline{C}}. \tag{7}$$

Since $\omega^2_{\widetilde{\mathcal{X}}/\overline{C}}$ is nef (Arakelov's theorem, see e.g. [**Deb82**]), its self-intersection is non-negative and we only have to check that $C \cdot \delta = \sum_{F \text{ sing.}} \Delta\chi_{\text{top}}(F)$. Both sides are additive, so we can check the contribution for each singular fiber and each node of such a fiber individually. A local equation $xy - t^n$ gives a contribution of n to the intersection number. To resolve the singularity we have to replace the node by a chain of $n-1$ rational curves. The Euler characteristic of a curve of genus $g-1$ with such a chain differs by n from the Euler characteristic of a smooth curve of genus g and this proves the claim. □

Slope bounds in general are studied in [**Xia87**]. Note that Xiao uses a related ratio, namely

$$\widetilde{s}(C) = \frac{\omega^2_{\widetilde{\mathcal{X}}/\overline{C}}}{\overline{C} \cdot \lambda} = 12 - s(C)$$

that he also calls slope. In our slope convention he obtains the following sharper bound.

THEOREM 3.4. *Let $f : \mathcal{X} \to C$ be a family of curves, giving rise to a curve $C \to M_g$. Then the slope satisfies the inequality*

$$0 \leq s(C) \leq 12 - 4\frac{g-1}{g}.$$

and the lower bound is attained if and only if every fiber is smooth and reduced.

A curve is called *trigonal*, if it admits a degree three map to a projective line. For curves of genus ≤ 4 all curves are trigonal, but from genus $g \geq 5$ on this is no longer the case. For $g = 5$ the locus of trigonal curves is the divisor BN_3^1 discussed in Section 3.3.

THEOREM 3.5 ([**SF00**]). *Suppose that $f : \mathcal{X} \to C$ is a family of trigonal curves. Then we have the slope bound*

$$s(C) \leq \frac{36(g+1)}{5g+1},$$

which is attained for certain families of trigonal curves all whose fibers are irreducible.

References: The recent survey by Farkas ([**Far09a**]) summarizes what is known on the slopes of divisors on the moduli space of curves and its consequences for the birational classification of $\mathcal{M}_g$. The slope conjecture appears first in [**HM98**], the first counterexample appears in [**FP05**].

One can also use Teichmüller curves to prove slope estimates for divisors. The bounds presently obtained in that way are as good (and as weak) as for 'moving' families, compare [**Che10a**] to [**Mor09**].

4. Variation of Hodge structures and real multiplication

The abstract concept of a variation of Hodge structure should be viewed as a formalism of how the cohomology of a variety or a family of varieties looks like. We indicate how the weight one situation, the most important for us, mimics the situation of (families of) abelian varieties. The reader may as well think of families of curves and their Jacobians. The advantage of the abstract concept is that one can handle multilinear operators (such as dual and tensor products) easily. Even if our main interest is weight one only, we need to consider endomorphisms of those (variations of) Hodge structures and thus need the general concept.

Hodge structures. For any field $K \subset \mathbb{R}$, we define a (weight k) *K-Hodge structure* on the K-vector space V to be a decomposition

$$V_{\mathbb{C}} := V \otimes_{\mathbb{R}} \mathbb{C} = \bigoplus_{p+q=k} V^{(p,q)}$$

into $\mathbb{C}$-vector spaces, such that $\overline{V^{(p,q)}} = V^{(q,p)}$. We say that V is a *$\mathbb{Z}$-Hodge structure*, if V is a $\mathbb{Q}$-Hodge structure and $V = \Lambda \otimes_{\mathbb{Z}} \mathbb{Q}$. (Sometimes Λ is called the *integral lattice*.) A *polarization* of a K-Hodge structure is a $\mathbb{C}$-valued bilinear form Q on $V_{\mathbb{C}}$, such that the generalization of the Riemann bilinear relations hold, i.e. $Q(V^{(p,q)}, V^{(r,s)}) = 0$ unless $p = s$ and $q = r$ and $i^{p-q}Q(\cdot, \bar{\cdot})$ is positive definite. For $\mathbb{Z}$-Hodge structures we require moreover, that the polarization is $\mathbb{Z}$-valued on Λ.

This definition is motivated by the fact that an abelian variety $A = \mathbb{C}^g/\Lambda$ gives rise to a polarized $\mathbb{Z}$-Hodge structure of weight one. In fact we define $V = \Lambda \otimes_{\mathbb{Z}} \mathbb{Q}$ and Hodge theory of complex tori states that $V = V^{(1,0)} \oplus V^{(0,1)}$ with $V^{(1,0)} = H^0(A, \Omega^1_A)$ and $V^{(0,1)} = H^1(A, \mathcal{O}_A) = \overline{V^{(1,0)}}$. The polarization is just the hermitian form and the required conditions are met because of the Riemann bilinear relations.

Conversely, given a polarized weight one $\mathbb{Z}$-Hodge structure (V, Λ, Q), we let $A = V^{(0,1)}/\Lambda$. The complex conjugation condition guarantees that this is a complex torus and the polarizations is a Hermitian form which is the first Chern class of a positive line bundle. Hence the complex torus is an abelian variety.

The following alternative viewpoint will generalize to families. Giving a Hodge structure is the same as giving a decreasing filtration

$$\ldots \supseteq F^p(V) \supseteq F^{p+1}(V) \supseteq \ldots$$

of $V_{\mathbb{C}}$, such that $V_{\mathbb{C}} = F^p(V) \oplus \overline{F^{k-p+1}}$. For the above decomposition one obtains the filtration by

$$F^p(V) = \bigoplus_{i \geq p} V^{(i,k-i)}$$

and conversely, the filtration determines the decomposition by

$$V^{(p,q)} = F^p(V) \cap \overline{F^q(V)}.$$

Period domains. From the filtration viewpoint it is obvious that the collection of all possible Hodge structures on a fixed vector space V form a complex subvariety $D^\vee$ of a product of Grassmann varieties. The polarization imposes a further positivity condition and the collection of all possible polarized Hodge structures form a domain D in $D^\vee$.

We specialize to the case of weight one. Then

$$D = \mathbb{H}_g = \{Z \in \mathbb{C}^{g \times g} : Z^T = Z \text{ and } \Im(Z) > 0\}$$

is the Siegel upper half space. Even more special, for $g = 1$, i.e. for polarized rank two weight one variations of Hodge structures the period domain is just the upper half plane $\mathbb{H}$.

Variation of Hodge structures The filtration viewpoint of Hodge structures generalizes to families so as to maintain the correspondence of weight one with abelian varieties. Recall that (for K a field or $\mathbb{Z}$) a K-local system $\mathbb{V}$ on a base B with fiber a K-vector space V is just a representation of $\pi_1(B) \to \mathrm{GL}(V)$. Equivalently, one may view a local system as a vector bundle over B whose transition functions are locally constant, or yet equivalently, as a vector bundle over B with a flat connection ∇. For families of curves (or abelian varieties), the local system will be given by the monodromy representation of $\pi_1(B)$ on the first cohomology of some fiber. Equivalently, the connection is the Gauss-Manin connection given by parallel transport of (e.g. singular) cohomology classes. Given a local system $\mathbb{V}$ on B, and suppose that the base is the complement of a normal crossing divisor Δ in a smooth, projective complex variety $\overline{B}$. Then there is a natural 'Deligne' extension of the vector bundle $\mathbb{V} \otimes \mathcal{O}_B$ to a vector bundle $\mathcal{V}$ on the whole base $\overline{B}$. We will not discuss the details of how to construct this extension and refer to [**Del70**, Chapter II.5] for a thorough definition.

For any field $K \subset \mathbb{R}$ and any complex ('base') manifold B, we define a K-*variation of Hodge structures* of weight k (VHS for short) to be a K-local system $\mathbb{V}$ together with a filtration of $\mathcal{V}$

$$F^\bullet(\mathcal{V}) = (\ldots \supseteq F^p(\mathcal{V}) \supseteq F^{p+1}(\mathcal{V}) \supseteq \ldots)$$

with the following properties: i) For every point $b \in B$ the stalks of the filtration form a weight k K-Hodge structure on the stalk $\mathbb{V}_b$. ii) Griffiths' transversality holds, i.e.

$$\nabla(F^p(\mathcal{V})) \subset F^{p-1}(\mathcal{V}) \otimes \Omega^1_{\overline{B}}.$$

We define a $\mathbb{Z}$-*variation of Hodge structures* to be a $\mathbb{Q}$-VHS $(\mathbb{V}_\mathbb{Q}, F^\bullet(\mathcal{V}))$ together with a $\mathbb{Z}$-local system $\mathbb{V}_\mathbb{Z}$ with the property that $\mathbb{V}_\mathbb{Z} \otimes_\mathbb{Z} \mathbb{Q} = \mathbb{V}_\mathbb{Q}$. A *polarization* of a K-VHS is a locally constant $\mathbb{C}$-bilinear map $Q : \mathbb{V}_\mathbb{C} \otimes \mathbb{V}_\mathbb{C} \to \mathbb{C}_B$ to the constant (rank one) local system $\mathbb{C}$ on B, whose stalks at every point $b \in B$ give a polarization of the induced Hodge structure on $\mathbb{V}_b$. Again, for $\mathbb{Z}$-variation of Hodge structures we require moreover, that the image $Q(\mathbb{V}_\mathbb{Z}, \mathbb{V}_\mathbb{Z})$ lies in the constant (rank one) local system $\mathbb{Z}_B$.

Recall the setup in Section 3.1 how we associated to a map $C \to M_g$ a family of curves $f : \mathcal{X} \to B$. The corresponding family of Jacobians has a weight one $\mathbb{Z}$-VHS that we can define explicitly as follows. Let X be a fiber over some point $b \in B$. The local system is given by the monodromy representation of $\pi_1(B)$ on $H^1(X, \mathbb{Z})$, in sheaf theory language as the higher direct image $R^1 f_* \mathbb{Z}$.

A map between VHS is a linear map of the underlying local systems that is compatible with the filtrations. The notion of VHS obviously admits all kinds of operations of multilinear algebra. In particular, the dual of a K-VHS of weight k is a K-VHS of weight $-k$. The tensor product of two K-VHS of weight k_1 and k_2 is a K-VHS of weight $k_1 + k_2$. The reader should keep in mind the particular case that $\mathbb{E}\mathrm{nd}(\mathbb{V}) = \mathbb{V}^\vee \otimes \mathbb{V}$ carries a weight zero VHS.

The most important and remarkable theorem is Deligne's semisimplicity result.

THEOREM 4.1 (Deligne ([**Del74**]),[**Sch73**, Theorem 7.25]). *Let B be complex manifold, complement of a normal crossing divisor Δ in an algebraic manifold $\overline{B}$. If $(\mathbb{V}, F^\bullet(\mathcal{V}), Q)$ is a polarized K-VHS over B, then monodromy representation*

$$\pi_1(B, b) \to \mathrm{GL}(\mathbb{V}_b)$$

is completely reducible into VHS, i.e. any $\mathbb{R}$-subrepresentation has a $\pi_1(B)$-invariant complement and inherits from $F^\bullet$ a filtration that makes it into a sub-$\mathbb{R}$-VHS.

One may want to decompose a local system into $\mathbb{C}$-irreducible pieces. Deligne defined in [**Del87**] the corresponding notion of $\mathbb{C}$-VHS and the semisimplicity theorem still holds. Since this is much less common, we chose to avoid this concept here at the cost of decomposing VHS over $\mathbb{R}$ only. This will be sufficient for the main properties of Teichmüller curves thanks to Theorem 2.8

In Section 5 we will prove a special instance of Theorem 4.1 with an application to Teichmüller curves. We give here the necessary prerequisites that we want to reduce the theorem to.

THEOREM 4.2 ([**Sch73**, Theorem 7.22]). *If B is a complex manifold, complement of a normal crossing divisor Δ in an algebraic manifold $\overline{B}$, and $(\mathbb{V}_K, F^\bullet(\mathcal{V}), Q)$ is a polarized K-VHS of weight k over B, then for any flat section $e = \sum_p e_p$ the Hodge components $e_p \in V^{(p,k-p)}$ are also flat.*

INGREDIENTS OF THE PROOF: Let ℓ be the least integer such that $e_p = 0$ for $p > \ell$. Then one uses the following properties of period domains and period mappings. The function $\varphi = i^{2\ell-k}Q(e_\ell, e_\ell)$ is plurisubharmonic for any VHS ([**Sch73**, Lemma 7.19], see also [**CMSP03**]). It is bounded above by a consequence of the Nilpotent orbit theorem, hence constant by the hypothesis on B. The curvature calculation that gives plurisubharmonicity can now be revisited to show that e_ℓ is flat. The proof concludes by induction on ℓ. □

Period mapping. Suppose that locally on $U \subset B$ we have trivialized the local system $\mathbb{V}$ underlying a VHS. Then we may associate with $b \in U$ the point in the period domain determined by the filtration $F^\bullet(\mathbb{V}_b)$. This defines a map to $U \to D$, if the VHS is polarized, or from the universal covering $\widetilde{B}$ of B to D. This map is called period map, it is known to be holomorphic ([**Gri68**]).

Concretely, for a rank two weight one VHS $\mathbb{L}$, choose a basis $\{\alpha, \beta\}$ of $\mathbb{L}^\vee$ locally on U (or on $\widetilde{B}$) and let $\omega(b)$ be a non-zero section of $\mathbb{L}^{(1,0)}$. Then the period map is given by

$$b \mapsto \langle \beta, \omega(b)\rangle / \langle \alpha, \omega(b)\rangle.$$

Less abstractly, take $\mathbb{L}$ to be the VHS of a family of elliptic curves, choose α, β generators of $H_1(E, \mathbb{Z})$, where E is the elliptic curve in some fiber. Then the contraction $\langle \alpha, \omega(b)\rangle$ is just the integration $\int_\alpha \omega(b)$ of the one-form along some loop representing α.

Hodge norm. In the section on Lyapunov exponents we want to measure sizes of cohomology classes in $H^1(X, \mathbb{R})$ where X is a curve (or maybe also an abelian variety). We can build a norm on $H^1(X, \mathbb{R})$ using the polarization Q as follows. Write $V = H^1(X, \mathbb{R}) \ni v = v^{(1,0)} + v^{(0,1)}$ and let C be the linear map defined as i^{p-q} on $V^{(p,q)}$. Then define $(v, w) = Q(Cv, \bar{w})$ and the *Hodge norm* is the associated norm $||v||^2 = (v, v)$.

Alternatively, for a curve X define the Hodge-* operator as $*v = i(\overline{v^{(1,0)}} - \overline{v^{(0,1)}})$. Then $(v, w) = Q(v, *w)$ is a scalar product on $H^1(X, \mathbb{R})$ and the associated norm

$$||v||^2 = (v, v) = \frac{i}{2} \int_X v \wedge *v.$$

is called the *Hodge norm* of v.

Period coordinates. As a motivation recall that the Torelli theorem states that the period map for a family of curves is locally an embedding outside the hyperelliptic locus ([**OS80**]). We may thus view a period matrix a 'coordinate system' for $\mathcal{M}_g$ outside the hyperelliptic locus, but we should keep in mind that the Torelli map is not (locally) onto, since $\dim \mathcal{M}_g < \dim \mathcal{A}_g$ for $g \leq 4$. We thus used quotation marks for this coordinate system. The period matrix entries use *all* holomorphic one-forms and all periods, but *only absolute* periods.

To contrast, for the strata $\Omega\mathcal{M}_g$ we do have *period coordinates* defined as follows. Fix locally around a given point (X, ω) a basis $\gamma_1, \ldots, \gamma_N$ of $H_1(X, Z(\omega), \mathbb{Z})$ and map a neighboring flat surface (Y, η) to $(\int_{\gamma_i} \eta)_{i=1}^N \in \mathbb{C}^N$. This map is indeed a local isomorphism (see [**Vee90**] for a proof using flat surface geometry, [**HM79**] and [**Möl09**] for algebraic proofs). We emphasize that these period coordinates use *only one* of the one-forms on X but *also relative* periods.

4.1. Hilbert modular varieties and the locus of real multiplication

Let $\mathcal{A}_g = \mathbb{H}_g/\mathrm{Sp}_{2g}(\mathbb{Z})$ be the moduli space of g-dimensional principally polarized Abelian varieties, where $\mathbb{H}_g$ is the $g(g+1)/2$-dimensional Siegel upper half space. Most abelian varieties only have the endomorphisms given by multiplication by an integer. A locus where the endomorphism ring is strictly bigger will play a central role when analyzing the VHS of a Teichmüller curve. We denote by $\mathcal{R}A_{\mathfrak{o}} \subset \mathcal{A}_g$ the *locus of Abelian varieties with real multiplication* by the order $\mathfrak{o}$, that we now define precisely.

Consider a totally real number field F of degree g. A *lattice in* F is a subgroup of the additive group of F isomorphic to a rank g free Abelian group. An *order* in F is a lattice which is also a subring of F containing the identity element. The ring of integers in F is the unique maximal order.

Let A be a principally polarized g-dimensional Abelian variety. We let $\mathrm{End}(A)$ be the ring of endomorphisms of A and $\mathrm{End}^0(A)$ the subring of endomorphisms such that the induced endomorphism of $H_1(A;\mathbb{Q})$ is self-adjoint with respect to the symplectic structure defined by the polarization.

Real multiplication by F on A is a monomorphism $\rho\colon F \to \mathrm{End}^0(A) \otimes_{\mathbb{Z}} \mathbb{Q}$. The subring $\mathfrak{o} = \rho^{-1}(\mathrm{End}(A))$ is an order in F, and we say that A has real multiplication by $\mathfrak{o}$.

The locus of real multiplication $\mathcal{R}A_{\mathfrak{o}}$ is the image of a union of Hilbert modular varieties, defined as follows. Choose an ordering $\iota_1, \ldots, \iota_g$ of the g real embeddings of F. We use the notation $x^{(i)} = \iota_i(x)$. The group $\mathrm{SL}_2(F)$ then acts on $\mathbb{H}^g$ by $A \cdot (z_i)_{i=1}^g = (A^{(i)} \cdot z_i)_{i=1}^g$, where $\mathrm{SL}_2(\mathbb{R})$ acts on the upper-half plane $\mathbb{H}$ by Möbius transformations in the usual way.

Given a lattice $M \subset F^2$, we define $\mathrm{SL}(M)$ to be the subgroup of $\mathrm{SL}_2(F)$ which preserves M. The *Hilbert modular variety* associated to M is

$$X(M) = \mathbb{H}^g/\mathrm{SL}(M).$$

Given an order $\mathfrak{o} \subset F$, we define

$$X_{\mathfrak{o}} = \coprod_M X(M),$$

where the union is over a set of representatives of all isomorphism classes of proper rank two symplectic $\mathfrak{o}$-modules. If $\mathfrak{o}$ is a maximal order, then every rank two symplectic $\mathfrak{o}$-module is isomorphic to $\mathfrak{o}\oplus\mathfrak{o}^\vee$ (this also holds if $g=2$; see [**McM07**]), so in this case $X_{\mathfrak{o}}$ is connected. In general, $X_{\mathfrak{o}}$ is not connected, as there are non-isomorphic proper symplectic $\mathfrak{o}$-modules; see the Appendix of [**BM**].

We now construct the map $X(M) \to \mathcal{A}_g$ in the simplified situation where $M = \mathfrak{o}\oplus\mathfrak{o}^\vee$. Pick a $\mathbb{Z}$-basis $\omega_1, \ldots, \omega_g$ of $\mathfrak{o}$ and let $B = (\iota_j(\omega_k))_{j,k=1}^g$ and $A = B^{-1}$. Then the map

$$\psi : (\tau_1, \ldots, \tau_g) \mapsto A \cdot \mathrm{diag}(\tau_1, \ldots, \tau_g) \cdot A^T, \quad \mathbb{H}^g \mapsto \mathbb{H}_g$$

is equivariant with respect to the action of $\mathrm{SL}(\mathfrak{o}\oplus\mathfrak{o}^\vee)$ on $\mathbb{H}^g$ and its Ψ-image on $\mathbb{H}_g$, where Ψ is defined as follows. For $a \in F$ let a^* be the diagonal matrix with entries $\iota_k(a)$, write

$$\mathrm{diag}(\gamma) = \begin{pmatrix} a^* & b^* \\ c^* & d^* \end{pmatrix} \quad \text{for} \quad \gamma = \begin{pmatrix} a & b \\ c & d \end{pmatrix} \in \mathrm{SL}(\mathfrak{o}\oplus\mathfrak{o}^\vee).$$

and we let

$$\Psi : \gamma \mapsto \operatorname{diag}(A, (A^T)^{-1}) \cdot \operatorname{diag}(\gamma) \cdot \operatorname{diag}(A, (A^T)^{-1})^{-1}, \quad \mathrm{SL}(\mathfrak{o} \oplus \mathfrak{o}^\vee) \to \mathrm{Sp}(2g, \mathbb{Z}).$$

Consequently ψ descends to the desired map $X(M) \to \mathcal{A}_g$. One easily checks that abelian varieties with the period matrix $(I_g \; A\operatorname{diag}(\tau_1, \ldots, \tau_g)A^T)$ or equivalently with period matrix $(B, \operatorname{diag}(B^T)^{-1})$ do indeed have real multiplication with $\mathfrak{o}$.

A criterion for real multiplication. The next theorem gives a criterion (see [**Del71**, Section 4.4]) how to detect from the decomposition of a weight one VHS that the corresponding family has real multiplication.

THEOREM 4.3. *Let $\mathbb{V}$ be the weight one VHS associated with a family of abelian varieties $f : \mathcal{A} \to C$.*

If the VHS $\mathbb{V}$ decomposes over $\mathbb{Q}$ as $\mathbb{V}_1 \oplus \mathbb{V}_2$, then the family of abelian varieties decomposes up to isogeny into two families of abelian varieties of dimension $\mathrm{rk}(\mathbb{V}_j)/2$.

If we have a decomposition

$$\mathbb{V}_L = (\oplus_{\sigma \in \mathrm{Gal}(L/\mathbb{Q})/\mathrm{Gal}(L/F)} \mathbb{L}^\sigma)$$

with the property that $\mathbb{L}^\sigma \cong \mathbb{L}^\tau$ if and only if $\sigma\tau^{-1}$ fixes F, then the family of abelian varieties has real multiplication by F.

PROOF. A homomorphism between two abelian varieties gives rise to a $\mathbb{C}$-linear map between their universal coverings and a homomorphism between their period lattices. Conversely, a homomorphism between the period lattices compatible with a $\mathbb{C}$-linear map of the universal coverings defines a homomorphism between two abelian varieties. Similar statements hold in families.

Let $\mathrm{End}(\mathbb{V}_\mathbb{Q})$ denote the global sections of the local system $\mathbb{E}\mathrm{nd}(\mathbb{V}_\mathbb{Q})$. We claim that an element of $\varphi \in \mathrm{End}(\mathbb{V}_\mathbb{Q}) \cap \mathbb{E}\mathrm{nd}(\mathbb{V})^{(0,0)}$ defines an element of $\mathrm{End}(\mathcal{A}) \otimes \mathbb{Q}$. In fact, a multiple $n\varphi$ will lie in $\mathrm{End}(\mathbb{V}_\mathbb{Z})$ and thus defines a self-map between the local system (of 'period lattices'). Lying in $\mathbb{E}\mathrm{nd}(\mathbb{V})^{(0,0)}$ says that $n\varphi$ preserves the graded pieces of the Hodge filtration. The universal covering of an abelian variety A can be identified with the tangent space at zero or with $H^1(A, \mathcal{O}_A)$. Consequently, the map on the graded pieces associated with $n\varphi$ defines a family of $\mathbb{C}$-linear maps between the family of universal coverings.

In the first case the map $\mathrm{id}_{\mathbb{V}_1}$ is certainly a global section of $\mathbb{E}\mathrm{nd}(\mathbb{V}_\mathbb{Q})$ that lies in $\mathbb{E}\mathrm{nd}(\mathbb{V})^{(0,0)}$. In the second case we map $a \in F$ to $\sum_{\sigma \in \mathrm{Gal}(L/\mathbb{Q})/\mathrm{Gal}(L/F)} \sigma(a) \cdot \mathrm{id}_{\mathbb{L}^\sigma}$. This endomorphism certainly lies in $\mathbb{E}\mathrm{nd}(\mathbb{V})^{(0,0)}$. It also lies in $\mathrm{End}(\mathbb{V}_L)$ and the action of $\mathrm{Gal}(L/\mathbb{Q})$ just renumbers the summands by the hypothesis on the $\mathbb{L}^\sigma$. Consequently, this endomorphism lies in $\mathrm{End}(\mathbb{V}_\mathbb{Q})$. □

4.2. Examples

We present some examples of families of curves whose VHS decomposes. We will see later that the examples are related to Teichmüller curves.

Cyclic coverings. Consider the family of curves $f : \mathcal{X} \to C$ with

$$\mathcal{X} : y^N = x^{a_1}(x-1)^{a_2}(x-t)^{a_3}$$

over $C = \mathbb{P}^1$ with coordinate t. Define $0 < a_4 < N$ such that $a_1 + a_2 + a_3 + a_4 \equiv 0 \mod N$. This is a family of covers of the projective line ramified over 4 points $0, 1, t, \infty$. The automorphism $\varphi(x, y) = (x, \zeta_N y)$ generates the Deck group of this covering. It acts on $H^1(X, \mathbb{C})$, $H^0(X, \Omega^1_X)$ and $H^1(X, \mathcal{O}_X)$ for every fiber

X of $\mathcal{X}$ and the decomposition of $H^1(X, \mathbb{C})$ commutes with parallel transport. Consequently, the VHS $\mathbb{V}$ decomposes over $\mathbb{C}$ (see the remark after Theorem 4.1) into the eigenspaces of φ. The quotient of $\mathcal{X}$ by the group generated by φ is a family of projective spaces $\mathbb{P}^1_x$ with coordinate x. Only when considered as family of projective spaces with 4 marked points $0, 1, t, \infty$, this is a non-trivial family over $\mathbb{P}^1_t$.

More concretely, one can explicitly write down a basis of holomorphic one-forms in the eigenspace with eigenvalue ζ_N^i as

$$\omega_j^i = \frac{x^{j-1}y^i dx}{x^{b_{1,i}}(x-1)^{b_{2,i}}(x-t)^{b_{3,i}}}$$

for $j = 1, \ldots, 2 - b(i)$, where $b_{k,i} = \lfloor ia_k/N \rfloor$ and where $b(i) = \sum_{k=1}^{4} \langle ia_k/N \rangle - 1$. Here $\langle \cdot \rangle$ denotes the fractional part. Similarly, one can represent elements in $H^1(X, \mathcal{O}_X)$ e.g. in Czech cocycles explicitly.

As a result, one obtains that for all i the eigenspaces $\mathbb{L}_i$ corresponding to the eigenvalue ζ_N^i is of rank two. The details appear in many places in the literature, e.g. [**Bou01**], [**BM10a**], [**EKZa**].

For special values of the covering parameters the curve $\mathcal{X}$ has additional automorphisms. E.g. take m, n odd and coprime, let $N = 2mn$ and $a_i = mn \pm m \pm n$. Then the $(\mathbb{Z}/2\mathbb{Z})^2$-action on $\mathbb{P}^1_x$ acting as double transpositions of the 4 branch points lifts to a $(\mathbb{Z}/2\mathbb{Z})^2$-action on $\mathcal{X}$. The VHS of the quotient family decomposes over the *reals* into rank two VHS $\mathbb{L}_i$, where the $\mathbb{L}_i$ are isomorphic to the $\mathbb{L}_i$ appearing in the VHS associated with $f : \mathcal{X} \to C$, but only subset of indexes i of $\{1, \ldots, N\}$ appears. Consequently, by Theorem 4.3 the quotient family has RM (on some part of its Jacobian). Details can be found in [**BM10a**].

5. Teichmüller curves

A *Teichmüller curve* $C \to \mathcal{M}_g$ is an algebraic curve in the moduli space of curves that is totally geodesic with respect to the Teichmüller metric.

We do not recall the definition of this metric since we only use two consequences. First, the Teichmüller metric on Teichmüller space T_g is the same as the Kobayashi metric explained below. Moreover, if $C \to \mathcal{M}_g$ is a Teichmüller curve the universal covering map is a map $\mathbb{H} \to T_g$ to Teichmüller space. This map is also totally geodesic with respect to the Teichmüller metric and those totally geodesic maps $\mathbb{H} \to T_g$ are called *Teichmüller discs*. The second property of the Teichmüller metric we us is that every Teichmüller disc is the $\mathrm{SL}_2(\mathbb{R})$-orbit of a flat surface (X, ω) or a half-translation surface (X, q) (see e.g. [**Hub06**, Chapter 5.3]). We say that the (X, ω) or (X, q) *generates* the Teichmüller curve. In the sequel we exclusively consider Teichmüller curves generated by flat surfaces (X, ω). For Teichmüller curves generated by a half-translation surface (X, q) one can pass to the canonical double covering and then apply the results below for some information about their trace fields, slopes etc.

Since C is an algebraic curve and $C \cong \mathbb{H}/\mathrm{SL}(X, \omega)$ ([**McM03**]), this implies that $\mathrm{SL}(X, \omega)$ is a lattice in $\mathrm{SL}_2(\mathbb{R})$. A surface (X, ω) with this property is called a lattice surface or, more frequently, a *Veech surface*.

Consequently, a Teichmüller curve is the image of the projection of an closed $\mathrm{SL}_2(\mathbb{R})$-orbit from $\Omega\mathcal{M}_g$ to $\mathcal{M}_g$. Also the converse holds:

THEOREM 5.1 (Smillie, Weiss, [**Vee95**], [**SW04**]). *If the* $\mathrm{SL}_2(\mathbb{R})$*-orbit of* (X,ω) *is closed then* (X,ω) *generates a Teichmüller curve.*

We emphasize that Teichmüller curves are closed in $\mathcal{M}_g$, but never closed in $\overline{\mathcal{M}}_g$, i.e. they are never compact curves, as we have seen in Proposition 2.4.

Motivation: Veech dichotomy. A flat surface satisfies *Veech dichotomy* or is *dynamically optimal* if the following property of the straight line flow ϕ_t^θ holds. For every fixed direction θ either ϕ_t^θ is uniquely ergodic or all trajectories of ϕ_t^θ are closed. (Here saddle connections also count as closed trajectories).

The fact that Veech surfaces are dynamically optimal is one of the key observations of [**Vee89**], but the proof has to be fixed concerning the unique ergodicity statement. See [**MT02**] for a complete proof. The property Veech dichotomy is not quite characterizing Veech surfaces: this holds for $g = 2$ ([**McM05b**]), but counterexamples exist for higher genus. The most frequently used consequence of Veech dichotomy is that on a Veech surface every direction that contains a saddle connection is indeed periodic.

We give examples of Teichmüller curves before we turn to their algebraic characterization.

5.1. Square-tiled surfaces and primitivity

A *square-tiled surface* is a flat surface (X,ω), where X is obtained as a covering of a torus ramified over one point only and ω is the pullback of the holomorphic one-form on the torus. ('Parallelogram tiled' would be slightly more accurate, but square-tiled has become standard terminology.) The affine group of the torus is $\mathrm{SL}_2(\mathbb{Z})$ and such coverings change the affine group only by a finite amount. For a more precise statement we introduce the following notions. Two subgroups Γ_1 and Γ_2 of $\mathrm{SL}_2(\mathbb{R})$ are called *commensurable*, if there is a subgroup Γ that has finite index both in Γ_1 and in Γ_2.

A *translation covering* $\pi : (X,\omega) \to (Y,\eta)$ is a covering $\pi : X \to Y$ of Riemann surfaces such that $\omega = \pi^*\eta$.

THEOREM 5.2 ([**GJ00**]). *Let* $\pi : (X,\omega) \to (Y,\eta)$ *be a translation covering. If* π *is branched only over* $Z(\eta)$ *or if* $g(Y) = 1$ *and* π *is branched over at most one point, then* $\mathrm{SL}(X,\omega)$ *and* $\mathrm{SL}(Y,\eta)$ *are commensurable.*

In view of Theorem 5.5 we restate that the trace field of a square-tiled surface is $\mathbb{Q}$. Square-tiled surfaces are a rich source of examples of Teichmüller curves. More precisely:

PROPOSITION 5.3. *For every* g *and each connected component of every stratum of* $\Omega\mathcal{M}_g$ *Teichmüller curves generated by square-tiled surfaces are dense (for the complex topology).*

PROOF. If a flat surface (X,ω) has all its period coordinates in $\frac{1}{N}\mathbb{Z}[i]$ for some N, then X admits a branched cover to the square torus branched over one point only. In fact, for any point $P \in X$ choose a path γ joining a zero of ω to X and map P to $N \cdot \int_\gamma \omega$. This map is well-defined by hypothesis.

Since period coordinates are indeed coordinates on every stratum, we may find a point with period coordinates in $\frac{1}{N}\mathbb{Z}[i]$ (for appropriately large N) in a neighborhood of every point of that stratum. $\square$

There is a convenient way to present square-tiled surfaces. In order to specify such a surface, with say d squares, it suffices to specify the monodromy of the covering. The fundamental group of a once-punctured torus is a free group on two generators. This monodromy is thus given by two permutations (π_u, π_r) on d letters, corresponding to going up and going to the right, respectively. The square-tiled surfaces drawn in Figure 2 is given in permutation representation by

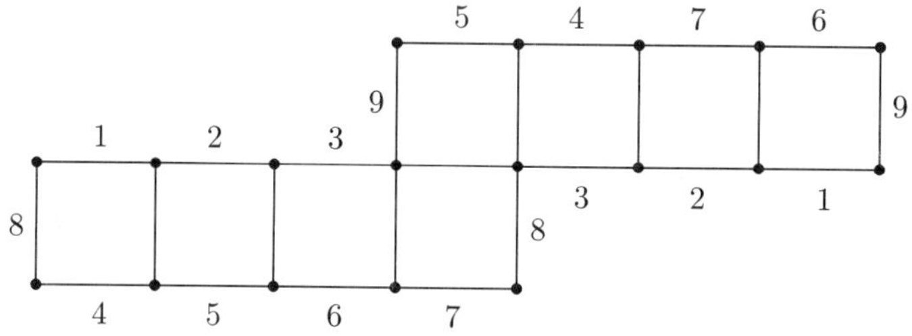

FIGURE 2. The 'eierlegende Wollmilchsau'

$(\pi_r = (1234)(5678), \pi_u = (1836)(2745))$. The name will be explained later.

Algebraic and geometric primitivity. From the point of view of classification of Teichmüller curves, square-tiled surfaces can be thought of as just one example plus a large amount of combinatorial decoration. One thus wants to classify Teichmüller curves generated from flat surfaces that do not stem from coverings. More precisely, we call a flat surface (X, ω) *geometrically primitive* if there is no translation covering $\pi : (X, \omega) \to (Y, \eta)$ with $g(Y) < g(X)$. A technical notion that is easy to check and a criterion of geometric primitivity is the following. We say that a Veech surface (X, ω) is *algebraically primitive*, if $g(X) = [K : \mathbb{Q}]$, where K is the trace field of $\mathrm{SL}(X, \omega)$. Examples of algebraically primitive Veech surfaces are the n-gons where n is an odd prime ([**Vee89**]).

If (X, ω) is algebraically primitive, then (X, ω) is geometrically primitive by Proposition 2.5. The converse does not hold in general. An infinite series of examples for this situation are given in [**McM06a**]. The surface in Figure 1 is an example of this situation if we perform the Thurston-Veech construction with multiplicities $d_n = (n, 1, n, 1, n, 1)$. We say that a Teichmüller curve is (algebraically) primitive, if a generating Veech surface is.

PROPOSITION 5.4. *The equivalence class of a flat surface (X, ω) given by the relation 'translation covering' contains an elliptic curve or a unique 'minimal' element, i.e. there is a flat surfaces (Y, η) such that all translation surfaces in the class of (X, ω) admit a translation covering to (Y, η).*

If the equivalence class contains an elliptic curve, i.e. if (X, ω) is square-tiled, the uniqueness of the minimal element does not hold, since we may always postcompose π by isogenies.

The major classification goal for Teichmüller curves seeks to classify those curves generated by the minimal representatives. The proof of this proposition is one of the corollaries in the next section.

5.2. The VHS of $\mathcal{T}$ curves

The reader should now recall the conventions for curves and flat surfaces that we set up in Section 3.1. Let $L/\mathbb{Q}$ be a Galois closure of $K/\mathbb{Q}$.

THEOREM 5.5 ([**Möl06b**, Prop. 2.4]). *Let B be a finite unramified cover of a Teichmüller curve generated by a flat surface (X,ω) and let $f:\mathcal{X}\to B$ be the universal family. Then the $\mathbb{R}$-variation of Hodge structures defined by B decomposes into sub-VHS*

$$R^1 f_*\mathbb{R} = (\oplus_{\sigma\in \mathrm{Gal}(L/\mathbb{Q})/\mathrm{Gal}(L/K)}\mathbb{L}^\sigma)\oplus \mathbb{M},$$

where $\mathbb{L}$ is the VHS with the standard 'affine group' representation of $\mathrm{SL}(X,\omega)\subset \mathrm{SL}_2(\mathbb{R})$, where $\mathbb{L}^\sigma$ are the Galois conjugates and where $\mathbb{M}$ is just some representation. Moreover, we have

$$2\deg(\mathbb{L}^{(1,0)}) = 2g(B)-2+|\Delta_B|.$$

There is also a converse of this, i.e. a characterization of Teichmüller curves.

THEOREM 5.6 ([**Möl06b**, Theorem 2.13]). *Let $f:\overline{\mathcal{X}}\to\overline{B}$ be a fibered surface and suppose that the VHS has a rank two sub-VHS $\mathbb{L}$, such that*

$$2\deg(\mathbb{L}^{(1,0)}) = 2g(B)-2+|\Delta_B|.$$

Then $B\to\mathcal{M}_g$ is a finite unramified covering of a Teichmüller curve $C\to\mathcal{M}_g$ generated by a flat surface (X,ω).

A sub-VHS $\mathbb{L}$ as in the statement of the theorem was called 'maximal Higgs' in [**Möl06b**]. Very few is known on which representations occur for $\mathbb{M}$ or what their numerical data (see the subsequent chapters) are. Note that for square-tiled surfaces $\mathbb{M}$ is almost the whole VHS!

We give two applications before we turn to the proof of these theorems.

PROOF OF PROPOSITION 5.4. Suppose that $f:\mathcal{X}\to B$ and $g:\mathcal{Y}\to B$ are two families of curves over a (finite covering of a) Teichmüller curve that arise from a translation covering $\pi:(X,\omega)\to(Y,\eta)$ of generating flat surfaces. Then the VHS associated with both with f and g contains the generating local system $\mathbb{L}$ and hence also its Galois conjugates. Note that the one-form ω on the family of Jacobians is zero on any complement to the abelian variety associated with $\oplus_\sigma\mathbb{L}^\sigma$.

Consequently, given (X,ω) we take the limit over all (ordered by inclusion) abelian subvarieties $A\subset \mathrm{Jac}(X)$ such that $\omega|_A=0$ of the normalization of the image Z of X in $\mathrm{Jac}(X)/A$. Obviously, we can provide Z with a one-form ω_Z, such that (X,ω) is a translation covering of (Z,ω_Z). Moreover any (Y,η) covers (Z,ω_Z) since we may take $A=\mathrm{Ker}(\mathrm{Jac}(X)\to\mathrm{Jac}(Y))$. □

COROLLARY 5.7. *The family of Jacobians associated with the fibered surface $f:\mathcal{X}\to B$ over a Teichmüller curve decomposes up to isogeny into two families of abelian varieties $g_1:\mathcal{A}_1\to B$ and $g_2:\mathcal{A}_2\to B$ of dimension $r=[K:\mathbb{Q}]$ and $g-r$ respectively. Moreover, $g_1:\mathcal{A}_1\to B$ has real multiplication by K.*

PROOF. Given the decomposition in Theorem 5.5 and the criterion Theorem 4.3 it suffices to verify the condition in the second statement of that criterion. Moreover, once we verify this condition for $\mathbb{L}$ it follows for the other $\mathbb{L}^\sigma$ by Galois conjugation.

If $\mathbb{L}\cong\mathbb{L}^\sigma$ then the traces of the underlying representation have to be fixed by σ, hence σ has to fix K. Conversely, suppose that $\mathbb{L}\ncong\mathbb{L}^\sigma$ for some σ fixing K. Then both $\mathbb{L}$ and $\mathbb{L}^\sigma$ appear in the VHS decomposition of $\mathbb{V}$. Let φ be some pseudo-Anosov element in $\mathrm{SL}(X,\omega)$ with dilatation λ and $t(\lambda)=\mathrm{tr}(D\varphi)=\lambda+\lambda^{-1}>2$. Both the fibers of $\mathbb{L}$ and of $\mathbb{L}^\sigma$ in $H^1(X,L)$ are in the kernel of $\varphi^*+\varphi^*-t\cdot\mathrm{id}\in$

$\mathrm{End}(H^1(X,L))$. Since the function $t(\lambda)$ is monotone for $\lambda \in (1,\infty)$, this implies that λ is a multiple eigenvalue of the action of φ on $H^1(X,L)$. But λ is known to be the largest eigenvalue of this action and simple by Perron-Frobenius (see e.g. [**McM03**, Theorem 5.3]]. This contradiction concludes the proof. □

Real multiplication and the classification problem. We give some indication of why real multiplication might be useful to classify Teichmüller curves and a warning why one should not be too optimistic. See Section 5.5 for more details on the classification problem.

Suppose for simplicity that we are interested in algebraically primitive Teichmüller curves only. The locus of real multiplication $\mathcal{RA}_\mathfrak{o}$ in $\mathcal{A}_g$ is the image of Hilbert modular varieties $X(M)$. They are of dimension g while $\dim \mathcal{A}_g = g(g+1)/2$. Since $\dim \mathcal{M}_g = 3g-3$, for large enough g the expected dimension of the intersection $X(M) \cap \mathcal{M}_g$ considered in $\mathcal{A}_g$ via the map defined in Section 4.1 resp. via the Torelli map is zero. Hence there should be no algebraically primitive Teichmüller curves for large genus – which is known to be false as is shown in the initial paper [**Vee89**] already.

On the other hand the union of all Hilbert modular varieties is Zariski dense in $\mathcal{A}_g$ by the Borel density theorem, so this explains why one should be cautious.

There is an analogous conjecture of Coleman stating that there are no Shimura curves (see Section 6.5) in $\mathcal{A}_g$ whose generic point lies in $\mathcal{M}_g$ for g large enough. Again, being a Shimura curve gives additional endomorphisms or at least additional Hodge cycles, so one can make a similar dimension heuristics. But there has not been much progress in the last decade on this question.

5.3. Proof of the VHS decomposition and real multiplication

The proof of Theorem 5.5 was given in [**Möl06b**] using $\mathbb{C}$-VHS and the corresponding $\mathbb{C}$-version of the Semisimplicity Theorem 4.1. Moreover, the fact that trace fields are totally real was deduced from that proof. We give here the proof for algebraically primitive Teichmüller curves, reducing the statement to Theorem 4.2. Theorem 2.8 allows us to stay entirely within the more well-known context of $\mathbb{R}$-VHS.

Proof (Algebraically primitive case): Let $\mathbb{V} = R^1 f_* \mathbb{R}$ be the $\mathbb{R}$-local system underlying the weight one VHS of the given family of curves. Since we deal with a Teichmüller curve, the family is generated as the $\mathrm{SL}_2(\mathbb{R})$-orbit of some flat surface (X,ω). That is, X is the fiber of f over some point $b \in B$ and in that fiber $\Re(\omega)$ and $\Im(\omega)$ generate a rank two real sub-vector space L of $\mathbb{V}_b = H^1(X,\mathbb{R})$. All the other fibers of f are the image of (X,ω) for some matrix $A \in \mathrm{SL}_2(\mathbb{R})$. All elements in $\gamma \in \pi_1(B)$ can be represented in the universal covering of B by a path from (X,ω) to $A_\gamma \cdot (X,\omega)$ for some matrix $A_\gamma \in \mathrm{SL}(X,\omega) \subset \mathrm{SL}_2(\mathbb{R})$. The matrix A_γ acts, by definition of the action of $\mathrm{SL}_2(\mathbb{R})$ by the standard representation on L. In particular it preserves the subspace L and we have found a local subsystem $\mathbb{L}$.

Now we use the simplifying additional hypothesis that $r = g$. Consequently, there is a collection of $\gamma_i \in \mathrm{SL}(X,\omega)$ such that the traces of γ_i generate a field of degree r over $\mathbb{Q}$. We denote by $\iota_1,\ldots,\iota_r$ the different embeddings of K into $\mathbb{R}$ and choose automorphisms σ_j of the Galois closure of $K/\mathbb{Q}$ such that $\sigma_j \circ \iota_1 = \iota_j$. Since $\mathbb{V}$ is defined over $\mathbb{Q}$ the σ_j-images of $\mathbb{L}$ are also K-local subsystems of $\mathbb{V}$. Since the standard two-dimensional representation of the lattice $\mathrm{SL}(X,\omega)$ is irreducible, so are the Galois conjugates and consequently, $\mathbb{L}^{\sigma_j} \cap \mathbb{L}^{\sigma_k} = 0$ for $j \neq k$.

To sum up, we know that

$$R^1 f_* \mathbb{R} = (\oplus_{j=1}^r \mathbb{L}^{\sigma_j})$$

as local systems. Said differently, we have a decomposition $V = H^1(X, \mathbb{R}) = (\oplus_{j=1}^r L_j)$ into $\pi_1(B)$-stable sub-vector spaces and we have to show that the Hodge filtration

$$0 \subset F^1(\mathcal{V}) \subset \mathcal{V} = (R^1 f_* \mathbb{R}) \otimes_{\mathbb{R}} \mathcal{O}_B$$

intersects each of the summands $\mathbb{L}_j \otimes_{\mathbb{R}} \mathcal{O}_B$ in a vector bundle. Dimensions of intersections of vector bundles are lower semicontinuous, and if the dimension is constant, such an intersection is again a vector bundle. Hence it suffices to show that at a general point (which we may suppose $b \in B$ to be)

$$L_j = (L_j \cap V^{(1,0)}) \oplus (L_j \cap V^{(0,1)}) \tag{8}$$

for all j, since then a dimension jump in a special fiber would lead to a dimension contradiction.

Consider the element $P_j \subset \operatorname{Hom}(V, V)$ that consists of projection to the subspace L_j composed with the inclusion. Since L_j is $\pi_1(B)$-invariant, P_j is a flat global section of the local system $\operatorname{Hom}(\mathbb{V}, \mathbb{V})$.

On the other hand, $\operatorname{Hom}(\mathbb{V}, \mathbb{V})$ carries a Hodge structure of weight zero and we may decompose P_j into its Hodge components

$$P_j = P_j^{(1,-1)} \oplus P_j^{(0,0)} \oplus P_j^{(-1,1)},$$

where the upper index indicates the shift in bidegree, if we consider P_j as endomorphism. Now we apply Theorem 4.2 to conclude that all the components, in particular $P_j^{(1,-1)}$ and $P_j^{(-1,1)}$ are also flat. The only flat global sections of $\oplus_{j=1}^r L_j$ are $\mathbb{C}^g$ acting diagonally, since the L_j (as $\pi_1(B)$-representations) are irreducible. In particular, no power of a non-zero global section vanishes. Since $(P_j^{(1,-1)})^2 = 0 = (P_j^{(-1,1)})^2$, these sections have to be zero. Hence $P_j = P_j^{(0,0)}$ and this implies (8), which we needed to show. □

Instead of fully proving the characterization we highlight one of the main arguments of the proof, the use of the Kobayashi metric in the next proposition. A fully self-contained proof will appear soon in a paper of A. Wright.

PROPOSITION 5.8. *Let $f : \overline{\mathcal{X}} \to \overline{B}$ be a fibered surface and suppose that the VHS has a rank two sub-VHS $\mathbb{L}$, such that the monodromy representation underlying $\mathbb{L}$ is the Fuchsian representation of $\pi_1(B) = \mathbb{H}/\Gamma$ in $\mathrm{SL}_2(\mathbb{R})$. Then $B \to \mathcal{M}_g$ is a finite unramified covering of a Teichmüller curve $C \to \mathcal{M}_g$.*

This is the only place in this text, where the Teichmüller metric actually appears. We use that the Teichmüller metric on Teichmüller space is indeed equivalent to the Kobayashi metric k_W, which is defined for any complex space W as follows.

We denote by d_Δ the Poincaré metric on the unit disc Δ and for all $x, y \in W$ we call a *chain* from x to y points $x_0, x_1, \ldots, x_n \in \Delta$ together with maps $f_i : \Delta \to W$ such that

$$f_1(x_0) = x, \quad f_j(x_j) = f_{j+1}(x_j),\ j = 1, \ldots, n-1, \quad f_n(x_n) = y.$$

Then

$$k_W(x, y) = \inf \sum_{i=1}^n d_\Delta(x_{i-1}, x_i),$$

where the infimum is over all chains from x to y.

From the definition already one can deduce that all holomorphic maps are distance non-increasing for the Kobayashi metric and that a composition of two holomorphic maps is a Kobayashi isometry only if the first map is a Kobayashi isometry, too.

PROOF. The VHS $\mathbb{L}$ gives rise to a period map p from the universal cover of B to the period domain of $\mathbb{L}$, i.e. $p : \mathbb{H} \to \mathbb{H}$. By definition, this map is equivariant with respect to the action of $\pi_1(B)$ on the domain and by the monodromy representation on its range. Since the image of the monodromy representation is Fuchsian, we may pass to the quotient map $\overline{p}$. Since the two quotient surfaces are isomorphic to B and since we may assume that $g(B) \geq 2$, Riemann-Hurwitz implies that $\overline{p}$ is an isomorphism. This implies that p is a Mobius transformation. On the other hand, p factors as

$$p : \mathbb{H} \to T_g \to \mathbb{H}_g \to \mathbb{H},$$

where the first two maps are the universal covering maps associated with the maps $C \to \mathcal{M}_g$ and with the Torelli map $\mathcal{M}_g \to \mathcal{A}_g$. The composition $\mathbb{H} \to \mathbb{H}_g$ is the period map for the full VHS $\mathbb{V}$, and since $\mathbb{L}$ is a factor of $\mathbb{V}$, we may represent the period map of $\mathbb{V}$ as a composition of this map and the projection onto one factor.

Finally, p is a composition of holomorphic maps and an isomorphism, hence an isometry for the Kobayashi metric. Consequently, the first map $\mathbb{H} \to T_g$ is an isometry for the Kobayashi metric, too. This implies that $B \to \mathcal{M}_g$ is a finite cover of a Teichmüller curve. □

5.4. Cusps and sections of $\mathcal{T}$ curves

A Teichmüller curve C is obtained as $\mathbb{H}/\mathrm{SL}(X,\omega)$, so the set of cusps Δ corresponds to the set of $\mathrm{SL}(X,\omega)$-conjugacy classes of maximal parabolic elements. In this section we first describe the stable curves associated with $\Delta_C \subset \overline{\mathcal{M}}_g$ and also neighborhoods of these points to perform intersection theory calculations. We then construct sections of the family of curves over C using the singularities $Z(\omega)$ and calculate their intersection number.

Let θ be a fixed direction of some parabolic element of $\mathrm{SL}(X,\omega)$. By conjugation we may suppose that θ is the horizontal direction. Then the geodesic $g_{-t}(X,\omega)$ runs into the cusp. We now describe how to obtain the surfaces along this geodesic. The horizontal direction is parabolic, in particular decomposes into maximal cylinders C_i of heights h_i and widths w_i for $i \in I$. Cut the surface X open along the core geodesics (at height $h_i/2$) of the cylinders. On the cut-open surface X^0 the original cylinders are decomposed into their top and bottom part $C_i^\top, C_i^\perp \subset X^0$. We define strips $U_i^\top \subset C_i^\top$ as the lower half (of height $h_i/4$) of the top part and $U_i^\perp \subset C_i^\perp$ as the top half (of height $h_i/4$) of the bottom part. For each i take a Z_i cylinder of height $(e^{2t} - 1/2)h_i$ and width w_i. Glue the top strip of height $h_i/4$ of Z_i to $U_i^\top$ and the bottom strip of height $h_i/4$ of Z_i to $U_i^\perp$ to form a flat surface $(Y_,\eta_t)$.

Obviously, the holomorphic map $z \mapsto e^{-t}z$ on each cylinder extends to a biholomorphism between the two Riemann surfaces $g_{-t}(X,\omega)$ and (Y_t,η_t).

Finally we recall that the topology from the complex structure of $\overline{\mathcal{M}}_g$ can be phrased in quasi-conformal language as follows. A basis of neighborhoods of the stable surface X_∞ consists of stable curves X together with ('smaller and smaller') compact sets V_X and V_∞ around the cusps of X and X_∞ such that there is a K-quasiconformal map $\varphi : X \setminus V_X \to V_\infty$ (with K close to one). This together with the previous description implies that $g_{-t}(X, \omega)$ converges to the stable curve where we replace each cylinder C_i of (X, ω) by two half-infinite cylinders. We summarize:

PROPOSITION 5.9. *The stable curves corresponding to the boundary points Δ_C of a Teichmüller curve $C \to \mathcal{M}_g$ are obtained by choosing a parabolic direction of a generating flat surface (X, ω) and replacing each cylinder by a pair of half-infinite cylinders whose points at $i\infty$ resp. at $-i\infty$ are identified.*

Since half-infinite cylinders are conformal to punctured discs, we may equivalently replace each cylinder by two punctured discs to obtain the (punctured) normalization of the stable curve. The stable curve itself is the obtained by adding the zeros to the punctured discs and identifying the corresponding pairs.
We defined a Teichmüller curve to be a curve in $\mathcal{M}_g$ and saw that Teichmüller curves stem from $\mathrm{SL}_2(\mathbb{R})$-orbits of flat surfaces in $\Omega\mathcal{M}_g$. If we quotient by scalar (of absolute value one) or equivalently by the $\mathrm{SO}_2(\mathbb{R})$-action we obtain a curve in $\mathbb{P}\Omega\overline{\mathcal{M}}_g$, more precisely in $\mathbb{P}\Omega\overline{\mathcal{M}}_g(\mu)$, where μ is the signature of a generating surface. We call this curve the *canonical lift* of the Teichmüller curve.

PROPOSITION 5.10. *Suppose that C is a Teichmüller curve generated by an abelian differential (X, ω) in $\Omega\mathcal{M}_g(\mu)$ and let μ' be a degeneration of the signature μ. Then the canonical lift of C to $\mathbb{P}\Omega\overline{\mathcal{M}}_g(\mu)$ is disjoint from $\mathbb{P}\Omega\overline{\mathcal{M}}_g(\mu')$.*

The holomorphic one-form given as the $\mathrm{SL}_2(\mathbb{R})$-image over each smooth fiber over a Teichmüller curve extends to a section ω_∞ of the relative dualizing sheaf (i.e. a stable one-form) for each singular fiber X^∞ over the closure of a Teichmüller curve.

PROOF. The claim is obvious over the interior of the moduli space. We only need to check the disjointness over the boundary. We may approach the boundary along a geodesic ray. In the construction of the limiting surface the open subset

$$H = X^0 \setminus \cup \bigcup_{i \in I} (U_i^\top \cup U_i^\perp),$$

is never touched as explained at the beginning of this section. This subset H contains the zeros of ω. Since the multiplicity of a zero is a local property, this implies the claim. □

COROLLARY 5.11. *Let X^∞ be a stable curve corresponding to a boundary point of a Teichmüller curve. Then X^∞ does not contain separating nodes. In particular $C \cdot \delta_i = 0$ for $i \geq 1$.*

For each irreducible component Y of X^∞ the number of zeros of ω_∞ is equal to $2g(Y) - 2 + n$, where n is the number of nodes of Z. In particular each irreducible component Y of X^∞ contains at least one zero of ω_∞.

PROOF. If a node was separating, then by the description of the degeneration a core curve γ_i of a cylinder C_i was separating. If this was true, view $X \setminus C_i = X^\top \cup X^\perp$ both as gluing of cylinders. For a translation structure saddle connections on the top of a cylinders have to be glued to the bottom of a cylinder. But the total length of top sides of cylinders of $X^\top$ exceeds the total length of bottom sides of cylinders in $X^\top$. It is thus impossible to form a closed surface $X^\top$ with boundary ∂C_i with the gluing rules just described. This contradiction proves the first statement.

The second statement is an immediate consequence of the degeneration description and the Gauss-Bonnet formula. □

Note that for a quadratic differential it is permitted to glue a saddle connection on the top of a cylinder to another one on the top of a cylinder. Consequently, the above proof does not apply to Teichmüller curves generated by quadratic differentials. They may indeed have degenerate fibers that are stable curves with separating nodes.

Sections defined by singularities. Let C be a Teichmüller curve generated by $(X,\omega) \in \Omega\mathcal{M}_g(m_1,\ldots,m_k)$. The $\mathrm{SL}_2(\mathbb{R})$-orbit of each singularity Z defines a section $S(Z)$ over the Teichmüller disc $\mathbb{H} = \mathrm{SO}_2(\mathbb{R})\backslash\mathrm{SL}_2(\mathbb{R})$. The affine group permutes the singularities. Since there are only finitely many of them, there is a finite index subgroup of Γ of $\mathrm{SL}(X,\omega)$ that fixes each of the singularities. We may take Γ even smaller, but still of finite index, and suppose that all the conditions of Section 3.1 are met for Γ. Let $B = \mathbb{H}/\Gamma \to C$ be a finite unramified cover of the Teichmüller curves such that the zero Z_i of order m_i defines a section σ_i (not only a multi-section) with image S_i of the pullback family $f : \overline{\mathcal{X}} \to \overline{B}$. Such a section can be used to define a lift of $\overline{B}$ to $\overline{\mathcal{M}}_{g,1}$.

PROPOSITION 5.12. *The section S_i has self-intersection number*

$$S_i^2 = \frac{-\chi}{2(m_i+1)},$$

where $\chi = 2g(\overline{B}) - 2 + |\Delta|$ and Δ is the set of cusps in $\overline{B}$. In particular the intersection number of $\overline{B}$ with $\omega_{i,\mathrm{rel}}$, which is by definition equal to $-S_i^2$, is given by

$$\overline{B}\cdot\omega_{i,\mathrm{rel}} = \frac{\overline{B}\cdot\lambda - (\overline{B}\cdot\delta)/12}{(m_i+1)\kappa_\mu},$$

where $\kappa_\mu = \frac{1}{12}\sum_{j=1}^k \frac{m_j(m_j+2)}{m_j+1}$.

PROOF. Let $\mathcal{L} \subset f_*\omega_{\widetilde{\mathcal{X}}/\overline{B}}$ be the ('maximal Higgs', see [**Möl06b**]) line bundle whose fiber over the point corresponding to $[X]$ is $\mathbb{C}\cdot\omega$, the generating differential of the Teichmüller curve. The property 'maximal Higgs' says by definition that

$$\deg(\mathcal{L}) = \chi/2. \tag{9}$$

Let S be the union of the sections $S_1,\ldots,S_k$. Pulling back the above inclusion to $\mathcal{X}$ gives an exact sequence

$$0 \to f^*\mathcal{L} \to \omega_{\widetilde{\mathcal{X}}/\overline{B}} \to \mathcal{O}_S\left(\sum_{j=1}^k m_jS_j\right) \to 0,$$

since the multiplicities of the vanishing locus of the generating differential of the Teichmüller curve are constant along the whole compactified Teichmüller curve. This implies that $\omega_{\widetilde{\mathcal{X}}/\overline{B}}$ is numerically equal to $f^*\mathcal{L}+\sum_{j=1}^k m_jS_j$. By the adjunction formula we get

$$S_i^2 = -\omega_{\mathcal{X}/\overline{B}}\cdot S_i = -m_iS_i^2 - \deg(\mathcal{L}),$$

since the intersection product of two fibers of f is zero. Together with (9) we thus obtain the desired self-intersection formula.

We write the Noether formula (6) as

$$12(\overline{B}\cdot\lambda) - (\overline{B}\cdot\delta) = \omega^2_{\widetilde{\mathcal{X}}/\overline{B}}$$

and calculate

$$\begin{aligned}
\omega^2_{\widetilde{\mathcal{X}}/\overline{B}} &= (f^*\mathcal{L} + \sum_{j=1}^{k} m_j S_j)^2 = \sum_{j=1}^{k} m_j^2 S_j^2 + 2\deg(\mathcal{L}) \sum_{j=1}^{k} m_j \\
&= \frac{\chi}{2} \sum_{j=1}^{k} \frac{-m_j}{m_j+1} + 2m_j = \frac{\chi}{2} \sum_{j=1}^{k} \frac{m_j(m_j+2)}{m_j+1} \\
&= -12(m_i+1)\kappa_\mu S_i^2.
\end{aligned}$$

Solving for $-S_i^2$ gives the claimed formula. □

In Proposition 2.3 we saw that in the interior of $\mathcal{M}_g$ a Teichmüller curve generated by a hyperelliptic curve always stays inside the hyperelliptic locus. By contraposition the argument implies that a Teichmüller curve generated by a non-hyperelliptic curve does not meet the hyperelliptic locus in $\mathcal{M}_g$. In boundary points the behavior is much more subtle. Still sometimes the dichotomy can still be proved. It will be important for non-varying results in Section 6.4.

PROPOSITION 5.13. *Let C be a Teichmüller curve generated by (X,ω) in $\Omega\mathcal{M}_g(\mu)$. Suppose that an irreducible degenerate fiber X^∞ over a cusp of C is hyperelliptic. Then X is hyperelliptic, hence the whole Teichmüller curve lies in the locus of hyperelliptic flat surfaces.*

Moreover, if $\mu \in \{(4),(3,1),(6),(5,1),(3,3),(3,2,1),(8),(5,3)\}$ and (X,ω) is not hyperelliptic, then no degenerate fiber of the Teichmüller curve is hyperelliptic.

The last conclusion does not hold for all strata. For instance, Teichmüller curves generated by a non-hyperelliptic flat surface in the stratum $\Omega\mathcal{M}_3(2,1,1)$ always intersect the hyperelliptic locus at the boundary. The proof is an intersection number argument similar to the non-varying results for Lyapunov exponents below. It can be found in [**CM11**].

PROOF. Suppose that the stable model X^∞ of the degenerate fiber is irreducible of geometric genus h with $(g-h)$ pairs of points (p_i,q_i) identified. A semi-stable model of X^∞ admits a degree two admissible cover of the projective line if and only if the normalization X_n of X^∞ is branched at $2h+2$ branch points over a main component with covering group generated by an involution ϕ and, moreover, for each of the $2(g-h)$ nodes there is a projective line intersecting X_n in p_i and $q_i = \phi(p_i)$ with two branch points.

In the flat coordinates of X_n given by ω, the surface consists of a compact surface X_0 with boundary of genus h and $2(g-h)$ half-infinite cylinders (corresponding to the nodes) attached to the boundary of X_0. We may define X_0 canonically by sweeping out the half-infinite cylinder at p_i (or q_i) with lines of slope equal to the residue (considered as element in $\mathbb{R}^2$) of ω at p_i until such a line hits a zero of ω, i.e. a singularity of the flat structure.

With this normalization, the above discussion shows that for irreducible stable curves the hyperelliptic involution exchanges the half-infinite cylinders corresponding to p_i and q_i and it defines an involution ϕ of X_0. As in the smooth case, ϕ acts as $-\mathrm{Id}$ on X_0.

To obtain smooth fibers over the Teichmüller curve (in a neighborhood of X^∞) one has to glue cylinders of finite (large) height in place of the half-infinite cylinders of appropriate ratios of moduli. The hypothesis on ϕ acting on X_0 and on the half-infinite cylinders implies that ϕ is a well-defined involution on the smooth curves. Moreover, ϕ has two fixed points in each of the finite cylinders and $2h+2$ fixed points on X_0, making $2g+2$ fixed points in total. This shows that the smooth fibers of the Teichmüller curve are hyperelliptic.

To complete the proof we have to consider the two-component degenerations for $\mu \in \{(3,1),(5,1),(5,3)\}$. In both cases, the hyperelliptic involutions can neither exchange the components (since the zeros are of different order) nor fix the components (since the zeros are of odd order).

For $\mu = (3,3)$ a hyperelliptic involution ϕ cannot fix the component, since 3 is odd. It cannot exchange the two components and exchange a pair of half-infinite cylinders that belong to different nodes, since ϕ could then be used to define a non-trivial involution for each component. This involution fixes the zeros and this contradicts that 3 is odd. If ϕ exchanges all pairs of half-infinite cylinders that belong to the same node, ϕ has two fixed points in each cylinder on the smooth 'opened up' surface. Now we can apply the same argument as in the irreducible case to conclude that the 'opened up' flat surfaces are hyperelliptic as well.

For $\mu = (3,2,1)$ a hyperelliptic involution can neither fix the component with the (unique) zero of order three, since 3 is odd, nor map it elsewhere, since the zeros are of different order. □

5.5. The classification problem of $\mathcal{T}$ curves: state of the art

One of the main questions of the theory is the classification of Teichmüller curves. In this section we summarize what is known today and what the open problems are. It also explains to which extent the VHS decomposition and real multiplication have so far been useful for solving the classification problem.

Genus two. In genus two the notion of primitive and algebraically primitive coincide. In the stratum $\Omega\mathcal{M}_2(2)$ an infinite series of primitive Teichmüller curves was found independently in [**Cal04**] and in [**McM03**]. This was shown to be the complete list of primitive Teichmüller curves in this stratum ([**McM05a**]). This is the only stratum where a complete classification is known, since square-tiled surfaces in this stratum have been classified in [**HL06b**] and also in [**McM05a**]. In the stratum $\Omega\mathcal{M}_2(1,1)$ primitive Teichmüller curves have been classified: There is only one example, the regular decagon with opposite sides identified. The proof in [**McM06b**] relies on a 'torsion' characterization of periodic points ([**Möl06a**]) on those Teichmüller curves, that in turn relies on the VHS decomposition in Theorem 5.5.

Infinite series for fixed genus. The only genera that are known to have an infinite number of primitive Teichmüller curves are $g = 2,3,4$. They are not algebraically primitive for $g = 3$ and $g = 4$. The construction of these examples is in [**McM06a**]. A classification of Teichmüller curves with quadratic trace field in these strata has recently been announced by Lanneau and D.-M. Nguyen.

Infinite series of (algebraically primitive) Teichmüller curves. Beside the examples in low genus $g \leq 4$, the only known primitive Teichmüller curves have a triangle group as affine group. These are the original examples of Veech ([**Vee89**]) and Ward, as well as a series of examples of primitive Teichmüller curves realizing all possible triangle groups as affine groups in [**BM10a**].

Finiteness results. In the hyperelliptic component of $\Omega\mathcal{M}_g(g-1, g-1)$ there are (for given g) only finitely many algebraically primitive Teichmüller curves. This is again a consequence of the 'torsion condition' ([**Möl08**]). These components are particular, since they have more than one zero to apply the criterion, but not too many zeros and moreover the hyperelliptic involution in order to reduce combinatorial complexity of the problem.

A more conceptual approach in order to exploit real multiplication for the classification of algebraically primitive Teichmüller curves was taken in [**BM**]. It combines the boundary behavior of the real multiplication locus in $\mathcal{M}_g$ with the 'torsion condition' to give finiteness in the stratum $\Omega\mathcal{M}_3(3, 1)$. Moreover, the real multiplication condition gives an algorithm and quite convincing numerical evidence for finiteness in the stratum $\Omega\mathcal{M}_3(4)^{\mathrm{hyp}}$.

Open problems. Finiteness for the number of algebraically primitive Teichmüller curves in the strata with many zeros seems likely to hold, following the torsion point approach in [**Möl06a**], but the combinatorics might be so difficult that essential new ideas are needed even for strata in low genus. On the other hand, the approach in [**BM**] should be extended to give finiteness at least for the strata $\Omega\mathcal{M}_g(2g-2)$.

If we consider square-tiled surfaces instead of algebraically primitive curves, the smallest open strata are $\Omega\mathcal{M}_2(1, 1)$ and $\Omega\mathcal{M}_3(4)$. The combinatorial approach of [**HL06b**] goes as follows. Take the 'obvious' combinatorial invariants and show that one can connect any pair of cusps by passing from one direction on a given flat surface to another. This approach might work also in both cases just mentioned, but the reader should be warned of the combinatorial complexity.

6. Lyapunov exponents

6.1. Motivation: Asymptotic cycles, deviations and the wind-tree model

Fix a generic surface (X, ω) in some stratum $\Omega\mathcal{M}_g(m_1, \ldots, m_k)$ and a generic point P on that surface. Consider a vertical straight line starting at P and close it up along a small horizontal slit I once this slit is hit for the first time as in Figure 3. We thus obtain a closed cycle $c_1 \in H_1(X, \mathbb{R})$.

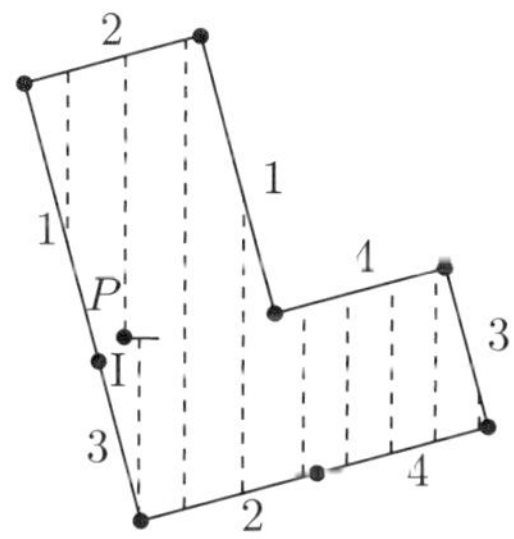

FIGURE 3. Straight line converging to the asymptotic cycle

We let $c_n \in H_1(X, \mathbb{R})$ be the cycle obtained by closing up the n-th hit on I. The limit

$$c = \lim_{n\to\infty} \frac{c_n}{||c_n||}$$

exists and is called asymptotic cycle. We are interested on the *deviation from this first order approximation.* For that purpose we provide $H_1(X, \mathbb{R})$ with the Hodge norm and let p_2 be the projection onto the orthogonal complement of $V_1 = \langle c \rangle$. Then the limit

$$\lambda_2 = \lim_{n\to\infty} \frac{\log ||p_2(c_n)||}{\log n}$$

exists. In fact, the projections $p_2(c_n)$ normalized to Hodge length one converge and we let V_2 be the subspace generated by the limit. Then we define p_3 to be the projection onto the orthogonal complement of V_1 and V_2 and let

$$\lambda_3 = \lim_{n\to\infty} \frac{\log ||p_3(c)||}{\log n}.$$

This process can be repeated to produce a full flag on $H_1(X, \mathbb{R})$.

For another example we consider the *wind-tree model* for the diffusion of gas molecules. In this model a particle drifts in billiard paths in the plane or the space and is reflected at randomly placed scatterers. Mathematically even the simplified model of scatterers places in a regular pattern at fixed positions is interesting. We restrict even further to the regular pattern being the lattice $\mathbb{Z}^2$ in the plane and the scatterers being boxes of side lengths (a, b) with $a, b \in (0, 1)$ centered at the lattice points as shown in Figure 4. For some (e.g. rational) values of parameters (a, b) there is a dense subset of S^1 such that trajectories in that given direction are periodic. Clearly, periodicity is a rare phenomenon. But still, for almost every direction the directional flow is recurrent.

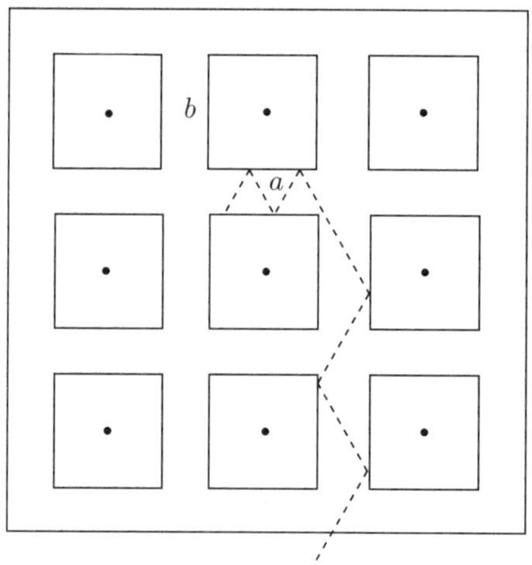

FIGURE 4. Wind-tree model

On the other hand, most trajectories make excursions further and further out. Let ϕ_t^θ denote the flow in the direction θ for time t. In fact, for every (a, b) there is some λ_2 such that for almost every direction θ and almost every starting point we have

$$\limsup_{t\to\infty} \frac{\log d(\phi_t^\theta(x), x)}{\log t} = \lambda_2.$$

The (non-)dependence of λ_2 on the parameters (a, b) is one of the main motivations of the notion 'non-varying' that we introduce below.

References: In both cases the answer to the problem is related to Lyapunov exponents as defined in the following section. The strata of $\Omega\mathcal{M}_g$ carry a finite invariant measure $\mu_{\rm gen}$ ([**Mas82**], [**Vee82**]) with support equal to the whole stratum and the λ_i that stem from the deviations are the Lyapunov exponents for the Teichmüller geodesic flow acting on the Hodge bundle over $\Omega\mathcal{M}_g$ with respect to $\mu_{\rm gen}$.

For the wind-tree model the λ_2 turns out to be the second Lyapunov exponent for the Haar measure supported on a Teichmüller curve in some cases, e.g. when a and b are rational, and for the Masur-Veech measure in the remaining cases.

The proof of these results will not be discussed here. It can be found in [**Zor94**] and [**Zor99**] (see also [**Zor06**]) for asymptotic cycles and in [**HLT**] and [**DHL**] for the wind-tree model.

6.2. Lyapunov exponents

We first state Oseledec's general theorem for the existence of Lyapunov exponents and then explain the instances we want to apply this theorem to.

THEOREM 6.1 (Oseledec). *Let $g_t : (M,\mu) \to (M,\mu)$ be a flow that acts ergodically on a space M with finite measure μ. Suppose that the action of $t \in \mathbb{R}^+$ lifts equivariantly to a flow also denoted by g_t on some measurable real vector bundle V on M. Suppose there exists a norm $||\cdot||$ on V (of course not supposed to be g_t-equivariant) such that for all $t \in \mathbb{R}^+$*

$$\int_M \log(1 + ||g_t||(m))\mu(m) < \infty,$$

where $||g_t||(m))$ denotes the operator norm at the point m induced by the map g_t and the norm on V.

Then there exist real constants $\widetilde{\lambda_1} > \cdots > \widetilde{\lambda_k}$ and a filtration

$$V = V_1 \supsetneq \cdots \supsetneq V_k \supset 0$$

by measurable vector subbundles such that, for almost all $m \in M$ and all $v \in V_m \setminus \{0\}$, one has

$$||g_t(v)|| = \exp(\widetilde{\lambda_i}t + o(t)),$$

where i is the maximal value such that $v \in (V_i)_m$. The $\widetilde{\lambda_i}$ and the V_i do not change if $||\cdot||$ is replaced by another norm of 'comparable' size (e.g. if one is a scalar multiple of the other).

Note that the $\widetilde{\lambda_i}$ and V_i are unchanged if we replace the support of μ by a finite unramified covering with a lift of the flow and the pullback of V.

From now on we adopt the convention to repeat the exponents $\widetilde{\lambda_i}$ according to the rank of V_i/V_{i+1} such that we will always have $2g$ of them, possibly some of them equal. The resulting sequence of numbers

$$\lambda_1 = \widetilde{\lambda_1} \geq \lambda_2 \geq \cdots \geq \lambda_{2g} = \widetilde{\lambda_k}$$

is called the spectrum of *Lyapunov exponents of g_t*. If V is a symplectic vector bundle, the spectrum is symmetric, i.e. $\lambda_{g+k} = -\lambda_{g-k+1}$.

We will use measures μ on $\Omega\mathcal{M}_g$ that will even be $\mathrm{SL}_2(\mathbb{R})$-invariant and such that the Teichmüller geodesic flow g_t acts ergodically. Let V be the restriction of the real Hodge bundle (i.e. the bundle with fibers $H^1(X,\mathbb{R})$) to the support M of μ. Let g_t be the lift of the geodesic flow to V via the Gauss-Manin connection. The norm on V will be the Hodge norm, the norm associated with the bilinear form Q defined in Section 4.

To sum up, the Lyapunov exponents for the Teichmüller geodesic flow on $\Omega\mathcal{M}_g$ measure the logarithm of the growth rate of the Hodge norm of cohomology classes during parallel transport along the geodesic flow. The reader may consult [**For06**] or [**Zor06**] for a more detailed introduction to this subject.

Most of our results will be about the *sum* of the top half of the Lyapunov exponents defined as

$$L = \sum_{i=1}^{g} \lambda_i.$$

Two sorts of measures. We will apply Oseledec's theorem in two instances. The first are Masur-Veech measures $\mu_{\rm gen}$ ([**Mas82**], [**Vee82**]) with support equal to the whole (hypersurface of flat surfaces of area one in) a connected component. These measures are constructed using period coordinates, giving the unit cube $(\mathbb{Z}[i])^N$ volume one. Since coordinate changes are in the symplectic group with integral coefficients, this is a well-defined normalization. The sum of Lyapunov exponents for these measure can be calculated by first calculating Siegel-Veech constants using [**EMZ03**] and then transferring the information using [**EKZb**]. We will denote by $L_{\Omega\mathcal{M}_g(m_1,\ldots,m_k)}$ the sum of Lyapunov exponents for the measure $\mu_{\rm gen}$ supported on the stratum $\Omega\mathcal{M}_g(m_1,\ldots,m_k)$. It is a combinatorially very involved procedure to actually compute these values.

When talking about Lyapunov exponents for Teichmüller curves we take μ to be the measure on the unit tangent bundle T^1B to a Teichmüller curve that stems from the Poincaré metric $g_{\rm hyp}$ on $\mathbb{H}$ with scalar curvature -4. This normalization is equivalent to require that our choice of scaling of the geodesic flow $g_t = \operatorname{diag}(e^t, e^{-t})$ has unit speed. It implies that the first Lyapunov exponent (corresponding to the subbundle $\Re\omega$ where (X,ω) is a generating flat surface of the Teichmüller curve) equals one. This normalization is consistent with the normalization of $\mu_{\rm gen}$ where also the first Lyapunov exponent is one. It is thus meaningful to compare Lyapunov exponents for μ and for the $\mu_{\rm gen}$ of the stratum the generating flat surface lies in.

In both cases, the integrability condition of Oseledec's theorem has to be verified. For $\mu_{\rm gen}$ this is done using a discretization and the language of matrix-valued cocycles in [**Zor99**], see also [**Zor06**]. For μ on a Teichmüller curve this can be deduced in continuous time from [**For02**, Lemma 2.1 and Corollary 2.2]. It holds in much greater generality as we shall soon see, since Lemma 6.10 below can also be used also here.

6.3. Lyapunov exponents for Teichmüller curves

The bridge between the 'dynamical' definition of Lyapunov exponents and the 'algebraic' method applied in the sequel is given by the following result. Note that if the VHS splits into direct summands one can apply Oseledec's theorem to the summands individually. The full set of Lyapunov exponents is the union (with multiplicity) of the Lyapunov exponents of the summands. Note that both sides of the equations of the following theorem are invariant when passing to finite unramified covers. We will thus use a convenient model of a fibered surface as explained in Section 3.1 instead of the Teichmüller curve itself. If one wants to evaluate the right hand side on the Teichmüller curve, one has to take into account orbifold degrees of line bundles.

THEOREM 6.2 ([**Kon97**], [**KZ**], [**BM10a**]). *If the VHS over the Teichmüller curve contains a sub-VHS* $\mathbb{W}$ *of rank* $2k$*, then the sum of the corresponding* k *non-negative Lyapunov exponents equals*

$$\sum_{i=1}^{k} \lambda_i^{\mathbb{W}} = \frac{2 \deg \mathbb{W}^{(1,0)}}{2g(\overline{B}) - 2 + |\Delta|},$$

where $\mathbb{W}^{(1,0)}$ *is the* $(1,0)$*-part of the Hodge-filtration of the vector bundle associated with* $\mathbb{W}$*. In particular, we have*

$$\sum_{i=1}^{g} \lambda_i = \frac{2 \deg f_* \omega_{\overline{\mathcal{X}}/\overline{B}}}{2g(\overline{B}) - 2 + |\Delta|}.$$

In particular, the first Lyapunov exponent of a Teichmüller curve is one by our normalization convention.

The following proof is from [**EKZb**]. This proof has three main ingredients. Instead of averaging over a g_t-orbit as stated in the definition of Lyapunov exponents one averages also over $\mathrm{SO}_2(\mathbb{R})$-orbits first and thus over the whole disc. This does not change the limit appearing in the definition of Lyapunov exponents, since the set where some other limit occurs has measure zero. Second, instead of taking the limit for any special vector we can (to determine the top Lyapunov exponent) average over all vectors or any subset whose intersection with the next filtration step has measure zero. Third, we rewrite a k-fold wedge-product of flat sections into a k-fold wedge-product of holomorphic sections plus contributions that are killed when taking $\partial\overline{\partial}$ in order to compute a curvature form. This last step works for the middle wedge power only and this is why this method only determines the sum of Lyapunov exponents. All the arguments are perfectly valid in the setting of Section 6.5 and justify Proposition 6.11.

We now give the details of this outline. Let

$$\Omega = \wedge^k(Q) : \wedge^{2k}\mathbb{W} \to \mathbb{C}$$

be the volume form on $\mathbb{W}$ induced by the Hodge inner product with values in the constant local system $\mathbb{C}$ on M, i.e.

$$\Omega(w_1 \wedge \ldots \wedge w_k) = \sum_{\sigma \in S_{2k}} \mathrm{sign}(\sigma) Q(w_{\sigma(1)}, w_{\sigma(2)}) \cdots Q(w_{\sigma(2k-1)}, w_{\sigma(2k)})$$

The volume form Ω can be extended $\mathcal{O}_M$-linearly to a volume form on the vector bundle $\wedge^{2k}\mathcal{W}$. For any section $L = v_1 \wedge \cdots \wedge v_k$ of $\Lambda^k(\mathbb{W})$ given by a decomposable vector we have a norm induced by the Hodge norm, that can be written in terms of the Hodge-* as

$$||L||^2 = \Omega(v_1 \wedge \cdots \wedge v_k \wedge *v_1 \wedge \cdots \wedge *v_k).$$

One checks that for any basis of holomorphic sections ω_i of $\mathbb{W}^{(1,0)}$ over some open subset U of B we have

$$(10) \quad ||L||^2 = \frac{|\Omega(\omega_1 \wedge \cdots \wedge \omega_k \wedge v_1 \wedge \cdots \wedge v_k)| \cdot |\Omega(v_1 \wedge \cdots \wedge v_k \wedge \bar{\omega}_1 \wedge \cdots \wedge \bar{\omega}_k)|}{|\Omega(\omega_1 \wedge \cdots \wedge \omega_k \wedge \bar{\omega}_1 \wedge \cdots \wedge \bar{\omega}_k)|}.$$

The numerator can also be written as

$$|\Omega(\omega_1 \wedge \cdots \wedge \omega_k \wedge \bar{\omega}_1 \wedge \cdots \wedge \bar{\omega}_k)| = \det(Q(\omega_i, \omega_j)_{i,j=1}^k).$$

LEMMA 6.3. *Given a section L of $\Lambda^k(\mathbb{W})$ as above over some open subset U of B, we have on U the equality of functions*

$$\Phi_k := \Delta_{\mathrm{hyp}}(\log(||L||)) = -\frac{1}{2}\Delta_{\mathrm{hyp}} \log|\det(Q(\omega_i,\omega_j)_{i,j=1}^k)|,$$

where Δ_{hyp} is the Laplacian for the hyperbolic metric along the Teichmüller disc.

PROOF. We apply Δ_{hyp} to the defining equation (10), calculate $||L||$ and obtain

$$\begin{aligned} 2\Phi_k &= \Delta_{\mathrm{hyp}} \log|\Omega(\omega_1\wedge\cdots\wedge\omega_k\wedge v_1\wedge\cdots\wedge v_k)| \\ &\quad + \Delta_{\mathrm{hyp}} \log|\Omega(v_1\wedge\cdots\wedge v_k\wedge\bar\omega_1\wedge\cdots\wedge\bar\omega_k)| - \Delta_{\mathrm{hyp}}\log|\det(Q(\omega_i,\omega_j)_{i,j=1}^k)|. \end{aligned}$$

Since the v_i are flat sections, the first summand on the right hand side is a holomorphic function and the second summand is an antiholomorphic function of the parameter on the Teichmüller disc. Hence they are harmonic and vanish after applying Δ_{hyp}. □

Let $\mathrm{Gr}_k(B)$ be the bundle whose fiber over b consist of the k-dimensional $\mathbb{R}$-subspaces of $(\mathbb{W})_b$. The Grassmannian carries a Haar measure which we denote by γ. We pull back this Grassmannian bundle to the unit tangent bundle $T_1B = \mathrm{SL}_2(\mathbb{R})/\Gamma$ of B. Within the Grassmannian bundle there is the subset of decomposable vectors $\mathrm{Gr}_k^{\mathrm{dec}}(T_1B)$ and this will be the set we are averaging over.

PROOF OF THEOREM 6.2. We first rewrite the sum of Lyapunov exponents as an integral over Φ_k in the following way. First, for almost every L in $\mathrm{Gr}_k^{\mathrm{dec}}(T_1B)$ we have

$$\sum_{i=1}^k \lambda_i = \lim_{T\to\infty}\frac{1}{T}\log||g_T(L)||$$

Together with the main theorem of calculus and additional averaging where the loci that give smaller contributions are of measure zero, we obtain the first line of (11). In the second line we do yet another circle averaging, in order to write the inner integral over a disc Δ_t of radius t. To pass to the third line we use a result from harmonic analysis stating that for any smooth rotation invariant function L we have

$$\frac{1}{2\pi}\frac{\partial}{\partial t}\int_0^{2\pi} L(t,\theta)d\theta = \frac{1}{2}\tanh(t)\frac{1}{\mathrm{vol}(\Delta_t)}\int_{\Delta_t}\Delta_{\mathrm{hyp}} L\, d\mu.$$

We are interpreting points c in Δ_t as elements in $\mathrm{SL}_2(\mathbb{R})$ by writing $c = g_t(r_\theta(b))$. Passing to the forth line we perform a change of variables and use the $\mathrm{SL}_2(\mathbb{R})$-invariance of μ. Passing to the fifth line we use the preceding lemma and get rid of the additional Grassmannian averaging. The sixth line needs a global bound ([**For06**]) to justify the change of integration order. Then we may take the limit of the integral over the tanh first.

(11)

$$\begin{aligned}
\operatorname{vol}(B)\sum_{i=1}^{k}\lambda_i &= \int_{\mathrm{Gr}_k^{\mathrm{dec}}(T_1B)} \lim_{T\to\infty}\frac{1}{T}\int_0^T \frac{d}{dt}\log\|g_t(L,b)\|dt\,d\mu(b)\,d\gamma(L)\\
&= \int_{\mathrm{Gr}_k^{\mathrm{dec}}(T_1B)} \lim_{T\to\infty}\frac{1}{T}\int_0^T \frac{1}{2\pi}\int_0^{2\pi}\frac{d}{dt}\log\|g_t(r_\theta L,b)\|d\sigma(\theta)\,dt\,d\mu(b)\,d\gamma(L)\\
&= \int_{\mathrm{Gr}_k^{\mathrm{dec}}(T_1B)} \lim_{T\to\infty}\frac{1}{T}\int_0^T \frac{\tanh(t)}{2\operatorname{vol}(\Delta_t)}\int_{\Delta_t}\Delta_{\mathrm{hyp}}\log\|c_*(L,b)\|d\mu(c)\,dt\,d\mu(b)\,d\gamma(L)\\
&= \int_{T_1B} \lim_{T\to\infty}\frac{1}{T}\int_0^T \frac{\tanh(t)}{2\operatorname{vol}(\Delta_t)}\int_{\Delta_t}-\frac{1}{2}\Delta_{\mathrm{hyp}}\log|\det Q(\omega_i,\omega_j)_{i,j=1}^k|(c_*(b))\,d\mu(c)\,dt\,d\mu(b)\\
&= \int_{T_1B} \lim_{T\to\infty}\frac{1}{T}\int_0^T \frac{\tanh(t)}{2\operatorname{vol}(\Delta_t)}\int_{\Delta_t}-\frac{1}{2}\Delta_{\mathrm{hyp}}\log|\det Q(\omega_i,\omega_j)_{i,j=1}^k|(b')\,)d\mu(c)\,dt\,d\mu(b')\\
&= \int_{T_1B}\Phi_k(b')\left(\lim_{T\to\infty}\frac{1}{T}\int_0^T\frac{\tanh(t)}{2\operatorname{vol}(\Delta_t)}\int_{\Delta_t}1\ d\mu(c)\right)d\mu(b')\\
&= \int_B \Phi_k(b)\,d\overline{\mu}(b),
\end{aligned}$$

where $\overline{\mu}$ denotes the direct image of μ under the projection $T_1B \to B$. Now, if $\Theta(\det(\mathbb{W}^{(1,0)}))$ denotes the curvature form of the line bundle $\det(\mathbb{W}^{(1,0)})$, this curvature form is represented by the differential form $-2\partial\overline{\partial}\log|\det(Q(\omega_i,\omega_j)_{i,j=1}^k)|$. Since $\operatorname{vol}(B) = \frac{\pi}{2}(2g(\overline{B}) - 2 + |\Delta|)$ we obtain

$$\begin{aligned}
\int_B \Phi_k(b)\,d\overline{\mu}(b) &= -\frac{1}{4}\int_B \Delta_{\mathrm{hyp}}\log|\det(Q(\omega_i,\omega_j)_{i,j=1}^k)|(b)\,d\overline{\mu}(b)\\
&= -\frac{1}{4}\int_B 4\partial\overline{\partial}\log|\det(Q(\omega_i,\omega_j)_{i,j=1}^k)|\,\frac{i}{2}dz\wedge\overline{dz} \qquad (12)\\
&= \frac{i}{2}\int_B[\Theta(\det(\mathbb{W}^{(1,0)}))] = \pi\deg(\mathbb{W}^{(1,0)}).
\end{aligned}$$

□

In the remainder of this section we give the bridge between the above formula for the sum of Lyapunov exponents and the slope for Teichmüller curves.

Recall that we denote the signature of a stratum of $\Omega\mathcal{M}_g$ by the tuple $\mu = (m_1,\ldots,m_k)$ where $\sum m_i = 2g-2$. Let κ_μ be the constant

$$\kappa_\mu = \frac{1}{12}\sum_{i=1}^{k}\frac{m_i(m_i+2)}{m_i+1}.$$

PROPOSITION 6.4. *Let $C \to \mathcal{M}_g$ be a Teichmüller curve generated by a flat surface in $\Omega\mathcal{M}_g(m_1,\ldots,m_k)$. Then knowing the slope is equivalent to knowing the sum of Lyapunov exponents, since the two quantities are related by the formula*

$$s(C) = 12 - \frac{12\kappa_\mu}{L(C)}. \qquad (13)$$

This is yet another consequence of the Noether formula and the self-intersection number of the sections S_i. Versions of the formula appear also in [**Che10b**, Thm. 1.8] and in [**EKZb**].

PROOF. As in the proof of Proposition 5.12 we deduce that

$$\frac{12(\overline{C}\cdot\lambda)-(\overline{C}\cdot\delta)}{12\kappa_\mu}=\frac{\chi}{2} \tag{14}$$

and Theorem 6.2 states that $L = 2(\overline{C}\cdot\lambda)/\chi$. Hence the relation (13) follows immediately. □

The preceding proposition and Xiao's bound (3.4) gives immediately the following upper bound.

COROLLARY 6.5. *Let $C \to \mathcal{M}_g$ be a Teichmüller curve generated by a flat surface in $\Omega\mathcal{M}_g(m_1,\dots,m_k)$. Then*

$$L(C) \leq \frac{g}{4(g-1)}\sum_{i=1}^{k}\frac{m_i(m_i+2)}{m_i+1}.$$

An explicit formula. For any given Teichmüller curve at a time the sum of Lyapunov exponents can be calculated, provided the Veech group can be calculated. This is a non-trivial condition, since at the time of writing there is still no deterministic algorithm to determine the Veech groups of the primitive Teichmüller curves in genus two (see Section 5.5), if the order D is large.

PROPOSITION 6.6. *Suppose the Teichmüller curve $C = \mathbb{H}/\mathrm{SL}(X,\omega)$ generated by (X,ω) has orbifold Euler characteristic χ and let Δ be the the set of cusps of C. For each $i \in \Delta$ let C_{ij} for $j \in J_i$ be the set of maximal cylinders of (X,ω) in the direction corresponding to the cusp. Suppose that the generator of the cusp stabilizer in $\mathrm{SL}(X,\omega)$ acts on C_{ij} as k_{ij}-fold Dehn twist. Then the sum of Lyapunov exponents can be calculated as follows.*

$$L(C) = \kappa_\mu + \frac{1}{6}\frac{\sum_{i\in\Delta}\sum_{j\in J_i}k_{ij}}{\chi}. \tag{15}$$

Note that the fraction is invariant under passing the finite covers unramified outside the cusps. Hence working with the orbifold Euler characteristic and the true Teichmüller curve or the finite cover introduced for technical reasons in Section 3.1 does not change the result.

The k_{ij} need not be integral. In fact in the square-tiled surface in Figure 2 the Veech group is equal to $\mathrm{SL}_2(\mathbb{Z})$. The generator of the stabilizer of the horizontal cusp produces a 1/4 Dehn twist on both horizontal cylinders, i.e. its 4-th power produces is simple Dehn twist on both horizontal cylinders.

PROOF. From the Noether formula in the version of (14) and $L(C) = 2(\overline{C}\cdot\lambda)/\chi$ we immediately obtain

$$L(C) = \kappa_\mu + \frac{1}{6}\frac{\overline{C}\cdot\delta}{\chi}.$$

We may suppose that we work over a base curve B as in Section 3.1 and suppose moreover that all the k_{ij} are integral by passing to yet another finite cover (if necessary) that we still denote by B. In the proof of Theorem 3.3 we checked that $\overline{B}\cdot\delta = \sum_{F\text{ sing.}}\Delta\chi_{\mathrm{top}}(F)$ and that each singularity $xy = t^n$ of the stable model of $f:\mathcal{X}\to B$ contributes n to this quantity. By Proposition 5.9 the degenerate fibers X^∞ at the point $\infty \in \Delta$ of the Teichmüller curve is obtained by replacing, in a given direction, the core curves of the cylinders by half-infinite cylinders attached at their points at infinity to from a node. It thus suffices to check that if a simple loop around ∞ makes and n-fold Dehn twist around a cylinder, the stable model locally looks like $xy = t^n$. This is the classical computation of the Picard-Lefschetz monodromy of surface singularity. □

For any given square-tiled surface (X, ω) calculating the Veech group is no problem, an explicit algorithm is given in [**Sch04**]. In this case the formula specializes as follows. Let X_i be the square-tiled surfaces in the $\mathrm{SL}_2(\mathbb{Z})$-orbit of (X, ω). For any square-tiled surface X_i we decompose the horizontal direction into maximal cylinders C_{ij} and denote by m_{ij} their moduli. Then

$$L(C) = \kappa_\mu + \frac{1}{|\mathrm{SL}_2(\mathbb{Z}) \cdot (X, \omega)|} \sum_{X_i \in \mathrm{SL}_2(\mathbb{Z}) \cdot (X,\omega)} \sum_{\substack{\text{all } C_{ij} \\ \text{cylinder of} X_i}} m_{ij}. \tag{16}$$

This formula is derived in [**EKZb**] as a consequence of their main theorem relating the sum of Lyapunov exponents to Siegel-Veech constants (for *any* $\mathrm{SL}_2(\mathbb{R})$-invariant measure that satisfies a technical regularity condition). We provide here a proof that is algebraic, but it works for Teichmüller curves only.

PROOF OF FORMULA (16). We have $6\chi = [\mathrm{SL}_2(\mathbb{Z}) : \mathrm{SL}(X, \omega)] = |\mathrm{SL}_2(\mathbb{Z}) \cdot (X, \omega)|$ for a square-tiled surface (X, ω). To check that the double sums in (15) and (16) are equal, it suffices to group coset representatives of $\mathrm{SL}_2(\mathbb{Z})/\mathrm{SL}(X, \omega)$ according to the cusps and to observe that $\begin{pmatrix} 1 & 1 \\ 0 & 1 \end{pmatrix}$ makes a m_{ij}-fold Dehn twist on a cylinder of modulus m_{ij}. □

This formula leaves open the need for a conceptual explanation which values occur for the sum of Lyapunov exponents. This is precisely the motivation for Section 6.4.

6.4. Non-varying properties for sums of Lyapunov exponents

We have seen that Teichmüller curves, in fact already those that are generated by square-tiled surfaces, are dense in each stratum. Indeed, on the level on sums of Lyapunov exponents we have the following convergence statement. Let C_d denote the union of all Teichmüller curves in a fixed stratum generated by square-tiled surfaces with d squares. We define the Lyapunov exponent $L(C_d)$ as the average of the Lyapunov exponents of the individual components weighted by the orbifold Euler characteristic (or hyperbolic volume) of the corresponding component.

PROPOSITION 6.7. *For $d \to \infty$ the weighted sum of Lyapunov exponents $L(C_d)$ of square-tiled surfaces in a component of a stratum of $\Omega\mathcal{M}_g$ converges to the sum of Lyapunov exponents $L_{\mu_{\mathrm{gen}}}$ for the measure μ_{gen} with support on the whole component.*

The proof of this fact is due to Eskin and is published in [**Che10b**, App. A]. For the proof the statement is translated into the language of Siegel-Veech constants (that we will not define here). This translation works only for the sum of Lyapunov exponents, and thus the corresponding statement for individual exponents is still an open problem.

Even with such a limiting behavior there is no reason to expect that this sum should be the same for all Teichmüller curves. A reason to appreciate such a phenomenon – if it holds – is the wind-tree example given above. It implies the independence of the escape rate from the side lengths of the scatterers.

We say that a connected component of a stratum is *non-varying*, if for all Teichmüller curves C in that component the sum of Lyapunov exponents $L(C)$ is the same. Such a non-varying phenomenon was observed numerically by Kontsevich and Zorich along with the initial observations on Lyapunov exponents for the Teichmüller geodesic flow ([**Kon97**]). Today, there are two types of non-varying results, one for low genus and one for hyperelliptic loci with completely different methods of proof. The low genus result uses a translation of the problem into algebraic geometry, in particular slope calculations, the hyperelliptic result uses the correspondence to Siegel-Veech constants.

THEOREM 6.8 ([**CM11**]). *For all strata in genus $g = 3$ but the principal stratum the sum of Lyapunov exponents is non-varying.*

For the strata with signature $(6)^{\mathrm{even}}$, $(6)^{\mathrm{odd}}$, $(5,1)$, $(3,3)$, $(3,2,1)$ and $(2,2,2)^{\mathrm{odd}}$ as well as for the hyperelliptic strata in genus $g = 4$ the sum of Lyapunov exponents is non-varying. For all the remaining strata, except maybe $(4,2)^{\mathrm{odd}}$ and $(4,2)^{\mathrm{even}}$, the sum of Lyapunov exponents is varying.

For the strata with signature $(8)^{\mathrm{even}}$, $(8)^{\mathrm{odd}}$ and $(5,3)$ as well as for the hyperelliptic strata in genus $g = 5$ the sum of Lyapunov exponents is non-varying. For all the other strata, except maybe $(6,2)^{\mathrm{odd}}$, the sum of Lyapunov exponents is varying.

THEOREM 6.9 ([**EKZb**]). *Hyperelliptic strata are non-varying. For a Teichmüller curve C generated by (X,ω) we have*

$$\begin{aligned} L(C) &= \frac{g^2}{2g-1} \quad \text{and} \quad s(C) = 8 + \frac{4}{g} \quad \text{if} \quad (X,\omega) \in \Omega\mathcal{M}_g^{\mathrm{hyp}}(2g-2), \\ L(C) &= \frac{g+1}{2} \quad \text{and} \quad s(C) = 8 + \frac{4}{g} \quad \text{if} \quad (X,\omega) \in \Omega\mathcal{M}_g^{\mathrm{hyp}}(g-1,g-1). \end{aligned} \tag{17}$$

The result of [**EKZb**] is a consequence of their main result relating sum of Lyapunov exponents to Siegel-Veech constants. It also gives non-varying statements in hyperelliptic loci. The case of genus two curves is a special case of this theorem and a proof appears in [**Bai07**] and in [**BM10b**].

In the left open cases of Theorem 6.8 computer experiments suggest that the strata should be non-varying, but the method of proof had no success so far, maybe due to our much smaller knowledge of divisor on the moduli space of spin curves compared to the $\mathcal{M}_g$. In most of the varying cases the method of proof of Theorem 6.8 gives interesting upper bounds for the sum of Lyapunov exponents. This can be translated e.g. into upper bounds on the escape rate of the 'wind-tree model' with other patterns of scatterers.

We now describe the idea of the proof of Theorem 6.8. Suppose that *all* Teichmüller curves C in a stratum $\Omega\mathcal{M}_g(\mu)$ are disjoint from a divisor D in $\overline{\mathcal{M}}_g$. Then $C \cdot D = 0$ for all these Teichmüller curves, hence $s(C) = s(D)$. Since the slopes are non-varying, so are the sums of Lyapunov exponents and the Siegel-Veech constants for these Teichmüller curves. In general, we also need to consider the moduli spaces of curves with marked points or spin structures. But the idea still relies on the non-intersection property of the Teichmüller curves with certain divisors on those moduli spaces.

We give the proof of the theorem in two instances. The first one is very simple and shows the general method. The second one is more involved. It shows that in general we have to work with the moduli space $\mathcal{M}_{g,n}$ rather than $\mathcal{M}_g$, since the slopes we expect for this stratum in $g = 4$ are smaller than the slope of any divisor in $\mathcal{M}_4$

In the case $\Omega\mathcal{M}_3(4)^{\text{odd}}$ the algorithm of [**EMZ03**] to calculate Siegel-Veech constants for components of strata can be translated using [**EKZb**] into $L_{(4)^{\text{odd}}} = 8/5$, hence if non-varying holds we expect by (13) any Teichmüller curve C in that stratum to have $s(C) = 9$.

PROOF OF THEOREM 6.8, CASE $\Omega\mathcal{M}_3(4)^{\text{odd}}$. For genus three, the connected components $\Omega\mathcal{M}_3(4)^{\text{odd}}$ and $\Omega\mathcal{M}_3(4)^{\text{hyp}}$ are not only disjoint in $\Omega\mathcal{M}_3$ they are also disjoint in $\Omega\overline{\mathcal{M}}_3$ since they are distinguished by a parity of a spin structure, which is known to be deformation invariant over all of $\overline{\mathcal{M}}_g$ for any g.

We need this property only for boundary points of Teichmüller curves, which is shown in Proposition 5.13. In any case, Teichmüller curves in this stratum do not intersect the closure of the hyperelliptic locus H in $\overline{\mathcal{M}}_3$. Recall the divisor class of H in (3). From $s(H) = 9$ and $C \cdot H = 0$, we obtain that $s(C) = 9$, hence $L(C) = 8/5$ for all Teichmüller curves in this stratum using (13). □

In the case $\Omega\mathcal{M}_4(3,3)^{\text{non-hyp}}$ we can calculate as above $L_{(3,3)^{\text{non-hyp}}} = 2$ hence if non-varying holds we expect for any Teichmüller curve C in that stratum $s(C) = 33/4$. Note that $s(C)$ is smaller than the slope $17/2$ of the divisor in $\mathcal{M}_4$ with the smallest slope ([**HM90**]).

Recall from Section 5.4 that we may define, after a finite unramified covering that does not change Lyapunov exponents, sections σ_i for $i_1, \ldots, i_k$ of the family $f : \overline{X} \to \overline{B}$ over the Teichmüller curve corresponding to the singularities of the generating holomorphic one-from ω. With the help of (a selection of) these section we may lift the Teichmüller curve to a map $\overline{B} \to \overline{\mathcal{M}}_{g,n}$ for some $n \leq k$.

PROOF OF THEOREM 6.8, CASE $\Omega\mathcal{M}_4(3,3)^{\text{non-hyp}}$. Let C be a Teichmüller curve generated by a flat surface (X, ω) in the stratum $\Omega\mathcal{M}_4(3,3)^{\text{non-hyp}}$, lifted to $B \to \overline{\mathcal{M}}_{4,2}$. Recall that $\text{Lin}_3^1 \subset \overline{\mathcal{M}}_{4,2}$ parametrizes pointed curves (X, p, q) that admit a g_3^1 with a section vanishing at p, q, r for some $r \in X$. We first want to show that C (or $\overline{B}$) does not intersect Lin_3^1.

Suppose that (X, p, q) is contained in the intersection of $\overline{B}$ with Lin_3^1. Since $\omega_X \sim \mathcal{O}_X(3p + 3q)$ and since being in Lin_3^1 implies $h^0(\mathcal{O}_X(p + q + r)) \geq 2$, by the Riemann-Roch theorem we know that $h^0(\mathcal{O}_X(2p + 2q - r)) \geq 2$. If $r \neq p, q$, then $h^0(\mathcal{O}_X(2p + 2q)) > 3$, hence $2p + q$ and $2q + p$ both admit g_3^1. Since X is not hyperelliptic the canonical image of X is contained in a quadric in $\mathbb{P}^3$. This quadric has at most two rulings (only one if the quadric is singular) and each g_3^1 corresponds to a ruling of the quadric. Consequently, both $2p + q$ and $2q + p$ define the same g_3^1, the one defined by the line connecting p, q. This implies $p \sim q$, a contradiction. If $r = p$ or q, again, $2p + q$ and $2q + p$ both admit g_3^1 and consequently C is hyperelliptic. But this stratum is non-hyperelliptic, and Proposition 5.13 yields the desired contradiction.

From $\overline{C} \cdot \text{Lin}_3^1 = 0$ together with Proposition 5.12 and $\kappa_{(3,3)} = 5/8$, the result follows immediately. □

The last calculation is the calculation for the second test curve in the proof of Proposition 3.2 read backwards.

6.5. Lyapunov exponents for general curves in $\mathcal{M}_g$ and in $\mathcal{A}_g$

The main point of this section is that all the formalism of Lyapunov exponents is also perfectly valid in this more general setting. For curves in $\mathcal{M}_g$ we show that calculating the sum is equivalent to the (interesting and partially understood) problem of computing slopes that we addressed in Section 3.4. Moreover, it raises the interesting question of identifying individual Lyapunov exponents for other curves but Teichmüller curves.

Let $f : \mathcal{A} \to C$ be a non-constant family of abelian varieties. Then the universal covering of C is the upper half plane and $C = \mathbb{H}/\Gamma$. We provide the unit tangent bundle T^1C to C with the metric μ that stems from Haar measure from $\mathrm{SL}_2(\mathbb{R})$ and with the geodesic flow g_t. In order to apply the calculation used in the proof of Theorem 6.2 we only need scalar curvature -4 for the correct relation between $\chi(C)$ and the $\mathrm{vol}(C)$ and we needed g_t the have geodesic unit speed, i.e. $g_t = \mathrm{diag}(e^t, e^{-t})$. We also use these conventions here.

The direct image $R^1 f_* \mathbb{R}$ is a local system $\mathbb{V}$ on C (with fiber $H^1(A, \mathbb{R})$ of dimension $2 \dim A$). We pull this local system back to T^1C and provide it with the Hodge metric. The key to get started is the following lemma.

LEMMA 6.10. *The lift of g_t to $\mathbb{V}$ with respect the μ and the Hodge norm satisfies the integrability hypothesis in Oseledec's theorem.*

PROOF. It suffices to give a global bound (i.e. independent of the abelian variety A) on the operator norm of $g_t(\cdot)$ on the (Hodge) norm one-ball in $H^1(A, \mathbb{R})$. To do so, it suffices to bound the derivative of the Hodge norm of any element in the (Hodge) norm one-ball in $H^1(A, \mathbb{R})$ in the direction of some $v \in T^1\mathcal{A}_g$. If $p : \mathbb{H} \to \mathbb{H}_g$ denotes the period map associated with f, this can be rephrased by bounding $||dp||$, where $||dp||$ is the norm induced by the Bergmann-Siegel metric and the Poincaré metric on $\mathbb{H}_g$ and $\mathbb{H}$ respectively. A generalization of the Schwarz-Pick Lemma (e.g. [**Roy80**, Theorem 2]) implies that $||dp|| \leq k/K$ where k and K are the curvatures of the metrics on domain and range of p respectively. With our normalization we have $k = K = -4$ and thus $||dp|| \leq 1$. □

Given the lemma, we may talk of the Lyapunov exponents λ_i of the geodesic flow on C. From the proof we deduce that

$$\lambda_i \leq 1 \tag{18}$$

for all i with our curvature normalization, generalizing the expectation from the case of Teichmüller curves.

As before, we take a subgroup $\Gamma_1 \subset \Gamma$ without elements of finite order and such that the monodromy around the cusps acts unipotently on the fibers. We let $B = \mathbb{H}/\Gamma_1$ and denote by $f : \mathcal{A} \to B$ the pullback of the family over C to B. Again, such a pullback does not change the spectrum of Lyapunov exponents.

The following proposition has been justified along with Theorem 6.2.

PROPOSITION 6.11. *Let $\mathbb{V}$ be the weight one VHS associated with the family of abelian varieties f of dimension g. Then then the sum of the g non-negative*

Lyapunov exponents equals

$$\sum_{i=1}^{g} \lambda_i^{\mathbb{V}} = \frac{2 \deg \mathbb{V}^{(1,0)}}{2g(\overline{B}) - 2 + |\Delta|},$$

where $\mathbb{V}^{(1,0)}$ *is the* $(1,0)$*-part of the Hodge-filtration of the vector bundle associated with* $\mathbb{V}$*.*

This proposition together with equation (18) implies the *Arakelov inequality*

$$\frac{2 \deg \mathbb{V}^{(1,0)}}{2g(\overline{B}) - 2 + |\Delta|} \leq g. \tag{19}$$

This inequality appears first in [**Fal83**], see also [**VZ04**] for some background and references to other versions.

Towards a characterization of Shimura curves. A *Shimura curve* is a curve $C \to \mathcal{A}_g$ that is obtained as the quotient $C = K\backslash G_{\mathbb{R}}/\Gamma \to U(g)\backslash \mathrm{Sp}_{2g}(\mathbb{R})/\mathrm{Sp}_{2g}(\mathbb{Z})$ induced by some inclusion of $\mathbb{Q}$-algebraic groups $G \to \mathrm{Sp}_{2g}$, where K is a maximal compact subgroup of $G_{\mathbb{R}}$ and where Γ is some arithmetic lattice. Equivalently, a Shimura curve is a locus of abelian varieties admitting additional 'endomorphisms'. We put 'endomorphisms' in quotation marks, since some Shimura curves are defined by the existence of endomorphisms but in general the presence of Hodge classes is the appropriate condition. We refer to the recent survey [**MO**] for details of the definition.

This is almost the same as requiring that $C \to \mathcal{A}_g$ is totally geodesic for the 'Bergmann-Siegel' Riemannian metric on $\mathcal{A}_g$, the unique (up to scale) Riemannian metric on the Hermitian symmetric domain $\mathbb{H}_g$ that is invariant under the action of $\mathrm{Sp}_{2g}(\mathbb{R})$. (Note that for $g > 1$ the Bergmann-Siegel metric is not the Kobayashi metric on $\mathbb{H}_g$. The latter is just a Finsler metric, not Riemannian for $g > 1$.) In fact, such a totally geodesic curve is a Shimura curve if and only if it contains a CM point. This obstruction is very minor. The only thing that can happen to produce a totally geodesic curve that is not Shimura is to take a product of a family of abelian varieties over a Shimura curve times a constant family consisting of a non-CM abelian variety. See again [**MO**] an the references therein for details.

We have the following partial characterization of Shimura curves.

PROPOSITION 6.12. *If the curve* $C \to \mathcal{A}_g$ *is totally geodesic for the Riemannian metric on* $\mathcal{A}_g$*, then the Lyapunov spectrum contains only the values* ± 1 *and zero.*

Conversely, if the Lyapunov spectrum of $C \to \mathcal{A}_g$ *contains only the values* ± 1*, then* C *is a Shimura curve.*

PROOF. If $C \to \mathcal{A}_g$ is totally geodesic, then we may change C into a Shimura curve without changing the Lyapunov spectrum. Consequently, we may apply [**Möl**, Theorem 1.2]. This theorem states that the VHS decomposes into a unitary part (that gives rise to zero Lyapunov exponents) and the standard Fuchsian representation of Γ (that gives rise to Lyapunov exponent ± 1) tensored with some unitary representation (that accounts for the multiplicity of ± 1).

Conversely, if the positive Lyapunov exponents are all one, then the VHS over C attains the upper bound in the Arakelov equality (19). Now the fundamental result of Viehweg and Zuo ([**VZ04**, Theorem 0.5]) implies that C is a Shimura curve (without unitary direct summand in the VHS). □

Open problem. Can one generalize Proposition 6.12 to characterize Shimura curves as those curves in $C \to \mathcal{A}_g$ with Lyapunov spectrum only zero and ± 1 ? That is, do the Lyapunov subspaces for zero and those with ± 1 span local subsystems?

6.6. Known results and open problems

There are many variants of the comparison problem of Lyapunov exponents for Teichmüller curves to those of the ambient strata. For the Masur-Veech measure it was shown by Avila and Viana that the Lyapunov spectrum is simple, thus solving the Kontsevich-Zorich conjecture ([**AV07**]). It is tempting to guess that for square-tiled surfaces with a large number of squares the Lyapunov spectrum of the corresponding Teichmüller curves (in a fixed connected component of a stratum) converges to the Lyapunov spectrum for the measure μ_{gen} supported on the whole connected component. Besides Proposition 6.7 no such statement is presently known. Is the Lyapunov spectrum for all but finitely many Teichmüller curves in a given stratum simple?

The nature of individual Lyapunov exponents is a wide open question. Are they related to characteristic classes? It seems, on the contrary, that partial sums of (non-zero) Lyapunov exponents are rational only if the Oseledec's bundle of this partial sum is a summand in the decomposition of the VHS. At least, there are presently no counterexamples.

Lyapunov exponents are hard to determine even numerically. The presently known and implemented algorithms are based on the first motivation (Zorich, Kontsevich), but they are exponentially slow (in desired accuracy) due the log in the definition of Lyapunov exponents. It would be interesting to have an alternative formulation that allows faster computation.

The stratum $\Omega\mathcal{M}_3(1,1,1,1)$ is one of smallest that does not have the non-varying property. With computer help one easily produces examples of Teichmüller curves in this stratum with

$$L(C) \in \{1, 3/2, 5/3, 7/4, 9/5, 11/6, 19/11, 33/19, 83/46, 544/297\}.$$

The value for the measure with support on the whole stratum is $L_{(1,1,1,1)} = 53/28$. What is the set of values that the sum of Lyapunov exponents for Teichmüller curves attains? At least, what is its set of accumulation points?

A Teichmüller curve has *maximally degenerate Lyapunov spectrum*, if $\lambda_1 = 1$ and $\lambda_{2g} = -1$ are the only two non-zero Lyapunov exponents. The 'eierlegende Wollmilchsau' in Figure 2 is one of the two known Teichmüller curves with maximally degenerate Lyapunov spectrum. Its name refers to that fact it has many different remarkable properties at the same time and serves ubiquitously as counterexamples to many naive conjectural properties of square-tiled surfaces. The eierlegende Wollmilchsau was discovered by [**HS08**] and [**For06**] independently. It is a cyclic cover

$$y^4 = x(x-1)(x-t) \quad \omega = \frac{dx}{y}$$

and also the other known Teichmüller curve with maximally degenerate Lyapunov spectrum ([**FM**]) is a cyclic cover

$$y^6 = x(x-1)(x-t) \quad \omega = \frac{dx}{y}.$$

In fact, a Teichmüller curve has maximally degenerate Lyapunov spectrum if and only if it is one of the two families, possibly with exceptions in strata in $g = 5$ ([**Möl**]). Does such an exception exist?

References: A good introduction to Lyapunov exponents with a lot of motivating examples is the survey by Zorich ([**Zor06**]). With (even) more emphasis on dynamics an introduction to Lyapunov exponents is given in the lecture notes of M. Viana ([**Via**]).

Bibliography

[AC87] E. Arbarello and M. Cornalba. The Picard groups of the moduli spaces of curves. *Topology*, 26(2):153–171, 1987. MR895568 (88e:14032)

[AV07] A. Avila and M. Viana. Simplicity of Lyapunov spectra: proof of the Zorich-Kontsevich conjecture. *Acta Math.*, 198(1):1–56, 2007. MR2316268 (2008m:37010)

[Bai07] M. Bainbridge. Euler characteristics of Teichmüller curves in genus two. *Geom. Topol.*, 11:1887–2073, 2007. MR2350471 (2009c:32025)

[BHPVdV04] W. P. Barth, K. Hulek, C. A. M. Peters, and A. Van de Ven. *Compact complex surfaces*, volume 4 of *Ergebnisse der Mathematik und ihrer Grenzgebiete. 3. Folge. A Series of Modern Surveys in Mathematics [Results in Mathematics and Related Areas. 3rd Series. A Series of Modern Surveys in Mathematics]*. Springer-Verlag, Berlin, second edition, 2004. MR2030225 (2004m:14070)

[BM] M. Bainbridge and M. Möller. The Deligne-Mumford compactification of the real multiplication locus and Teichmüller curves in genus three, *Acta Math.*, 208(2):1–92. MR2910796

[BM10a] I. Bouw and M. Möller. Teichmüller curves, triangle groups, and Lyapunov exponents. *Ann. of Math. (2)*, 172(1):139–185, 2010. MR2680418 (2012b:32020)

[BM10b] I. I. Bouw and M. Möller. Differential equations associated with nonarithmetic Fuchsian groups. *J. Lond. Math. Soc. (2)*, 81(1):65–90, 2010. MR2580454 (2011e:32014)

[Bou01] I. I. Bouw. The p-rank of ramified covers of curves. *Compositio Math.*, 126(3):295–322, 2001. MR1834740 (2002e:14045)

[Cal04] K. Calta. Veech surfaces and complete periodicity in genus two. *J. Amer. Math. Soc.*, 17(4):871–908, 2004. MR2083470 (2005j:37040)

[Che10a] D. Chen. Covers of elliptic curves and the moduli space of stable curves. *J. Reine Angew. Math.*, 649:167–205, 2010. MR2746470 (2012b:14056)

[Che10b] D. Chen. Square-tiled surfaces and rigid curves on moduli spaces, 2010, *Adv. Math.*, 228(2):1135–1162. MR2822219 (2012k:14038)

[CM11] D. Chen and M. Moeller. Non-varying sums of Lyapunov exponents for Abelian differentials in low genus, 2011, arXiv: math.AG/1104.3932.

[CMSP03] J. Carlson, S. Müller-Stach, and C. Peters. *Period mappings and period domains*, volume 85 of *Cambridge Studies in Advanced Mathematics*. Cambridge University Press, Cambridge, 2003. MR2012297 (2005a:32014)

[Deb82] O. Debarre. Inégalités numériques pour les surfaces de type général. *Bull. Soc. Math. France*, 110(3):319–346, 1982. With an appendix by A. Beauville. MR688038 (84f:14026)

[Del70] P. Deligne. *Équations différentielles à points singuliers réguliers*. Lecture Notes in Mathematics, Vol. 163. Springer-Verlag, Berlin, 1970. MR0417174 (54:5232)

[Del71] P. Deligne. Théorie de Hodge. II. *Inst. Hautes Études Sci. Publ. Math.*, (40):5–57, 1971. MR0498551 (58:16653a)

[Del74] P. Deligne. Théorie de Hodge. III. *Inst. Hautes Études Sci. Publ. Math.*, (44):5–77, 1974. MR0498552 (58:16653b)

[Del87] P. Deligne. Un théorème de finitude pour la monodromie. In *Discrete groups in geometry and analysis (New Haven, Conn., 1984)*, volume 67 of *Progr. Math.*, pages 1–19. Birkhäuser Boston, Boston, MA, 1987. MR900821 (88h:14013)

[DHL] V. Delecroix, P. Hubert, and S. Leliévre. Diffusion for the periodic wind-tree model, arXiv: 1107.1810.

[EKZa] A. Eskin, M. Kontsevich, and A. Zorich. Lyapunov spectrum of square-tiled cyclic covers, *J. Mod. Dyn.*, 5(2):319–353, 2011. MR2820564 (2012h:37073)

[EKZb] A. Eskin, M. Kontsevich, and A. Zorich. Sum of Lyapunov exponents of the Hodge bundle with respect to the Teichmüller geodesic flow, work in progress.

[EMZ03] A. Eskin, H. Masur, and A. Zorich. Moduli spaces of Abelian differentials: the principal boundary, counting problems, and the Siegel-Veech constants. *Publ. Math. Inst. Hautes Études Sci.*, 97:61–179, 2003. MR2010740 (2005b:32029)

[Fal83] G. Faltings. Arakelov's theorem for abelian varieties. *Invent. Math.*, 73(3):337–347, 1983. MR718934 (85m:14061)

[Far09a] G. Farkas. Birational aspects of the geometry of M_g. In *Surveys in differential geometry. Vol. XIV. Geometry of Riemann surfaces and their moduli spaces*, volume 14 of *Surv. Differ. Geom.*, pages 57–110. Int. Press, Somerville, MA, 2009. MR2655323 (2011g:14065)

[Far09b] G. Farkas. Koszul divisors on moduli spaces of curves. *Amer. J. Math.*, 131(3):819–867, 2009. MR2530855 (2010f:14030)

[FM] G. Forni and C. Matheus. An example of a Teichmüller disk in genus 4 with degenerate Kontsevich-Zorich spectrum, arXiv:math.DS/0810.0023.

[For02] G. Forni. Deviation of ergodic averages for area-preserving flows on surfaces of higher genus. *Ann. of Math. (2)*, 155(1):1–103, 2002. MR1888794 (2003g:37009)

[For06] G. Forni. On the Lyapunov exponents of the Kontsevich-Zorich cocycle. In *Handbook of dynamical systems. Vol. 1B*, pages 549–580. Elsevier B. V., Amsterdam, 2006. MR2186248 (2010c:37004)

[FP05] G. Farkas and M. Popa. Effective divisors on M_g, curves on $K3$ surfaces, and the slope conjecture. *J. Algebraic Geom.*, 14(2):241–267, 2005. MR2123229 (2006a:14043)

[GJ00] E. Gutkin and C. Judge. Affine mappings of translation surfaces: geometry and arithmetic. *Duke Math. J.*, 103(2):191–213, 2000. MR1760625 (2001h:37071)

[Gri68] P. A. Griffiths. Periods of integrals on algebraic manifolds. II. Local study of the period mapping. *Amer. J. Math.*, 90:805–865, 1968. MR0233825 (38:2146)

[HL06a] P. Hubert and E. Lanneau. Veech groups without parabolic elements. *Duke Math. J.*, 133(2):335–346, 2006. MR2225696 (2007d:37055)

[HL06b] P. Hubert and S. Lelièvre. Prime arithmetic Teichmüller discs in $\mathcal{H}(2)$. *Israel J. Math.*, 151:281–321, 2006. MR2214127 (2008f:37073)

[HLT] P. Hubert, S. Leliévre, and S. Troubetzkoy. The Ehrenfest wind-tree model: periodic directions, recurrence, diffusion, *J. Reine Angew. Math.*, 656:233–244, 2011. MR2818861 (2012f:37079)

[HM79] J. Hubbard and H. Masur. Quadratic differentials and foliations. *Acta Math.*, 142(3-4):221–274, 1979. MR523212 (80h:30047)

[HM82] J. Harris and D. Mumford. On the Kodaira dimension of the moduli space of curves. *Invent. Math.*, 67(1):23–88, 1982. With an appendix by William Fulton. MR664324 (83i:14018)

[HM90] J. Harris and I. Morrison. Slopes of effective divisors on the moduli space of stable curves. *Invent. Math.*, 99(2):321–355, 1990. MR1031904 (91d:14009)

[HM98] J. Harris and I. Morrison. *Moduli of Curves*, volume 187 of *Graduate Texts in Mathematics*. Springer-Verlag, New York, 1998. MR1631825 (99g:14031)

[HS08] F. Herrlich and G. Schmithüsen. An extraordinary origami curve. *Math. Nachr.*, 281(2):219–237, 2008. MR2387362 (2008k:14059)

[Hub06] J. H. Hubbard. *Teichmüller theory and applications to geometry, topology, and dynamics. Vol. 1.* Matrix Editions, Ithaca, NY, 2006. Teichmüller theory, With contributions by Adrien Douady, William Dunbar, Roland Roeder, Sylvain Bonnot, David Brown, Allen Hatcher, Chris Hruska and Sudeb Mitra, With forewords by William Thurston and Clifford Earle. MR2245223 (2008k:30055)

[Kon97] M. Kontsevich. Lyapunov exponents and Hodge theory. In *The mathematical beauty of physics (Saclay, 1996)*, volume 24 of *Adv. Ser. Math. Phys.*, pages 318–332. World Sci. Publishing, River Edge, NJ, 1997. MR1490861 (99b:58147)

[KZ] M. Kontsevich. and A. Zorich. Lyapunov exponents and Hodge theory, arXiv:hep-th/9701164.

[KZ03] M. Kontsevich and A. Zorich. Connected components of the moduli spaces of Abelian differentials with prescribed singularities. *Invent. Math.*, 153(3):631–678, 2003. MR2000471 (2005b:32030)

[Log03] A. Logan. The Kodaira dimension of moduli spaces of curves with marked points. *Amer. J. Math.*, 125(1):105–138, 2003. MR1953519 (2003j:14035)

[Mas82] H. Masur. Interval exchange transformations and measured foliations. *Ann. of Math. (2)*, 115(1):169–200, 1982. MR644018 (83e:28012)

[McM03] C. McMullen. Billiards and Teichmüller curves on Hilbert modular surfaces. *J. Amer. Math. Soc.*, 16(4):857–885, 2003. MR1992827 (2004f:32015)

[McM05a] C. McMullen. Teichmüller curves in genus two: Discriminant and spin. *Math. Ann.*, 333(1):87–130, 2005. MR2169830 (2006h:32011)

[McM05b] C. McMullen. Teichmüller curves in genus two: the decagon and beyond. *J. Reine Angew. Math.*, 582:173–199, 2005. MR2139715 (2006a:32017)

[McM06a] C. McMullen. Prym varieties and Teichmüller curves. *Duke Math. J.*, 133(3):569–590, 2006. MR2228463 (2007a:32018)

[McM06b] C. McMullen. Teichmüller curves in genus two: torsion divisors and ratios of sines. *Invent. Math.*, 165(3):651–672, 2006. MR2242630 (2007f:14023)

[McM07] C. McMullen. Dynamics of $SL_2(\mathbb{R})$ over moduli space in genus two. *Ann. of Math. (2)*, 165(3):397–456, 2007. MR2299738 (2008k:32035)

[MO] B. Moonen and F. Oort. The Torelli locus and special subvarieties., Preprint, August 2010. Submitted to the Handbook of Moduli (G. Farkas and I. Morrisson, eds).

[Möl] M. Möller. Shimura and Teichmüller curves, *J. Mod. Dyn.*, 5(1):1-32, 2011. MR2787595 (2012f:14055)

[Möl06a] M. Möller. Periodic points on Veech surfaces and the Mordell-Weil group over a Teichmüller curve. *Invent. Math.*, 165(3):633–649, 2006. MR2242629 (2007e:14012)

[Möl06b] M. Möller. Variations of Hodge structures of a Teichmüller curve. *J. Amer. Math. Soc.*, 19(2):327–344 (electronic), 2006. MR2188128 (2007b:32026)

[Möl08] M. Möller. Finiteness results for Teichmüller curves. *Ann. Inst. Fourier (Grenoble)*, 58(1):63–83, 2008. MR2401216 (2010d:14037)

[Möl09] M. Möller. Affine groups of flat surfaces. In *Handbook of Teichmüller theory. Vol. II*, volume 13 of *IRMA Lect. Math. Theor. Phys.*, pages 369–387. Eur. Math. Soc., Zürich, 2009. MR2497782 (2011a:32023)

[Mor09] I. Morrison. Mori theory of moduli spaces of stable curves, 2009, in preparation, available from the authors' web page.

[MT02] H. Masur and S. Tabachnikov. Rational billiards and flat structures. In *Handbook of dynamical systems, Vol. 1A*, pages 1015–1089. North-Holland, Amsterdam, 2002. MR1928530 (2003j:37002)

[OS80] Frans Oort and Joseph Steenbrink. The local Torelli problem for algebraic curves. In *Journées de Géometrie Algébrique d'Angers, Juillet 1979/Algebraic Geometry, Angers, 1979*, pages 157–204. Sijthoff & Noordhoff, Alphen aan den Rijn, 1980. MR605341 (82i:14014)

[Roy80] H. L. Royden. The Ahlfors-Schwarz lemma in several complex variables. *Comment. Math. Helv.*, 55(4):547–558, 1980. MR604712 (82i:32049)

[Sch73] W. Schmid. Variation of Hodge structure: the singularities of the period mapping. *Invent. Math.*, 22:211–319, 1973. MR0382272 (52:3157)

[Sch04] G. Schmithüsen. An algorithm for finding the Veech group of an origami. *Experiment. Math.*, 13(4):459–472, 2004.

[SF00] Zvezdelina E. Stankova-Frenkel. Moduli of trigonal curves. *J. Algebraic Geom.*, 9(4):607–662, 2000. MR1775309 (2001h:14031)

[SW04] J. Smillie and B. Weiss. Minimal sets for flows on moduli space. *Israel J. Math.*, 142:249–260, 2004. MR2085718 (2005g:37067)

[Thu88] W. P. Thurston. On the geometry and dynamics of diffeomorphisms of surfaces. *Bull. Amer. Math. Soc. (N.S.)*, 19(2):417–431, 1988. MR956596 (89k:57023)

[Vee82] W. A. Veech. Gauss measures for transformations on the space of interval exchange maps. *Ann. of Math. (2)*, 115(1):201–242, 1982. MR644019 (83g:28036b)

[Vee89] W. Veech. Teichmüller curves in moduli space, Eisenstein series and an application to triangular billiards. *Invent. Math.*, 97(3):553–583, 1989. MR1005006 (91h:58083a)

[Vee90] W. Veech. Moduli spaces of quadratic differentials. *J. Analyse Math.*, 55:117–171, 1990. MR1094714 (92e:32014)

[Vee95] W. A. Veech. Geometric realizations of hyperelliptic curves. In *Algorithms, fractals, and dynamics (Okayama/Kyoto, 1992)*, pages 217–226. Plenum, New York, 1995. MR1402493 (98f:14022)

[Via] M. Viana. IMPA lecture notes on Lyapunov exponents, available on http://w3.impa.br/~viana.

[Vor96] Ya. B. Vorobets. Plane structures and billiards in rational polygons: the Veech alternative. *Uspekhi Mat. Nauk*, 51(5(311)):3–42, 1996. MR1436653 (97j:58092)

[VZ04] E. Viehweg and K. Zuo. A characterization of certain Shimura curves in the moduli stack of abelian varieties. *J. Differential Geom.*, 66(2):233–287, 2004. MR2106125 (2006a:14015)

[Xia85] G. Xiao. *Surfaces fibrées en courbes de genre deux*, volume 1137 of *Lecture Notes in Mathematics*. Springer-Verlag, Berlin, 1985. MR872271 (88a:14042)

[Xia87] G. Xiao. Fibered algebraic surfaces with low slope. *Math. Ann.*, 276(3):449–466, 1987. MR875340 (88a:14046)

[Zor94] A. Zorich. Asymptotic flag of an orientable measured foliation on a surface. In *Geometric study of foliations (Tokyo, 1993)*, pages 479–498. World Sci. Publishing, River Edge, NJ, 1994. MR1363744 (97f:58102)

[Zor99] A. Zorich. How do the leaves of a closed 1-form wind around a surface? In *Pseudoperiodic Topology*, volume 197 of *Amer. Math. Soc. Transl. Ser. 2*, pages 135–178. Amer. Math. Soc., Providence, RI, 1999. MR1733872 (2001c:57019)

[Zor06] A. Zorich. Flat surfaces. In *Frontiers in Number Theory, Physics and Geometry. Volume 1: On random matrices, zeta functions and dynamical systems*, pages 439–586. Springer-Verlag, Berlin, 2006. MR2261104 (2007i:37070)

Introduction to arithmetic mapping class groups

Makoto Matsumoto

IAS/Park City Mathematics Series
Volume 20, 2011

Introduction to arithmetic mapping class groups

Makoto Matsumoto

Introduction

The main goal of these lectures is to explain why the absolute Galois group $\mathrm{Gal}(\overline{\mathbb{Q}}/\mathbb{Q})$ and its action on the profinite completion of the fundamental group $\Pi_{g,n}$ of a (g,n)-surface appear naturally in the study of the mapping class group $\mathrm{MCG}_{g,n}$. A second goal is to explain (in Lecture 5) how this Galois action can be used to bound the size of the image of $\mathrm{MCG}_{g,n}$ in the group of automorphisms of the Lie algebraization of $\Pi_{g,n}$ by giving *Galois obstructions* to the surjectivity of the higher Johnson homomorphisms.

These studies are based on Grothendieck's theory of *algebraic fundamental groups* of algebraic varieties (Lecture 1). We consider only the case where X is an algebraic variety over a field $K \subset \mathbb{C}$ where the associated complex analytic variety X^{an} is connected.

As is classically known, the topological fundamental group of X^{an} is isomorphic to the automorphism group of a universal covering $\tilde{X^{\mathrm{an}}} \to X^{\mathrm{an}}$. Since the map $\tilde{X^{\mathrm{an}}} \to X^{\mathrm{an}}$ is not necessarily algebraic when $\pi_1(X^{\mathrm{an}})$ is infinite, X^{an} does not always have a universal covering in the category of algebraic varieties. In the Grothendieck's theory, the universal covering of X is replaced by the system of all the finite unramified (more precisely, *etale*) coverings of X. Let x be a geometric point of X. The algebraic fundamental group of X with base point x is defined to be the group of automorphisms of the corresponding tower of pointed finite etale coverings:

$$\pi_1^{\mathrm{alg}}(X,x) := \operatorname*{proj\,lim}_{\lambda\in\Lambda} \mathrm{Aut}(Y_\lambda/X)^{\mathrm{op}},$$

where Y_λ runs through the isomorphism classes of finite etale Galois coverings of X equipped with a geometric point $y \in Y_\lambda$ above x. Each finite group $\mathrm{Aut}(Y_\lambda/X)^{\mathrm{op}}$ is equipped with discrete topology. Then, the above projective limit is a closed subgroup of the direct product of these finite groups. Thus, $\pi_1^{\mathrm{alg}}(X,x)$ is a topological group, called a *profinite* topological group.

The algebraic fundamental group has the following three fundamental properties, which will be explained in the lectures:

Graduate School of Mathematical Sciences, the University of Tokyo, Tokyo 153 8914 Japan
E mail address: matumoto@ms.u-tokyo.ac.jp

The author would like to thank Professor Richard Hain for all the helps related to these notes, and Mr. Kazuaki Miyatani as my Teaching Assistant. The author also thanks the anonymous referees who gave many important comments. Remark 5.17 is to answer a question from a referee. This study is partially supported by JSPS Grant-in-Aid for Scientific Research (A) No. 19204002 and No. 23244002.

(1) A comparison theorem between topological fundamental groups and algebraic fundamental groups: Let $X \otimes_K \overline{K}$ denote X considered as a variety over $\overline{K}$, where $\overline{K}$ is the algebraic closure of K in $\mathbb{C}$. Then, there is a natural isomorphism

$$\pi_1(X^{\mathrm{an}}, x)^{\widehat{}} \cong \pi_1^{\mathrm{alg}}(X \otimes \overline{K}, x),$$

where $\pi_1(X^{\mathrm{an}}, x)^{\widehat{}}$ is the profinite completion of $\pi_1(X^{\mathrm{an}}, x)$.

(2) The algebraic fundamental group of $\operatorname{Spec} K$, a single point with its function field K, is isomorphic to the absolute Galois group of K:

$$\pi_1^{\mathrm{alg}}(\operatorname{Spec}(K), b) \cong \operatorname{Gal}(\overline{K}/K),$$

where b is a geometric point of $\operatorname{Spec} K$ corresponding to $K \to \overline{K}$.

(3) There is a short exact sequence of algebraic fundamental groups:

$$1 \to \pi_1^{\mathrm{alg}}(X \otimes \overline{K}, x) \to \pi_1^{\mathrm{alg}}(X, x) \to \pi_1^{\mathrm{alg}}(\operatorname{Spec} K, b) \to 1.$$

The last short exact sequence intertwines the topology of X^{an} with the arithmetic of K. The left-hand group is isomorphic to the profinite completion of the fundamental group of the associated complex manifold X^{an}. It is called the geometric part of the middle group and depends only on the homotopy type of X^{an}. The right-hand group is isomorphic to the absolute Galois group $\operatorname{Gal}(\overline{K}/K)$ of K. It is of an "arithmetic nature," and is independent of X^{an}. The middle group intertwines these two. It is the algebraic (or *arithmetic*) fundamental group of X.

If X is the moduli stack $\mathcal{M}_{g,n}$ of (g,n)-curves over $\mathbb{Q}$, then the above sequence becomes

$$1 \to \widehat{\mathrm{MCG}_{g,n}} \to \mathrm{AMCG}_{g,n} \to \operatorname{Gal}(\overline{\mathbb{Q}}/\mathbb{Q}) \to 1,$$

where $\mathrm{MCG}_{g,n}$ is the mapping class group of n-ordered pointed genus g Riemann surface, $\widehat{\mathrm{MCG}_{g,n}}$ denotes its profinite completion, and $\mathrm{AMCG}_{g,n}$ is the *arithmetic mapping class group*, which is defined as $\pi_1^{\mathrm{alg}}(\mathcal{M}_{g,n})$.

In topology, if one has a smooth family of Riemann surfaces $F \to B$ over a connected base space B, then one has a *geometric monodromy representation*

$$\pi_1(B, b) \to \operatorname{Out}(\pi_1(F_b)).$$

An analogous construction for the universal family of (g,n)-curves over $\mathcal{M}_{g,n}$ yields a representation

$$\mathrm{AMCG}_{g,n} \to \operatorname{Out}(\widehat{\Pi_{g,n}}),$$

where $\Pi_{g,n}$ denotes the fundamental group of a genus g Riemann surface minus n points. Its restriction to the geometric part $\widehat{\mathrm{MCG}_{g,n}}$ comes from the standard representation $\mathrm{MCG}_{g,n} \to \operatorname{Out}(\Pi_{g,n})$ as follows. Since the profinite completion is a functor, the above representation extends to $\mathrm{MCG}_{g,n} \to \operatorname{Out}(\widehat{\Pi_{g,n}})$. Because $\widehat{\Pi_{g,n}}$ is topologically finitely generated, $\operatorname{Out}(\widehat{\Pi_{g,n}})$ is proved to be a profinite group, and by the universality of profinite completion, $\mathrm{MCG}_{g,n} \to \operatorname{Out}(\widehat{\Pi_{g,n}})$ factors through a unique continuous group homomorphism

$$\widehat{\mathrm{MCG}_{g,n}} \to \operatorname{Out}(\widehat{\Pi_{g,n}}),$$

and this is the restriction of $\mathrm{AMCG}_{g,n} \to \operatorname{Out}(\widehat{\Pi_{g,n}})$ to its geometric part. Thus, both mapping class groups and the Galois group $G_{\mathbb{Q}}$ are represented in $\operatorname{Out}(\widehat{\Pi_{g,n}})$. As explained in Lecture 4, their relation is subtle and interesting.

There are many other related interesting topics, such as Grothendieck-Teichmüller groups and Grothendieck's anabelian conjectures, which the author has chosen not mention at all due to lack of space and time.

LECTURE 1

Algebraic fundamental groups

1.1. The classic topological fundamental group and universal covering

Let $\mathcal{X}$ be an arcwise connected topological space, and x be a point on $\mathcal{X}$. Then, the (classical) fundamental group of $\mathcal{X}$ with base point x is defined by

$$\pi_1(\mathcal{X}, x) := \{\text{paths from } x \text{ to } x\}/(\text{homotopy with both ends fixed to } x).$$

We define the composition $\gamma \circ \gamma'$ of the loops γ and γ' at x to be the path that first traverses γ' and then γ.[1]

DEFINITION 1.1. Let $\mathcal{X}$ be a locally connected topological space. A continuous homomorphism $f : \mathcal{Y} \to \mathcal{X}$ between topological spaces is an *unramified covering* if it is surjective and for any $x \in \mathcal{X}$ there is an open neighborhood U such that $f : f^{-1}(U) \to U$ when restricted to any connected component of $f^{-1}(U)$ is a homeomorphism.

An arc-wise connected and simply connected unramified covering of $\mathcal{X}$, if exists, is called a *universal covering* of $\mathcal{X}$.

A universal covering of $\mathcal{X}$ is known to exist and be unique up to isomorphism as coverings under a weak assumption on $\mathcal{X}$ (arcwise connected, locally arcwise connected and semi-locally simply connected).

Fix a universal covering $\widetilde{\mathcal{X}} \to \mathcal{X}$ and a point $\tilde{x} \in \widetilde{\mathcal{X}}$ over $x \in \mathcal{X}$. A loop γ in X based at x lifts uniquely to a path $\gamma|_{\tilde{x}}$ in $\widetilde{\mathcal{X}}$ that starts at $\tilde{x}$. Denote its terminal point by $\tilde{x}_\gamma$. It lies over x and depends only on the homotopy class of γ. The universality of $\widetilde{\mathcal{X}}$ shows that there is a unique automorphism $\tilde{\gamma}$ of $\widetilde{\mathcal{X}}$ that maps $\tilde{x}$ to $\tilde{x}_\gamma$.

THEOREM 1.2. *The assignment $\gamma \mapsto \tilde{\gamma}$ gives a group isomorphism*

$$\pi_1(\mathcal{X}, x) \to \mathrm{Aut}(\widetilde{\mathcal{X}}/\mathcal{X})^{op}.$$

Here, G^{op} means a group with the same underlying set with G but the composition is given by

$$g_1 \circ_{G^{op}} g_2 := g_2 \circ g_1.$$

(Check: We want show

$$\tilde{\gamma_1}\tilde{\gamma_2} = \widetilde{\gamma_2\gamma_1}.$$

Consider a unique lift of the composition of two paths $\gamma_2\gamma_1$. This lift factors through $\tilde{x}_{\gamma_1}$, i.e.,

$$\tilde{x} \overset{\gamma_1|_{\tilde{x}}}{\to} \tilde{x}_{\gamma_1} \overset{\gamma_2|_{\tilde{x}_{\gamma_1}}}{\to} \tilde{x}_{\gamma_2\gamma_1} = \widetilde{\gamma_2\gamma_1}(\tilde{x}).$$

[1] Most topologists compose paths in the opposite order.

The second path $\gamma_2|_{\tilde{x}_{\gamma_1}}$ is characterized as a unique lift of γ_2 starting at $\tilde{x}_{\gamma_1} = \gamma_1(\tilde{x})$, hence coincides with the image by $\tilde{\gamma_1}$ of the path $\gamma_2|_{\tilde{x}}$ that has the same property. Its terminal point is the image by $\tilde{\gamma_1}$ of the terminal $\tilde{x}_{\gamma_2} = \tilde{\gamma_2}(\tilde{x})$, hence $\tilde{\gamma_1}\tilde{\gamma_2}(\tilde{x})$.

Thus, comparing the last point, we have that the assignment is a homomorphism. It is surjective since $\tilde{\mathcal{X}}$ is arcwise connected, and injective since simply-connected.)

1.2. The algebraic fundamental group for varieties over $\mathbb{C}$

To define an analogue of the topological fundamental group for algebraic varieties (such as those over finite fields), one has to overcome the problem that there may be no obvious analogue of a non-constant path of real dimension one in such varieties. Fortunately there is a reasonable analogue of the universal covering, and it is this which yields the algebraic fundamental group.

For non-algebraic geometers:

A smooth *affine algebraic variety* X over $\mathbb{C}$ is a nonempty subset of $\mathbb{C}^n$ that is the set of common zeroes of a set of n-variable polynomials, which is smooth and connected.[2] The coordinates $x_1, \ldots, x_n$ of $\mathbb{C}^n$ will be regarded as functions from X to $\mathbb{C}$. They are called the coordinate functions of X. A *rational function* on X is a partially defined function from X to $\mathbb{C}$ that is defined by a rational function $f(x_1, \ldots, x_n)/g(x_1, \ldots, x_n)$ where $g(x_1, \ldots, x_n)$ does not vanish identically on X. Such a rational function is defined away from the zero locus of g on X. Rational functions on X form a field, which is called the field of rational functions of X and is denoted by $\mathbb{C}(X)$.

Morphisms from X to Y are those mappings defined by polynomial functions of the coordinates:

$$y_i = f_i(x_1, \ldots, x_n) \quad \text{for each coordinate } y_i \text{ of } Y.$$

Affine algebraic varieties almost suffice for the purposes of this lecture. We only note that, roughly speaking, an *algebraic variety* is defined by glueing together a finite number of affine algebraic varieties via local isomorphisms. The field of rational functions of a variety is the same as the field of rational functions of one of its affine open subsets.

For example, in $\mathbb{C}^2$, the equation $xy = 1$ defines an affine variety X, which is nothing but $\mathbb{C} - \{0\}$; an isomorphism is given by $(x, y) \in X \mapsto x \in \mathbb{C} - \{0\}$ with inverse given by $x \mapsto (x, x^{-1})$. The x is a regular (i.e., entire) function on X, and $y = 1/x$ is another regular function on X. On the other hand, $1/(x-1)$ is a rational function, which has a pole at $x = 1$ on X. In this example, for any integer N, an unramified covering $X \to X$ is defined by $x \mapsto x^N$. Since the (topological) fundamental group of X is isomorphic to $\mathbb{Z}$, these are all the isomorphism classes of finite unramified coverings of X. However, a universal covering is given by $\exp : \mathbb{C} \to \mathbb{C} - \{0\}$, which is not an algebraic morphism.

A rough idea of algebraic fundamental groups is to use a system of all unramified *algebraic* coverings instead of a universal covering.

1.3. Algebraic fundamental groups by multivalued functions

One way to consider all the algebraic coverings in a unified way is to use multivalued analytic functions, as follows.

---

[2] Throughout these notes, all algebraic varieties will be assumed to be smooth.

DEFINITION 1.3. Let X be an algebraic variety over $\mathbb{C}$, and x be a point on it. Let M_x denote the field of the germs of algebraic meromorphic functions at x on X (in the complex analytic sense) such that the germ can be meromorphically continued to a multivalued meromorphic function along all paths in X.[3]

Let us temporarily denote by $\mathcal{M}_x$ the set of all germs of meromorphic functions at x on X which admit meromorphic continuation to a multivalued function over X. That is, they are germs of meromorphic functions at x on X that can be meromorphically continued along all paths in X. Here, $h \in \mathcal{M}_x$ is *algebraic* if it satisfies a polynomial equation with coefficients in $\mathbb{C}(X)$. Thus, $M_x \subset \mathcal{M}_x$ is the subfield of algebraic elements. Take $h \in M_x$. By assumption, for any $\gamma \in \pi_1(X,x)$, there is a meromorphic continuation $\gamma(h) \in \mathcal{M}_x$. By meromorphic continuation of the minimal polynomial of h, $\gamma(h)$ has the same minimal polynomial, and hence is in M_x. In particular, the set $\{\gamma(h) \mid \gamma \in \pi_1(X,x)\}$ is finite (i.e., h is finitely multivalued), since its cardinality does not exceed the degree of the minimal polynomial. Since any γ only permutes these functions, each symmetric polynomial of these functions is a single-valued meromorphic function on X, which is algebraic over $\mathbb{C}(X)$, and hence it is in $\mathbb{C}(X)$.[4] Hence, all the algebraic conjugates of h over $\mathbb{C}(X)$ are realized in M_x (i.e. $M_x/\mathbb{C}(X)$ is a normal algebraic extension, thus a Galois extension).

DEFINITION 1.4. The algebraic fundamental group $\pi_1^{\text{alg}}(X,x)$ is defined by

$$\pi_1^{\text{alg}}(X,x) := \text{Aut}(M_x/\mathbb{C}(X)).$$

Here, the right-hand side denotes the group of automorphisms of the field M_x that act trivially on $\mathbb{C}(X)$. Since this field extension is Galois, the right-hand side is the Galois group $G(M_x/\mathbb{C}(X))$ of the (typically infinite) extension $M_x/\mathbb{C}(X)$.

By meromorphic continuation, a path from x to x yields an automorphism of M_x that is trivial on $\mathbb{C}(X)$. Since this automorphism depends only on the homotopy class of γ, we have a natural homomorphism

$$\pi_1(X,x) \to \pi_1^{\text{alg}}(X,x).$$

By definition, a field extension M/L is a *Galois extension* if M is a union of finite normal separable extensions L_λ/L $(\lambda \in \Lambda)$. It is standard in the (infinite) Galois theory that

$$G(M/L) = \operatorname*{proj\,lim}_{\lambda \subset \Lambda} G(L_\lambda/L).$$

Thus, the Galois group of an infinite field extension is a projective limit of finite groups, namely a *profinite group*.

DEFINITION 1.5. A group isomorphic to a projective limit of finite groups is called a profinite group. It is a compact topological group with the topology induced from the discrete topology on the finite groups.[5]

[3]A multivalued meromorphic function on X can be considered as a meromorphic function on a universal covering $\widetilde{X}$ of X.

[4]We leave a proof of the last claim to the readers.

[5]Compactness follows from Tychonoff's Theorem because the projective limit is a closed subset of the product.

1.4. Comparison theorem

How much of $\pi_1(X,x)$ is detected in $\pi_1^{\mathrm{alg}}(X,x)$? The answer is that, the latter detects all the finite quotients of the former.

Fix a universal covering $\tilde{X}$ of X and a point $\tilde{x} \in \tilde{X}$ above x. Let N be a normal finite index subgroup of $\pi_1(X,x)$. Then, $Y := \tilde{X}/N \to X$ is a Galois covering with $\mathrm{Aut}(Y/X)^{op} \cong \pi_1(X,x)/N$. The image y of $\tilde{x}$ in Y is a distinguished base point of Y. Conversely, given a pointed Galois covering $(Y,y) \to (X,x)$, one recovers N as the stabilizer of y.

THEOREM 1.6 (Grothendieck's Riemann-Existence-Theorem). [**10**, XII]. *If X is an algebraic variety over $\mathbb{C}$, then every finite unramified covering (in the sense of topology) Y of X has a (unique) structure of a complex algebraic variety such that $Y \to X$ is an algebraic morphism.*

Let Y be as above. Fix a point $y \in Y$ above $x \in X$. Then the germ at y of each element of $\mathbb{C}(Y)$ is an element of M_x. Thus each choice of a point $y \in Y$ that lies over x gives an inclusion $\mathbb{C}(Y) \subset M_x$. Since Y is a Galois covering, $\mathbb{C}(Y)/\mathbb{C}(X)$ is a Galois extension with Galois group $\pi_1(X,x)/N$. Thus we have a homomorphism

$$\pi_1(X,x) \to \pi_1(X,x)/N \cong G(\mathbb{C}(Y)/\mathbb{C}(X)).$$

Each $h \in M_x$ is the germ at y of a single-valued meromorphic function on some finite unramified covering $(Y,y) \to (X,x)$. Thus, M_x is the union of $\mathbb{C}(Y)$ as (Y,y) runs over the finite Galois coverings. By taking the limit over N, we have

THEOREM 1.7 (Comparison Theorem). *There are natural isomorphisms*

$$\begin{aligned} \pi_1(X,x)^{\widehat{}} &:= \operatorname{proj\,lim}_{N} \pi_1(X,x)/N \\ &\cong \operatorname{proj\,lim}_{(Y,y)} G(\mathbb{C}(Y)/\mathbb{C}(X)) = G(M_x/\mathbb{C}(X)) = \pi_1^{\mathrm{alg}}(X,x). \end{aligned}$$

The left-hand group is called the *profinite completion* of the discrete group $\pi_1(X,x)$:

DEFINITION 1.8. Let G be a (discrete) group. The normal subgroups N with finite index (i.e. G/N is a finite group) constitute a directed preorder. The profinite completion of G is defined by

$$G^{\widehat{}} := \operatorname{proj\,lim}_{N} G/N.$$

EXAMPLE 1.9. The free cyclic group $\mathbb{Z}$ has its profinite completion

$$\mathbb{Z}^{\widehat{}} = \operatorname{proj\,lim}_{N \in \mathbb{Z}, N \geq 1} (\mathbb{Z}/N\mathbb{Z}).$$

By Chinese Remainder Theorem, we have an isomorphism

$$\mathbb{Z}^{\widehat{}} \cong \prod_{\ell:\text{prime}} \mathbb{Z}_\ell,$$

where $\mathbb{Z}_\ell := \operatorname{proj\,lim}_n \mathbb{Z}/(\ell^n)$ is the pro-ℓ completion of $\mathbb{Z}$.

1.5. Algebraic fundamental group (over $K \subset \mathbb{C}$)

So far, the algebraic fundamental group is merely the profinite completion of the classical fundamental group. It has a much richer structure when we consider algebraic varieties defined over a subfield K of $\mathbb{C}$, such as $K = \mathbb{Q}$. An algebraic variety X defined over K (or a K-variety) is an algebraic variety whose defining polynomials have coefficients in K. When we want to emphasize that X is a K-variety, we will write X_K. Thus, $X_{\mathbb{C}}$ means the complex algebraic variety with the same defining polynomials as X_K. Morphisms over K of K-varieties are those morphisms represented by polynomials with coefficients in K. The rational function field $K(X)$ of X_K is the field of rational functions on $X_{\mathbb{C}}$ which can be written as rational functions with coefficients in K of the coordinate functions of X.

In our definition of a K-variety $X_{\mathbb{C}}$ is assumed to be connected.[6] Let $\overline{K} \subset \mathbb{C}$ be the algebraic closure of K in $\mathbb{C}$. We may consider $X_{\overline{K}}$ and $\overline{K}(X)$. It is not obvious, but the connectivity of $X_{\mathbb{C}}$ implies that $\overline{K}(X) \cong K(X) \otimes_K \overline{K}$ and that

$$G(\overline{K}(X)/K(X)) \cong G(\overline{K}/K).$$

DEFINITION 1.10. Let X be a K-algebraic variety and $x \in X_{\mathbb{C}}$. Then one has the field M_x defined above for $X_{\mathbb{C}}$. Let M_x^{alg} be the algebraic closure of $\overline{K}(X)$ in M_x. Define

$$\pi_1^{\text{alg}}(X_K, x) := G(M_x^{\text{alg}}/K(X)).$$

We need to show that the extension $M_x^{\text{alg}}/K(X)$ is normal. This requires some effort: an element $h \in M_x^{\text{alg}}$ is contained in $\mathbb{C}(Y_{\mathbb{C}})$ for some $Y_{\mathbb{C}}$, and then in $\overline{K}(Y_{\overline{K}})$ by algebraicity. Since there are only finitely many polynomials involved in the definition of $Y_{\overline{K}}$, $h \in L(Y_L)$ for a finite Galois extension L of K. We can consider for each $\sigma \in G(L/K)$, a function $h^\sigma \in L(Y_L^\sigma)$ in an L-variety Y_L^σ obtained by changing all the coefficients (in both defining polynomials and the expression of h in terms of coordinates) in L by σ. These h^σ and their meromorphic continuations give all the conjugate elements of h over $K(X)$.

THEOREM 1.11. [**10**, §7, Ch.XIII] *If $Y \to X$ is a finite unramified covering of complex algebraic variety, and if X is defined over $\overline{K}$, then there is a model $Y_{\overline{K}} \to X_{\overline{K}}$ over $\overline{K}$, such that $Y_{\mathbb{C}} \to X_{\mathbb{C}}$ is isomorphic to the original covering. If $\mathbb{C}(Y)/\mathbb{C}(X)$ is Galois, so is $K(Y)/\overline{K}(X)$, and their Galois groups coincide.*

COROLLARY 1.12. *If X is a K-variety, then there are natural isomorphisms*

$$\pi_1^{\text{alg}}(X_{\mathbb{C}}, x) = G(M_x/\mathbb{C}(X)) \simeq G(M_x^{\text{alg}}/\overline{K}(X)) = \pi_1^{\text{alg}}(X_{\overline{K}}, x).$$

The fundamental theorem of Galois theory for infinite extensions gives the short exact sequence

$$1 \to G(M_x^{\text{alg}}/\overline{K}(X)) \to G(M_x^{\text{alg}}/K(X)) \to G(\overline{K}(X)/K(X)) \to 1.$$

Identifying Galois groups with the algebraic fundamental groups, we obtain:

THEOREM 1.13. *If X is a K-algebraic variety (with $X_{\mathbb{C}}$ connected), then there is a short exact sequence*

$$1 \to \pi_1^{\text{alg}}(X_{\overline{K}}, x) \to \pi_1^{\text{alg}}(X_K, x) \to G(\overline{K}/K) \to 1. \tag{1.1}$$

[6]For algebraic geometers: we are thus considering only geometrically connected varieties.

The left-hand group is the profinite completion of the topological fundamental group of $X_{\mathbb{C}}$; it is called the geometric part of the middle group. The right-hand group is the absolute Galois group of K. The middle group thus combines topology of $X_{\mathbb{C}}$ with the arithmetic of K.

A lift $\tilde{\sigma} \in \pi_1^{\text{alg}}(X_K, x)$ of an element σ of $G(\overline{K}/K)$ acts on $\pi_1^{\text{alg}}(X_{\overline{K}}, x)$ by conjugation $\tilde{\sigma}(-)\tilde{\sigma}^{-1}$. Any other lift of σ is of the form $\tilde{\sigma}' = \tilde{\sigma}\gamma$ with $\gamma \in \pi_1^{\text{alg}}(X_{\overline{K}}, x)$. The difference of the actions of σ and σ' on $\pi_1^{\text{alg}}(X_{\overline{K}}, x)$ is the inner automorphism given by conjugating by γ. We thus have a representation

$$(1.2)\quad \rho_X : G(\overline{K}/K) \to \operatorname{Out}(\pi_1^{\text{alg}}(X_{\overline{K}}, x)) := \operatorname{Aut}(\pi_1^{\text{alg}}(X_{\overline{K}}, x))/\operatorname{Inn}(\pi_1^{\text{alg}}(X_{\overline{K}}, x))$$

of the Galois group of K in the group of *continuous* outer automorphisms of $\pi_1^{\text{alg}}(X_{\overline{K}}, x)$.

From now on, let us denote $G_K := G(\overline{K}/K)$.

1.6. The case $\mathbb{C} - \{0\}$

Consider the case where $X_{\mathbb{Q}}$ is the zero locus of $xy - 1 \in \mathbb{Q}[x, y]$. The map $t \mapsto (t, 1/t)$ defines an isomorphism $\mathbb{C} - \{0\} \cong X_{\mathbb{C}}$. Take the base point to be the point that corresponds to $x = 1$, which we denote by 1. Thus, we consider M_1^{alg} and $\pi_1^{\text{alg}}(X, 1)$ in the following. For each positive integer N, let t_N be the germ at $x = 1$ of a holomorphic function on X at $x = 1$ that satisfies $(t_N)^N = x$ and $t_N(1) = 1$. Thus, t_N is a fixed branch of $x^{1/N}$ at $x = 1$; it admits analytic continuation along all paths in X and thus defines a multivalued holomorphic function on X. Thus $t_N \in M_1^{\text{alg}}$. On the other hand, because the fundamental group of X is isomorphic to $\mathbb{Z}$, the coverings $Y_N := \mathbb{C} - \{0\} \to \mathbb{C} - \{0\}$ given by $\alpha \mapsto \alpha^N$ are all the possible unramified coverings of X.

Let K be a subfield of $\mathbb{C}$ as above. Since t_N generate the function field $\overline{K}(Y_N)$ over $\overline{K}$, we know that M_1^{alg} is generated, as a field over $\overline{K}$, by $\{t_N : N \geq 1\}$. That is, the inclusions $\overline{K}(t_N) \hookrightarrow M_1^{\text{alg}}$ induce an isomorphism $M_1^{\text{alg}} \cong \cup_{N \in \mathbb{N}} \overline{K}(t_N)$.

The morphism $\pi_1(X, 1) \to G(M_1^{\text{alg}}/\overline{K}(X))$ is described as follows: Take a standard generator $\gamma \in \pi_1(X, 1)$ (i.e., a path that circles the origin once, counter clockwise). Then the analytic continuation of t_N along γ is given by

$$\gamma(t_N) = \zeta_N t_N$$

where $\zeta_N := \exp(2\pi\sqrt{-1}/N)$ is an N-th root of unity; this is because the above choice gives explicitly

$$t_N = \exp(\frac{1}{N} \log x)$$

where the branch of $\log x$ around $x = 1$ is taken so that $\log 1 = 0$. This determines the image of γ in the Galois group $G(M_1^{\text{alg}}/\overline{K}(X))$, which we shall also denote by γ.

Our final task is to compute the action of $G(\overline{K}/K)$ on γ. Let $\sigma \in G(\overline{K}/K) \cong G(\overline{K}(X)/K(X))$. A lift $\tilde{\sigma} \in G(M_1^{\text{alg}}/K(X))$ can be chosen as follows: for each N, let $\tilde{\sigma}$ act on a rational function of t_N with coefficients in $\overline{K}$ by letting σ act on its coefficients. Its conjugate $\tilde{\sigma}\gamma\tilde{\sigma}^{-1} \in G(M_1^{\text{alg}}/\overline{K}(X))$ acts on t_N by

$$t_N \overset{\tilde{\sigma}^{-1}}{\longmapsto} t_N \overset{\gamma}{\longmapsto} \zeta_N t_N \overset{\tilde{\sigma}}{\longmapsto} \sigma(\zeta_N) t_N.$$

Define $\chi(\sigma)_N \in (\mathbb{Z}/N)^\times$ by

$$\sigma(\zeta_N) = \zeta_N^{\chi(\sigma)_N}.$$

By taking the projective limit of the χ_N, we obtain a homomorphism

$$\chi : G_K \to (\mathbb{Z}^\frown)^\times,$$

which is called the *cyclotomic character*. When $K = \mathbb{Q}$, the irreducibility of the cyclotomic polynomials implies that the cyclotomic character is surjective.

For all N, the action of $\sigma(\gamma)$ and $\gamma^{\chi(\sigma)}$ on each t_N agree. It follows that

$$\sigma(\gamma) = \gamma^{\chi(\sigma)} \in \pi_1^{\mathrm{alg}}(X_{\overline{K}}, 1) \cong \mathbb{Z}^\frown.$$

Thus, we have proved that the homomorphism

$$\rho_X : G_K \to \mathrm{Aut}\, \pi_1^{\mathrm{alg}}(X_{\overline{K}}, 1)$$

takes σ to the automorphism $\gamma \mapsto \gamma^{\chi(\sigma)}$.

1.7. Fundamental group of moduli spaces

The central object in this PCMI summer school is the moduli space $\mathcal{M}_{g,n\mathbb{C}}$ that classifies all the (holomorphic) isomorphic classes of n-pointed compact Riemann surfaces of genus g. Recall that throughout in this note we assume that $2g-2+n > 0$.

Let $\Sigma_{g,n}$ be a compact oriented surface of genus g with n marked points. Let $\mathrm{MCG}_{g,n} := \pi_0(\mathrm{Diff}^+\Sigma_{g,n})$ denote its mapping class group. As explained in other lecture series, Teichmüller theory constructs the Teichmüller space $\mathcal{T}_{g,n}$, which is homeomorphic to a ball of real $6g - 6 + 2n$ dimension and on which $\mathrm{MCG}_{g,n}$ acts properly discontinuously. The moduli space $\mathcal{M}_{g,n\mathbb{C}}$ is the quotient:

$$\mathcal{M}_{g,n\mathbb{C}} := \mathcal{T}_{g,n}/\mathrm{MCG}_{g,n}.$$

If $n > 2g + 2$, then the action of $\mathrm{MCG}_{g,n}$ on $\mathcal{T}_{g,n}$ is fixed point free. In this case $\mathcal{T}_{g,n}$ is the universal covering of $\mathcal{M}_{g,n\mathbb{C}}$, and we have an isomorphism

$$\pi_1(\mathcal{M}_{g,n\mathbb{C}}) \cong \mathrm{MCG}_{g,n}.$$

Results of Deligne-Mumford [**8**] and Knudsen [**20**] imply that $\mathcal{M}_{g,n}$ is an algebraic variety over $\mathbb{Q}$. Thus, we can define a short exact sequence of algebraic fundamental groups

$$1 \to \pi_1^{\mathrm{alg}}(\mathcal{M}_{g,n\mathbb{C}}) \to \pi_1^{\mathrm{alg}}(\mathcal{M}_{g,n\mathbb{Q}}) \to G_{\mathbb{Q}} \to 1. \tag{1.3}$$

The middle group is the *arithmetic mapping class group* of (g, n)-surfaces (or (g, n)-curves, since a Riemann surface is a complex curve), denoted by $\mathrm{AMCG}_{g,n}$. Thus, we have the short exact sequence

$$1 \to \mathrm{MCG}^\frown_{g,n} \to \mathrm{AMCG}_{g,n} \to G_{\mathbb{Q}} \to 1.$$

For general (g, n) satisfying $2g - 2 + n > 0$, Deligne and Mumford [**8**] (and Knudsen [**20**] for $n > 0$) constructed $\mathcal{M}_{g,n}$ as an object of the (subsequently named) category of Deligne-Mumford stacks. Roughly speaking, a Deligne-Mumford stack is (locally) a quotient of an algebraic variety X by a finite group G. If the finite group G acts fixed point freely on X, then the quotient X/G is just the quotient in the category of varieties. When there are fixed points, we keep the data of $X \to X/G$, which is regarded as an unramified covering. Deligne-Mumford stacks are the analogue of orbifolds in the category of varieties.

The author will not define Deligne-Mumford stacks, but only mentions the fact that algebraic fundamental groups of Deligne-Mumford stacks can be defined and that the short exact sequence (1.3) holds for all (g, n) satisfying $2g - 2 + n > 0$. Details can be found in [**29**].

Monodromy representation on fundamental groups

2.1. Classical topological case

We return to the topological setting. Let $f : \mathcal{F} \to \mathcal{B}$ be a locally trivial fiber bundle in the category of topological spaces. Take a point $b \in \mathcal{B}$, let $\mathcal{F}_b$ be the fiber over b, and take a point $x \in \mathcal{F}_b$ (hence $b = f(x)$). Then one has the homotopy exact sequence

$$\cdots \to \pi_2(\mathcal{B}) \to \pi_1(\mathcal{F}_b, x) \to \pi_1(\mathcal{F}, x) \to \pi_1(\mathcal{B}, f(x)) \to \pi_0(\mathcal{F}_b, x) \to \cdots .$$

If $\pi_0(\mathcal{F}_b, x)$ is trivial and $\pi_1(\mathcal{F}_b, x) \to \pi_1(\mathcal{F}, x)$ is injective, then this becomes the short exact sequence

$$1 \to \pi_1(\mathcal{F}_b, x) \to \pi_1(\mathcal{F}, x) \to \pi_1(\mathcal{B}, f(x)) \to 1. \tag{2.1}$$

Recall that the outer automorphism group of a group G is defined as the quotient

$$\mathrm{Out}(G) := \mathrm{Aut}(G)/\,\mathrm{Inn}(G)$$

of the automorphism group of G by the subgroup of inner automorphisms.

Suppose that $\gamma \in \pi_1(\mathcal{B}, f(x))$. Choose a lift $\tilde{\gamma} \in \pi_1(\mathcal{F}, x)$ of γ. This acts on $\pi_1(\mathcal{F}_b, x)$ by conjugation. The ambiguity of the action of this lift is an inner automorphism of $\pi_1(\mathcal{F}_b, x)$. Hence we have a well-defined homomorphism

$$\pi_1(\mathcal{B}, b) \to \mathrm{Out}(\pi_1(\mathcal{F}_b, x)). \tag{2.2}$$

It is called the *outer monodromy representation.*

2.2. Arithmetic/algebraic case

The outer Galois representation (1.2) is an analogue of the monodromy representation (2.2). To explain this necessitates a very brief discussion of schemes. Here I do not explain the definition, but instead state some of their basic properties.

- The category of schemes contains the category of algebraic varieties.
- For a commutative ring A with unit, there is an associated scheme $\operatorname{Spec} A$. One should think of A as the ring of functions on $\operatorname{Spec} A$. The functor Spec is a contravariant fully faithful functor from the category of commutative rings with unit to the category of schemes.
- A K-variety X has a natural morphism $X \to \operatorname{Spec} K$.
- Fibered products exist in the category of schemes.
- For a field L and a scheme X, a morphism $\operatorname{Spec} L \to X$ is called an L-*rational point* of X. If L is algebraically closed, an L-rational point is called a *geometric point* of X.

An *affine scheme* is a scheme of the form $\operatorname{Spec} A$. General schemes are obtained by glueing together affine schemes. A scheme is locally noetherian if it is covered by affine schemes $\operatorname{Spec} A$ where A is a noetherian ring.

FACT 2.1 (algebraic fundamental groups for schemes).

(1) For a connected locally noetherian scheme X and a geometric point x of X, one can define a profinite group $\pi_1^{\mathrm{alg}}(X, x)$. If X is a K-variety, where $K \subset \mathbb{C}$, and $x \in X^{\mathrm{an}}$, it is (canonically) isomorphic to the one defined in Definition 1.10.

(2) This fundamental group is functorial. That is, a morphism $f : (X, x) \to (Y, f(x))$ of pointed schemes induces a morphism of profinite groups $\pi_1(f) : \pi_1^{\mathrm{alg}}(X, x) \to \pi_1^{\mathrm{alg}}(Y, f(x))$.

(3) If $\overline{K}$ is an algebraic closure of K and x is the geometric point $\operatorname{Spec} \overline{K} \to \operatorname{Spec} K$, then there is a canonical isomorphism

$$\pi_1(\operatorname{Spec} K, x) \cong G(\overline{K}/K) =: G_K.$$

We have the following analogy. The structure morphism $X \to \operatorname{Spec} K$ is an analogue of a topological fiber bundle $\mathcal{F} \to \mathcal{B}$. A geometric point $b : \operatorname{Spec} \overline{K} \to \operatorname{Spec} K$ is an analogue of a point $b \in \mathcal{B}$. The fiber X_b over $b : \operatorname{Spec} \overline{K} \to \operatorname{Spec} K$ is the fiber product $X_b := X_{\overline{K}} \to \operatorname{Spec} \overline{K}$ in the category of schemes. Take a geometric point x in the fiber $X_{\overline{K}}$. Now $\pi_1^{\mathrm{alg}}(X_{\overline{K}}, x)$ (respectively $\pi_1^{\mathrm{alg}}(X_K, x)$, $G(\overline{K}/K) = \pi_1(\operatorname{Spec} K, b)$) is an analogue to $\pi_1(\mathcal{F}_b, x)$ (respectively $\pi_1(\mathcal{F}, x)$, $\pi_1(\mathcal{B}, b)$), and the short exact sequence of the algebraic fundamental groups (1.1) is an analogue to the topological one (2.1). The method defining the outer Galois representations (1.2) is the same with the method defining the monodromy representations (2.2), based on these short exact sequences.

LECTURE 3

Arithmetic mapping class groups

3.1. Moduli of curves and moduli stack

If we work in the category of schemes, an appropriate notion of "family of (g,n)-curves" is described as follows:

DEFINITION 3.1. Let $C^* \to B$ be a proper smooth morphism whose fibers are geometrically connected curves of genus g. Let $s_1, s_2, \dots, s_n : B \to C^*$ be n sections where the n images of B in C^* do not intersect. The complement

$$C := (C^* - \cup_{i=1}^n s_i(B))$$

over B is called a family of (g,n)-curves. We always assume the hyperbolicity of the curve; namely that $2g-2+n>0$.

You don't need to keep the details in the above definition. For our purpose, it suffices to know that a K-variety

$$X \to \operatorname{Spec} K$$

is a (g,n)-curve if the corresponding complex analytic variety is a compact Riemann surface of genus g with n distinct K-rational points removed.

Let (g,n) be non-negative integers satisfying $2g-2+n>0$. We want to find a universal family $\mathcal{C}_{g,n} \to \mathcal{M}_{g,n}$ of (g,n)-curves over $\mathbb{Q}$ with the following universal property (called the fine moduli property): For any family $C \to B$ of (g,n)-curves over $\mathbb{Q}$, there exists a unique pair of morphisms of schemes $B \to \mathcal{M}_{g,n}$ and $C \to \mathcal{C}_{g,n}$ such that the square

$$\begin{array}{ccc} C & \to & \mathcal{C}_{g,n} \\ \downarrow & & \downarrow \\ B & \to & \mathcal{M}_{g,n} \end{array}$$

is a fiber product.[1]

A difficulty is that except when $n > 2g+2$, there is no such a universal family However, if we enlarge the category of schemes to the category of Deligne-Mumford stacks, such family $\mathcal{C}_{g,n} \to \mathcal{M}_{g,n}$ does exist [8], [20]. In fact, this family is isomorphic to the projection

$$\mathcal{M}_{g,n+1} \to \mathcal{M}_{g,n}$$

obtained by forgetting the $n+1$-st point.[2]

It is worth stating the universality for a (g,n)-curve $C \to \operatorname{Spec} K$ for a field K (of characteristic 0, since we consider the moduli stack over $\mathbb{Q}$): There is a unique K-rational point $\operatorname{Spec} K \to \mathcal{M}_{g,n}$ of $\mathcal{M}_{g,n}$ such that $C \to \operatorname{Spec} K$ is the pullback

[1] More precisely, there is a family $\mathcal{C}^*_{g,n} \to \mathcal{M}_{g,n}$ with n disjoint sections, and $C^* \to \mathcal{C}^*_{g,n}$ pull backs these universal sections to the sections in C^*.

[2] An important fact, that we shall not need in these lectures, is that all of these stacks and morphisms are defined over $\mathbb{Z}$.

of the universal curve along $\operatorname{Spec} K \to \mathcal{M}_{g,n}$. Thus, the isomorphism classes of (g, n)-curves defined over K are in one-to-one correspondence to K-rational points of $\mathcal{M}_{g,n}$. (Strictly speaking, we need the notion of stack to deal with curves with non-trivial automorphisms in the above bijection.)

The homotopy exact sequence for $\mathcal{M}_{g,n} \to \operatorname{Spec} \mathbb{Q}$ is described in (1.3). The following is the homotopy exact sequence for the family $\mathcal{C}_{g,n} \to \mathcal{M}_{g,n}$.

THEOREM 3.2. *Let b be a geometric point of $\mathcal{M}_{g,n}$ and C_b be the (g, n)-curve obtained as the fiber of the universal (g, n) curve over b. If x is a geometric point of C_b, there there is a short exact sequence*

$$1 \to \pi_1^{\text{alg}}(C_b, x) \to \pi_1^{\text{alg}}(\mathcal{C}_{g,n}, x) \to \pi_1^{\text{alg}}(\mathcal{M}_{g,n}, b) \to 1.$$

For a proof, see [**29**] and [**23**, Lemma 2.7].

3.2. Universal monodromy representation

As already stated, the algebraic fundamental group of $\mathcal{M}_{g,n}$ is called the *arithmetic mapping class group* for (g, n)-curves, and denoted by $\mathrm{AMCG}_{g,n}$. This terminology may be justified by the following facts. Under the assumptions of Theorem 3.2, we have the *universal monodromy representation*

$$\rho_{g,n} : \mathrm{AMCG}_{g,n} = \pi_1^{\text{alg}}(\mathcal{M}_{g,n}, b) \to \operatorname{Out} \pi_1^{\text{alg}}(C_b, x),$$

which fits into the following diagram:

$$\begin{array}{ccccccccc} 1 & \to & \pi_1(\mathcal{M}_{g,n}^{\text{an}}, b)\widehat{\ } & \to & \pi_1^{\text{alg}}(\mathcal{M}_{g,n}, b) & \to & \mathrm{Gal}(\overline{\mathbb{Q}}/\mathbb{Q}) & \to & 1 \\ & & & & \downarrow \rho_{g,n} & & & & \\ & & & & \operatorname{Out}(\pi_1^{\text{alg}}(C_b, x)). & & & & \end{array} \tag{3.1}$$

The fact that $\pi_1(C_b^{\text{an}}, x)$ is finitely generated implies that $\operatorname{Out}(\pi_1(C_b^{\text{an}}, x)\widehat{\ })$ is a profinite group. The homomorphism

$$\mathrm{MCG}_{g,n} \to \operatorname{Out}(\pi_1(C_b^{\text{an}}, x)) \to \operatorname{Out}(\pi_1(C_b^{\text{an}}, x)\widehat{\ })$$

thus induces a homomorphism

$$\rho_{g,n}^{\text{geom}} : \widehat{\mathrm{MCG}}_{g,n} \to \operatorname{Out}(\pi_1(C_b^{\text{an}}, x)\widehat{\ }).$$

Takayuki Oda [**29**] showed that

$$\pi_1(\mathcal{M}_{g,n} \otimes \overline{\mathbb{Q}}, b) \cong \widehat{\mathrm{MCG}}_{g,n},$$

and that the restriction

$$\rho_{g,n}|_{\widehat{\mathrm{MCG}}_{g,n}} : \widehat{\mathrm{MCG}}_{g,n} \to \operatorname{Out}(\pi_1^{\text{alg}}(C_b, x))$$

of $\rho_{g,n}$ to the geometric part is isomorphic to $\rho_{g,n}^{\text{geom}}$ described as above, induced from the standard outer action

$$\mathrm{MCG}_{g,n} \to \operatorname{Out}(\pi_1(C_b^{\text{an}}, x))$$

of the mapping class group.

To summarize: the geometric part of $\mathrm{AMCG}_{g,n}$ is $\widehat{\mathrm{MCG}}_{g,n}$. Its action on the fundamental group of the curve comes from pure topology.

LECTURE 4

Topology versus arithmetic

A basic question in the theory of arithmetic fundamental groups is how the arithmetic part (i.e., the Galois part) and the geometric part are intertwined.

4.1. The profinite case

DEFINITION 4.1. Let $\Pi_{g,n}$ denote the (classical) fundamental group of (g,n)-curve. It has generators $\alpha_1, \beta_1, \ldots, \alpha_g, \beta_g, \gamma_1, \ldots, \gamma_n$ that satisfy the single relation

$$[\alpha_1, \beta_1] \cdots [\alpha_g, \beta_g]\gamma_1 \cdots \gamma_n = 1.$$

Consider the case $(g,n) = (0,3)$. Since the projective line with three marked and labeled points is rigid, $\mathcal{M}_{0,3} = \operatorname{Spec} \mathbb{Q}$. (This follows from that for any commutative ring R, a proper smooth curve of genus 0 over R with an R-rational point is isomorphic to a projective line over R, and for every ordered triple of distinct R-rational points in a projective line can be taken to $(0,1,\infty)$ by a unique element in $\mathrm{PGL}_2(R)$, which is the automorphism group of the projective line). In the complex analytic case, the universal family is trivial. However, in the arithmetic context, the universal family $\mathcal{C}_{0,3} \to \mathcal{M}_{0,3}$ is $\mathbb{P}^1 - \{0,1,\infty\} \to \operatorname{Spec} \mathbb{Q}$, and we have the monodromy representation

$$\rho_{0,3} : \mathrm{AMCG}_{0,3} = G_{\mathbb{Q}} \to \mathrm{Out}(\widehat{\Pi_{0,3}})$$

The right-hand side is the outer automorphism group of the profinite completion of a free group with two generators. Note that, in the $(0,3)$ case, $\mathrm{AMCG}_{0,3}$ has no geometric part, only a large arithmetic part.

The following striking result of Belyĭ is fundamental. Remarkably, its proof is elementary.

THEOREM 4.2 (Belyĭ [**3**]). *The outer action $\rho_{0,3}$ is injective.*

Note that $G_{\mathbb{Q}}$ is an infinite (and thus uncountable) profinite group. It is considered to be mysterious. So the above embedding into a (relatively) understandable group is interesting.

Belyĭ's result has a natural generalization to all hyperbolic curves. Throughout in this section, let K be a finite extension of $\mathbb{Q}$. Let C be a geometrically connected (g,n)-curve defined over K, where $2g-2+n > 0$. As in (1.2), we have an associated outer Galois representation

$$\rho_C : \mathrm{Gal}(\overline{K}/K) \to \mathrm{Out}(\pi_1(C^{\mathrm{an}}, x)^\wedge).$$

THEOREM 4.3. *If $2g-2+n > 0$, then ρ_C is injective.*

This is a generalization of Belyĭ's Theorem. It is proved by reducing to Belyĭ's Theorem. The case $n \geq 1$ was proved by the author in [**22**]. The general case was recently settled by Y. Hoshi and S. Mochizuki in [**13**].

The image of ρ_C has trivial intersection with the image of $\mathrm{MCG}^{\wedge}_{g,n}$, as stated below. By the universality, $C \to \operatorname{Spec} K$ is a pullback of the universal curve $\mathcal{C}_{g,n} \to \mathcal{M}_{g,n}$ along a morphism $a : \operatorname{Spec} K \to \mathcal{M}_{g,n}$. Let b be a geometric point of $\mathcal{M}_{g,n}$ over a. We have

$$\pi_1(a) : G_K \to \pi_1(\mathcal{M}_{g,n}, b)$$

such that the composition $\rho_{g,n} \circ \pi_1(a)$ is ρ_C, where $\rho_{g,n}$ is the universal monodromy as in §3.2.

THEOREM 4.4. *If $2g-2+n>0$, then the intersection of the image of*

$$\rho_C : G_K \to \operatorname{Out}(\pi_1(C^{\mathrm{an}}, x)^{\wedge})$$

with the image of

$$\rho_{g,n} : \mathrm{MCG}^{\wedge}_{g,n} \to \operatorname{Out}(\pi_1(C^{\mathrm{an}}, x)^{\wedge})$$

is trivial.

The case $n \geq 1$ was proved by A. Tamagawa and the author in [**23**]. The general case was proved by Hoshi and Mochizuki in [**13**].

REMARK 4.5. Y. Iijima [**16**] recently showed the injectivity of the outer representation on the profinite mapping class groups

$$\rho : G_{\mathbb{Q}} \to \operatorname{Out}(\mathrm{MCG}^{\wedge}_{g,n})$$

for $3g-3+n>0$ ($3g-3+n$ is the dimension of the moduli stack). This is another direction of generalization of Belyĭ's Theorem, from $\mathbb{P}^1 - \{0,1,\infty\} = \mathcal{C}_{0,3} = \mathcal{M}_{0,4}$ to $\mathcal{M}_{g,n}$.

4.2. The pro-ℓ case

In many arithmetic problems it is natural to focus one's attention on one prime number. To that end, we fix a prime number ℓ.

DEFINITION 4.6. Let G be a discrete group (viewed as a topological group with the discrete topology) or a profinite group. Its pro-ℓ completion $G^{(\ell)}$ is defined by

$$G^{(\ell)} := \operatorname*{proj\,lim}_{N} G/N,$$

where N runs over the open normal subgroups of G where G/N is a finite group of order a power of ℓ.

The pro-ℓ completion of $\mathbb{Z}$ is denoted by $\mathbb{Z}_\ell$. Actually, the projective system $\mathbb{Z}/(\ell^n)$, $(n = 1, 2, \ldots)$, is a system of finite rings, and thus $\mathbb{Z}_\ell$ has a structure of topological ring. It is an integral domain, and its fraction field is denoted by $\mathbb{Q}_\ell$.

Since every ℓ-group is nilpotent, $G^{(\ell)}$ is a pronilpotent group. Since there is a surjective homomorphism $G^{\wedge} \to G^{(\ell)}$ whose kernel is characteristic (namely, stabilized by any automorphism of $G^{\wedge}$), we have a homomorphism

$$\operatorname{Aut}(G^{\wedge}) \to \operatorname{Aut}(G^{(\ell)}).$$

Consequently, an action of a group (e.g., a Galois group) on $G^{\wedge}$ induces an action on $G^{(\ell)}$. The analogues of the results above in the pro-ℓ case, stated below, are surprisingly different.

Let K be a finite extension of $\mathbb{Q}$ and let $C \to \operatorname{Spec} K$ be a geometrically connected (g,n)-curve. The outer Galois representation (1.2) composed with

$$\operatorname{Out}(\Pi^{\wedge}_{g,n}) \to \operatorname{Out}(\Pi^{(\ell)}_{g,n})$$

is called a *pro-ℓ outer representation*

$$\rho_C^{(\ell)} : \mathrm{Gal}(\overline{K}/K) \to \mathrm{Out}(\Pi_{g,n}^{(\ell)}).$$

Similarly, we have the universal pro-ℓ representation

$$\rho_{g,n}^{(\ell)} : \pi_1(\mathcal{M}_{g,n}) \to \mathrm{Out}\,\Pi_{g,n}^{(\ell)}. \tag{4.1}$$

As seen previously, by the universality, C corresponds to a K-rational point $a : \mathrm{Spec}\,K \to \mathcal{M}_{g,n}$, and we see that $\rho_C^{(\ell)}$ factors as $\rho_{g,n}^{(\ell)} \circ \pi_1(a)$. Thus, the image of $\rho_{g,n}^{(\ell)}$ in $\mathrm{Out}\,\Pi_{g,n}^{(\ell)}$ contains the image of $\rho_C^{(\ell)}$.

In the pro-ℓ case, in contrast to the profinite case, there are many curves C for which the image of the Galois group and the image of the mapping class group intersect maximally.

THEOREM 4.7 ([**23**]). *There are infinitely many pairs (K, a) where K is a finite extension of $\mathbb{Q}$ and a : $\mathrm{Spec}\,K \to \mathcal{M}_{g,n}$ (or equivalently, infinitely many isomorphism classes of (g, n) curves $C \to \mathrm{Spec}\,K$ when K runs through all the finite extensions of $\mathbb{Q}$ in $\mathbb{C}$) for which the image of*

$$\rho_C^{(\ell)} : \mathrm{Gal}(\overline{K}/K) \to \mathrm{Out}\,\Pi_{g,n}^{(\ell)}$$

contains the image of

$$\rho_{g,n}^{(\ell)} : \mathrm{MCG}^{\widehat{}} \to \mathrm{Out}\,\Pi_{g,n}^{(\ell)}.$$

This says that for many curves, the action by any element of MCG is realized as an action by an element of the absolute Galois group. The proof uses Hilbert's Irreducibility Theorem.

LECTURE 5

The conjectures of Oda and Deligne-Ihara

In general, profinite groups are interesting, but difficult to deal with. Often, it is useful to replace a profinite group by a "not so non-abelian" invariant, such as its pro-ℓ completion or by the associated Lie-algebra.

One way to obtain a Lie algebra from a group is to take its Malcev completion. Here, rather than explain Malcev completion, I will use the graded Lie algebra obtained from a central filtration of the pro-ℓ completion of the group.

5.1. Weight filtration and the higher Johnson homomorphisms

In this subsection, we treat only discrete groups with the discrete topology. In particular, we recall the higher Johnson homomorphisms. (The pro-ℓ analogue is discussed in the next subsection.)

Let C be a complex (g, n)-curve. Identify its discrete fundamental group with $\Pi_{g,n}$.

DEFINITION 5.1 (Weight filtration on $\Pi_{g,n}$). Define a filtration

$$\Pi_{g,n} = W_{-1}\Pi_{g,n} \supset W_{-2}\Pi_{g,n} \supset W_{-3}\Pi_{g,n} \supset \cdots$$

on $\Pi_{g,n}$ by inductively by $W_{-1}\Pi_{g,n} := \Pi_{g,n}$,

$$W_{-2}\Pi_{g,n} := \langle [\Pi_{g,n}, \Pi_{g,n}], \gamma_1, \gamma_2, \ldots, \gamma_n \rangle_{norm},$$

and

$$W_{-N}\Pi_{g,n} := \langle [W_{-i}\Pi_{g,n}, W_{-j}\Pi_{g,n}] | i + j = N \rangle_{norm}$$

when $N \geq 3$. Here $\langle\ \rangle_{norm}$ denotes the normal subgroup generated by the elements inside $\langle\ \rangle$, $[\ ,\ ]$ denotes the commutator product, and the γ_i are the elements in the presentation given in Definition 4.1. When it is obvious from the context, we write W_{-N} for $W_{-N}\Pi_{g,n}$.

It is easy to check that $\mathrm{gr}_{-j}(\Pi_{g,n}) := W_{-j}/W_{-j-1}$ is abelian, and that it is central in $\Pi_{g,n}/W_{-j-1}$ for each $j \geq 1$. In other words, $W_\bullet$ is the fastest decreasing central filtration such that W_{-2} contains $\gamma_1, \ldots, \gamma_n$. It is known that each gr_{-j} is a free $\mathbb{Z}$-module of finite rank ([**21**] for $n = 0$, [**19**] for $n > 0$). This notion of the weight filtration came from the study of the mixed Hodge structure on the fundamental groups by Richard Hain [**11**], but was independently introduced by Ihara [**14**] when $(g, n) = (0, 3)$. Forcing γ_i to have weight -2 is natural from the *"yoga" of weights* (see an expository paper [**7**] for the cohomology case; weight filtrations appear in Galois representations, Hodge theory and the theory of motives).

Define

$$\mathrm{Gr}\,\Pi_{g,n} := \oplus_{i=1}^{\infty} \mathrm{gr}_{-i}(\Pi_{g,n}).$$

This is a free graded $\mathbb{Z}$-module, which has the structure of a graded Lie algebra with the bracket

$$\mathrm{gr}_{-i} \otimes \mathrm{gr}_{-j} \to \mathrm{gr}_{-i-j}, \quad x \otimes y \mapsto [x, y] := xyx^{-1}y^{-1}$$

induced by the commutator. The following presentation of the Lie algebra is known [**21**], [**19**], [**1**]:

THEOREM 5.2. *The graded Lie algebra* $\mathrm{Gr}\,\Pi_{g,n}$ *is generated by the* $2g$ *generators* $A_1, B_1, \ldots, A_g, B_g$ *of weight* -1 *and* n *generators* $C_1, \ldots, C_n$ *of weight* -2*. These satisfy the single homogeneous relation*

$$[A_1, B_1] + \cdots + [A_g, B_g] + C_1 + \cdots + C_n = 0.$$

DEFINITION 5.3 (Induced filtration). Let $\Gamma := \mathrm{Aut}(\Pi_{g,n})$. We define the induced filtration

$$\Gamma \supset I_0\Gamma \supset I_{-1}\Gamma \supset I_{-2}\Gamma \cdots$$

by

$$\sigma \in I_{-j}\Gamma \Leftrightarrow \text{for any } k \in \mathbb{N} \text{ and } x \in W_{-k}(\Pi_{g,n}),\ \sigma(x)x^{-1} \in W_{-k-j}(\Pi_{g,n}) \text{ holds.}$$

This latter condition is equivalent to the condition that

$$\sigma(\alpha_i)\alpha_i^{-1},\ \sigma(\beta_i)\beta_i^{-1} \in W_{-1-j}(\Pi_{g,n}) \text{ and } \sigma(\gamma_i)\gamma_i^{-1} \in W_{-2-j}(\Pi_{g,n}).$$

Note that if $n > 1$, then Γ is strictly larger than $I_0\Gamma$, since $\alpha_1, \ldots, \beta_g, \gamma_1$ is a subset of free generators and there is an automorphism of $\Pi_{g,n}$ that maps γ_1 to α_1 (thus an element of weight -2 to weight -1). In most articles, such automorphisms are excluded by restricting to a subgroup $\Gamma^* \subset \Gamma = \mathrm{Aut}\,\Pi_{g,n}$ which preserves the conjugacy classes of the subgroup generated by γ_i for each cusp i.

DEFINITION 5.4. Let (Γ, I) be a filtration on a group.

- If there is a group homomorphism $f : \Gamma \to \Gamma'$, then we define the pushout filtration I' on Γ' by $I'_j(\Gamma') = f(I_j(\Gamma))$.
- If there is a group homomorphism $h : \Gamma'' \to \Gamma$, then we define the pullback filtration I'' on Γ'' by $I''_j(\Gamma'') = h^{-1}(I_j(\Gamma))$.

LEMMA 5.5. *In the above definition, for any integer* j *for which* I_{-j} *is defined,*

- f *induces a surjective homomorphism*

$$\mathrm{gr}_{-j}\Gamma \to \mathrm{gr}_{-j}\Gamma'$$

- h *induces an injective homomorphism*

$$\mathrm{gr}_{-j}\Gamma'' \to \mathrm{gr}_{-j}\Gamma$$

- *If* $\Gamma'' \xrightarrow{h} \Gamma \xrightarrow{f} \Gamma'$ *is exact and if* $I_{-j-1}\Gamma$ *is normal in* $I_{-j}\Gamma$*, then we have a short exact sequence*

$$0 \to \mathrm{gr}_{-j}\Gamma'' \to \mathrm{gr}_{-j}\Gamma \to \mathrm{gr}_{-j}\Gamma' \to 0.$$

Since $\mathrm{Aut}(\Pi_{g,n})$ has the induced filtration I, we can pushout the filtration to $\mathrm{Out}(\Pi_{g,n})$, and then pullback to $\mathrm{MCG}_{g,n}$ through $\mathrm{MCG}_{g,n} \to \mathrm{Out}(\Pi_{g,n})$. By an abuse of notation, we denote all of these filtrations by I.

DEFINITION 5.6. Let R be a commutative ring. A Lie algebra L over R (or R-Lie algebra) is a Lie algebra which has a structure of an R-module, with its Lie-product being R-bilinear. (In this and the next sections, we treat only the cases where $R = \mathbb{Z}$ or $R = \mathbb{Q}$. Later in Section 5.3, we treat cases where $R = \mathbb{Z}_\ell$ or $R = \mathbb{Q}_\ell$). Let L be a Lie algebra over R. Its derivation Lie algebra (over R) is defined by

$$\mathrm{Der}(L) := \{D \in \mathrm{End}_R(L) \mid D[X,Y] = [DX,Y] + [X,DY]\}.$$

It is a Lie subalgebra of the Lie algebra of R-linear endomorphism (as an R-module, not as a Lie algebra) ring $\mathrm{End}_R(L)$ of L (Lie bracket is given by $[D,D'] = DD' - D'D$). The derivation D is determined by its action on a generating set of L (generating as an R-Lie algebra). For any $X \in L$, the assignment $Y \mapsto [X,Y]$ is a derivation of L, called an *inner derivation.* The Lie ideal of $\mathrm{Der}(L)$ consisting of the inner derivations is denoted by $\mathrm{Inn}(L)$. The outer derivation Lie algebra of L is defined by

$$\mathrm{OutDer}(L) := \mathrm{Der}(L)/\mathrm{Inn}(L).$$

Let $L := \oplus_{m\in\mathbb{Z}} L_m$ be a graded Lie algebra over R. Suppose that L is generated by finitely many generators as a Lie algebra over R. Then, $\mathrm{Der}(L)$ is a graded Lie algebra over R: A homogeneous derivation of degree j is a derivation that maps degree i part of L to degree $i+j$ part of L for any i. We denote the set of such derivations by $\mathrm{Der}(L)_j \subset \mathrm{Der}(L)$. Then we have a projection $\mathrm{Der}(L) \to \mathrm{Der}(L)_j$ for each j. The image of $\mathrm{Der}(L)$ in $\prod_{j\in\mathbb{Z}} \mathrm{Der}(L)_j$ lies in $\oplus_{j\in\mathbb{Z}}\mathrm{Der}(L)_j$ because L is finitely generated, and hence we have

$$\mathrm{Der}(L) = \oplus_{j\in\mathbb{Z}}\mathrm{Der}(L)_j.$$

If L is a graded and finitely generated Lie algebra over R, then $\mathrm{OutDer}(L)$ is also a graded Lie algebra over R, whose degree m-th part is $\mathrm{Der}(L)_m/\mathrm{Inn}(L_m)$.

PROPOSITION 5.7.

- *The graded group* $\mathrm{Gr}_{\le -1}\mathrm{MCG}_{g,n} := \bigoplus_{j=1}^{\infty} \mathrm{gr}_{-j}(\mathrm{MCG}_{g,n})$ *is a graded Lie algebra over* $\mathbb{Z}$, *which is free as a* $\mathbb{Z}$ *module.*
- *The group homomorphism* $\mathrm{MCG}_{g,n} \to \mathrm{Out}\,\Pi_{g,n}$ *induces an injective homomorphism of graded* $\mathbb{Z}$*-Lie algebras*

$$\tau : \mathrm{Gr}_{\le -1}\mathrm{MCG}_{g,n} \to (\mathrm{Out}\,\mathrm{Der}(\mathrm{Gr}\,\Pi_{g,n}))_{\le -1}$$

given by mapping $\sigma \in I_{-j}\mathrm{MCG}_{g,n}$ *to the derivation that takes* $x \in W_{-k}\Pi_{g,n}$ *to* $\sigma(x)x^{-1} \in W_{-k-j}\Pi_{g,n}$

A representation $G \to \mathrm{Aut}(H)$ of one Lie group on another induces a Lie algebra homomorphism $\mathrm{Lie}(G) \to \mathrm{Der}(\mathrm{Lie}(H))$. Proposition 5.7 is an analogue for discrete groups. This kind of results is well-known to combinatorial group theorists. In the above context, it was proved by Morita [**25**] (with $n = 1$ and for $\mathrm{Aut}\,\Pi_{g,1}$) in the discrete case, and by Asada and Kaneko in the pro-ℓ case ([**2**] $n = 0$, [**19**] $n > 0$, and [**1**] in the remaining cases; see Section 5.3 for the pro-ℓ version).

For the cases of $n \le 1$, each graded piece τ_{-j} of the morphism τ in Proposition 5.7 is studied by D. Johnson (see a survey [**18**]).

DEFINITION 5.8 (Higher Johnson homomorphisms). The homomorphism

$$\tau_{-j} : \mathrm{gr}_{-j}\mathrm{MCG}_{g,n} \to (\mathrm{Out}\,\mathrm{Der}(\mathrm{Gr}\,\Pi_{g,n}))_{-j}$$

is called the j-th Johnson homomorphism.

5.2. Higher Johnson homomorphisms

For simplicity, in this subsection we assume that $n = 0$. In this case (and also the case $n = 1$), the weight filtration coincides with the lower central filtration:

$$W_{-n}\Pi_{g,0} = [\Pi_{g,0}, [\Pi_{g,0}, [\cdots, \Pi_{g,0}]]] \quad (n\text{-times}) .$$

The induced filtration $I_{-N}(\mathrm{MCG}_{g,0})$ on $\mathrm{MCG}_{g,0}$ is given by

$$I_{-N} = \{\gamma \in \mathrm{MCG}_{g,0} \mid \exists \text{ lift } \tilde{\gamma} \in \mathrm{Aut}\,\Pi_{g,0} \text{ of } \gamma \text{ that acts trivially on } \Pi_{g,0}/W_{-N-1}\}.$$

Thus $I_{-1}\mathrm{MCG}_{g,0}$ is the kernel of

$$\mathrm{MCG}_{g,0} \to \mathrm{Aut}\,\Pi_{g,0}^{ab} = \mathrm{Aut}\,H_1(\Sigma_g, \mathbb{Z}).$$

That is, it is the Torelli group. For an element $\sigma \in I_{-1}\mathrm{MCG}_{g,0}$, its image by the first Johnson homomorphism $\tau_{-1}(\sigma)$ is given by the associated derivation: $\alpha_i \in W_{-1}\Pi_{g,0}/W_{-2}\Pi_{g,0} \mapsto (\tilde{\sigma}(\alpha_i))\alpha_i^{-1} \in W_{-2}\Pi_{g,0}/W_{-3}\Pi_{g,0}$ (and similarly for β_i) modulo the inner derivation of grade -1. The image by τ_{-1} of this σ in $\mathrm{Out}\,\mathrm{Der}(\mathrm{Gr}\,\Pi_{g,0})_{-1}$ is trivial if and only if there is a lift $\tilde{\sigma}$ of σ such that $(\tilde{\sigma}(\alpha_i))\alpha_i^{-1} \in W_{-3}\Pi_{g,0}$. That is, $\tau_{-1}(\sigma) = 1$ if and only if $\sigma \in I_{-2}\mathrm{MCG}_{g,0}$. Thus, τ_{-1} is an injective homomorphism

$$\tau_{-1} : I_{-1}\mathrm{MCG}_{g,0}/I_{-2}\mathrm{MCG}_{g,0} \to (\mathrm{Out}\,\mathrm{Der}\,\mathrm{Gr}\,\Pi_{g,0})_{-1}.$$

It is the original Johnson homomorphism. Similarly, for $j \geq 1$, the j-th Johnson homomorphism

$$\tau_{-j} : \mathrm{gr}_{-j}\mathrm{MCG}_{g,0} \to (\mathrm{Out}\,\mathrm{Der}\,\mathrm{Gr}\,\Pi_{g,0})_{-j}$$

is defined and is injective.

THEOREM 5.9. *If $g \geq 3$, then τ_{-1} is an isomorphism (Johnson* [**17**]*) and the cokernel of τ_{-2} is a finite 2-torsion group (Morita* [**24**]*).*

The following shows that higher Johnson homomorphisms τ_{-m} for odd integers $m \geq 3$ has a cokernel of positive rank ([**25**] for the case $n = 1$, see [**26**, Section 6.1] for the case $n = 0$).

THEOREM 5.10 (Morita Trace). *When $g \geq 3$, there is a graded $\mathbb{Z}$-Lie algebra homomorphism*

$$MT : \mathrm{OutDer}\,\mathrm{Gr}(\Pi_{g,0}) \to \bigoplus_{m\geq 3,\ odd} \mathrm{Sym}^m H_1,$$

(called the Morita trace), where the right hand side is considered as an abelian Lie algebra. This is a homomorphism as $\mathrm{Sp}(g, \mathbb{Z})$-modules, becomes surjective when tensoring with $\mathbb{Q}$, and $\mathrm{Gr}_{\leq -1}\,\mathrm{MCG}_{g,n}$ is contained in the kernel of the Morita trace.

Here, $\mathrm{Sym}^m H_1$ denotes the m-th symmetric product of $H_1 := \Pi_{g,0}^{ab}$. The construction of the Morita trace is based on a combinatorial group theoretic argument (Fox free differentials). This theorem implies that the cokernel of the m-th Johnson homomorphism for odd $m \geq 3$ has $\mathbb{Z}$-rank at least $\dim(\mathrm{Sym}^m H_1)$.

In the next subsection, we exhibit a different type of obstruction to the surjectivity of the Johnson homomorphisms. It arises from $G(\overline{\mathbb{Q}}/\mathbb{Q})$.

5.3. The conjectures of Deligne-Ihara and Oda

There are constructions in the pro-ℓ setting analogous to those above. The main consequence of this extended construction is that the image of $\mathrm{AMCG}_{g,n} \to \mathrm{Out}\,\Pi_{g,n}^{(\ell)}$ (denoted by $A_{g,n}$) is strictly larger than the image of $\widehat{\mathrm{MCG}}_{g,n}$ (denoted by $M_{g,n}$), and that the free $\mathbb{Z}_\ell$-ranks of the graded quotients of $A_{g,n}/M_{g,n}$ are independent of (g,n) if $2g-2+n>0$. A further consequence of this and Francis Brown's recent solution [**4**] of the Deligne-Ihara conjecture is that the direct sum of the graded quotients of $A_{g,n}/M_{g,n}$ (which is a $\mathbb{Z}_\ell$-Lie algebra), after taking $\otimes_{\mathbb{Z}_\ell}\mathbb{Q}_\ell$, is isomorphic to a certain countably generated free Lie algebra. These "Galois obstructions" give strong bounds on the size of the image of the Johnson homomorphisms that do not appear to have a topological explanation.

DEFINITION 5.11. Let $\Pi_{g,n}^{(\ell)}$ be the pro-ℓ completion of $\Pi_{g,n}$. It is the pro-ℓ group with the same set of generators and relations among the generators as in the discrete case, which was given in Definition 4.1.

Definition 5.1 extends directly to the pro-ℓ case. The only difference is that we take the topological closure at each step. Thus,

$$W_{-1}\Pi_{g,n}^{(\ell)} := \Pi_{g,n}^{(\ell)},\ W_{-2}\Pi_{g,n}^{(\ell)} := \langle[\Pi_{g,n}^{(\ell)},\Pi_{g,n}^{(\ell)}],\gamma_1,\gamma_2,\ldots,\gamma_n\rangle_{norm},$$

and

$$W_{-N}\Pi_{g,n}^{(\ell)} := \langle[W_{-i}\Pi_{g,n}^{(\ell)},W_{-j}\Pi_{g,n}^{(\ell)}]|i+j=N\rangle_{norm}$$

when $N\geq 3$, where $\langle\ \rangle_{norm}$ denotes the *topological closure of* the normal subgroup generated by the elements inside $\langle\ \rangle$.

Define

$$\mathrm{Gr}\,\Pi_{g,n}^{(\ell)} := \oplus_{i=1}^{\infty}\mathrm{gr}_{-i}(\Pi_{g,n}^{(\ell)}).$$

The following comparison theorem with discrete case is known [**2**], [**19**], [**1**]:

THEOREM 5.12. *The associated graded* $\mathrm{Gr}\,\Pi_{g,n}^{(\ell)}$ *is a graded* $\mathbb{Z}_\ell$*-Lie algebra which is free as a* $\mathbb{Z}_\ell$*-module. It is generated by* $2g$ *generators* $A_1,B_1,\ldots,A_g,B_g$ *of weight* -1 *and* n *generators* $C_1,\ldots,C_n$ *of weight* -2 *which satisfy the single homogeneous relation*

$$[A_1,B_1]+\cdots+[A_g,B_g]+C_1+\cdots+C_n=0.$$

The pullback to $\Pi_{g,n}$ *of* $W_{-N}\Pi_{g,n}^{(\ell)}$ *via* $\Pi_{g,n}\to\Pi_{g,n}^{(\ell)}$ *coincides with the weight filtration defined in the discrete case (Definition* 5.1*), and induces an isomorphism*

$$(\mathrm{Gr}\,\Pi_{g,n})\otimes_{\mathbb{Z}}\mathbb{Z}_\ell \cong \mathrm{Gr}\,\Pi_{g,n}^{(\ell)}.$$

Consequently,

$$(\mathrm{OutDer}\,\mathrm{Gr}\,\Pi_{g,n})\otimes_{\mathbb{Z}}\mathbb{Z}_\ell \cong \mathrm{OutDer}\,\mathrm{Gr}\,\Pi_{g,n}^{(\ell)}.$$

Now $\mathrm{Aut}(\Pi_{g,n}^{(\ell)})$ has the induced filtration I defined in exactly the same manner as in the discrete case Definition 5.3. By pushing-out I along $\mathrm{Aut}(\Pi_{g,n}^{(\ell)})\to\mathrm{Out}(\Pi_{g,n}^{(\ell)})$ and pulling-back along the homomorphism $\rho_{g,n}^{(\ell)}:\mathrm{AMCG}_{g,n}\to\mathrm{Out}(\Pi_{g,n}^{(\ell)})$ (cf. (4.1)), we obtain an induced filtration (denoted by the same letter I) on $\mathrm{AMCG}_{g,n}$. The short exact sequence (1.3) then gives a pullback filtration on $\widehat{\mathrm{MCG}}_{g,n}$ and a pushout filtration on $G(\overline{\mathbb{Q}}/\mathbb{Q})$. One has the short exact sequence

$$0\to\mathrm{Gr}_{\leq -1}\widehat{\mathrm{MCG}}_{g,n}\to\mathrm{Gr}_{\leq -1}\mathrm{AMCG}_{g,n}\to\mathrm{Gr}_{\leq -1,(g,n)}G(\overline{\mathbb{Q}}/\mathbb{Q})\to 0. \tag{5.1}$$

The filtration on $G(\overline{\mathbb{Q}}/\mathbb{Q})$ possibly depends on (g,n). This dependence is indicated in the subscript. We also have the homomorphism

$$\mathrm{Gr}_{\leq -1}\,\mathrm{AMCG}_{g,n} \to \mathrm{OutDer}(\mathrm{Gr}\,\Pi_{g,n}^{(\ell)}),$$

which is injective by Lemma 5.5.

Consider the commutative diagram with exact rows

$$\begin{array}{ccccccccc}
0 & \to & \mathrm{Gr}_{\leq -1}\,\widehat{\mathrm{MCG}}_{g,n} & \to & \mathrm{Gr}_{\leq -1}\,\mathrm{AMCG}_{g,n} & \to & \mathrm{Gr}_{\leq -1,(g,n)}\,G(\overline{\mathbb{Q}}/\mathbb{Q}) & \to & 0 \\
 & & \| & & \downarrow & & \downarrow & & \\
0 & \to & \mathrm{Gr}_{\leq -1}\,\widehat{\mathrm{MCG}}_{g,n} & \to & \mathrm{OutDer}(\mathrm{Gr}\,\Pi_{g,n}^{(\ell)})_{\leq -1} & \to & \text{Cokernel} & \to & 0 \\
 & & \uparrow & & \| & & \uparrow & & \\
0 & \to & (\mathrm{Gr}_{\leq -1}\,\mathrm{MCG}_{g,n})\otimes\mathbb{Z}_\ell & \to & \mathrm{OutDer}(\mathrm{Gr}\,\Pi_{g,n})_{\leq -1}\otimes\mathbb{Z}_\ell & \to & \text{Cokernel}' & \to & 0.
\end{array}$$

The upper middle vertical arrow is injective (Lemma 5.5), and hence so is the upper right vertical arrow. Since the bottom left morphism (in the bottom row) comes from the higher Johnson homomorphism (Definition 5.8) $\tau : \mathrm{Gr}_{\leq -1}\,\mathrm{MCG}_{g,n} \to \mathrm{OutDer}(\mathrm{Gr}\,\Pi_{g,n})_{\leq -1}$, the rank of each graded piece of Cokernel′ over $\mathbb{Z}_\ell$ is equal to that of the cokernel of τ. Since the morphism Cokernel′ to Cokernel is surjective, the rank of Cokernel′ is no less than that of $\mathrm{Gr}_{\leq -1,(g,n)}\,G(\overline{\mathbb{Q}}/\mathbb{Q})$. In sum, $\mathrm{Gr}_{\leq -1,(g,n)}\,G(\overline{\mathbb{Q}}/\mathbb{Q})$ gives an obstruction to the surjectivity of the Johnson homomorphisms. A conjecture of Oda (cf. [**30**]) claims that this obstruction is independent of (g,n).

CONJECTURE 5.13 (Takayuki Oda). *The pushout filtration $I_{g,n}$ on $G(\overline{\mathbb{Q}}/\mathbb{Q})$ is independent of (g,n) when $2g-2+n>0$.*

In particular, this conjecture implies that the cokernel of

$$\mathrm{Gr}_{\leq -1}(\widehat{\mathrm{MCG}}_{g,n}) \to Gr_{\leq -1}\mathrm{AMCG}_{g,n}$$

should be independent of (g,n). In particular, it should coincide with the filtration obtained in the case $(g,n)=(0,3)$.

After taking $\otimes_{\mathbb{Z}_\ell}\mathbb{Q}_\ell$, the conjecture is true:

THEOREM 5.14. *As a Lie algebra, the cokernel of*

$$\mathrm{Gr}_{\leq -1}(\widehat{\mathrm{MCG}}_{g,n})\otimes_{\mathbb{Z}_\ell}\mathbb{Q}_\ell \to \mathrm{Gr}_{\leq -1}\,\mathrm{AMCG}_{g,n}\otimes_{\mathbb{Z}_\ell}\mathbb{Q}_\ell$$

is independent of (g,n) whenever $2g-2+n>0$.

The case $n \geq 1$ was proved around 1994 by H. Nakamura, Y. Ihara, R. Ueno, and N. Takao [**28, 27, 15**]. The general case $n \geq 0$ was proved in 2010 by N. Takao [**31**].

Returning to the case $(g,n)=(0,3)$, we have an induced filtration $I_\bullet G_{\mathbb{Q}}$ from $G_{\mathbb{Q}} \to \mathrm{Out}\,\Pi_{0,3}^{(\ell)}$.

CONJECTURE 5.15 (Deligne-Ihara). *The Lie algebra $\mathrm{Gr}_{\leq -1,(0,3)}(G_{\mathbb{Q}})\otimes_{\mathbb{Z}_\ell}\mathbb{Q}_\ell$ is a free Lie algebra with one free generator in $\mathrm{gr}_{-2(2d+1)}G_{\mathbb{Q}}$ for each $d \geq 1$.*

Ihara [**14**] proved that $\mathrm{Gr}_{-2(2d+1)}\,G_{\mathbb{Q}}$ has rank at least one by producing a specific non-zero element. He also showed that the commutator of any two of them does not vanish. Hain and the author [**12**] proved that this Lie algebra is generated by Ihara's generators.

In 2010, Francis Brown proved this conjecture [**4**]. This conjecture and Theorem 5.14 determine the rank of the each graded piece of the cokernel appeared in

Theorem 5.14, and hence giving an obstruction to the surjectivity of Johnson homomorphisms. This obstruction (on the even weights) is orthogonal to the Morita trace (on the odd weights).

The existence of such Galois obstructions was conjectured by Takayuki Oda, and its non-triviality was proved by H. Nakamura [**27**] and the author [**22**] independently. Recently, more obstructions to surjectivity have been found by Conant-Kassabov-Vogtmann [**6**] and Enomoto-Satoh [**9**].

Note that the target $\operatorname{Out}\operatorname{Der}(\operatorname{Gr}(\Pi_{g,0}))_{-m}$ of the m-th Johnson homomorphism has an explicit description due to Morita [**25**, Lemma 4.5], which is given below.[1]

Denote $\operatorname{gr}_{-1}\Pi_{g,0}$ by H. It is naturally isomorphic with the first homology of the genus g reference surface and is a free $\mathbb{Z}$-module of rank $2g$. Fix a basis $A_1, B_1, \ldots, A_g, B_g$ of H that is symplectic with respect to the intersection form.

PROPOSITION 5.16. *For all $m \geq 1$, there is a natural map*

$$\operatorname{Der}(\operatorname{Gr}\Pi_{g,0})_{-m} \to \operatorname{Hom}(H, \operatorname{gr}_{-m-1}\Pi_{g,0}),$$

defined by taking $D \in \operatorname{Der}(\operatorname{Gr}\Pi_{g,0})_{-m}$ to its restriction to H. This homomorphism is injective, since H generates $\operatorname{Gr}\Pi_{g,0}$ as an Lie algebra. Its image is the set of linear maps D that satisfy

$$\sum_{i=1}^{g}[D(A_i), B_i] + [A_i, D(B_i)] = 0.$$

Under the identification $\operatorname{Hom}(H, \operatorname{gr}_{-m-1}\Pi_{g,0}) \cong H \otimes \operatorname{gr}_{-m-1}\Pi_{g,0}$ given by the intersection pairing, the image corresponds to the kernel of the bracket mapping

$$H \otimes \operatorname{gr}_{-m-1}\Pi_{g,0} \xrightarrow{[,]} \operatorname{gr}_{-m-2}\Pi_{g,0}.$$

Note that this also works with $m = 0$. The graded quotient $\operatorname{Out}\operatorname{Der}(\operatorname{Gr}\Pi_{g,0})_{-m}$ of $\operatorname{Out}\operatorname{Der}(\operatorname{Gr}\Pi_{g,0})$ can be computed as the quotient of $\operatorname{Der}(\operatorname{Gr}\Pi_{g,0})_{-m}$ by $\operatorname{Inn}\operatorname{gr}_{-m}(\Pi_{g,0})$.

REMARK 5.17. In the commutative diagram consisting of the three short exact sequences in the previous page, the homomorphism from the left term in the bottom row to the left term in the middle row

$$(\operatorname{Gr}_{\leq -1}\mathrm{MCG}_{g,n}) \otimes \mathbb{Z}_\ell \to \operatorname{Gr}_{\leq -1}\widehat{\mathrm{MCG}}_{g,n}$$

is interesting. By Lemma 5.5 and Theorem 5.12, the morphism can be shown to be injective. For $(g,n) = (1,1)$, $\mathrm{MCG}_{1,1} \cong \mathrm{SL}(2,\mathbb{Z})$ faithfully acts on H_1, so the left hand side is 0, but the right hand side is proved to have countably infinitely many generators, using theory of continuous relative Malcev completion (in preparation by Hain and the author), and hence the homomorphism is far from surjective.

For $g \geq 2$ and $g = 0$, the morphism is proved to be an isomorphism by Asada [**1**, Theorem B].

[1] To be precise, [**25**] deals with $(g,1)$-case. In his survey [**26**, §5], Morita treats $(g,0)$-case and denotes $\operatorname{Out}\operatorname{Der}(\operatorname{Gr}(\Pi_{g,0}))_{-m}$ by $\mathfrak{h}_g(m)$.

APPENDIX

Algebraic fundamental groups via fiber functors

The purpose of this appendix is to describe the "standard" method of defining the algebraic fundamental group in terms of etale coverings and fiber functors. To make the appendix independent and self-contained, some material from the lectures is duplicated.

A.1. The classic topological fundamental group

Let $\mathcal{X}$ be an arcwise connected topological space, and x be a point on $\mathcal{X}$. Then, the (classical) fundamental group of $\mathcal{X}$ with base point x is defined by

$$\pi_1(\mathcal{X}, x) := \{\text{paths from } x \text{ to } x\}/(\text{homotopy with } x \text{ fixed}).$$

We adopt the convention that the composition $\gamma \circ \gamma'$ of two paths is the path that first goes along γ' and then γ. (Topologists generally use the reverse ordering.)

DEFINITION A.1. Let $\mathcal{X}$ be a locally connected topological space. A continuous homomorphism $f : \mathcal{Y} \to \mathcal{X}$ between topological spaces is an *unramified covering* if each $x \in \mathcal{X}$ has an open neighborhood U such that the restriction $f : f^{-1}(U) \to U$ to each connected component of $f^{-1}(U)$ is a homoomorphism.

The unramified covering of $\mathcal{X}$ form a category that we denote by $\mathcal{X}_{et}$ (et is for étale). Its objects are the unramified covering $f : \mathcal{Y} \to \mathcal{X}$; a morphisms from $f : \mathcal{Y} \to \mathcal{X}$ to $f' : \mathcal{Y}' \to \mathcal{X}$ is a continuous mapping $g : \mathcal{Y} \to \mathcal{Y}'$ such that $f' \circ g = f$. Note that we do not assume the connectivity of $\mathcal{Y}$.

Let $(\mathcal{X}, x)$ be a pair consisting of a topological space $\mathcal{X}$ and a point $x \in \mathcal{X}$. The category of pointed unramified coverings of $(\mathcal{X}, x)$ has objects pairs $(\mathcal{Y}, y)$ and an unramified covering $g : \mathcal{Y} \to \mathcal{X}$ where $g(y) = x$; its morphisms are base point preserving maps of coverings. This category is denoted by $(\mathcal{X}, x)_{et}$.

Under a mild assumption on $\mathcal{X}$ (arcwise connected, locally arcwise connected and semi-locally simply connected), the category $(\mathcal{X}, x)_{et}$ has an initial object $(\widetilde{\mathcal{X}}, \tilde{x}) \to (\mathcal{X}, x)$ — namely a pointed universal covering. It is characterized by the properties that $\widetilde{\mathcal{X}} \to \mathcal{X}$ is connected and simply connected unramified covering of $\mathcal{X}$. It is also characterized by its universal property: for any pointed unramified covering $(\mathcal{Y}, y) \to (\mathcal{X}, x)$, there is a unique morphism $(\widetilde{\mathcal{X}}, \tilde{x}) \to (\mathcal{Y}, y)$.

A direct connection between $\pi_1(\mathcal{X}, x)$ and the universal covering $\widetilde{\mathcal{X}} \to \mathcal{X}$ is given by the isomorphism

$$\pi_1(\mathcal{X}, x) \cong \operatorname{Aut}(\widetilde{\mathcal{X}}/\mathcal{X})^{op}, \tag{A.1}$$

which depends on x. It is defined as follows: For unramified covering $f : \mathcal{Y} \to \mathcal{X}$ and $x \in \mathcal{X}$, let $F_x(\mathcal{Y}) := f^{-1}(x)$ be the inverse image of x. It is called the *fiber* of $\mathcal{Y}$ over x.

If $g : \mathcal{Y} \to \mathcal{Y}'$ is a morphism in $\mathcal{X}_{et}$, we have a map $F_x(\mathcal{Y}) \to F_x(\mathcal{Y}')$. This results in a functor

$$F_x : \mathcal{X}_{et} \to \mathbf{Sets}$$

to the category of sets, which is called the *fiber functor* at x.

Let $\gamma \in \pi_1(\mathcal{X}, x)$ be a path, and $(\mathcal{Y}, y)$ be a pointed unramified covering of $(\mathcal{X}, x)$. Then there is a unique lift of γ to $\mathcal{Y}$ that starts from y. Denote its end point by $\gamma_{\mathcal{Y}}(y)$. This gives, for every $\mathcal{Y} \in \mathcal{X}_{et}$, a mapping

$$\gamma_{\mathcal{Y}} : F_x(\mathcal{Y}) \to F_x(\mathcal{Y})$$

and a group homomorphism

$$\pi_1(\mathcal{X}, x) \to \operatorname{Aut} F_x,$$

where $\operatorname{Aut} F_x$ means the group of invertible natural transformations from F_x to itself.

This group homomorphism is actually an isomorphism: the universality of $\widetilde{\mathcal{X}}$ implies that there is a bijection

$$\operatorname{Hom}_{\mathcal{X}_{et}}(\widetilde{\mathcal{X}}, Y) \xrightarrow{\simeq_{\tilde{x}}} F_x(Y)$$

where an element of the left-hand side corresponds to the image of $\tilde{x}$ by that element.[1] This means that the functor F_x is representable by $\widetilde{\mathcal{X}}$. Recall that G^{op} denotes the group G with same underlying set but with the opposite multiplication:

$$g_1 \circ_{G^{op}} g_2 := g_2 \circ g_1.$$

Yoneda's Lemma implies that $\tilde{x}$ determines an isomorphism

$$\operatorname{Aut} F_x \xrightarrow{\simeq_{\tilde{x}}} \operatorname{Aut}(\widetilde{\mathcal{X}})^{op}.$$

If we trace the image of $\gamma \in \pi_1(\mathcal{X}, x)$, we obtain the automorphism

$$\tilde{\gamma} : \widetilde{\mathcal{X}} \to \widetilde{\mathcal{X}}, \quad \tilde{\gamma} : \tilde{x} \mapsto \gamma_{\widetilde{\mathcal{X}}}(\tilde{x}).$$

This is the image of γ under the composite

$$\pi_1(\mathcal{X}, x) \to \operatorname{Aut} F_x \to \operatorname{Aut}(\widetilde{\mathcal{X}})^{\mathrm{op}},$$

which is the standard isomorphism of the fundamental group with (the opposite of) the automorphism group of the universal covering. It follows that $\pi_1(\mathcal{X}, x)$ is isomorphic to the automorphism group of F_x.

The fundamental group controls the category of unramified coverings:

THEOREM A.2. *Let $\mathcal{X}$ be a connected topological space that has a universal covering. For each $x \in \mathcal{X}$, there is an equivalence of categories*

$$F_x : \mathcal{X}_{et} \to \{\pi_1(\mathcal{X}, x)\text{-}sets\}.$$

---

[1] The notation $\xrightarrow{\simeq_{\tilde{x}}}$ means that the choice of $\tilde{x}$ determines the isomorphism.

A.2. Schemes and algebraic fundamental groups

There is a notion of *algebraic fundamental groups* by Grothendieck [**10**]. These groups are defined for *schemes* satisfying some mild conditions (connected and locally noetherian). We do not go into the definition of the category of schemes. (See §2.2 for a very brief overview.)

Grothendieck's definition uses "unramified coverings" in the category of schemes. The appropriate generalization of both unramified *finite* coverings of a topological space (Example A.8 below) and finite separable extensions of a field (Example A.7 below) is the notion of finite etale coverings [**10**, I].

First we state the definition in the language of schemes.

DEFINITION A.3. A morphism $f : Y \to X$ locally of finite type between locally noetherian schemes is etale at $y \in Y$ if the induced morphism of local rings $f^* : \mathcal{O}_{X,f(y)} \to \mathcal{O}_{Y,y}$ is flat, and

$$\mathcal{O}_{X,f(y)}/m_{X,f(y)} \to \mathcal{O}_{Y,y}/(f^*(m_{X,f(y)})\mathcal{O}_{Y,y})$$

is a finite separable extension of fields.

We say f is etale if it is etale at every point on Y. If f is finite and etale and if X is connected, then $f : Y \to X$ is said to be a finite etale covering.

Suppose that X is connected. Let us define X_{et} to be the category of the finite etale coverings of X.

DEFINITION A.4. (Algebraic fundamental group, see SGA1[**10**, V].)

Let X be a connected locally noetherian scheme. Let Ω be an algebraically closed field, and $x : \operatorname{Spec}\Omega \to X$ be a geometric point. Then

$$F_x : (Y, p_Y : Y \to X) \mapsto p_Y^{-1}(x) := \{y : \operatorname{Spec}\Omega \to Y : f \circ y = x\}$$

defines a functor

$$F_x : X_{et} \to \mathbf{FinSets}$$

to the category of finite sets (called a *fiber functor*). We define the algebraic fundamental group by

$$\pi_1(X, x) := \pi_1(X_{et}, F_x) := \operatorname{Aut} F_x.$$

As in the case of topological spaces, there is the notion of connectedness of schemes: a scheme is connected if it is not a coproduct of two nonempty schemes. As in the topological case, $Y \to X \in X_{et}$ is connected if and only if $\pi_1(X, x)$ acts on the fiber $F_x(Y)$ transitively. A connected finite etale cover $Y \to X$ is a Galois covering, if $\operatorname{Aut}(Y/X)$ acts transitively on $F_x(Y)$. A *pointed covering* $(Y, y) \to (X, x)$ is a finite etale covering with a geometric point y over x being specified. We consider the category of pointed Galois coverings (Y, y). Since there is at most one morphism between two pointed connected coverings of (X, x), the isomorphism classes of pointed Galois coverings of (X, x) form a directed preorder Λ.

PROPOSITION A.5. *If X is a connected locally Noetherian scheme, there is a natural isomorphism*

$$\pi_1(X, x) \cong \operatorname*{proj\,lim}_{(Y,y)\in\Lambda} \operatorname{Aut}(Y/X)^{op}.$$

The above proposition follows from the following observation. There is no universal covering in X_{et}. On the other hand, the functor

$$\operatorname{inj\,lim}_{(Y,y)\in\Lambda} \operatorname{Hom}(Y, -)$$

is isomorphic to the fiber functor F_x, and its automorphism group becomes the right-hand side of the above proposition.

Note that $\operatorname{Aut}(Y/X)$ is a finite group, which we regard as a topological group with the discrete topology. Taking the projective limit above in the category of topological groups we conclude that $\pi_1(X, x)$ is a compact topological group. Such an inverse limit of finite groups is called a *profinite group.*

Let X be a connected locally noetherian scheme and $x : \Omega \to X$ be a geometric point. Set $\Pi := \pi_1(X, x) = \operatorname{Aut} F_x$. A finite set (viewed as a topological space with the discrete topology) with a continuous Π-action is called a Π-set. The category of finite Π-sets will be denoted Π-**FinSets**. For each finite etale covering Y of X, $F_x(Y)$ is a finite Π-set.

THEOREM A.6 (Galois Theory). *The fiber functor F_x*

$$F_x : X_{et} \to \Pi\text{-}\mathbf{FinSets}$$

is an equivalence of categories [**10**, V]. *A finite etale covering $Y \to X$ is connected if and only if $F_x(Y)$ is a transitive Π-set.*

The following examples of algebraic fundamental groups should suffice for the lectures.

EXAMPLE A.7. Let K be a field. A finite etale covering of $\operatorname{Spec} K$ is isomorphic to

$$(\coprod_{i\in I} \operatorname{Spec} L_i) \to \operatorname{Spec} K$$

where $K \hookrightarrow L_i$ are finite separable field extensions of K and i runs over a finite set I. This is connected if and only if I is a singleton. A Galois cover (in the sense defined above) is $\operatorname{Spec} L \to \operatorname{Spec} K$ where $K \to L$ is a finite Galois extension of field K (i.e., a finite normal separable extension), as shown below.

A geometric point $x : \operatorname{Spec} \Omega \to \operatorname{Spec} K$ corresponds to an embedding $K \hookrightarrow \Omega$ of K into an algebraically closed field. A point y over x in $\operatorname{Spec} L$ is an embedding $L \hookrightarrow \Omega$ whose restriction to K is the embedding $x : K \hookrightarrow \Omega$ that corresponds to x. By definition of etale, L/K is a finite separable extension. By definition of a Galois covering, $\operatorname{Aut}(L/K)$ acts transitively on the set of embeddings of $L \to \Omega$ that extend x, which is a characterization of a normal extension.

Fix a geometric point $x : \operatorname{Spec} \Omega \to \operatorname{Spec} K$. Denote the separable closure of K in Ω by K^{sep}. Finite Galois extensions L of K in K^{sep} form a directed preorder Λ. By Proposition A.5

$$\pi_1(\operatorname{Spec} K, x) = \pi_1((\operatorname{Spec} K)_{et}, F_x) = \operatorname{Aut}(F_x) \cong \operatorname{proj\,lim}_{L\in\Lambda} \operatorname{Gal}(L/K),$$

Where $\operatorname{Gal}(L/K)$ denotes the usual Galois group — viz., the automorphism group of L that is trivial on K. The right-hand side (i.e. the projective limit) is actually isomorphic to the automorphism group of K^{sep} over K. This group, denoted by G_K or $\operatorname{Gal}(K^{\text{sep}}/K)$, is called the absolute Galois group of K. It is a profinite group.

EXAMPLE A.8. Let $K \subset \mathbb{C}$ be an algebraically closed field. Let X be a scheme of finite type over K. Let X^{an} be the corresponding complex analytic set. Then, a morphism $f : Y \to X$ is finite etale if and only if its analytification $f^{\mathrm{an}} : Y^{\mathrm{an}} \to X^{\mathrm{an}}$ is finite and unramified. Details can be found in [**10**, XII].

One advantage to using fiber functors to define algebraic fundamental groups is that it makes the functoriality of π_1 clear. For a finite etale covering $Y \to X$ and a morphism $f : Z \to X$ where Z is a locally noetherian scheme, the fiber product $Y \times_X Z \to Z$ is finite and etale. Thus, we have a functor

$$f^* : X_{et} \to Z_{et}.$$

For each geometric point z on Z, there is an isomorphism of fiber functors

$$F_{f(z)} = F_z \circ f^*.$$

It follows that $\mathrm{Aut}\, F_z$ acts on the right-hand side. We thus have a homomorphism

$$\pi_1(Z, z) = \mathrm{Aut}\, F_z \to \mathrm{Aut}\, F_{f(z)} = \pi_1(X, f(z)).$$

REMARK A.9. In Definition 1.10 we defined the algebraic fundamental group by

$$\pi_1^{\mathrm{alg}}(X_K, x) := \mathrm{Aut}\, F,$$

where

$$F(Y) := \mathrm{Hom}_{K(X)}(K(Y), M_x).$$

There is a fixed isomorphism $F_x \to F$ which is defined by assigning to a point $y \in Y$ over x the embedding of $K(Y)$ into M_x obtained by identifying a neighbourhood of $y \in Y$ with a neighbourhood of $x \in X$. Conversely, an embedding of $K(Y) \to M_x$ determines a discrete valuation ring in $K(Y)$ and thus a point $y \in Y$ that lies over x. Specifically, take the embedding to the inverse image in $K(Y)$ of the subring of M_x consisting of germs that are regular (i.e., holomorphic) at x; its maximal ideal is the ideal of functions that vanish at x. This determines a geometric point on Y by the valuative criterion of properness of $Y \to X$.

Bibliography

[1] M. Asada, On the filtration of topological and pro-ℓ mapping class groups of punctured Riemann surfaces, J. Math. Soc. Japan 48 (1996), 13–36. MR1361545 (97h:57001)

[2] M. Asada and M. Kaneko, On the automorphism groups of some pro-ℓ fundamental groups, Advanced studies in Pure Math. 12 1987, 137–159. MR948240 (89h:20053)

[3] G.V. Belyĭ, *On Galois extensions of a maximal cyclotomic field*, Math. USSR Izv. **14** (1980), 247–256. MR534593 (80f:12008)

[4] F. Brown, Mixed Tate Motives over Spec(Z), Annals of Math., 175 (2012), 949–976. MR2993755

[5] J.W.S. Cassels and A. Fröhlich, Eds., "Algebraic Number Theory", Academic Press 1967. MR0215665 (35:6500)

[6] J. Conant, M. Kassabov, and K. Vogtmann, Hairy graphs and the unstable homology of Mod(g, s), Out(F_n) and Aut(F_n), arXiv:1107.4839v2 [math.AT].

[7] P. Deligne, "Poids dans la cohomologie des variétés algébriques," Proceedings of the International Congress of Mathematicians (Vancouver, B. C., 1974), Vol. 1, pp. 79–85. Canad. Math. Congress, Montreal, Que., 1975. MR0432648 (55:5634)

[8] P. Deligne and D. Mumford, The irreducibility of the space of curves of given genus, Publ. Math. I.H.E.S 36 (1969) 75–110. MR0262240 (41:6850)

[9] N. Enomoto and T. Satoh, New series in the Johnson cokernels of the mapping class groups of surfaces, arXiv:1012.2175v3 [math.RT].

[10] A. Grothendieck and Mme M. Raynaud, "Revêtement Étales et Groupe Fondamental (SGA 1)", Lecture Notes in Math. **224**, Springer-Verlag, 1971.

[11] R. Hain, The geometry of the mixed Hodge structure on the fundamental group, Algebraic geometry, Bowdoin, 1985 (Brunswick, Maine, 1985), 247–282, Proc. Sympos. Pure Math., 46, Part 2, Amer. Math. Soc., Providence, RI, 1987. MR927084 (89g:14010)

[12] R. Hain and M. Matsumoto, Weighted completion of Galois groups and Galois actions on the fundamental group of $P^1 - \{0, 1, \infty\}$, Compositio Mathematicae 139-2 (2003), 119–167. MR2025807 (2005c:14031)

[13] Y. Hoshi and S. Mochizuki, On the combinatorial anabelian geometry of nodally nondegenerate outer representations, Hiroshima Mathematical Journal 41 (2011) 275–342. MR2895284

[14] Y. Ihara, Profinite braid groups, Galois representations and complex multiplications, Ann. of Math., 123 (1986), 43–106. MR825839 (87c:11055)

[15] Y. Ihara and H. Nakamura, On deformation of maximally degenerate stable marked curves and Oda's problem, J. Reine Angew. Math. 487 (1997), 125–151. MR1454262 (98j:14040)

[16] Y. Iijima, Galois action on mapping class groups, Master's Thesis at Hiroshima University, 2012.

[17] D. Johnson, An abelian quotient of the mapping class group $\mathcal{T}_g$. Math. Ann. 249 (1980), 225–242. MR579103 (82a:57008)

[18] D. Johnson, A survey of the Torelli group, Contemporary Math. 20 (1983), 165–179. MR718141 (85d:57009)

[19] M. Kaneko, Certain automorphism groups of pro ℓ fundamental groups of punctured Riemann surfaces, J. Fac. Sci. Univ. Tokyo Sect. IA. Math. 36 (1989), 363–372. MR1015004 (90j:14019)

[20] F.F. Knudsen, The projectivity of the moduli space of stable curves II: The stacks $M_{g,n}$, Math. Scad. 52 (1983), 161-199. MR702953 (85d:14038a)

[21] J. Labute, On the descending central series of groups with a single defining relation, J. Algebra, 14 (1970), 16-23. MR0251111 (40:4342)

[22] M. Matsumoto, Galois representations on profinite braid groups on curves, J. Reine. Angew. Math. 474 (1996), 169–219. MR1390695 (97h:14043)

[23] M. Matsumoto and A. Tamagawa, Mapping-Class-group action versus Galois action on profinite fundamental groups, American Journal of Mathematics 122 1017–1026 (2000). MR1781929 (2002a:14022)

[24] S. Morita, On the structure of the Torelli group and the Casson invariant. Topology 30 (1991), 603–621, MR1133875 (92i:57016)

[25] S. Morita, Abelian quotients of subgroups of the mapping class group of surfaces, Duke. Math. J. (1993), 699–726. MR1224104 (94d:57003)

[26] S. Morita, Structure of the mapping class groups of surfaces: a survey and a prospect, in: Geometry & Topology Monographs Volume 2: Proceedings of the Kirbyfest, 349–406, 1999. MR1734418 (2000j:57039)

[27] H. Nakamura, Coupling of universal monodromy representations of Galois- Teichmüller modular groups, Math. Ann. 304 (1996), 99–119. MR1367885 (97a:14026)

[28] H. Nakamura, N. Takao, R. Ueno, Some stability properties of Teichmüller modular function fields with pro-ℓ weight structures. Math. Ann. 302 (1995), no. 2, 197–213. MR1336334 (96h:14041)

[29] T. Oda, *Etale homotopy type of the moduli spaces of algebraic curves*, in "Geometric Galois Actions 1", London Math. Soc. Lect. Note Ser. 242 (1997) 85–95. MR1483111 (2000a:14030)

[30] T. Oda, The universal monodromy representations on the pro-nilpotent fundamental groups of algebraic curves, Mathematische Arbeitstagung (Neue Serie) 9-15 Juni 1993, Max-Planck-Institute for Mathematics preprint 93-57. MR1296709 (95i:22023)

[31] N. Takao, Braid monodromies on proper curves and pro-ℓ Galois representations, Journal of the Institute of Mathematics of Jussieu, 2011 (online). MR2862378 (2012k:14032)

Published Titles in This Series

20 **Benson Farb, Richard Hain, and Eduard Looijenga, Editors,** Moduli Spaces of Riemann Surfaces, 2013

19 **Hongkai Zhao, Editor,** Mathematics in Image Processing, 2013

18 **Cristian Popescu, Karl Rubin, and Alice Silverberg, Editors,** Arithmetic of L-functions, 2011

17 **Jeffery McNeal and Mircea Mustaţă, Editors,** Analytic and Algebraic Geometry, 2010

16 **Scott Sheffield and Thomas Spencer, Editors,** Statistical Mechanics, 2009

15 **Tomasz S. Mrowka and Peter S. Ozsváth, Editors,** Low Dimensional Topology, 2009

14 **Mark A. Lewis, Mark A. J. Chaplain, James P. Keener, and Philip K. Maini, Editors,** Mathematical Biology, 2009

13 **Ezra Miller, Victor Reiner, and Bernd Sturmfels, Editors,** Geometric Combinatorics, 2007

12 **Peter Sarnak and Freydoon Shahidi, Editors,** Automorphic Forms and Applications, 2007

11 **Daniel S. Freed, David R. Morrison, and Isadore Singer, Editors,** Quantum Field Theory, Supersymmetry, and Enumerative Geometry, 2006

10 **Steven Rudich and Avi Wigderson, Editors,** Computational Complexity Theory, 2004

9 **Brian Conrad and Karl Rubin, Editors,** Arithmetic Algebraic Geometry, 2001

8 **Jeffrey Adams and David Vogan, Editors,** Representation Theory of Lie Groups, 2000

7 **Yakov Eliashberg and Lisa Traynor, Editors,** Symplectic Geometry and Topology, 1999

6 **Elton P. Hsu and S. R. S. Varadhan, Editors,** Probability Theory and Applications, 1999

5 **Luis Caffarelli and Weinan E, Editors,** Hyperbolic Equations and Frequency Interactions, 1999

4 **Robert Friedman and John W. Morgan, Editors,** Gauge Theory and the Topology of Four-Manifolds, 1998

3 **János Kollár, Editor,** Complex Algebraic Geometry, 1997

2 **Robert Hardt and Michael Wolf, Editors,** Nonlinear partial differential equations in differential geometry, 1996

1 **Daniel S. Freed and Karen K. Uhlenbeck, Editors,** Geometry and Quantum Field Theory, 1995